Nanomaterials for the Life Sciences Volume 5
Nanostructured Thin Films and Surfaces

Edited by Challa S. S. R. Kumar

Further Reading

Kumar, C. S. S. R. (Ed.)
Nanotechnologies for the Life Sciences (NtLS)
10 Volume Set
2007
ISBN: 978-3-527-31301-3

Kumar, C. S. S. R. (Ed.)
Nanomaterials for the Life Sciences (NmLS)
Book Series, 10 Volumes

Vol. 1
Metallic Nanomaterials
2009
ISBN: 978-3-527-32151-3

Vol. 2
Nanostructured Oxides
2009
ISBN: 978-3-527-32152-0

Vol. 3
Mixed Metal Nanomaterials
2009
ISBN: 978-3-527-32153-7

Vol. 4
Magnetic Nanomaterials
2009
ISBN: 978-3-527-32154-4

Vol. 5
Nanostructured Thin Films and Surfaces
2010
ISBN: 978-3-527-32155-1

Nanomaterials for the Life Sciences
Volume 5

Nanostructured Thin Films and Surfaces

Edited by
Challa S. S. R. Kumar

WILEY-VCH Verlag GmbH & Co. KGaA

The Editor

Dr. Challa S. S. R. Kumar
CAMD
Louisiana State University
6980 Jefferson Highway
Baton Rouge, LA 70806
USA

■ All books published by Wiley-VCH are carefully produced. Nevertheless, authors, editors, and publisher do not warrant the information contained in these books, including this book, to be free of errors. Readers are advised to keep in mind that statements, data, illustrations, procedural details or other items may inadvertently be inaccurate.

Library of Congress Card No.:
applied for

British Library Cataloguing-in-Publication Data
A catalogue record for this book is available from the British Library.

Bibliographic information published by the Deutsche Nationalbibliothek
The Deutsche Nationalbibliothek lists this publication in the Deutsche Nationalbibliografie; detailed bibliographic data are available on the Internet at <http://dnb.d-nb.de>.

© 2010 WILEY-VCH Verlag GmbH & Co. KGaA, Weinheim

All rights reserved (including those of translation into other languages). No part of this book may be reproduced in any form – by photoprinting, microfilm, or any other means – nor transmitted or translated into a machine language without written permission from the publishers. Registered names, trademarks, etc. used in this book, even when not specifically marked as such, are not to be considered unprotected by law.

Composition Toppan Best-set Premedia Ltd., Hong Kong
Printing betz-druck GmbH, Darmstadt
Bookbinding Litges & Dopf GmbH, Heppenheim
Cover Design Schulz Grafik-Design, Fußgönheim

Printed in the Federal Republic of Germany
Printed on acid-free paper

ISBN: 978-3-527-32155-1

Contents

Preface *XIII*
List of Contributors *XVII*

1 **Polymer Thin Films for Biomedical Applications** *1*
Venkat K. Vendra, Lin Wu, and Sitaraman Krishnan
1.1 Introduction *1*
1.2 Biocompatible Coatings *2*
1.2.1 Protein-Repellant Coatings *2*
1.2.1.1 PEGylated Thin Films *2*
1.2.1.2 Non-PEGylated Hydrophilic Thin Films *3*
1.2.1.3 Thin Films of Hyperbranched Polymers *4*
1.2.1.4 Multilayer Thin Films *5*
1.2.2 Antithrombogenic Coatings *6*
1.2.2.1 Surface Chemistry and Blood Compatibility *6*
1.2.2.2 Membrane-Mimetic Thin Films *7*
1.2.2.3 Heparin-Mimetic Thin Films *8*
1.2.2.4 Clot-Lyzing Thin Films *8*
1.2.2.5 Polyelectrolyte Multilayer Thin Films *9*
1.2.2.6 Polyurethane Coatings *9*
1.2.2.7 Vapor-Deposited Thin Films *10*
1.2.3 Antimicrobial Coatings *10*
1.2.3.1 Cationic Polymers *10*
1.2.3.2 Nanocomposite Polymer Thin Films Incorporating Inorganic Biocides *11*
1.2.3.3 Antibiotic-Conjugated Polymer Thin Films *12*
1.2.3.4 Biomimetic Antibacterial Coatings *13*
1.2.3.5 Thin Films Resistant to the Adhesion of Viable Bacteria *13*
1.3 Coatings for Tissue Engineering Substrates *14*
1.3.1 PEGylated Thin Films *15*
1.3.2 Zwitterionic Thin Films *15*
1.3.3 Thin Films of Hyperbranched Polymers *16*
1.3.4 Polyurethane Coatings *17*
1.3.5 Polysaccharide-Based Thin Films *18*

1.3.6	Polyelectrolyte Multilayer Thin Films	18
1.3.7	Temperature-Responsive Polymer Coatings	25
1.3.8	Electroactive Thin Films	27
1.3.9	Other Functional Polymer Coatings	28
1.3.10	Multilayer Thin Films for Cell Encapsulation	30
1.3.11	Patterned Thin Films	31
1.4	Polymer Thin Films for Drug Delivery	33
1.5	Polymer Thin Films for Gene Delivery	36
1.6	Conclusions	39
	References	39
2	**Biofunctionalization of Polymeric Thin Films and Surfaces**	**55**
	Holger Schönherr	
2.1	Introduction: The Case of Biofunctionalized Surfaces and Interfaces	55
2.2	Polymer-Based Biointerfaces	58
2.2.1	Requirements for Biofunctionalized Polymer Surfaces	58
2.2.2	Surface Modification Using Functional Polymers and Polymer-Based Approaches	63
2.2.2.1	Grafting of Polymers to Surfaces	64
2.2.2.2	Polymer Brushes by Surface-Initiated Polymerization	66
2.2.2.3	Physisorbed Multifunctional Polymers	72
2.2.2.4	Multipotent Covalent Coatings	74
2.2.2.5	Plasma Polymerization and Chemical Vapor Deposition (CVD) Approaches	75
2.2.3	Surface Modification of Polymer Surfaces, and Selected Examples	79
2.2.3.1	Coupling and Bioconjugation Strategies	79
2.2.3.2	Interaction with Cells	86
2.2.3.3	Patterned Polymeric Thin Films in Biosensor Applications	89
2.3	Summary and Future Perspectives	91
	References	92
3	**Stimuli-Responsive Polymer Nanocoatings**	**103**
	Ana L. Cordeiro	
3.1	Introduction	103
3.2	Stimuli-Responsive Polymers	105
3.2.1	Polymers Responsive to Temperature	105
3.2.2	Polymers Responsive to pH	107
3.2.3	Dual Responsive/Multiresponsive Polymers	108
3.2.4	Intelligent Bioconjugates	109
3.2.5	Responsive Biopolymers	110
3.3	Polymer Films and Interfacial Analysis	112
3.4	Applications	114
3.4.1	Release Matrices	114
3.4.2	Cell Sheet Engineering	115

3.4.3	Biofilm Control	*122*
3.4.4	Cell Sorting	*123*
3.4.5	Stimuli-Modulated Membranes	*123*
3.4.6	Chromatography	*124*
3.4.7	Microfluidics and Laboratory-on-a-Chip	*128*
3.5	Summary and Future Perspectives	*130*
	Acknowledgments	*131*
	References	*131*
4	**Ceramic Nanocoatings and Their Applications in the Life Sciences**	*149*
	Eng San Thian	
4.1	Introduction	*149*
4.2	Magnetron Sputtering	*150*
4.3	Physical and Chemical Properties of SiHA Coatings	*151*
4.4	Biological Properties of SiHA Coatings	*154*
4.4.1	In Vitro Acellular Testing	*154*
4.4.2	In Vitro Cellular Testing	*160*
4.5	Future Perspectives	*168*
4.6	Conclusions	*170*
	References	*171*
5	**Gold Nanofilms: Synthesis, Characterization, and Potential Biomedical Applications**	*175*
	Shiho Tokonami, Hiroshi Shiigi, and Tsutomu Nagaoka	
5.1	Introduction	*175*
5.2	Preparation of Various AuNPs	*177*
5.3	Functionalization of AuNPs and their Applications through Aggregation	*178*
5.4	AuNP Assemblies and Arrays	*182*
5.4.1	AuNP Assemblies Structured on Substrates	*182*
5.4.2	AuNP Assembly on Biotemplates	*184*
5.4.3	AuNP Arrays for Gas Sensing	*186*
5.4.4	AuNP Arrays for Biosensing	*188*
5.5	Conclusions	*195*
	References	*195*
6	**Thin Films on Titania, and Their Applications in the Life Sciences**	*203*
	Izabella Brand and Martina Nullmeier	
6.1	Introduction	*203*
6.2	Titanium in Contact with a Biomaterial	*204*
6.3	Lipid Bilayers at the Titania Surface	*206*
6.3.1	Formation of Lipid Bilayers on the Titania Surface	*207*
6.3.1.1	Spreading of Vesicles on a TiO_2 Surface: Comparison to a SiO_2 Surface	*207*

6.3.2	Interactions: Lipid Molecule–Titania Surface *213*
6.3.3	Structure and Conformation of Lipid Molecules in the Bilayer on the Titania Surface *216*
6.3.3.1	Structure of Phosphatidylcholine on the Titania Surface *217*
6.4	Characteristics of Extracellular Matrix Proteins on the Titania Surface *226*
6.4.1	Collagen Adsorption on Titania Surfaces *227*
6.4.1.1	Morphology of Collagen Adsorbed on an Oxidized Titanium Surface *228*
6.4.1.2	Adsorption of Collagen on a Hydroxylated Titania Surface *228*
6.4.1.3	Morphology and Structure of Collagen Adsorbed on a Calcified Titania Surface *229*
6.4.1.4	Conclusions *231*
6.4.1.5	Structure of Collagen on the Titania Surface: Theoretical Predictions *232*
6.4.2	Fibronectin Adsorption on the Titania Surface *234*
6.4.2.1	Morphology of Fibronectin Adsorbed on the Titania Surface *235*
6.4.2.2	Fibronectin–Titania Interactions *236*
6.4.2.3	Structure of Fibronectin Adsorbed onto the Titania Surface *238*
6.4.2.4	Atomic-Scale Picture of Fibronectin Adsorbed on the Titania Surface: Theoretical Predictions *240*
6.4.2.5	Conclusions *241*
6.5	Conclusions *242*
	Acknowledgments *244*
	References *244*

7 Preparation, Characterization, and Potential Biomedical Applications of Nanostructured Zirconia Coatings and Films *251*
Xuanyong Liu, Ying Xu, and Paul K. Chu

7.1	Introduction *251*
7.2	Preparation and Characterization of Nano-ZrO_2 Films *251*
7.2.1	Cathodic Arc Plasma Deposition *251*
7.2.2	Plasma Spraying *254*
7.2.3	Sol–Gel Methods *255*
7.2.4	Electrochemical Deposition *257*
7.2.5	Anodic Oxidation and Micro-Arc Oxidation *260*
7.2.6	Magnetron Sputtering *262*
7.3	Bioactivity of Nano-ZrO_2 Coatings and Films *263*
7.4	Cell Behavior on Nano-ZrO_2 Coatings and Films *267*
7.5	Applications of Nano-ZrO_2 Films to Biosensors *269*
	References *273*

8 Free-Standing Nanostructured Thin Films *277*
Izumi Ichinose

8.1	Introduction *277*
8.2	The Roles of Free-Standing Thin Films *277*

8.2.1	Films as Partitions	277
8.2.2	Nanoseparation Membranes	279
8.2.3	Biomembranes	281
8.3	Free-Standing Thin Films with Bilayer Structures	282
8.3.1	Supported Lipid Bilayers and "Black Lipid Membranes"	282
8.3.2	Foam Films and Newton Black Films	283
8.3.3	Dried Foam Film	284
8.3.4	Foam Films of Ionic Liquids	286
8.4	Free-Standing Thin Films Prepared with Solid Surfaces	289
8.5	Free-Standing Thin Films of Nanoparticles	290
8.6	Nanofibrous Free-Standing Thin Films	292
8.6.1	Electrospinning and Filtration Methods	292
8.6.2	Metal Hydroxide Nanostrands	293
8.6.3	Nanofibrous Composite Films	295
8.6.4	Nanoseparation Membranes	298
8.7	Conclusions	299
	References	300

9	**Dip-Pen Nanolithography of Nanostructured Thin Films for the Life Sciences** 303	
	Euiseok Kim, Yuan-Shin Lee, Ravi Aggarwal, and Roger J. Narayan	
9.1	Introduction	303
9.2	Dip-Pen Nanolithography	304
9.2.1	Important Parameters	305
9.2.2	Applications of DPN	306
9.3	Direct and Indirect Patterning of Biomaterials Using DPN	310
9.3.1	Background	310
9.3.2	Direct Patterning	311
9.3.3	Indirect Patterning	314
9.4	Applications of DPN for Medical Diagnostics and Drug Development	316
9.4.1	General Methods of Nano/Micro Bioarray Patterning	317
9.4.2	Virus Array Generation and Detection Tests	318
9.4.3	Diagnosis of Allergic Disease	320
9.4.4	Cancer Detection Using Nano/Micro Protein Arrays	321
9.4.5	Drug Development	323
9.4.6	Lab-on-a-Chip Using Microarrays	323
9.5	Summary and Future Directions	325
	References	325

10	**Understanding and Controlling Wetting Phenomena at the Micro- and Nanoscales** 331	
	Zuankai Wang and Nikhil Koratkar	
10.1	Introduction	331
10.2	Wetting and Contact Angle	333
10.3	Design and Creation of Superhydrophobic Surfaces	335

10.3.1	Design Parameters for a Robust Composite Interface 335
10.3.2	Creation of Superhydrophobic Surfaces 335
10.3.3	Superhydrophobic Surfaces with Unitary Roughness 336
10.3.4	Superhydrophobic Surfaces with Two-Scale Roughness 336
10.3.5	Superhydrophobic Surfaces with Reentrant Structure 338
10.4	Impact Dynamics of Water on Superhydrophobic Surfaces 339
10.4.1	Impact Dynamics on Nanostructured MWNT Surfaces 340
10.4.2	Impact Dynamics on Micropatterned Surfaces 342
10.5	Electrically Controlled Wettability Switching on Superhydrophobic Surfaces 344
10.5.1	Reversible Control of Wettability Using Electrostatic Methods 344
10.5.2	Electrowetting on Superhydrophobic Surfaces 345
10.5.3	Novel Strategies for Reversible Electrowetting on Rough Surfaces 348
10.6	Electrochemically Controlled Wetting of Superhydrophobic Surfaces 350
10.6.1	Polarity-Dependent Wetting of Nanotube Membranes 350
10.6.2	Mechanism of Polarity-Dependent Wetting and Transport 352
10.6.3	Potential Applications of Electrochemically Controlled Wetting and Transport 354
10.7	Summary and Future Perspectives 356
10.7.1	Future Perspectives 357
	Acknowledgments 357
	References 358

11 Imaging of Thin Films, and Its Application in the Life Sciences 363
Silvia Mittler

11.1	Introduction 363
11.2	Thin Film Preparation Methods 364
11.2.1	Dip-Coating 364
11.2.2	Spin-Coating 364
11.2.2	Langmuir–Blodgett (LB) Films 364
11.2.4	Self-Assembled Monolayers 366
11.2.5	Layer-by-Layer Assembly 366
11.2.6	Polymer Brushes: The "Grafting-From" Approach 367
11.3	Structuring: The Micro- and Nanostructuring of Thin Films 368
11.3.1	Photolithography 368
11.3.2	Ion Lithography and FIB Lithography 369
11.3.3	Electron Lithography 369
11.3.4	Micro-Contact Printing and Nanoimprinting (NIL) 369
11.3.5	Near-Field Scanning Methods 370
11.3.6	Other Methods 371
11.4	Imaging Technologies 371
11.4.1	The Concept of Total Internal Reflection 371
11.4.2	The Concept of Waveguiding 372
11.4.3	Brewster Angle Microscopy (BAM) 373

11.4.4	Resonant Evanescent Methods	*375*
11.4.4.1	Surface Plasmon Resonance Microscopy	*375*
11.4.4.2	Waveguide Resonance Microscopy	*376*
11.4.4.3	Surface Plasmon Enhanced Fluorescence Microscopy	*378*
11.4.4.4	Waveguide Resonance Microscopy with Electro-Optical Response	*379*
11.4.5	Nonresonant Evanescent Methods	*379*
11.4.5.1	Total Internal Reflection Fluorescence (TIRF) Microscopy	*379*
11.4.5.2	Waveguide Scattering Microscopy	*380*
11.4.5.3	Waveguide Evanescent Field Fluorescence Microscopy (WEFFM)	*381*
11.4.5.4	Confocal Raman Microscopy and One- and Two-Photon Fluorescence Confocal Microscopy	*382*
11.5	Application of Thin Films in the Life Sciences	*383*
11.5.1	Sensors	*384*
11.5.2	Surface Functionalization for Biocompatibility	*384*
11.5.3	Drug Delivery	*385*
11.5.4	Bioreactors	*386*
11.5.5	Cell-Surface Mimicking	*387*
11.6	Summary	*389*
	References	*390*

12 Structural Characterization Techniques of Molecular Aggregates, Polymer, and Nanoparticle Films *397*
Takeshi Hasegawa

12.1	Introduction	*397*
12.2	Characterization of Ultrathin Films of Soft Materials	*398*
12.2.1	X-Ray Diffraction Analysis	*398*
12.2.2	Infrared Transmission and Reflection Spectroscopy	*402*
12.2.3	Multiple-Angle Incidence Resolution Spectrometry (MAIRS)	*406*
12.2.3.1	Theoretical Background of MAIRS	*406*
12.2.3.2	Molecular Orientation Analysis in Polymer Thin Films by IR-MAIRS	*408*
12.2.3.3	Analysis of Metal Thin Films	*411*
	References	*415*

Index *419*

Preface
Vol 5 (Nano-structured Films, Surfaces and Coatings for Life Sciences)

A number of books have covered topics related to microscopic films, surfaces and coatings. There are very few books focusing on nanostructured films, surfaces and coatings. However, to date there has not been a single book completely devoted to applications of nanostructured films, surfaces and coatings in life sciences. To fill this gap, 21 nanotechnology specialists have come together in bringing out the fifth volume of the NmLS series entitled, "Nano-structured Films, Surfaces and Coatings for Life Sciences." A big "thank you" to all the authors for having joined in this exciting effort. Of course, Wiley-VCH publishing team, as always, played a crucial role in timely publication of the book with uncompromised quality. I am grateful for each and every one who ensured that this book reaches the reader as soon as possible.

The book covers wide ranging type of films, from polymers to ceramics and metals, providing information on their preparation, biofunctionalization, imaging and understanding of the intrinsic phenomenon. The book begins with a chapter on polymer thin films and coatings emphasizing biocompatible coatings for implants, tissue engineering, drug delivery and gene therapy. In this chapter, materials, processing and physicochemical properties of coatings have been highlighted followed by delineating polymer film's protein-adsorption and cell adhesion characteristics. With the increase in application of polymeric films, a plethora of very successful functionalization and modification strategies for the polymer surface have been developed in the past decades. The second chapter reviews the chemistry and surface engineering involved in the fabrication of polymer-based bio-functional surfaces and bio-interfaces. It is now well established that polymers can be designed in such a way that they can respond to changes in the surrounding environment. In the third chapter, polymers especially biopolymers and intelligent polymer-bioconjugates, which are responsive specifically to temperature and pH have been presented.

The next four chapters deal with ceramic, gold, titania and zirconia films and coatings respectively. The fourth chapter describes novel, nanostructured silicon-substituted hydroxyapatite (SiHA) coatings of varying Si levels exhibiting excellent bioactivity in comparison with uncoated substrates and HA-coated substrates. In the fifth chapter, fabrication of biosensors based on gold nanofilms is discussed. The array sensing system, described in detail, is useful in quantification of organic gases, antigens, and DNA with special emphasis placed on the DNA assay. To gain medical and technological advantages, it is essential to recognize all processes occurring at the implant's surface in greatest detail possible. Underscoring this point, the sixth chapter, **Thin films on titania and their applications in life sciences**, aims to provide information about biological interfaces with titania surfaces; mimicking structural features in organisms. Apart from titania, Zirconia (ZrO_2) coatings and films are finding useful applications as bio-coatings, femoral heads, orthopedic implants, and biosensors due to their excellent mechanical strength, thermal stability, chemical inertness, lack of toxicity, and affinity for the groups containing oxygen. Therefore, the seventh chapter is completely dedicated to the study of nanostructured ZrO_2 coatings and films prepared by different techniques. In addition, details about their surface morphology, microstructure, crystallite size, phase composition, and biomedical characterization are also presented.

Unlike the completely supported coatings and films, there is yet another class of films called "free-standing films." A free-standing thin film is defined as a thin film in which at least a part of the film is not in contact with a support material. Such films are gaining a lot of importance due to their potential application in the development of nano-separation membranes for small molecules. These novel technqiues are providing solutions for a number of long standing energy and environmental problems. In this context, the eighth chapter brings out new fabrication methods for nanostructured free-standing thin films providing a number of examples to illustrate how they can help in separating two phases and control material transfer between those phases. Patterning of films and arrays especially biopatterning has made tremendous progress thanks to nanolithography methods such as Dip-pen nanolithography (DPN). Since its discovery, DPN has been found to be a simple, flexible tool that can be used, directly or indirectly, in order to fabricate bioarrays and patterned biofilms. In the ninth chapter, you will find an up to date summary of dip-pen nanolithography process and its applications in life science. You will also find recent approaches to fabricate bioarrays including direct and indirect DPN, important process parameters, and several applications with examples.

The wetting phenomena is a classical and key issue in surface engineering and hence the next chapter explores this fundamental concept in surface science. This study is crucial as the phenomenon is not only ubiquitous in our everyday lives, but also plays an important role in many technological processes as well as in many biological systems. In the tenth chapter, you will learn about the fundamental physical mechanisms of wetting, contact angle, wetting transition, dynamic wetting and impact, the design/creation of superhydrophobic surfaces and the active control of wettability and wettability switching. The penultimate chapter, **Imaging of Thin Films and its Application in Life Sciences**, summarizes the fabrication, pattering, microscopy and life science application of organic thin films on metal and glass substrates. Especially, the chapter is invaluable to the readers interested in imaging of thin films as it focuses on optical microscopy technologies, mainly based on evanescent technology and provides an overview of the near field methods and electron microscopy. Continuing on the same theme of characterization of thin films, the final chapter underscores the importance of surfaces and interfaces in providing a platform for chemical reactions and material interactions; key to understand properties of functionalized materials. In this chapter, the characterization techniques using vibrational spectroscopy for polymeric thin films are explained first. This will be followed with an introduction to a new spectroscopic tool, namely, multiple-angle incidence resolution spectrometry (MAIRS), for analyzing plasmon polarization.

With the publication of this book, we are exactly half way through the NmLS series and I am excited to let you know that the rest of the five volumes have already been planned. The titles of sixth to eighth volumes are semiconductor nanomaterials for life sciences, biomimetic and bio-inspired nanomaterials for life sciences, and nanocomposites for life sciences respectively. The volumes 6-8 are already in print and I am pleased to see the timely efforts of all in bringing these unique books to the reader. As I write this preface, volumes 9-10 have also completed the design process and will be in print early next year. While the ninth volume is entitled carbon nanomaterials for life sciences, the tenth and the final volume is on application of organic nanomaterials in life sciences.

As the fledgling field of nanomaterials for life sciences is poised to take off in a big way, I would like to thank all the authors, readers and supporters of this new field of sceintific endevour for all their support, enouragement and inspiration. May the new year and years to come help us to continue our team effort to make a difference in the scientific world.

Baton Rouge   *Challa S.S.R.Kumar*

List of Contributors

Ravi Aggarwal
North Carolina State University
Department of Biomedical
Engineering and
Department of Materials Science
and Engineering
Raleigh, NC 27606
USA

Izabella Brand
Carl von Ossietzky University of
Oldenburg
Center of Interface Science (CIS)
and Department of Pure and
Applied Chemistry
26111 Oldenburg
Germany

Paul K. Chu
City University of Hong Kong Tat
Department of Physics &
Materials Science
Kowloon
Hong Kong
China

Ana L. Cordeiro
Leibniz Institute of Polymer
Research Dresden
Max Bergmann Center of
Biomaterials Dresden
Hohe St. 6
01069 Dresden
Germany

Takeshi Hasegawa
Tokyo Institute of Technology
Department of Chemistry
Graduate School of Science and
Engineering
2-12-1 Ookayama
Meguro-ku
Tokyo 152-8551
Japan

Izumi Ichinose
National Institute for Materials
Science (NIMS)
Organic Nanomaterials Center
1-1 Namiki
Tsukuba 305-0044
Japan
Japan Science and Technology
Agency
CREST
5 Sanbancho
Chiyoda-ku
Tokyo 102-0075
Japan

Euiseok Kim
North Carolina State University
Edward P. Fitts Department of
Industrial and Systems
Engineering
Raleigh, NC 27606
USA

Nikhil Koratkar
Rensselaer Polytechnic Institute
Department of Mechanical
Aerospace and Nuclear
Engineering
110 8th Street
Troy, NY 12180
USA

Sitaraman Krishnan
Clarkson University
Department of Chemical and
Biomolecular Engineering
Potsdam, NY 13699
USA

Yuan-Shin Lee
North Carolina State University
Edward P. Fitts Department of
Industrial and Systems
Engineering
Raleigh, NC 27606
USA

Xuanyong Liu
Shanghai Institute of Ceramics
Chinese Academy of Sciences
1295 Dingxi Road
Shanghai 200050
China

Silvia Mittler
The University of
Western Ontario
Department of Physics
and Astronomy
London
ON N6A 3K7
Canada

Tsutomu Nagaoka
Osaka Prefecture University
Frontier Science Innovation
Center
1-2 Gakuen-cho
Naka-ku
Sakai
Osaka 599-8570
Japan

Roger J. Narayan
North Carolina State University
Department of Biomedical
Engineering and
Department of Materials Science
and Engineering
Raleigh, NC 27606
USA

Martina Nullmeier
Carl von Ossietzky University of
Oldenburg
Center of Interface Science (CIS)
and Department of Pure and
Applied Chemistry
26111 Oldenburg
Germany

Holger Schönherr
University of Siegen
Department of Physical
Chemistry
Adolf-Reichwein-Str. 2
57068 Siegen
Germany

Hiroshi Shiigi
City University of Hong Kong Tat Chee Avenue
Department of Physics & Materials Science
Kowloon
Hong Kong
China

Osaka Prefecture University
Frontier Science Innovation Center
1-2 Gakuen-cho
Naka-ku
Sakai
Osaka 599-8570
Japan

Eng San Thian
National University of Singapore
Department of Mechanical Engineering
9 Engineering Drive 1
Singapore 117576
Singapore

Shiho Tokonami
Osaka Prefecture University
Frontier Science Innovation Center
1-2 Gakuen-cho
Naka-ku
Sakai
Osaka 599-8570
Japan

Venkat K. Vendra
Clarkson University
Department of Chemical and Biomolecular Engineering
Potsdam, NY 13699
USA

Zuankai Wang
Rensselaer Polytechnic Institute
Department of Mechanical Aerospace and Nuclear Engineering
110 8th Street
Troy, NY 12180
USA

Lin Wu
Clarkson University
Department of Chemical and Biomolecular Engineering
Potsdam, NY 13699
USA

Ying Xu
Shanghai Institute of Ceramics
Chinese Academy of Sciences
1295 Dingxi Road
Shanghai 200050
China

1
Polymer Thin Films for Biomedical Applications
Venkat K. Vendra, Lin Wu, and Sitaraman Krishnan

1.1
Introduction

Modern medicine uses a variety of synthetic materials and devices to treat medical conditions and diseases. Biomedical devices such as coronary stents, vascular grafts, heart valves, blood bags, blood oxygenators, renal dialyzers, catheters, hip prostheses, knee prostheses, intraocular lenses, contact lenses, cochlear implants, and dental implants have definitely played an important role in transforming lives and improving the quality of living. Advances in protein-based drugs, gene therapy, targeted drug delivery, and tissue engineering have the potential to revolutionize contemporary medicine. Artificial skins to treat burn victims, artificial pancreas for people with diabetes, and cardiac patches to regenerate cardiac muscle damaged by a heart attack, no longer seem far-fetched, because of developments in tissue engineering. Thus, a wide range of synthetic materials are used to evaluate, treat, augment or replace any tissue, organ or function of the body. "Biomaterial" is a term used to categorize such materials and devices that directly "interact" with human tissues and organs [1]. The interactions may involve, for example, platelet aggregation and blood coagulation in the case of blood-contacting devices, immune response and foreign body reactions around biomaterials or devices implanted in the body, or more desirably, structural and functional connection between the implant and the host tissue (this is termed osseointegration in the case of dental and orthopedic implants).

Biomaterials interact with biological systems through their surfaces. It is, therefore, vitally important to control the surface properties of a biomaterial so that it integrates well with host tissues – that is, to make the material "biocompatible" [2]. Organic thin films and coatings, particularly those of polymers, are very attractive as biomaterial coatings because they offer great versatility in the chemical groups that can be incorporated at the surface (to control tissue–biomaterial interactions); the coatings also have mechanical properties that are similar to soft biological tissues. The relative ease of processing is another reason for the extensive interest in organic thin films. Biomaterial surfaces can be coated with polymers using simple techniques such as dip-coating, spray-coating, spin-coating, or solvent casting. Coating techniques involving the chemical grafting of molecules onto the biomaterial surface are also available. Nanothin coatings based on self-assembled monolayers (SAMs), surface-tethered polymers (polymer brushes), or multilayer coatings based on layer-by-layer assembly offer precise control on the location and orientation of chemical groups and biomolecules on the surface of the coating.

In this chapter we discuss polymer thin films and coatings that have potential biomedical applications. Three main areas are covered: (i) biocompatible coatings for implants (e.g., protein-repellant coatings, antithrombogenic coatings that can prevent blood coagulation around implants, and antibacterial coatings); (ii) polymer thin films for tissue engineering; and (iii) polymer thin films for drug delivery and gene therapy. Emphasis is placed on the material and processing aspects of the coating, the physico-chemical properties of the coating, and its protein-adsorption and cell-adhesion characteristics.

1.2
Biocompatible Coatings

1.2.1
Protein-Repellant Coatings

Protein adsorption on biomaterial surfaces plays a critical role in determining cellular events at the tissue–implant interface. The adsorption of plasma proteins onto the surface of a blood-contacting implant can trigger a cascade of chemical reactions, leading to the formation of a blood clot surrounding the implant. The blood protein factor XII is known to be activated by negatively charged surfaces such as glass, and some polymers such as poly(vinyl chloride) [3] may initiate the intrinsic pathway of blood coagulation. Protein adsorption can also activate the host foreign-body response, by switching on the complement system. The adhesion of neutrophils and macrophages, which are associated with the host foreign-body response, can subject the biomaterial surface to attack by destructive enzymes, superoxide anions, and hydrogen peroxide [4]. Finally, protein adsorption can promote bacterial colonization on the implant surface, that in time will require the reoperation of infected implants. There is, therefore, a great interest in developing biomedical coatings that can resist protein adsorption.

1.2.1.1 PEGylated Thin Films
The ability of hydrophilic surfaces, especially coatings based on poly(ethylene glycol) (PEG), to resist protein adsorption has been demonstrated in several studies [5–14]. PEG surfaces seem to be the benchmark in protein-adsorption studies, because of their exceptional resistance to protein adsorption. Silane-based PEG SAMs were found to maintain their integrity after sterilization in an autoclave using 20 psi steam at 120 °C for 1 h. Strategies to bind PEG molecules to the surfaces of metallic implants, such as the use of PEG conjugates with the adhesive

amino acid L-2,4-dihydroxyphenylalanine (DOPA) [15–17], or with the cyanobacterial iron chelator anachelin [18], have been reported.

Amphiphilic block copolymer coatings containing PEGylated and fluoroalkyl moieties have shown to have extremely low forces of interaction with protein molecules (C.J. Wienman et al., unpublished results). Novel architectures, such as multi-armed molecules, have been used to prepare monolayer coatings on surfaces, which decreased the protein adsorption [19]. Hydrogel thin films were prepared on amine-functionalized surfaces by spin-coating of a six-armed, star-shaped poly(ethylene glycol-ran-propylene glycol) polymer, with crosslinkable peripheral isocyanate groups [20]. Crosslinking of the star-shaped molecules occurred by reaction between the isocyanate groups in an aqueous environment to form urea linkages. The coatings prevented adsorption of avidin (over a pH range of 5 to 9.5), and were stable over several months when stored under ambient conditions. A multicomponent coating formulation to obtain robust, readily functionalized PEG-based thin films has been reported by Lochhead and coworkers for imparting resistance to protein adsorption, fibroblast cell adhesion, and bacterial adhesion [21]. The formulation consisted of: (i) the active component, NHS–PEG-aminosilane (NHS = N-hydroxysuccinimide); (ii) the matrix-forming component, polyoxyethylene sorbitan tetraoleate; and (iii) the molecular crosslinker, 6-azido-sulfonylhexyltriethoxy silane. Solutions of these three components in dimethylsulfoxide (DMSO) were spin-coated onto glass, silicon, or tissue culture grade polystyrene (TCPS) substrates and thermally cured at 0.1 mmHg pressure for 75 min. The coatings showed significant inhibition of fibrinogen and lysozyme adsorption, microbial adhesion, and fibroblast cell adhesion. The *in situ* crosslinked PEG-based coatings could be functionalized with an RGD (Arg-Gly-Asp) peptide to promote cell adhesion for tissue engineering applications. PEGylated surfaces were also made biofunctional using latent aldehyde groups that were used to covalently tether proteins and bioligands [22]. Similarly, PEGylated polymer brushes were functionalized with cell-adhesive proteins to promote the osseointegration of bone and dental implants [23].

Coatings prepared using the polysaccharide, chitosan (CS) that was grafted with PEG side chains, showed good resistance to the adsorption of fibrinogen and bovine serum albumin (BSA) [24]. In the chitosan polymer, 72% of the sugar units were deacetylated, and 56% of all sugar units carried a $-(CH_2CH_2O)_{44}CH_3$ side chain. SAMs of 16-mercaptodecanoic acid on gold were used as substrates for the PEGylated chitosan coatings. The coating adhered to the substrate via electrostatic as well as covalent coupling between the primary amine groups in chitosan and carboxylic acids in the SAM.

1.2.1.2 Non-PEGylated Hydrophilic Thin Films

A variety of hydrophilic thin films have been investigated as alternatives to PEGylated coatings for biomedical applications [25, 26]. There is significant interest in developing biomedical coatings that have better thermal and oxidative stability than PEG, but with protein-repellency comparable to that of PEG. A recent review highlighted some developments in the design and synthesis of protein-resistant polymer coatings, and provided a discussion of the mechanism of antifouling activity [27]. Pulsed plasma-deposited poly(N-acryloylsarcosine methyl ester) coatings were shown to be resistant towards the adsorption of fibrinogen and lysozyme [28], while hydrophilic tertiary amine oxide surfaces were found to compare favorably with PEGylated coatings in preventing nonspecific protein adsorption [29]. Hydrogel thin films of poly(N,N'-methylene bisacrylamide) showed less than $0.5\,\text{ng}\,\text{cm}^{-2}$ adsorption of BSA. The highly hydrophilic polyacrylamide was chemically tethered to the hydrophobic substrate of an alkyl thiol SAM using a photoinitiator, benzophenone, which induced crosslinking with the substrate via hydrogen abstraction [30]. Zwitterionic polymers have found renewed interest as protein-resistant coatings. These polymers attempt to mimic the excellent resistance to protein adsorption and hemocompatibility of the zwitterionic phosphorylcholine group that is a major component of cell membranes [31]. Thin film coatings of zwitterionic SAMs and polymer brushes, of phosphobetaines, carboxybetaines and sulfobetaines [32], have shown excellent resistance to protein adsorption [33–39]. In addition to their protein repellency, the carboxylic acid groups of carboxybetaine thin films have been used to covalently immobilize cell-adhesive peptides for tissue engineering applications [37]. Interpenetrating polymer networks of polyurethanes and zwitterionic polymers combined the desirable mechanical properties of polyurethane with the antifouling properties of the zwitterionic polymer [40]. In a novel approach, zwitterionic amino acids were used to prepare thin film coatings on gold-coated glass slides [41]. SAMs of N-3-mercaptopropylamino acids, prepared from the 19 natural amino acids, were investigated for their ability to resist the nonspecific adsorption of proteins. When the SAMs were exposed to a solution of BSA ($76\,\text{mg}\,\text{ml}^{-1}$) in phosphate-buffered saline (PBS) for 20 min, the concentration of nonspecifically bound proteins ranged from approximately $400\,\text{ng}\,\text{cm}^{-2}$, with polar and ionic amino acids, to approximately $800\,\text{ng}\,\text{cm}^{-2}$, with amino acids of increased hydrophobicity. The nonspecific adsorption of BSA increased in the following order: Asp < Asn < Ser < Met < Glu < Gln < Thr < Gly < His < Cys < Arg < Phe < Trp < Val < Pro < Ile < Leu < Ala < Tyr. Thin-film coatings prepared using poly(N-substituted glycine) (polypeptoids) exhibited significant reductions in the adsorption of lysozyme, fibrinogen, and serum proteins [42]. Poly(2-methyl-2-oxazoline), a peptide-like polymer, has been used to prepare surfaces with a protein adsorption below a level of $2\,\text{ng}\,\text{cm}^{-2}$ [43]. The polyoxazoline surfaces had protein repellency comparable to that of the best PEG-based coatings.

1.2.1.3 Thin Films of Hyperbranched Polymers

Hyperbranched polymers with hydrophilic groups have attracted interest because of their similarity to the antifouling glycocalyx (extracellular matrix; ECM) of cells [44–47]. Dendritic polyglycerols (cf. Figure 1.1) combine the characteristics of highly protein-resistant polymers, namely a highly flexible aliphatic polyether segments, hydrophilic surface groups, and a highly branched architecture [49]. Fibrinogen adsorption on these surfaces was comparable to that on a PEGylated SAM of $HS(CH_2)_{11}(OCH_2CH_2)_3OH$, and better than the dextran-coated surfaces which

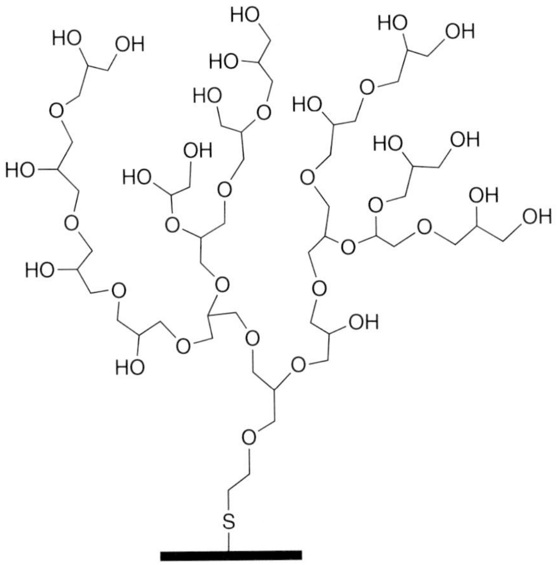

Figure 1.1 A dendritic polyglycerol monolayer. Adapted with permission from Ref. [48]; © 2008, American Chemical Society.

have been used for decades as low-protein-binding substrates [50]. Self-assembled monothiol-terminated hyperbranched polyglycerols on gold surfaces have shown a better resistance to the adsorption of BSA and immunoglobulin adsorption than the SAMs of linear PEG thiols [48]. Dendritic or hyperbranched hydrophilic polymers will result in a highly hydrated surface, which is an important characteristic of most antibiofouling surfaces; however, similar hyperbranched polyglycerols have shown a lower resistance to bacterial adhesion than linear PEG molecules [51]. The combined effects of steric repulsion, which is expected to be higher for linear architectures because of higher polymer flexibility, and the packing density of hydrophilic groups, which will be higher for dendritic architectures; this plays a critical role in determining the protein and cell repellency of the dendritic coatings. In contrast, Jiang and coworkers have argued that conformational flexibility is not required for protein resistance, and that only hydration played a dominant role in surface resistance to nonspecific protein adsorption [52]. Thus, the role of steric repulsion and polymer conformation on resistance to protein adsorption requires further investigation.

1.2.1.4 Multilayer Thin Films

Polyelectrolyte multilayer (PEM) thin films have been widely explored as functional coatings in biomedical engineering, particularly in tissue engineering and gene/drug delivery. These coatings are, however, less protein-resistant than uncharged hydrophilic coatings such as PEG, or zwitterionic coatings such as those containing phosphorylcholine groups [53]. The interactions of proteins with PEMs have been characterized using human serum albumin and poly(sodium

4-styrenesulfonate) (PSS)/poly(allyl amine hydrochloride) (PAH) [(PSS/PAH)$_n$] multilayer thin films [54]. The protein adsorbed onto the multilayer coatings, regardless of whether PAH or PSS was the terminal layer. On PSS-ending multilayers, the human serum albumin adsorption was limited to a dense monolayer, whereas on PAH-ending multilayers protein films with thicknesses exceeding several-fold the native protein dimension could be formed. Protein interaction with PEMs was found to be electrostatic in origin. PEM coatings with balanced charges have shown resistance to protein adsorption. By controlling the amounts of the cationic poly(2-aminoethyl methacrylate hydrochloride) and anionic poly(2-carboxyethyl acrylate), the net charge (as estimated by direct force measurements) of the PEM thin film could be minimized; this minimum in net charge corresponded to a minimum in protein adsorption on the polyelectrolyte blends [55].

In order to make (PAH/PSS)$_n$ multilayer films resistant to protein adsorption, PEM coatings terminated with the anionic PSS polymer were additionally coated with a terminal layer of poly(L-lysine)-*graft*-poly(ethylene glycol) (PLL-g-PEG) [56]. The PEG grafts imparted protein resistance to the polyelectrolyte coatings. Protein adsorption from full serum on the PEGylated surfaces of (PAH/PSS)$_n$(PLL-g-PEG) multilayers (n = 1–4) was three orders of magnitude lower in comparison to (PAH/PSS)$_n$ surfaces that did not contain the PEG grafts. The layer-by-layer (LBL) assembly of multiarm PEG with reactive vinylsulfone end groups and dithiothreitol (DTT) was found to be resistant to protein adsorption and cell adhesion [57]. The vinylsulfone end groups of PEG reacted with thiol groups on DTT through a Michael-type reaction, producing a thin, crosslinked PEG coating. The resistance to cell adhesion then increased with an increase in the number of layers in the coating. For the same number of layers, multilayers prepared with eight-arm PEG molecules were more resistant to cell adhesion than multilayers containing four-arm PEG. An RGD-containing peptide, acetyl-GCGYGRGDSPG-NH$_2$, could be covalently immobilized on the surface, through reaction between the vinylsulfone end groups of PEG and the cysteine residue in the peptide, to enhance cell attachment by binding cell-surface receptors of the integrin family.

Polyelectrolytes bearing zwitterionic groups have been used in PEM coatings to impart resistance to protein adsorption [58, 59]. The polyelectrolytes shown in Figure 1.2 were obtained by modification of poly(L-glutamic acid) (PGA), poly(acrylic acid) (PAA), and poly(L-lysine) (PLL) [58]. An earlier study had also reported that fibronectin was adsorbed onto poly(allylamine hydrochloride)-terminated and Nafion™-terminated polyelectrolyte multilayer coatings, but fibronectin adsorption was low on multilayers terminated with a poly(acrylic acid-*co*-3-[2-(acrylamido)-ethyl dimethylammonio]propane sulfonate) copolymer that contained the zwitterionic sulfobetaine monomer [60].

1.2.2
Antithrombogenic Coatings

1.2.2.1 Surface Chemistry and Blood Compatibility
The blood compatibility of coatings is strongly influenced by chemical groups present at its surface. Sperling *et al.* found that leukocytes did not adsorb onto a

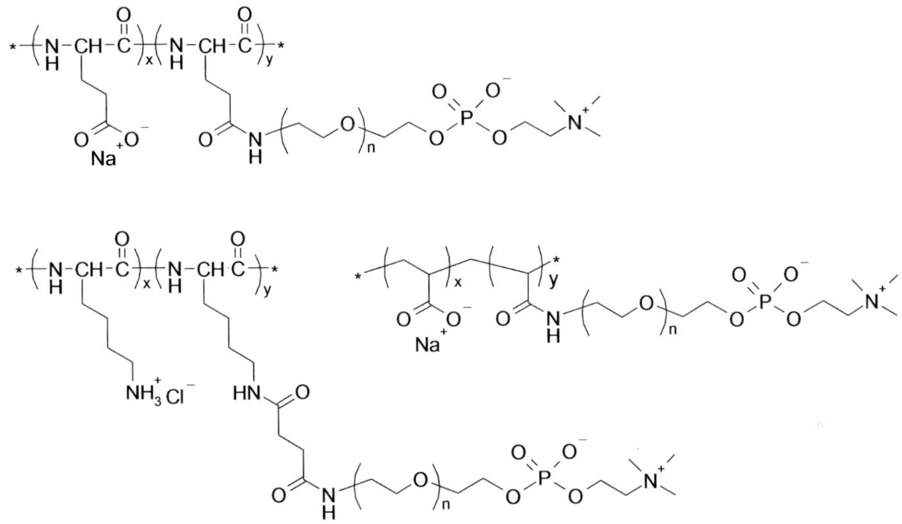

Figure 1.2 Polyelectrolytes bearing zwitterionic moieties. Adapted with permission from Ref. [58]; © 2009 Wiley-VCH Verlag.

–CH$_3$-terminated alkanethiol SAM, but their adhesion was greatly enhanced on surfaces with –OH groups [61]. The opposite was detected for the adhesion of platelets. A strong correlation between the activation of the complement system and the adhesion of leukocytes with the content of –OH groups was observed. Complement activation was also scaled with the amount of –COOH groups at the surface. Rodrigues et al. showed that fibrinogen adsorption decreased linearly with an increase of –OH groups on a SAM surface [62]. Platelet adhesion and activation were also seen to decrease with an increase of surface hydrophilicity. The adsorption of plasma albumin onto the surfaces passivated the surfaces and lowered platelet adhesion.

1.2.2.2 Membrane-Mimetic Thin Films

Chung et al. screened SAMs of different phospholipids molecules, on gold, for fibrinogen adsorption and platelet adhesion [63]. It was found that, of the bromoethylphosphorate-, phosphorylcholine-, phosphorylethanolamine-, and hydroxyl-terminated SAMs, the phosphorylcholine-terminated SAMs showed the best antifouling properties. Carboxybetaine-based SAMs and polymers showed not only a very low fibrinogen adsorption but also a very low platelet adhesion. Moreover, the poly(carboxybetaine methacrylate) polymer also exhibited anticoagulant activity and increased the clotting time of blood, which made it a promising candidate for coating blood-contacting devices and implants [64, 65]. Ito and coworkers have developed a new type of copolymer coating composed of L-histidine, a zwitterion, and n-butyl methacrylate, a hydrophobic moiety [66]. Polystyrene surfaces coated with the copolymer were found to have a significantly low nonspecific adsorption of proteins and adhesion of cells in comparison with BSA-passivated surfaces.

1.2.2.3 Heparin-Mimetic Thin Films

Heparin is a highly-sulfated, anionic, polysaccharide (5–25 kDa) (cf. Figure 1.12) that can bind to the blood protein antithrombin through ionic interactions, and result in a several-fold acceleration of the rate at which antithrombin inactivates clotting factors such as thrombin. The enzyme thrombin plays a key role in the coagulation cascade by cleaving fibrinogen to produce fibrin monomers; thrombin also increases platelet–platelet adhesion and stimulates platelet activation and degranulation. Hence, the inactivation of thrombin (by heparin-bound antithrombin) will inhibit blood coagulation. Surface-tethered heparin is also known to suppress platelet adhesion, complement activation and protein adsorption [67]. Ayres et al. found that polymer brushes containing sulfated carbohydrate repeat units, which resembled surface-tethered heparin, resulted in significantly longer plasma recalcification clotting times than with nonsulfated polysugar polymer brush surfaces used as a control [67]. The sulfated brushes also reduced the production of complement factor products C3a, C4a, and C5a, in comparison to the control surfaces. Other polysaccharide-based glycocalyx-mimicking polymer coatings that reduce platelet adhesion and improve blood compatibility have also been described [68, 69].

1.2.2.4 Clot-Lyzing Thin Films

The PEGylation of poly(dimethyl siloxane) (PDMS) surfaces conferred resistance to nonspecific protein adsorption. Moreover, the incorporation of free ε-amino groups on the surface, by using PEG–lysine conjugates, rendered the surface capable of dissolving fibrin clots because of adsorption of the fibrinolytic protein, plasminogen, from blood plasma [70]. Similar studies in the past had shown that polyurethane surfaces coated with a lysine-derivatized acrylamide polymer dissolved fibrin clots by a ready conversion of the adsorbed plasminogen to plasmin in the presence of tissue-plasminogen activator (TPA) [71]. The design of these lysine-based anticlotting coatings is based on the fact that surfaces incorporating a high density of lysine residues, in which the ε-amino groups are free, are capable of selective adsorption of plasminogen from blood plasma (up to a level of $1.2\,\mu g\,cm^{-2}$, corresponding to a compact monolayer of plasminogen), and virtually no other proteins [72]. In contrast, control surfaces that contained either no lysine, or lysine in which the ε-amino group was not available, adsorbed only very small amounts of plasminogen, and were unable to prevent clot formation.

Nanocomposite fibrinolytic coatings were obtained by tethering proteolytic enzymes to the surfaces of carbon nanotubes (CNTs), which were then dispersed in poly(methyl methacrylate) (PMMA) [73]. Enzymes, such as serine protease subtilisin Carlsberg (SC) and trypsin, were loaded onto the CNTs by physisorption. The extent of nonspecific protein adsorption on these biocatalytic films was 95% lower compared to the enzyme-free film. The incorporation of a fibrinolytic enzyme into the coating resulted in a lowering of fibrinogen fouling by 92%. Clot-lyzing coatings such as these could potentially prevent thrombosis in stents and other blood-contacting implants [74].

Local nitric oxide (NO) release from polymeric surfaces can potently inhibit platelet adhesion and activation, making the surface resistant to clot formation [75, 76]. A low-leaching, NO-generating PEM thin film comprising sodium alginate (ALG) and organoselenium-modified polyethyleneimine was prepared by LBL assembly [77]. The thin films were deposited on biomedical-grade polymer substrates (such as silicone rubber tubings and polyurethane catheters), and produced NO even after prolonged contact with sheep whole blood. The multilayers allowed endogenous S-nitrothiols such as S-nitrosoglutathione (GSNO) and thiol-reducing agents such as glutathione (GSH), to diffuse through the polymer matrix and reach the organoselenium sites, where the catalytic decomposition of GSNO to NO occurred. The LBL coatings showed a very low catalyst leaching.

1.2.2.5 Polyelectrolyte Multilayer Thin Films

There is an increasing interest in preparing antithrombogenic polymer thin films using the LBL assembly technique [78–90]. When PEM coatings of chitosan and dextran sulfate (DS) were prepared on poly(tetramethylene adipate-*co*-terephthalate) membranes [91], it was found that coatings with dextran sulfate as the outermost layer could resist platelet adhesion and human plasma fibrinogen adsorption. A LBL assembly approach has also been used for the *in vivo* repair of damaged blood vessels by forming a multilayer coating of anionic hyaluronic acid (HA) and cationic chitosan on the arterial walls [92]. Chitosan, with its excellent bioadhesive properties on negatively charged surfaces (such as those presented by the damaged arterial lumen), was deposited as the first layer to ensure a strong adhesion of the coating. The growth of blood clot on damaged arterial surfaces was significantly inhibited by the $(CS/HA)_n$ multilayers (87% reduction in platelet adhesion). The incorporation of L-arginine (which is known to inhibit monocyte and platelet adhesion) into the multilayer resulted in a 91% reduction in platelet adhesion compared to the unprotected damaged arteries.

1.2.2.6 Polyurethane Coatings

Novel polyurethane coatings that incorporated hyaluronic acid as a chain extender during polyurethane synthesis have been reported [93]. The surface hydrophilicity was increased with an increase in hyaluronic acid content. Ultimately, a 20-fold decrease in platelet adhesion was identified due to the inclusion of hyaluronic acid, when compared to polyurethane that did not contain the polysaccharide. The polyurethane–hyaluronic acid coatings were cytocompatible and supported endothelial cell adhesion and viability. The surfaces of the poly(ester urethane) guiding catheters were dip-coated with an amphiphilic conjugate of stearyl poly(ethylene oxide) (SPEO) with 4,4′-methylene diphenyl diisocyanate [94]. In order to improve adhesion of the SPEO–diisocyanate surface-modifying additive to the poly(ester urethane) substrate, a "film-building additive" was used in addition to the SPEO–diisocyanate conjugate, while preparing the coating formulation. The film-building additive was a poly(ether urethane), Pellethane® 2363-80AE (Dow Chemical Co.), a polytetramethylene glycol-based polyurethane elastomer.

The resultant coated surfaces resisted blood clotting much more effectively than did the uncoated polyurethane.

1.2.2.7 Vapor-Deposited Thin Films

Chemical vapor deposition (CVD) represents another attractive technique for preparing polymer thin film coatings for biomedical applications. Polymer coatings of various [2.2] paracyclophane (PCP) derivatives were codeposited in controlled ratios by CVD, and the multifunctional coatings evaluated for their biocompatibility and antithrombotic properties [95]. The functionalized PCPs were polymerized into poly(p-xylylenes) during the deposition process.

1.2.3
Antimicrobial Coatings

Of the three million cases of central venous catheter insertions per year in the USA, 30% result in infection-related mortality. This occurs because pathogens (e.g., bacteria) are inadvertently transferred from the skin or air into the wound site during the surgical insertion of implants. Bacterial infection also represents a serious complication in the case of orthopedic implants. When dealing with an infected fibrous capsule (through which antibiotics cannot easily penetrate), or with infections resulting from antibiotic-resistant bacterial strains, surgical removal of the implanted biomaterial very often becomes necessary. It follows that biomedical coatings that could resist bacterial adhesion and colonization would help to prevent implant-associated bacterial infections.

1.2.3.1 Cationic Polymers

Many bactericidal coatings are based on the membrane-disrupting activity of quaternary ammonium, phosphonium, or pyridinium groups [96–98]. Klibanov and coworkers found that surface-tethered poly(4-vinyl-N-hexyl pyridinium bromide) chains were highly effective against Gram-positive bacteria such as *Staphylococcus aureus* and *Staphylococcus epidermis*, as well as the Gram-negative bacteria *Pseudomonas aeruginosa* and *Escherichia coli* [99]. The antibacterial activity was found to depend on the molecular weight of the tethered polymer chains; thus, it was proposed that only sufficiently long and flexible chains would be able to penetrate the bacterial cell walls and disrupt the cell membrane. In contrast, Isquith *et al.* reported that even monolayers of 3-(trimethoxysilyl)-propyldimethyloctadecyl ammonium chloride exhibited antibacterial activity while chemically bonded to a variety of surfaces [100]. The antibacterial activity of these coatings was due to the presence of the surface-bonded molecules, and not to the slow release of membrane-disrupting molecules from the surface. By using spray-coated coatings of quaternized polystyrene-*block*-poly(4-vinylpyridine) block copolymers, the molecular weight was found not to be a limiting factor of antibacterial activity [101], as even a polymer with a relatively low-molecular-weight pyridinium block showed a high bactericidal activity. Similar results were obtained in recent studies involving poly(butylmethacrylate-*co*-Boc-aminoethyl methacrylate) polymer brushes of controlled layer thicknesses and grafting density [102]. Here, the bactericidal efficiency of these surfaces was shown to be independent of the polymer layer thickness, within the range of surfaces studied.

The surface concentration of pyridinium groups is critical in determining bactericidal efficiency [101, 103]. Antibacterial activity was higher when the surface concentration of the quaternary nitrogen [characterized using X-ray photoelectron spectroscopy (XPS) and near edge X-ray absorption fine structure (NEXAFS)] was higher [101]. The bactericidal effect was also higher when a semifluorinated alkyl bromide, $F(CF_2)_8(CH_2)_6Br$, was used for quaternization along with n-hexyl bromide. The fluorinated side chains increased the surface concentration of the high-surface energy pyridinium rings, and this resulted in an enhanced antibacterial activity. Hydrophobic interactions of the highly nonpolar fluoroalkyl groups with the bacterial cell membrane may also have contributed to the antibacterial effect. The extensive literature on antibacterial polymers and coatings based on cationic polymers is discussed elsewhere [101]. Polymeric surface modifiers (PSMs) with soft blocks comprising semifluorinated ($-CH_2OCH_2CF_3$) and 5,5-dimethylhydantoin or alkyl ammonium side groups were found to have good biocidal properties [104–106]. The near-surface amide groups of the hydantoin side groups were converted to antibacterial chloramide groups using hypochlorite. The surfaces of conventional polyurethane (isophorone diisocyante/1,4-butanediol-derived hard block and poly(tetramethylene oxide) soft block) blended with PSM were resistant to both, Gram-positive S. aureus and Gram-negative P. aeruginosa and E. coli. Indeed, only 1.6 wt% of the PSM was found to be sufficient to completely kill P. aeruginosa within a 15 min period. Polyurethane coatings have also been rendered antibacterial by a UV-induced surface-initiated polymerization of 4-vinylpyridine from the polyurethane surface [107].

Hydrophobic fluoroalkyl groups enhanced the bactericidal activity of pyridinium and ammonium polymer coatings [101, 104]. Interestingly, the incorporation of a hydrophilic comonomer such as poly(ethylene glycol)-methyl ether methacrylate or hydroxyethyl methacrylate was also found to increase the antibacterial activity of 4-vinyl-N-hexyl pyridinium coatings [108].

1.2.3.2 Nanocomposite Polymer Thin Films Incorporating Inorganic Biocides

The antibacterial activity of silver has been used to prepare bactericidal thin films for biomedical surfaces. Hybrids of silver particles with highly branched amphiphilically modified polyethyleneimines (PEIs) were found to be bactericidal [109]. Liposomes loaded with silver ions were embedded in poly(L-lysine)/hyaluronic acid multilayer thin films. The controlled release of encapsulated $AgNO_3$ from the coating resulted in a 4-log reduction in the number of viable E. coli cells in contact with the coating [110]. Rubner and coworkers prepared antibacterial coatings based on hydrogen-bonded multilayers containing in situ-synthesized silver nanoparticles, and found these coatings to be efficient against both Gram-positive and Gram-negative bacteria [111]. The same authors also reported a dual-function antibacterial coating with both quaternary ammonium salts and silver [112]. The coatings were prepared by a LBL assembly of poly(allylamine hydrochloride) and poly(acrylic acid); the polymer layers were then coated with silica nanoparticles that were later functionalized with quaternary ammonium silane. The silver nanoparticles were created *in situ* in the polymer layer.

Antibacterial surfaces with a wide range of surface wettability (water contact angles of 30–140°) have been prepared using hydrolytically stable N-alkylmethoxysilane pyridinium polymers [113]. Antibacterial agents such as silver bromide nanoparticles and triiodide ions were also incorporated into these coatings. An electrochemical deposition technique was used to prepare thin films of silver/polymer nanocomposites on stainless steel surfaces; here, the polymer matrix consisted of an inert poly(ethyl acrylate), a macroinitiator of controlled radical polymerization poly(2-phenyl-2-(2,2,6,6-tetramethylpiperidin-1-yloxy)ethyl acrylate), or poly(8-quinolinylacrylate) (P8QA) that has broad antibacterial activity and complexation ability toward metal ions [114]. The silver-containing, electro-grafted, acrylic polymer films showed significant bactericidal activity against *S. aureus*, with the P8QA-based coatings showing the highest antibacterial activity. Antibacterial coatings were also prepared on stainless steel substrates by first depositing an adhesion-promoter layer of hexamethyldisiloxane using a low-pressure plasma technique, followed by the plasma-deposition of ethylene diamine polymer [115]. The PEI coatings were quaternized with alkyl halides to impart antibacterial activity.

The two naturally occurring polymers, alginate and gelatin, have each been used to prepare antibacterial, biodegradable polymer coatings [116]. These coatings were loaded with silver nanoparticles to impart antibacterial activity. Moreover, the pH-responsive swelling/shrinking behavior, resulting from the alginate carboxylic acid groups, could potentially be used for the controlled storage and release of biomolecules. Surface immobilization with bioactive proteins was also shown to be possible. The catechol-functionalized polyelectrolytes were found to form stable LBL assemblies on a variety of substrates, such as poly(tetrafluoroethylene) (PTFE), polyethylene, poly(ethylene terephthalate), and polycarbonate [117]. The catechol groups were used to bind the polyelectrolytes to the substrate, and also for the *in situ* deposition of silver nanoparticles, imparting an antibacterial activity to the multilayer film.

PDMS substrates have been coated with titanium dioxide thin films by liquid-phase deposition, from water, under near-ambient conditions [118]. Such coatings reduced the adhesion of both Gram-positive and Gram-negative bacteria. Moreover, the bacterial adhesion was further reduced if the TiO_2 over-layer was irradiated with UV light before introduction of the bacteria. Antibacterial coatings have also been prepared using chitosan/heparin multilayer thin films that consisted of embedded TiO_2 or silver nanoparticles [119].

1.2.3.3 Antibiotic-Conjugated Polymer Thin Films

Atom transfer radical polymerization (ATRP) was used to grow poly(2-hydroxyethyl methacrylate) polymer brushes on titanium surfaces. For this, the pendent hydroxyl groups were converted into carboxyl or amine groups, which were used to covalently immobilize antibiotics such as gentamicin and penicillin. These coatings showed a significant decrease in the viability of *S. aureus* [120]. Surfaces of PTFE grafted with penicillin also showed antibacterial activity [121]. Poly(dimethylaminomethyl styrene), which is cationic by virtue of protonation, was coated onto substrates using an initiated chemical vapor deposition (iCVD) method, and found to be very effective against Gram-negative *E. coli* and Gram-positive *B. subtilis* [122]. Other biocompatible coatings and antimicrobial coatings have also been prepared using iCVD (for a review, see Ref. [123]).

1.2.3.4 Biomimetic Antibacterial Coatings

Facially amphiphilic polymers that mimic the physico-chemical properties of natural host defense peptides have been found to have excellent antibacterial activity coupled with selectivity, which in turn makes them promising candidates for imparting antibacterial activity to biomedical surfaces [124]. Etienne *et al.* prepared antifungal coatings by embedding the antifungal peptide chromofungin into PEM thin films [125]. Despite being embedded in the film, the chromofungin was able interact with the membrane of the fungi and demonstrate antimicrobial activity. The antifungal coatings did not exhibit any cytotoxicity towards eukaryotic cells (e.g., human gingival fibroblast cells). The antibacterial peptide, defensin, was also embedded in PEM films to impart bactericidal activity to the coatings [126]. In this way, copolymer brushes of 2-(2-methoxyethoxy)ethyl methacrylate and hydroxyl-terminated oligo(ethylene glycol) methacrylate were grafted with the natural antibacterial peptide, magainin I, and found to be effective against different strains of Gram-positive bacteria [127].

1.2.3.5 Thin Films Resistant to the Adhesion of Viable Bacteria

Nonbiocidal coatings have also been investigated for their ability to resist bacterial colonization. Coatings of "surfactant polymers" with a structure consisting of a poly(vinyl amine) backbone and hydrophilic PEG and hydrophobic *n*-hexyl grafts, were successful in suppressing bacterial adhesion on biomaterial surfaces [128]. The PEG packing density and hydration thickness were found to be critical in determining the ability of the coating to shield the surface against bacterial interactions. PEMs prepared using poly(L-lysine) and poly(L-glutamic acid)-*graft*-poly(ethylene glycol) were similarly found to drastically reduce both protein adsorption and bacterial adhesion [129].

Poly(L-lysine)-*graft*-poly(ethylene glycol) polymers, functionalized with bioligands such as RGD, were adsorbed from aqueous solutions onto negatively charged metal oxide surfaces, reducing protein adsorption as well as the adhesion of *S. aureus*, *S. epidermidis*, *S. mutans*, and *P. aeruginosa* to titanium surfaces [130]. The RGD-functionalized thin films selectively allowed cells such as fibroblasts to attach, which makes them useful as coatings for biomedical implants that can mediate the adhesion of host cells to the implant surface, but do not allow bacterial colonization. Zwitterionic poly(sulfobetaine methacrylate) polymer brushes resulted in a more than 90% reduction in the adhesion of viable bacterial cells relative to glass controls [131].

SAMs presenting methyl, L-gulonamide (a sugar alcohol tethered with an amide bond), and triethylene glycol were tested for resistance to *E. coli* biofilm formation

[132]. The triethylene glycol-terminated SAM was the most resistant. PEM coatings on glass, prepared using chitosan and hyaluronic acid, lowered the attachment density of E. coli by approximately 80% if the films were sufficiently thick (~300 nm) and hydrated (with low rigidity) [133]. The attachment density did not depend on the terminal layer, and was similar on both $(CS/HA)_{10}$ and $(CS/HA)_{10}CS$ films. However, if the LBL assembly was performed at a lower ionic strength (0.01 M NaCl instead of 0.15 M NaCl), then thinner coatings were obtained (~120 nm for a $(CS/HA)_{20}$ multilayer film), the surfaces were more rigid, and the bacterial adhesion was greater.

Lichter et al. prepared PEM thin films comprised of poly(allylamine hydrochloride) and poly(acrylic acid), with dry a thickness of about 50 nm, and found that the adhesion of viable S. epidermis, a Gram-positive bacterium, correlated positively with the stiffness of the polymeric substrates [134]. The elastic moduli of the hydrated films was controlled using the pH-modulation of the extent of ionic crosslinking in the thin films, and ranged over two orders of magnitude (from 0.8 to 80.4 MPa). The colony density was lowest on the most compliant films, with $E \sim 0.8$ MPa; similar trends were observed for Gram-negative E. coli. Thus, the mechanical stiffness of biomedical coatings could play an important role in regulating the adhesion and subsequent colonization of viable bacteria.

UV-induced grafting/polymerization of poly(N-vinyl-2-pyrrolidone) onto poly(ethylene terephthalate) (PET) surfaces, in an aqueous medium, prevented colonization of the surface by S. aureus [135]. The covalent immobilization of silk sericin on poly(methacrylic acid)-functionalized titanium surfaces was found to significantly reduce the adhesion of S. aureus and S. epidermis, while promoting the adhesion, proliferation, and alkaline phosphatase activity of osteoblasts [136]. The hydrophilic poly(methacrylic acid) lowered bacterial adhesion, while the silk sericin protein enhanced osteoblast attachment and proliferation. Such thin film coatings on titanium surfaces could potentially be used to prevent the bacterial infection of bone implants, while promoting osseointegration.

Puskas et al. reviewed the biomedical applications of polyisobutylene-based biomaterials [137]. The arborescent (randomly branched, tree-like) polyisobutylene-polystyrene block copolymers (PIB-PS) surfaces showed a greatly reduced attachment of a common uropathogenic species, E. coli 67, compared to medical-grade silicone rubber (SIL-K™). The relatively hydrophobic PIB-PS surfaces resulted in a strong binding of the adsorbed proteins; hence, when the surfaces of both PIB-PS and SIL-K™ were coated with a 29 kDa neutrophil protein, p29, further significant reductions in uropathogen attachment were observed (approximately 90% on PIB-PS and 60% on SIL-K™).

1.3
Coatings for Tissue Engineering Substrates

Polymers such as polystyrene (e.g., TCPS substrates) and PDMS [138] have frequently been used as substrates for cell culture and tissue engineering applications. A variety of thin-film coating strategies have been developed to impart biocompatibility and biofunctionality to these substrates. The different techniques available for the immobilization of bioactive molecules onto surfaces have been reviewed by Goddard *et al.* [139]. Here, we will highlight polymeric thin films that have been used to control protein adsorption and cell adhesion onto tissue engineering substrates.

1.3.1
PEGylated Thin Films

Poly(ethylene glycol) is used extensively in biomedical surface modification. In a review, Krsko and Libera have discussed strategies for controlling the interaction of cells and proteins with PEG-based coatings [140]. The different methods for attaching PEG molecules to a surface include:

- The use of SAMs of short oligomers.

- The adsorption of triblock copolymers such as poly(ethylene oxide)-*block*-poly(propylene oxide)-*block*-poly(ethylene oxide), known as Pluronic™.

- Surface-grafted PEGylated brushes.

- Thin-film hydrogels obtained by chemical crosslinking techniques that include irradiation with ionizing radiation, such as high-energy electron-beam or gamma rays.

Mrksich and coworkers have reported a strategy for the controlled, irreversible immobilization of adhesion proteins on biologically inert surfaces of PEG-terminated SAMs [141].

Low-friction surfaces for biomedical applications were obtained by functionalizing PDMS surfaces with PEG–DOPA–lysine conjugates [142]. The DOPA (L-3,4-dihydroxyl-L-phenylalanine) and lysine peptide mimics of mussel adhesive proteins resulted in a strong attachment of the PEGylated thin films to PDMS. The resultant surfaces had an extremely low friction coefficient (~0.03) compared to bare PDMS (~0.98), although a lowering of the friction coefficient occurred only when DOPA was bound to lysine. Modification with PEG–DOPA did not have any effect on the friction coefficient. Rather, lysine played a critical role in lowering the friction coefficient.

1.3.2
Zwitterionic Thin Films

Poly(carboxybetaine methacrylate) homopolymer brushes (of 10–15 nm thickness) were prepared using surface-initiated ATRP [143]. The zwitterionic nature of the thin-film resulted in a well-hydrated surface, and prevented any nonspecific adsorption of fibrinogen, lysozyme and human chorionic gonadotropin proteins. The carboxylic acid side groups were also used to selectively immobilize fibronectin, after which the surfaces showed a good adhesion and spreading of aortic endothelial cells. These dual-function polymer brushes could potentially be used as coatings for tissue engineering substrates and biosensors. The effects of incorporating cationic charges into zwitterionic polymers (based on phosphorylcholine; PC) on biocompatibility, specific cell–surface interactions and cytotoxicity have also been studied [144]. The cationic comonomer, choline methacrylate (CMA), was used to introduce cationic groups into the zwitterionic polymer. While phosphorylcholine $[Me_3^+N-CH_2-CH_2-O-P(=O)(O^-)OH]$ is a zwitterion, with no net charge, choline $[Me_3^+N-CH_2-CH_2-OH]$ is cationically charged. In general, although protein- and cell-repellant surfaces are necessary for biocompatibility, the ability to promote cell adhesion and growth is important in the surface modification of biomaterials such as vascular grafts. The cationic surfaces promoted both cell adhesion and growth; indeed, the presence of a cationic charge in the PC-based coatings led to a significant increase in the adsorption of different proteins, as well as in the adhesion of fibroblasts, epithelial cells, granulocytes, and mononuclear cells. The surfaces remained noncytotoxic up to 30 mol% of the cationic moiety. The coating stability in PBS, however, was greatly reduced when the CMA content exceeded 15 mol%, evidently because of an interaction of the ions in PBS with the cationic groups in the coating.

1.3.3
Thin Films of Hyperbranched Polymers

Fluoropolymer surfaces have been functionalized with thin films of a hyperbranched glycopolymer for biocompatibility. The glycopolymer was prepared by the atom transfer radical copolymerization of 2-(2-bromopropionyloxy)ethyl acrylate inimer (initiator + monomer) and a sugar-carrying acrylate, 3-O-acryloyl-1,2:5,6-di-O-isopropylidene-α-D-glucofuranoside, and was grafted onto the fluorocarbon substrate using low-pressure argon plasma [145]. The hyperbranched glycopolymer thin films were found to promote fibronectin adsorption and human umbilical vein endothelial cell (HUVEC) adhesion, which would make them useful as coatings for tissue engineering substrates. Spin-coated thin films of sorbitol-containing polyesters, which were synthesized via a one-pot, lipase-catalyzed condensation polymerization, were found to elicit fibroblast 3T3 cell behavior similar to that of a biocompatible poly(ε-caprolactone) control [146]. The polymer could be functionalized with biological molecules such as oligopeptides or oligosaccharides by attachment to the hydroxyl groups of sorbitol. Moreover, the lipase-catalyzed polymerization avoided the use of potentially toxic catalysts.

The adhesion and proliferation of human corneal epithelial cells (HCECs) on thin film coatings of hydroxyl-terminated aliphatic polyester dendrons were compared to the interaction of these cells with hydroxyl-terminated PEG SAMs on gold [147]. Whilst little or no HCEC adhesion was observed on the PEG SAMs, the HCEC proliferation was increased exponentially on the dendronized surfaces. The attachment density increased in line with an increase of the generation number of the dendrimer. When the peripheral hydroxyl groups of the dendronized surfaces were further reacted with methoxy-terminated PEG chains, HCEC adhesion was significantly reduced. Two facts became evident from this study. First, the surface hydrophilicity alone does not confer cell repellency to a surface. The hydroxyl-terminated PEG SAMs and hydroxyl-terminated dendrimers were both hydrophilic and hydrated, but the latter was cell-adhesive and the former cell-repellant. Second, the dendronized surfaces induced the adsorption of adhesive proteins (e.g., fibronectin) that are secreted by the cells, and thus promoted cell adhesion. When the peripheral hydroxyl groups of the dendrimers were PEGylated, however, the cell density was even lower than that of the PEG SAMs. Consequently, the high density of PEG groups at the surface caused the surface to become repulsive to both proteins and cells.

1.3.4
Polyurethane Coatings

By using combinations of five different polyols, six different isocyanates, and nine different chain extenders, a library of 120 polyurethanes has been synthesized and evaluated for cell adhesion, using microarray-based assays [148–150]. The polyurethanes which showed the highest number of adhered primary renal tubular epithelial cells contained 4,4′-methylenebis(phenylisocyante) (MDI) as the diisocyanate, and poly(tetramethylene glycol) (PTMG) as the diol [148]. Clearly, such polymer chemistry will influence factors such as surface wettability, surface roughness, and coating modulus, all of which will in turn affect cell adhesion. Similarly, those polyurethanes that successfully bound immature bone marrow dendritic cells also contained MDI as the diisocyanate, PTMG (250 or 1000 Da) as the diol, and either propyleneglycol, 1,4-butanediol, or no chain extender [149]. The immobilization of dendritic cells, which play a key role in the initiation of immune response by acting as antigen-presenting cells, is important in the development of vaccines, particularly those against tumors. It is important to identify efficient substrates for the immobilization of these cells in their immature state, as maturation can affect their ability to capture antigens by phagocytosis. The polyurethane coatings were also screened for applications such as human skeletal progenitor cell isolation and surface modification of tissue engineering scaffolds aimed at enhancing skeletal cell growth and differentiation. Among 120 polyurethanes analyzed in the present study for their ability to bind to skeletal progenitor cells from human bone marrow, only four exhibited high binding affinities for STRO-1+ cells from human bone marrow. Each of these high-affinity polyurethanes was hydrophilic ($\theta_w \sim 30°$), contained PEG (2000 or 900 Da) as the polyol, contained either MDI or 1,4-phenylene diisocyanate (PDI) as the diisocyanate, and either 1,4-butanediol or no chain extender. Alperin *et al.* used solvent-cast thin films of a biodegradable polyurethane to prepare cardiac grafts for heart tissue regeneration [151]. In myocardial infarction, a macroscopic area of the heart muscle tissue is damaged due to an inadequate supply of blood. As cardiomyocytes are terminally differentiated cells, they are unable to regenerate heart tissue after infarction. Hence, embryonic stem cell-derived cardiomyocytes were seeded onto polyurethane films, and coated with ECM proteins such as laminin or collagen type

IV to promote cell adhesion; the result was an elastomeric film that could respond to contractile forces produced by the cultured cardiomyocytes.

1.3.5
Polysaccharide-Based Thin Films

The presence of specific biomolecules such as chitosan or O-carboxymethylchitosan (OCMCS) at the surfaces of tissue-engineering substrates was found to be as important as surface-wettability or charge in influencing cell–surface interactions [152]. The surface-bound free radicals, which were generated by the treatment of a PET surface with argon plasma, were used to grow poly(acrylic acid) (PAA) brushes from the PET surface. Both, CS and OCMCS were covalently immobilized on the PAA brush using a carbodiimide-mediated reaction of the acid groups on the surface with amine groups of CS or OCMCS. The distinctly different morphologies of smooth muscle cells on the CS and OCMCS surfaces were attributed to specific interactions of OCMCS with the cell membrane. The OCMCS-modified surfaces were also found to be protein-repellant, with excellent antithrombogenic properties.

By using an *in vitro* cell culture of human mesenchymal stem cells and *in vivo* subcutaneous implantation into mice, it was found that covalently immobilized collagen coatings clearly improved the cytocompatibility of stainless steel implants [153]. The stainless steel surfaces were first coated with a 200 nm-thick tantalum coating (using magnetron sputtering from a tantalum target in argon atmosphere), followed by a 70 nm-thick tantalum oxide coating (by introducing oxygen during the sputtering process), to improve the corrosion resistance of stainless steel. These tantalum oxide-modified stainless steel surfaces were functionalized with aminopropyl triethoxy silane, and further activated by immersing the surfaces in N,N'-disulfosuccinimidyl suberate. Collagen molecules were covalently tethered to these activated surfaces, and chemically crosslinked using a carbodiimide chemistry. Such chemical crosslinking greatly improved the resistance against biodegradation and mechanical stability of the bioactive coating. The collagenous layer would also enhance cell adhesion and integration of the biomaterial with the surrounding tissue.

1.3.6
Polyelectrolyte Multilayer Thin Films

Currently, there is an active interest in using multilayer polymer thin films for tissue engineering, which is evident from the number of reports in this area; comprehensive reviews on this topic are available, produced by the groups of Schlenoff and Kotov [154, 155]. Hubbell and coworkers were among the first to investigate the interaction of cells with LBL-assembled PEM thin films [156]. By using human fibroblast cells, which are known to spread aggressively on most surfaces when cultured in serum-containing medium (the medium would thus contain proteins such as fibronectin to promote cell attachment), Hubbell's group showed that poly(L-lysine)/alginate multilayers could prevent the fibroblasts from spreading. The LBL assemblies were shown to be stable in cell culture medium for over 24 h and to remain attached to the substrate under fluid flow, even at a wall shear rate of $1000\,s^{-1}$. Additional tissue engineering applications of PEM assemblies were proposed by Ogier and coworkers, who investigated the cell adherence, viability, phenotype expression and inflammatory response [via tumor necrosis factor-α (TNF-α) and interleukin (IL)-8 secretion] of human osteoblast-like SaOS-2 cells and human periodontal ligament cells, on multilayer thin films [157–159]. Except for PEI, which was cytotoxic, poly(sodium 4-styrenesulfonate), poly(allylamine hydrochloride), poly(L-glutamic acid) and poly(L-lysine) were all shown to be biocompatible, which suggested that multilayer coatings prepared using these polyelectrolytes would be suitable for implant coatings. Subsequently, Chluba et al. immobilized a peptide hormone, α-melanocortin, on $(PGA/PLL)_n$ multilayer assembly by covalently binding the hormone to PLL forming the outer layer [160]. Chluba's group showed the immobilized hormone to be as biologically active as the free hormone. It follows that the covalent immobilization of growth factors on biomaterial surfaces may have important tissue engineering applications [161–165].

Boura et al. investigated the possibility of using PEM thin films as coatings to improve the biocompatibility of small-diameter vascular grafts. Specifically, the group determined the ability of these coatings to support and maintain a confluent layer of healthy HUVECs [166]. In these studies, the $(PSS/PAH)_n$ and poly(L-glutamic acid)/poly(D-lysine) $(PGA/PDL)_n$ multilayers were shown to be noncytotoxic, nor to alter the phenotype of the endothelial cells. The PEMs also showed a higher initial cell attachment compared to polyelectrolyte monolayers. Cell growth on these multilayer thin films was similar to that on TCPS. The $(PSS/PAH)_n$ multilayers showed excellent biocompatibility and a greater growth and adhesion of HUVECs than did the $(PGA/PDL)_n$ multilayers.

Mendelsohn et al. have reported the influence of processing conditions on the cell-adhesive properties of $(PAH/PAA)_n$ PEM thin films [167]. The degree of ionization of the polyelectrolytes (i.e., the relative number of NH_3^+ versus NH_2 groups for PAH, $pK_a \sim 9$, and the number of COO^- versus $COOH$ groups for PAA, $pK_a \sim 5$), as well as the crosslink density (i.e., the number of ionic bonds, $COO^- \cdots NH_3^+$) was tuned using the deposition pH conditions. When PAH and PAA were both deposited from solution at pH 6.5 (denoted as 6.5/6.5 PAH/PAA), both polymers were fully charged molecules and formed thin, flat layers because of the high ionic crosslink density (cf. Figure 1.3). The film swelled by only ~115% of its original dry height in PBS. In the 7.5/3.5 PAH/PAA multilayers, both PAH and PAA were partially ionized and adsorbed in loop-rich conformations, forming thick layers with a high degree of internal charge pairing. The multilayers did not possess well-blended surfaces, and the chemical groups of the last-deposited polymer dominated the surface. In the 2.0/2.0 PAH/PAA multilayers, both the interior and the surface of the film were enriched by PAA chains, irrespective of the outermost

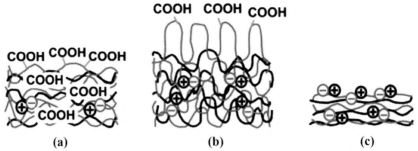

Figure 1.3 Schematics of the (a) 2.0/2.0, (b) 7.5/3.5, and (c) 6.5/6.5 PAH/PAA [poly(allyl amine hydrochloride)/poly(acrylic acid)] multilayer assemblies, shown with PAA as the outermost layer. Reproduced with permission from Ref. [167]; © 2003, American Chemical Society.

layer. The extent of ionic crosslinking was low because most of the PAA groups existed in their uncharged, protonated COOH state. The film swelled by almost 400% of its original thickness.

The PEM thin films were investigated for their *in vitro* interactions with a highly adhesive murine fibroblast cell line. Prior to seeding with the fibroblast cells, which were suspended in normal serum-containing media, the coatings were sterilized with 70% (v/v) ethanol (this sterilization did not affect the mechanical integrity of the films). The 2.0/2.0 PAH/PAA films completely resisted attachment of the NR6WT fibroblast cells, and were not cytotoxic; however, it was found that at least 15 layers were required to create a surface that resisted cell attachment. In contrast, cell attachment was observed on the TCPS control and the 6.5/6.5 and 7.5/3.5 PAH/PAA films. Interestingly, the 6.5/6.5 and 7.5/3.5 PAH/PAA multilayers were always cell-adhesive, irrespective of whether PAH or PAA was the last layer adsorbed; similarly, the 2.0/2.0 multilayers were always cell-resistant, irrespective of the last polyion deposited. Both, the cytophilic 7.5/3.5 and the cytophobic 2.0/2.0 PAH/PAA films, readily adsorbed model proteins from a solution in PBS (Figure 1.4). However, in comparison to an uncoated gold surface, the PEMs showed a lower adsorption of the predominantly anionic protein fibrinogen. All of the PEM coatings adsorbed the highly cationic lysozyme, regardless of their net surface charge. Notably, the cell-resistant 2.0/2.0 PAH/PAA coatings adsorbed more of each protein than did the 7.5/3.5 system. A similar study, on the influence of the type of the outermost layer, the presence of proteins, and the number of layers in the film on cell interactions, was reported earlier by Richert *et al.* [168].

Schneider *et al.* have proposed that (PLL/PGA)$_n$ PEMs grafted with sugar molecules (e.g., mannose) could be used not only as nonviral vectors but also as cell-adhesive substrates in tissue engineering [169]. PGA was selected as the terminal layer in this study because it showed a higher cell viability compared to PLL. Specific interactions were identified of the primary chondrocytes with the glycated thin films, because of which the cells adhered well to these films. On the other

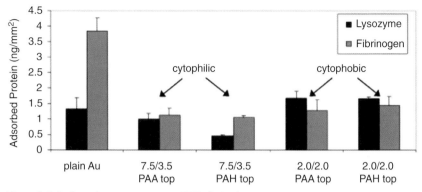

Figure 1.4 Surface plasmon resonance (SPR)-derived adsorption data for lysozyme and fibrinogen on an uncoated gold surface and on gold coated with 10 to 11 layers of the cytophilic 7.5/3.5 or 14 to 15 layers of the cytophobic 2.0/2.0 PAH/PAA multilayer system. Reproduced with permission from Ref. [167]; © 2003, American Chemical Society.

hand, chondrosarcoma cells did not grow well on the mannose-grafted film. Moreover, while cell adhesion was strongly influenced by the mannose, the effect of lactose was much less obvious. The specific interaction of primary chondrocytes with the surface was attributed to the large number of mannose receptor transmembrane proteins present at the cell surfaces. Such preferential adhesion of primary cells to a biomaterial surface, when compared to that of tumor cells, would be an important factor for improving the biocompatibility of implanted prostheses following surgical ablation.

Wittmer et al. studied protein adsorption and HUVEC attachment on fibronectin-terminated PEM thin films consisting of PLL and dextran sulfate [170]. It was observed that fibronectin, which enhanced cell adhesion, was adsorbed in an irreversible manner and to a greater extent on the positively charged and less hydrated PLL-terminated films, than on the DS-terminated films. The adsorbed fibronectin subsequently promoted cell spreading. Moreover, positively charged PLL-terminated films showed a greater degree of cell-spreading than with negatively charged DS-terminated films. Fibronectin adsorption on the LBL assembly resulted in a lower film hydration, a higher surface charge, and also enhanced cell spreading on the thin film.

Menu and coworkers used PEM thin films of PSS and PAH to coat the luminal side of cryopreserved human umbilical arteries, in order to promote re-endothelialization, so that the coated arteries could be used as vascular grafts [171]. The internal walls of the de-endothelialized arteries were coated with a (PAH/PSS)$_3$PAH film by sequential injection of the PAH and PSS solutions, with a 15 min incubation period followed by a 15 min rinse period after each injection. The biomechanical properties of the LBL-coated umbilical arteries were similar to those of fresh arteries. Notably, the PEM coating greatly promoted endothelialization of the coated surface via differentiation of endothelial progenitor cells into mature endothelial cells. A confluent endothelial cell monolayer was formed within two weeks, while previous fibronectin-coated surfaces required about two months to achieve confluence [172].

Wittmer et al. determined the factors affecting the attachment and function of hepatic cells on multilayer nanofilms formed by LBL assembly [173]. The group also studied the role of chemical crosslinking after LBL assembly. Both, biopolymers [e.g., chitosan and alginate (ALG)] and synthetic polymers (e.g., PAH and PSS) were used to prepare the multilayer films. The types of polymer which comprised the multilayer film played an important role in the bioresponse. Although, none of the pure polysaccharide films promoted attachment and growth of the human hepatocellular carcinoma (HepG2) cells, one polysaccharide–polypeptide multilayer, which was composed of PLL and ALG, promoted a strong attachment. Whilst the overall film charge was found to be unimportant in influencing cell behavior, the film terminal layer had a quite strong influence on hepatic cell attachment and growth. Multilayers which terminated with the anionic PSS, in (PAH/PSS)$_n$ thin films, promoted HepG2 attachment and growth. Likewise, a cationic terminal layer, in a PLL/ALG assembly, also resulted in confluent culture of the HepG2 cells. Film rigidity, which was engineered by chemical crosslinking of the layers, was found also to affect cell response; indeed, HepG2 attachment and growth was significantly enhanced by chemical crosslinking (and therefore film rigidity). The (PAH/PSS)$_n$ films, crosslinked (PLL/ALG)$_n$ films, and crosslinked PLL/PGA films with a terminal PLL layer, were each identified as the most promising candidates for *in vivo* human liver tissue engineering applications.

Ren et al. observed that the initial adhesion and proliferation of skeletal muscle cells (C2C12 cells), on 1 μm-thick PEM films, and their differentiation into myotubes, depended on the stiffness of the film [174]. The surface elastic moduli, E, were measured using atomic force microscopy (AFM) nanoidentation experiments, and were varied by varying the crosslink density of the films. Stiff films ($E > 320$ kPa) of crosslinked poly(L-lysine)/hyaluronic acid (PLL/HA) multilayers promoted the formation of focal adhesions and enhanced proliferation, whereas soft films were not favorable for cell anchoring, spreading, or proliferation. The crosslinked (PLL/HA)$_n$ films did not require specific protein or ligand precoating to promote cell adhesion. In fact, the crosslinked films were very hydrophilic (water contact angle, $\theta_w < 10°$), and showed a low adsorption (~100 ng cm^{-2}) of proteins from fetal bovine serum (FBS). Interestingly, the un-crosslinked PEM films were moderately hydrophobic and showed a high adsorption of FBS proteins (~2000 ng cm^{-2}).

Sallolum et al. studied the effect of surface charge, film thickness, hydrophobicity, and the presence of zwitterionic groups, on the adhesion and spreading of vascular smooth muscle cells on different PEM coatings [175]. Polyelectrolytes such as PAA, poly(methacrylic acid)-*block*-poly(ethylene oxide) (PMA-*b*-PEO), PSS, a perfluorosulfonate ionomer (Nafion™), PAH, poly(diallyldimethylammonium chloride) (PDADMA), poly(2-vinylpyridine)-*block*-poly(ethylene oxide) that was 86% quaternized with methyl iodide (PM2VP-*b*-PEO), and poly(4-vinylpyridine) that was 45% quaternized with 1H,1H,2H,2H-perfluorooctyl iodide (PFPVP), were used. The fluorinated polyelectrolytes were found to promote cell adhesion. In general, hydrophobic polyelectrolyte film surfaces, regardless of their formal charge, were more cytophilic than hydrophilic surfaces. Moreover, the number of multilayers had no effect on cell adhesion and growth. Thin films prepared from a copolymer of acrylic acid and (3-[2-(acrylamido)-ethyldimethyl ammonio]propane sulfonate) (AEDAPS) were used to study the effect of zwitterionic groups. Cell adhesion decreased with an increase in the fraction of AEDAPS in the copolymer. Interestingly, the negatively charged surfaces of (PM2VP-b-PEO/PMA-b-PEO)$_2$ showed a greater cell spreading ability than the positively charged surfaces of (PM2VP-b-PEO/PMA-b-PEO)$_2$(PM2VP-b-PEO). The images of rat aortic smooth muscle A7r5 cells cultured on the diblock surfaces reported by these authors, however, indicated that more cells had settled on the positively charge surface, as expected. A micropatterning of the cells could be achieved by stamping Nafion™ on the PAA–PAEDAPS copolymer.

Wu et al. found that PEMs prepared using HA and poly(allylamine hydrochloride) or collagen (COL), supported neural cell adhesion, neurite elongation, and neural network formation [176]. These films, when deposited onto amino-functionalized glass slides, were found to be cytocompatible with hippocampal and cortical neurons. The hippocampal neurons preferred the (HA/PAH)$_n$ films, while the cortical neurons preferred the (HA/COL)$_n$ films. Neurite outgrowth could not be simply correlated to the terminal layer, and was also found to depend on the number of bilayers. Nadiri et al. have reported that bone morphogenetic proteins (BMP) and the BMP antagonist ("Noggin"), which were embedded in poly(L-glutamic acid)/poly(L-lysine) PEM thin films, could be used to induce or inhibit cell death. Such a control on cell apoptosis could find applications in tissue repair, and in the specific "shaping" of artificial organs [177]. PEM thin films have also been prepared using proteins. Haynie et al. have reviewed the biomedical applications of polypeptide multilayer films [178].

Moby et al. used the LBL technique to coat the luminal surfaces of expanded PTFE tubes with PEM thin films composed of PEI, PSS, and PAH [179]. The PEI(PSS/PAH)$_3$ coatings promoted endothelial cell adhesion and resulted in a healthy confluent cell monolayer formation. The cell viability on the multilayer thin films was greatly improved compared to the nonmodified PTFE surface. The presence of a confluent endothelial layer is necessary for the successful replacement of diseased vessels by synthetic vascular grafts.

He et al. have used LBL assemblies of polycations such as PEI or CS, with polyanions such as gelatin or laminin (LN) to coat silicon microelectrode arrays of neural implants [180]. Neural implants are used for the in vivo recording of neural activity, or for stimulating neurons with electrical impulses from an external source. The insertion of rigid metal electrodes into soft neural tissue triggers the formation of a scar around the metal electrode, which electrically insulates the electrode from the neurons. Coatings that can prevent scar formation and promote the adhesion of neurons can avoid electrical isolation of the electrode due to scar tissue. In vitro experiments showed that the (PEI/LN)$_n$ coatings promoted the

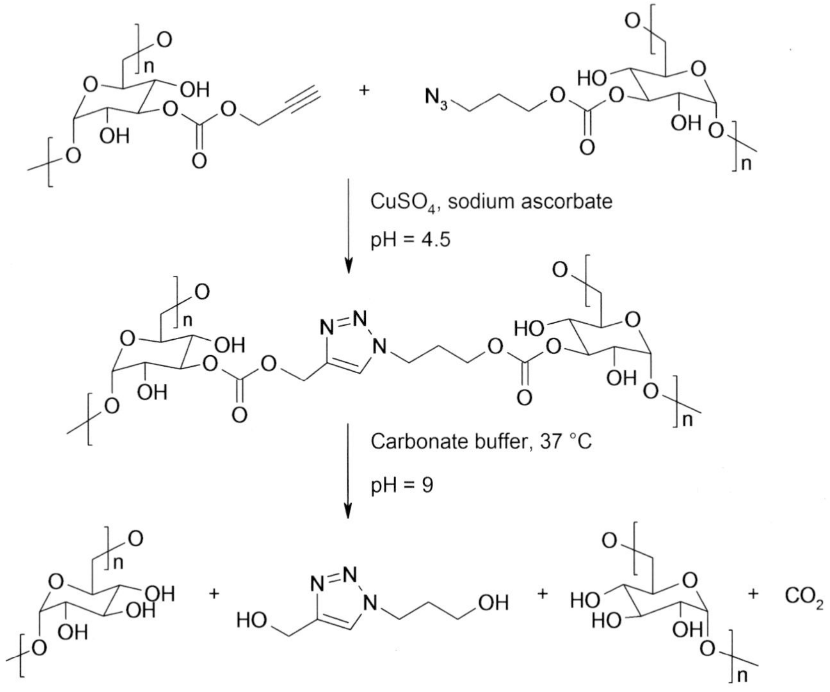

Figure 1.5 1,3-Dipolar click cycloaddition reaction between dextran-propargyl carbonate and dextran-azidopropyl carbonate to from triazole ring linkages between the layers. Hydrolysis of the carbonate esters that linked the triazole ring and dextrose resulted in a disintegration of the multilayered film. Adapted with permission from Ref. [183]; © 2008, Wiley-VCH Verlag.

adhesion and differentiation of chick cortical neurons, without increasing the impedance of the electrodes. Single-walled CNT PEMs have also been proposed as biocompatible platforms for neuroprosthetic implants [181].

Lee et al. [182] prepared tissue-engineering scaffolds using inverted colloidal crystals. For this, the internal surfaces of the scaffolds were coated with clay/poly(diallyl dimethylammonium chloride) LBL multilayers to enhance cell adhesion. Cocultures of adherent and nonadherent cells were obtained using these scaffolds, which were fabricated with the goal of an *in vitro* replication of the differentiation microenvironments, or niches, of hematopoietic stem cells.

De Geest et al. have prepared polyelectrolyte-free, polymeric multilayer films containing alkyne- and azide-functionalized dextrans using the LBL assembly technique [183]. The interlayer crosslinking was achieved via the triazole linkages formed by the Huisgen 1,3-dipolar cycloaddition reaction between the alkyne and azide groups (Figure 1.5). The coatings were biodegradable as a result of the hydrolysis of carbonate ester links present in the polymers. Such biodegradable

1.3.7
Temperature-Responsive Polymer Coatings

Thin films and coatings of temperature-responsive polymers, such as poly(N-isopropylacrylamide) (PNIPAAm), have been widely investigated in the area of cell sheet engineering [184]. This process, which involves tissue reconstruction from cell sheets rather than from single cells, was developed to overcome the limitations of tissue reconstruction using biodegradable scaffolds or by the injection of cell suspensions (Figure 1.6) [185–187]. In this approach, temperature-responsive polymers are covalently grafted onto tissue culture dishes, which allows various types of cell to adhere and proliferate at a temperature above the lower critical solution temperature (LCST) of the grafted polymer (a state wherein the surface is hydrophobic) [188]. The cells detach spontaneously when the temperature is lowered below the LCST, because of spontaneous hydration of the grafted polymer chains [189]. Cell detachment from the thermally responsive surfaces was a result of active cellular metabolic processes triggered by surface-wettability changes [190, 191]. A covalently grafted layer of PNIPAAm of about 20 nm thickness allows thermally responsive cell adhesion and detachment [192]. The confluent cells can be harvested noninvasively as single, contiguous cell sheets with intact cell–cell junctions

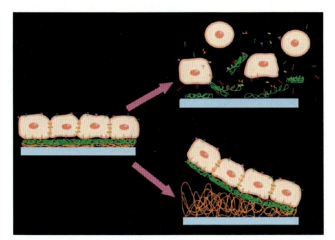

Figure 1.6 Cell sheet harvesting. Trypsin degrades the deposited extracellular matrix (ECM; green), as well as the membrane proteins, so that confluent, monolayer cells are harvested as single cells (upper right). The temperature-responsive polymer (orange), covalently immobilized on the dish surface, hydrates when the temperature is reduced; this decreases the interaction with the deposited ECM. All the cells connected via cell–cell junction proteins are harvested as a single, contiguous cell sheet, without the need for proteolytic enzymes (lower right). Reproduced with permission from Ref. [185]; © 2004, Elsevier.

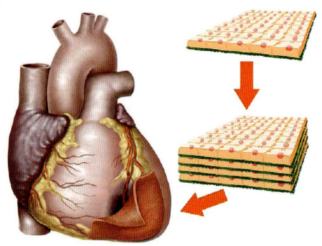

Figure 1.7 Cardia patch cell sheet engineering. Cardia myocyte sheets are harvested from temperature-responsive culture dishes. Four cell sheets are then stratified and transplanted to ischemic hearts as cardia patches. Reproduced with permission from Ref. [185]; © 2004, Elsevier.

and a deposited ECM. The low-temperature liftoff from PNIPAAm surfaces is less damaging to the ECM proteins than enzymatic digestion and mechanical dissociation methods of cell harvesting [193]. A cell sheet composed of different types of cell (patterned cocultured cell sheets) can be obtained by patterning the substrate with thermoresponsive polymers with different LCSTs [194]. Double-layered cell sheets can also be engineered in this way [195]. Corneal epithelial cell sheets and retina pigment epithelial cell sheets for ocular surface regeneration [196–199], periodontal ligament cell sheets for regenerating the connective tissue that attaches a tooth to the alveolar bone, urothelial cell sheets for bladder augmentation, and cardiomyocyte sheets for engineering electrically communicative, pulsatile, three-dimensional (3-D) cardiac constructs (Figure 1.7) [200–202], have each been obtained using cell sheet engineering.

Patterned surfaces have been used to produce heterotypic cell cocultures by covalently grafting PNIPAAm onto the tissue culture surfaces [203, 204]. PNIPAAm brushes were grown from TCPS surfaces by polymerization of the N-isopropylacrylamide (NIPAAm) monomer upon exposure to an electron beam through a patterned metal mask. The patterned surfaces were used for the culture of hepatocytes. When the temperature was reduced below 37 °C, hepatocytes became selectively detached from the PNIPAAm regions of the patterned substrate and were replaced by endothelial cells that were seeded and cocultured (with the remaining hepatocytes) at 37 °C.

Okajima and coworkers reported the details of a thin film coating that could regulate cell adhesion by controlling the potassium ion concentration in the cell

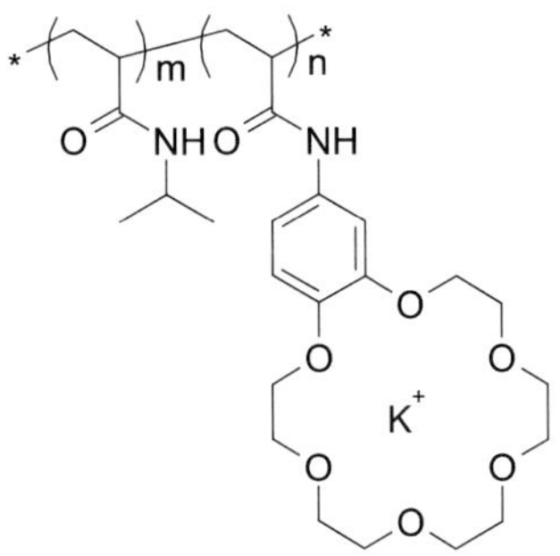

Figure 1.8 Structure of the NIPAAm-co-BCAAm polymer. Upon capturing a potassium ion, the copolymer of the thermoresponsive polymer NIPAAm and the crown ether polymer BCAAm swells and releases adhered cells. Adapted with permission from Ref. [206]; © 2005, American Chemical Society.

culture medium [205]. For this, polyethylene films were grafted with a copolymer of NIPAAm and benzo-18-crown-6-acrylamide (BCAAm) (see Figure 1.8), using radical copolymerization of the monomers on argon plasma-treated polyethylene substrates. In a cell culture medium, a complex formation of the pendant crown ether of the grafted polymer with potassium ions in the medium caused an increase in the LCST of the copolymer, and a switch to a more wettable hydrophilic state at 37 °C. This allowed the cell sheets to be detached from the substrate, without using proteolytic enzymes or changing the cell culture temperature. In this way, Okajima and colleagues were able to develop a substrate that could sense and selectively release dead cells in the culture [206]. Living cells concentrate potassium ions internally, and release them when they die; the polymer thin film coating was able to detect, locally, any potassium ions released from the dead cells, which were then selectively removed.

Cell-adhesive RGDS (Arg-Gly-Asp-Ser) peptides were immobilized on a temperature-responsive poly(N-isopropylacrylamide-co-2-carboxyisopropylacrylamide) copolymer grafted onto TCPS dishes [207, 208]. These surfaces facilitated both the adhesion and spreading of HUVECs and bovine aortic endothelial cells at 37 °C. The spread cells were seen to detach spontaneously from the surfaces when the temperature was lowered below the LCST of the polymer. In this way, the binding of cell integrin receptors located on cell membranes to immobilized RGDS located on cell culture substrates could be reversed simply by using a mild temperature stimulus, without enzymatic or chemical treatments. As these surfaces can be used to culture cells under serum-free conditions, they would be suited to applications where the use of animal-derived materials (e.g., serum) may not be desirable for reasons of cost and/or safety.

1.3.8
Electroactive Thin Films

Yeo and Mrksich prepared cell culture substrates that could be triggered to release tethered ligands by the oxidation or reduction of electroactive linkage groups [209]. The surface immobilization of ligands was carried out on two types of monolayer. The first type consisted of a maleimide group tethered to an electroactive quinone

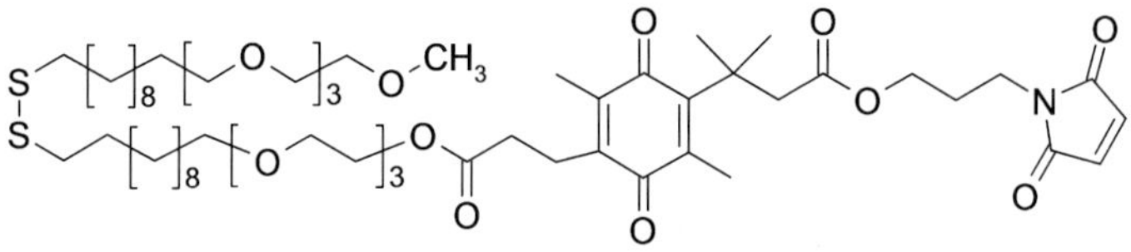

Figure 1.9 Redox molecules for preparing self-assembled monolayer (SAM) coatings for cell sheet engineering. Adapted with permission from Ref. [209]; © 2006, American Chemical Society.

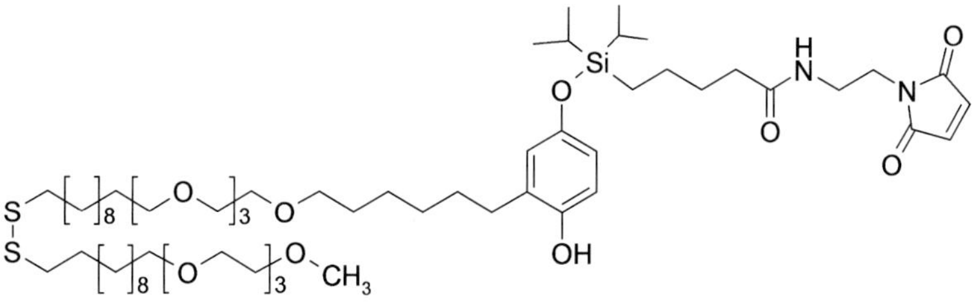

ester, while the second type consisted of a maleimide group tethered to an electroactive O-silyl hydroquinone moiety (Figure 1.9). An RGD-containing peptide (CGRGDS) was immobilized on the monolayer surfaces by reaction of the maleimide groups with the terminal cysteine residue of the peptide. In the former type of monolayer, the electrochemical reduction of the quinone to hydroquinone was followed by a cyclization reaction, to give a lactone and release the RGD ligand. In the latter type of monolayer, the electrochemical oxidation of O-silyl hydroquinone to benzoquinone resulted in a hydrolysis of the silyl ether and the selective release of RGD ligands. Swiss 3T3 fibroblast cells that had adhered to the RGD-presenting monolayers could be electrically triggered to release from the surface by applying an electrical potential to the monolayer. The same group also demonstrated that such electrochemical strategies could be used to release cells from surfaces in a selective and noninvasive manner, and may also be useful in directing stem cell differentiation, maturation, and function. The applications of other stimulus-responsive surfaces in areas such as biofouling, cell culture, tissue engineering and regenerative medicine have been reviewed recently by Mendes [210].

1.3.9
Other Functional Polymer Coatings

Fluoroalkyl groups, with a relatively low surface energy, can be used to produce a surface enrichment of bioactive molecules in a coating [211, 212]. Santerre and coworkers have used bioactive fluorinated surface modifiers to deliver vitamin E antioxidants [213] and cell-adhesive RGD peptides [214] to the surfaces of polycarbonate polyurethanes. Poly(trivinyltrimethylcyclotrisiloxane) thin films synthesized by iCVD were seen to show promise as electrical insulating coatings for neural implants (and as an alternative to the currently used Parylene-C coatings) on the basis of their high resistivity, hydrolytic stability, pin-hole-free smooth surface morphology, and biocompatibility [215].

Karp et al. found that spin-coated thin films of poly(DL-lactide-co-glycolide) (PLGA) of <100 nm thickness supported the formation of a bone matrix when seeded with rat bone marrow cells in an α-minimal essential medium. A confluent 0.5 μm-thick cement line, comprising a collagen-free layer of calcium hydroxyapatite, was formed on the PLGA surface despite the fact that the acidic products formed by the degradation of PLGA would be expected to dissolve calcium hydroxyapatite. The cement line served as a scaffold for the assembly of mineralized collagen, and would (potentially) connect new bone to the old bone surface in bone implants [216].

Biologically active dopants, such as growth factors, have been used in conductive polymer coatings. When nerve growth factor (NGF) was incorporated into electrochemically deposited polypyrrole and poly(3,4-ethylene dioxythiophene) thin films, the PC-12 cells adhered to the NGF-modified substrates and extended neurites [217]. Subsequently, NGF was shown to increase the conductivity and lower the impedance of the conducting polymer films, which can be used as coatings for electrodes that interface with neurons.

Li et al. prepared carboxylic acid gradients on the surfaces of PET films, and found that neurite outgrowths could be guided by the chemical gradient on the surface [218]. The exposure of PET surfaces to UV light resulted in the formation of surface peroxides that could be used to polymerize surface-located acrylic acid. The gradients were created by subjecting different areas of the substrate to different durations of UV exposure; neurite growth was shown to occur preferentially along the direction of decreasing –COOH density. Jhaveri et al. prepared 300 μm-thick hydrogel coatings on the surfaces of neural implants via the photopolymerization of lysine-conjugated 2-hydroxyethyl methacrylate and an ethylene glycol dimethacrylate crosslinker [219]. The coating was used to encapsulate and supply nerve growth factor (NGF) to the dorsal root ganglion (DRG) neurons in cell-culture experiments. By comparison with bath-applied NGF, a controlled release of NGF from the hydrogel coatings resulted in significantly longer neuronal processes.

Conformal coatings of hydrogel microparticles have been used to attenuate not only biofouling, leukocyte adhesion and activation, but also adverse host responses in biomedical and biotechnological applications [220]. Bridges et al. prepared thin films of poly(N-isopropyl acrylamide) hydrogel microparticles crosslinked with ethylene glycol diacrylate by using a spin-coating process. These particles were then grafted covalently, by exposure to UV light, onto aminobenzophenone-tethered poly(ethylene terephthalate) surfaces in order to obtain biocompatible coatings [221].

1.3.10
Multilayer Thin Films for Cell Encapsulation

Krol et al. created conformal coatings on individual human pancreatic islets using a polyelectrolyte LBL assembly [222]. In the islet transplantation approach for the treatment of diabetes, the islets must be encapsulated in semipermeable microcapsules so as to protect the donor cells from the host immune system, while allowing the transport of glucose, insulin, and other nutrients. Polyelectrolyte multilayer coatings seem to show promise in achieving these goals. Alternatively, the islet of Langerhans cells were deposited with multilayers of cationic PAH or poly(diallyldimethyl ammonium chloride), and anionic PSS. The surface charge of the islets was used as a binding site for the polyions, while the functionality of the encapsulated islets and permeability of the capsules were characterized by determining insulin release at different glucose levels. The coated islets accounted for about 40% of the insulin release by uncoated islets, under stimulation with a low concentration (3.3 M) of glucose. Both, the polymer nature and the molecular weight played important roles in the release behavior of the coated islets. Wilson et al. achieved pancreatic islet microencapsulation by an LBL assembly of PLL with biotinylated PEG side chains (PPB), and streptavidin (SA) (Figure 1.10) [223]. By

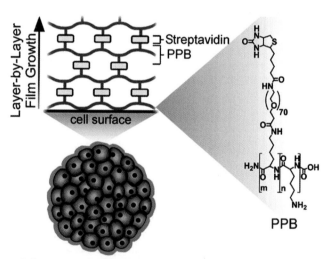

Figure 1.10 Nanothin conformal islet coatings via layer-by-layer (LBL) deposition of poly(L-lysine) with biotinylated PEG grafts (PPB), and streptavidin (SA). The PPB interacts electrostatically with negatively charged cell surfaces, facilitating the binding of SA. Unoccupied biotin binding sites of immobilized SA allow a second layer of PPB to be added, thereby enabling the incorporation of a second SA layer. This process may be repeated to generate thin films assembled via alternating deposition of PPB and SA. Reproduced with permission from Ref. [223]; © 2008, American Chemical Society.

controlling the extent of grafting with biotin–PEG, it was possible to avoid polycation-mediated cell death. The islets could be coated with (PPB/SA)$_8$ multilayer films without any loss of viability or function, and the coated islets also performed comparably to untreated controls *in vivo*. Pancreatic islets have also been encapsulated in photopolymerized PEG diacrylate hydrogel coatings using a novel apparatus [224].

1.3.11
Patterned Thin Films

Cell–biomaterial interactions in tissue engineering are influenced not only by topographical features comparable to cell size (1–100 μm) [225], but also by the nanoscale details of the biomaterial surface [226–228]. Surfaces can provide physical and chemical guidance to growth and alignment of cells such as neurons, Schwann cells, epithelial cells, and bone-derived cells [229]. The "contact guidance" or mechanical cues provided by grooves on a polymer surface – and also the chemical cues offered by micropatterns of molecules that promote or prevent cell adhesion – have attracted great interest in biomedical engineering.

Cheng *et al.* used a microheater array to pattern single or multiple types of cell onto plasma-polymerized NIPAAm thermoresponsive coatings [230]. Site-specific cell adhesion was achieved by temperature-controlled polymer conformation and surface wettability. By using a nanoembossing technique, Mills *et al.* prepared patterned poly(lactic acid) (PLA) structures, with dimensions much smaller than the size of an individual cell [231, 232]. Free-standing films of topographically patterned PLA were obtained by embossing a PLA film with microstructured or nanostructured silicon master patterns; feature dimensions ranging from tens of micrometers down to hundreds of nanometers, covering areas up to 1 cm^2, could be produced using this method (Figure 1.11). Such surfaces could, in principle, be used to examine the effects of local interactions of surface topography with cell surfaces. The optical clarity of the embossed films depended mainly on the surface roughness, which could be significantly decreased simply by sandwiching the film against an unstructured silicon master. Optically transparent patterned films are useful in studies employing optical microscopy. For example, Fernandez *et al.* produced optically transparent nanostructured chitosan thin sheets using soft lithography [233]. The technique here consisted of forming a film of chitosan on a topographically patterned silicon master, and obtaining a free-standing film by peeling off the coating from the silicon mold, after having evaporated off the solvent. This step was facilitated by a surface modification of the mold with a nonadhesive silane. One problem with this approach was the poor mechanical stability of the topographically patterned polymer film, when the polymer was not elastomeric in nature.

Feinberg *et al.* have fabricated microtopographies in PDMS elastomers by using micromachined silicon wafers having patterned geometries [227]. In order to form these microtopographies, the PDMS films were cast and cured on the patterned silicon substrates. The adhesion and growth of porcine vascular endothelial cells

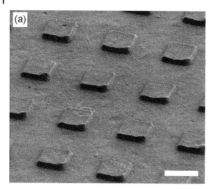 

Figure 1.11 Scanning electron microscopy images of: (a) 25 μm², 500 nm-tall square posts with a 10 μm period; and (b) 25 μm-diameter, 500 nm-tall round posts with a 5 μm period nanoimprinted into a freestanding sheet of poly(lactic acid). Reproduced with permission from Ref. [231]; © 2005, Wiley Periodicals, Inc.

(ECs) on the topographically patterned surfaces were affected by the interaction of factors such as surface chemistry (i.e., whether or not the PDMS surface was treated with a radiofrequency glow discharge argon plasma to make the surface hydrophilic, and whether or not the surfaces were coated with the protein fibronectin that promotes cell adhesion), the elastic modulus of the coating, and the topography (i.e., the height and spacing of the microridges at the surface).

One interesting application of micropatterned polymer films in tissue engineering is the use of a biodegradable thin and thick film scaffolds prepared from a blend of poly(L-lactic acid-co-glycolic acid) and poly(hydroxybutyrate-co-hydroxyvaleric acid) [229]. The films were prepared by casting a solution of the polymer blend onto a micropatterned silicon template that incorporated 21 μm- and 42 μm-wide grooves on the surfaces (with 20 μm ridge width and depth). The polymer film was detached from the template by immersing the coated template in distilled water, after which the topographically patterned film was seeded with photoreceptor cells. The effects of not only physical constraint (i.e., the grooves) but also the surface chemistry (chemical cues from laminin, a noncollagenous adhesive glycoprotein for neuronal cells) on cell adhesion, survival, and alignment of the photoreceptor cells were investigated. The rod and cone photoreceptor cells showed a clear preference for grooves on the surface rather than ridges, which highlighted the possibility of reconstituting rod–cone mosaics through the use of patterned scaffold surfaces. Such micropatterned thick/thin film scaffolds have the potential to deliver photoreceptor cells to the subretinal space of patients with blinding retinal diseases.

Shi and coworkers prepared biodegradable, topographically patterned thin films with unidirectional grooves that were, for example, 150 nm high and 1 μm wide, using a holographic diffraction grating as templates. For the nonphotolithographic approach, grooves with a sinusoidal cross-section were created by casting a solution of PLA onto the template, followed by peeling off the dried film from the template [234, 235]. The neurites of cells from chick sympathetic ganglia were found to align parallel to the grooves, and to be longer on patterned films than on unpatterned controls. Such well-controlled contact guidance of neurites has practical applications in nerve regeneration and reconnection, for the treatment of nerve injury. Libera and coworkers have described the fabrication and use of sub-micron-sized cell-repulsive PEG hydrogels patterned on an otherwise cell-adhesive substrate, to enable (in selective fashion) the growth of neurons and neuronal processes, but to repel astrocytes [236]. The approach of Libera *et al.* was based on differences in the shapes and sizes of the two cells. The *axons* are high aspect-ratio neuronal processes with diameters on the order of 1 µm and lengths exceeding centimeters, whereas the star-shaped *astrocytic glial cells* are substantially larger than 1 µm in size. When the hydrogel patterns were sufficiently closely spaced, the neurites could grow on the adhesive surface between the hydrogels, whilst the astrocytes were unable to adhere. One potential application of this concept might be to engineer an implantable nerve-guidance device that would selectively enable regrowing axons to bridge a spinal cord injury, without interference from the glial scar.

1.4
Polymer Thin Films for Drug Delivery

Leugen *et al.* have discussed several examples where polyelectrolyte LBL assemblies that are functionalized by embedded proteins, peptides or drugs, could control cell activation or act as local drug delivery systems [237]. Among the early reports on the pH-dependent deconstruction of PEM thin films for controlled release application are those of Hammond and coworkers [238, 239]. Here, hydrolytically degradable LBL thin films were prepared using a degradable poly(β-amino ester) as the cationic polymer, and a series of model therapeutic polysaccharides (e.g., heparin, low-molecular-weight heparin, chondroitin sulfate) that contain a large number of anionic sulfate groups. These degradable multilayer films were capable of both parallel and serial release of multiagents [239]. "Barrier" layers consisting of covalently crosslinked PEMs were used to block the interlayer diffusion of the model drugs. Dextran sulfate (a diffusing polyelectrolyte) and heparin (a nondiffusing polyelectrolyte) were used as model macromolecular drugs (Figure 1.12). This classification, as "diffusing" and "nondiffusing", was based on the interlayer diffusion characteristics of the polyelectrolytes in the PEM assemblies. Diffusing polysaccharides (e.g., many polypeptides and polysaccharides) rapidly diffuse throughout LBL architectures during assembly, and this results in poorly organized, blended structures. In contrast, nondiffusing polyelectrolytes (e.g., most synthetic, strong polyelectrolytes) do not diffuse across layers [239]. It was found that a covalently crosslinked barrier layer composed of nondiffusing polyelectrolytes would prevent the interlayer diffusion of model drugs, and could be used to tailor, in precise fashion, the sequential release of these drugs.

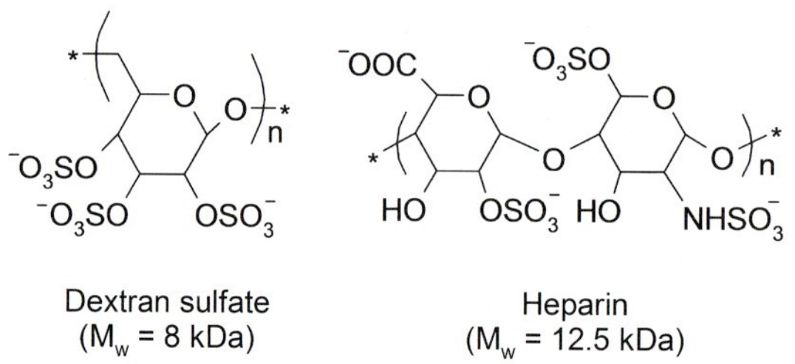

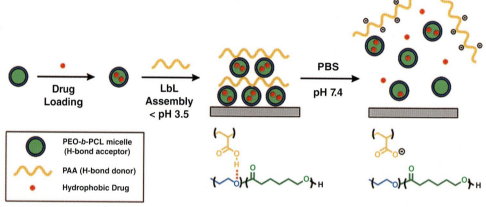

Figure 1.12 Chemical structures of dextran sulfate (a diffusing polyelectrolyte) and heparin (a nondiffusing polyelectrolyte) [239].

Figure 1.13 Schematic representation of the hydrogen-bonding LBL assembly of block copolymer micelles for hydrophobic drug delivery vehicles from surfaces. Reproduced with permission from Ref. [241]; © 2008, American Chemical Society.

Protein delivery from hydrolytically degradable and biocompatible LBL films has also been investigated by the Hammond research group [240]. The embedded protein molecules were found to retain 100% functionality after release from the multilayer thin films, showing that the processing conditions in LBL assembly were sufficiently gentle to avoid protein denaturation. Here, micelle-containing LBL films were prepared by the integration of biodegradable block copolymer micelles as nanosized carriers for hydrophobic drugs within LBL films, using hydrogen-bonding interactions as the driving force for assembly [241, 242] (Figure 1.13). For this, PAA was the H-bond donor, while biodegradable poly(ethylene oxide)-*block*-poly(ε-caprolactone) (PEO-*b*-PCL) was the H-bond acceptor. In this way, free-standing 3.1 μm-thick micelle LBL films of (PEO-*b*-PCL/PAA)$_{60}$ were isolated. This approach is useful for the surface delivery of hydrophobic and neutral drugs, which are difficult to encapsulate directly in PEMs. As an example, when the hydrophobic antibacterial drug, triclosan, was loaded into the micelles, the drug-loaded LBL film was found to release significant amounts of triclosan to inhibit the growth of S. aureus. Notably, the thermal crosslinking of PAA retarded drug release to the surrounding medium, enabling a sustained release which persisted for several days.

Addison et al. have demonstrated that stimulus-responsive block copolymer micelles can be used as triggerable delivery systems when incorporated within multilayer films deposited on polystyrene latex particles [243]. This approach of using core–shell architectures for the encapsulation and release of actives was pioneered by Caruso and others [244]. Cationic, pH-responsive micelles of poly[2-(dimethylamino)ethyl methacrylate-*block*-poly(2-(diethylamino)ethyl methacrylate)] (PDMA-*b*-PDEA) micelles and anionic poly(sodium 4-styrene sulfonate) polymer were deposited on the surface of anionic polystyrene latex particles using the LBL technique. The block copolymer micelles can be loaded with bioactive molecules such as a hydrophobic small-molecule drug. The block copolymer micelles were found to retain their micelle structure at pH 9.3, with very little release of the hydrophobic actives; however, at pH 4 the micelles underwent a transition to a polymer brush-like structure, resulting in a rapid release of the active agents.

Erel-Unal and Sukhishvili have reported the construction of hydrogen-bonded hybrid polymer multilayers comprising of poly(N-vinylcaprolactam) (PVCL)/poly(L-aspartic acid) (PLAA) bilayers with a critical disintegration pH of ~3.3, and poly(N-vinylcaprolactam) (PVCL)/tannic acid (TA) bilayers with a critical disintegration pH of 9.5 [245]. These authors have proposed that such biodegradable thin films could be used for the pH-responsive release of active molecules for future biomedical applications. PEMs have also been used for microencapsulation in drug delivery. For example, de Geest et al. coated dextran-based hydrogels with $(PSS/PAH)_n$ PEM thin films, using the LBL technique, to obtain pH-responsive self-rupturing microcapsules for both protein and drug delivery [246].

Schneider et al. investigated the release of model drugs – sodium diclofenac (an anti-inflammatory drug) and paclitaxel (an anticancer drug) – from covalently crosslinked CS/HA and $(PLL/HA)_n$ PEM thin films. When both crosslinked and uncrosslinked PEMs were compared, the crosslinked films were found to have the desired combination of properties – namely, mechanical resistance, biodegradability, and bioactivity. Paclitaxel for example, was found to remain active when loaded in crosslinked $(PLL/HA)_n$ films, and this led to a dramatic decrease in human colonic adenocarcinoma cell viability over a three-day period.

Chen et al. prepared a drug-eluting coronary stent to treat coronary arterial stenosis by coating the stent with LBL assemblies of collagen and sirolimus, an immunosuppressant drug. The collagen layers were chemically crosslinked using genipin, a naturally occurring crosslinking agent, so as to control the sirolimus release rate [247]. During use, a balloon expansion of the coated stent could be achieved without causing the coatings to crack or peel away from the stent wire.

Jewell et al. have reported that, by conjugating cationic protein transduction domains to therapeutic proteins, the extent to which the proteins were internalized by the cells could be increased [248]. Most likely, LBL assemblies of therapeutic functionalized proteins with PSS would allow both spatial and temporal control over the delivery of proteins to the cells and tissues.

Currently, CVD represents a convenient, single-step synthesis of high-quality polymer thin films on a variety of substrates, including microparticles and nanoparticles [249, 250]. For example, Lau et al. used iCVD to synthesize methacrylic acid copolymer thin film coatings for encapsulating drug microcrystals (<100 µm) for controlled release in the gastrointestinal tract. The thin film coatings showed an abrupt transition in swelling (from 5% to 30%) when the pH rose from 5 to 6.5 that would allow the drug to be protected while in the acidic environment of the stomach, but released on entering the more alkaline small intestine.

Thermoplastic elastomers, such as polystyrene-*block*-polyisobutylene-*block*-polystyrene, are easy to process and relatively stable in biological environments. This renders them attractive as a materials for the construction of, or the coating of, biomedical devices and implants. Recently, Ranade et al. have discussed the possible use of a polystyrene-*block*-polyisobutylene-*block*-polystyrene copolymer as a matrix for paclitaxel delivery from Boston Scientific's TAXUS™ coronary stent [251].

1.5
Polymer Thin Films for Gene Delivery

Polymer thin films that support cell adhesion and incorporate plasmid DNA for sustained release are of great interest in gene therapy and tissue engineering [252–256]. Among the several approaches available for gene delivery, DNA–polycation multilayer thin films have shown much promise as nonviral vectors for localized and sustained transfection, both *in vitro* and *in vivo*. The incorporation of DNA into multilayered films was first reported in 1993 by Lvov, Decher and Sukhorukov [257]. Lynn and coworkers have demonstrated this approach using a LBL assembly of a poly(β-amino ester) and a plasmid DNA encoding for enhanced green fluorescent protein (EGFP) [252–255]. The release of DNA occurred by hydrolytic degradation of the poly(β-amino ester) matrix. COS-7 line fibroblast cells began to express the protein after contacting these 100 nm-thick multilayer films. The experimentally observed increase in EGFP expression was consistent with the decrease in average film thickness, because of release of the embedded plasmid DNA, and correlated well with the DNA release profile. A sustained release over a period of about 31 h could be achieved in this way, although the majority of the plasmid was released from the film during the first 16 h of incubation. Approximately 19 ± 8% of the total number of cells expressed the protein after 48 h of contact with coated slides. The rate of DNA release could be controlled by tailoring the hydrophobicity of the poly(β-amino ester) (Figure 1.14a) [255]. In a similar approach, Lu et al. have discussed controlled release of DNA using LBL-assembled multilayer thin films composed of a cationic polymer poly(2-aminoethyl propylene phosphate) (PPE-EA) [258] and plasmid DNA [259]. The biodegradable

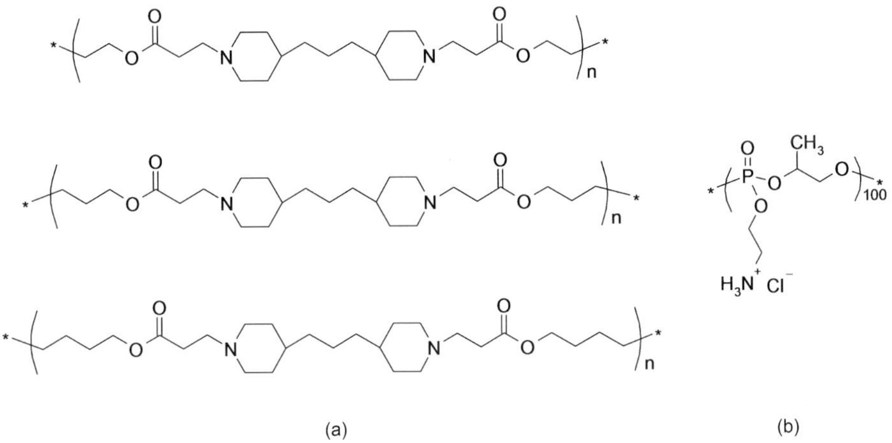

1.5 Polymer Thin Films for Gene Delivery | 37

Figure 1.14 Biodegradable polymers for preparing DNA–polycation multilayer thin films. (a) Poly(β-amino ester)s with different hydrophobicities [255]; (b) Poly(2-aminoethyl propylene phosphate) [258, 259].

polyphosphoester (see Figure 1.14b) degraded upon incubation in PBS, and provided a local and sustained delivery of bioactive plasmid DNA for up to two months. The multilayer thin films were found to be cytocompatible with osteoblast cells, and a sustained expression of GFP by the cells was detected for up to 20 days. About 47% of the cells were transfected after 10 days of contact with multilayer films with PPE-EA as the terminal layer. The polyphosphoester has a greater hydrolytic stability than poly(β-amino ester)s, and may be more suitable for long-term plasmid delivery in gene-induced tissue-engineering applications.

In a different approach towards preparing degradable PEM films, Lynn and coworkers have reported a new class of ester-functionalized "charge-shifting" polyamines [260]. Here, PAH was treated with an excess of methyl acrylate to obtain a cationic polymer with "charge-shifting" ester side chains (Figure 1.15). DNA-containing polyelectrolyte multilayers were prepared using this polymer. A gradual hydrolysis of the ester-functionalized side chains introduced carboxylate groups and reduced the net charge of the polymer, and resulted in film erosion and release of the entrapped DNA. The rate of release could be controlled by varying the degree of substitution of the amine groups with the methyl acrylate side chains.

Charge-shifting polymers were used to fabricate PEM films that were stable at neutral pH, but eroded over a period of several days at pH ~5 [261]. The addition of citraconic anhydride to poly(allyl amine) resulted in an anionic carboxylate-functionalized polymer that readily converted to the cationic poly(allyl amine) in an acidic environment. It was proposed that this approach could lead to a significant expansion of the range of different cationic agents (e.g., cationic proteins,

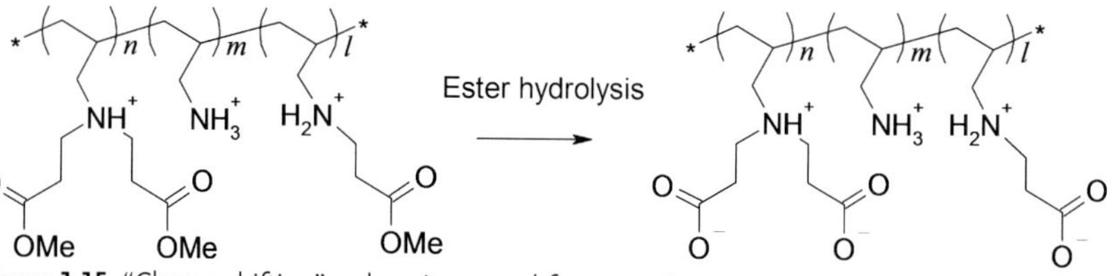

Figure 1.15 "Charge-shifting" polyamines used for preparing multilayer thin films incorporating DNA, for controlled release. Adapted with permission from Ref. [260]; © 2008, Wiley-VCH Verlag.

peptides, polymers, nanoparticles) that can be released or delivered from surfaces using PEMs.

Zhang et al. have investigated the transfection ability and intracellular DNA pathway of gene-delivery systems based on (PLL/HA)$_n$ thin films [262]. For this, plasmid DNA was complexed with PLL, β-cyclodextrin (CD) or β-cyclodextrin-grafted PLL (PLL-CD), in solution. Subsequently, (PLL/HA)$_5$ coatings were incubated with the plasmid DNA complexes for 90 min, followed by the assembly of another multilayer film of (PLL/HA)$_5$. When the films had been dried and sterilized by exposure to UV light, the transfection efficiency was found to be higher when the DNA complexes were delivered from the multilayer system than from solution. This higher efficiency was achieved because of an efficient internalization in the cytoplasm (through a nonendocytic pathway) and, subsequently, in the nuclei of the transfected cells. In contrast, internalization in solution was via endocytosis, when the complexes were trapped in endosomes and lysosomes. Degradation in the lysosomes resulted in a lower transfection efficiency.

Wang et al. demonstrated the release of DNA from an LBL-assembled thin film of DNA with an inorganic compound that had a relatively low binding affinity for DNA [263]. Zr^{4+}/DNA multilayer films were prepared through the LBL assembly of zirconyl chloride octahydrate ($ZrOCl_2 \cdot 8H_2O$) and sodium salt of fish sperm DNA. The addition of a chelator (e.g., sodium citrate) cleaved the electrostatic/coordinate covalent linkages between Zr^{4+} and the phosphate groups of the DNA, and this resulted in disintegration of the film and a release of DNA. The ability of the chelators to disassemble the multilayer film varied, in general, as sodium citrate > sodium tartarate > sodium fluoride > sodium acetate. About 97% of the (Zr^{4+}/DNA)$_{11}$ film was released within 6 h in a sodium citrate solution (5 mM), compared to 86% in sodium tartarate and 70% in sodium fluoride, for the same duration. Only about 17% of the film was released from the surface in 15 h when sodium acetate was used.

Sakai et al. coated electrospun fibrous mats of PLA with LBL thin films of PEI/plasmid-DNA multilayers. Because of the large surface-to-volume ratio, and flexibility, these mats could function as improved substrates for gene therapy and tissue engineering [264].

1.6
Conclusions

Among all the coatings and thin films discussed so far, those coatings based on PEG and zwitterionic polymers are exceptional in their ability to resist protein adsorption. The strong interaction of these coatings with water molecules is evidently the reason for their antibiofouling properties. These hydrophilic coatings are also successful in imparting blood compatibility and preventing blood coagulation, as well as rendering the biomaterial inert to the body's immune responses. Strategies to promote cell adhesion and proliferation on these otherwise non-adhesive coatings are available so that the coated implant can integrate with the host tissue. PEGylated and zwitterionic polymer coatings, when functionalized with reactive groups such as carboxylic acid, amino or aldehyde, have been used to tether biomolecules such as growth factors to facilitate tissue integration. Novel polymeric materials, including polymers with branched architectures, are currently being investigated and compared with PEGylated coatings for their protein resistance and biocompatibility. The influence of polymer chemistry and polymer conformation on biological interactions is an ongoing area of research. Among the different coating techniques available for preparing polymer thin films, the LBL assembly technique, employing polyelectrolytes, is currently a highly active area of research, the main reasons for this being the processing simplicity and ability to form conformal coatings. The CVD approach shares some of these traits. Stimulus-responsive coatings—and especially thermally responsive coatings based on poly(N-isopropylacrylamide)—have shown much promise for harvesting cells in the form of sheets for novel tissue-engineering applications. The use of polymer thin films as nonviral vectors for localized gene delivery from surfaces represents another highly interesting area of biomedical research with great potential.

References

1 Williams, D.F. (1999) *The Williams Dictionary of Biomaterials*, Liverpool University Press, Liverpool.

2 Vadgama, P. (2005) Surface biocompatibility. *Annual Report on the Progress of Chemistry, Section C*, **101**, 14–52.

3 Hong, J., Larsson, A., Ekdahl, K.N., Elgue, G., Larsson, R. and Nilsson, B. (2001) Contact between a polymer and whole blood: sequence of events leading to thrombin generation. *Journal of Laboratory and Clinical Medicine*, **138**, 139–45.

4 Brodbeck, W.G., Patel, J., Voskerician, G., Christenson, E., Shive, M.S., Nakayama, Y., Matsuda, T., Ziats, N.P. and Anderson, J.M. (2002) Biomaterial adherent macrophage apoptosis is increased by hydrophilic and anionic substrates *in vivo*. *Proceedings of the National Academy of Sciences of the United States of America*, **99**, 10287–92.

5 Lee, J.H., Lee, H.B. and Andrade, J.D. (1995) Blood biocompatibility of poly(ethylene oxide) surfaces. *Progress in Polymer Science*, **20**, 1043–79.

6 Ostuni, E., Chapman, R.G., Holmlin, R.E., Takayama, S. and Whitesides, G.M. (2001) A survey of structure-property relationships of surfaces that resist the

adsorption of protein. *Langmuir*, **17**, 5605–20.

7 Chapman, R.G., Ostuni, E., Liang, M.N., Meluleni, G., Kim, E., Yan, L., Pier, G., Warren, H.S. and Whitesides, G.M. (2001) Polymeric thin films that resist the adsorption of proteins and the adhesion of bacteria. *Langmuir*, **17**, 1225–33.

8 Zhang, F., Kang, E.T., Neoh, K.G., Wang, P. and Tan, K.L. (2001) Surface modification of stainless steel by grafting of poly(ethylene glycol) for reduction in protein adsorption. *Biomaterials*, **22**, 1541–8.

9 Herrwerth, S., Eck, W., Reinhardt, S. and Grunze, M. (2003) Factors that determine the protein resistance of oligoether self-assembled monolayers – Internal hydrophilicity, terminal hydrophilicity, and lateral packing density. *Journal of the American Chemical Society*, **125**, 9359–66.

10 Lee, S.-W. and Laibinis, P.E. (1998) Protein-resistant coatings for glass and metal oxide surfaces derived from oligo(ethylene glycol)-terminated alkytrichlorosilanes. *Biomaterials*, **19**, 1669–75.

11 Feller, L.M., Cerritelli, S., Textor, M., Hubbell, J.A. and Tosatti, S.G.P. (2005) Influence of poly(propylene sulfide-block-ethylene glycol) di- and triblock copolymer architecture on the formation of molecular adlayers on gold surfaces and their effect on protein resistance: a candidate for surface modification in biosensor research. *Macromolecules*, **38**, 10503–10.

12 Ma, H., Li, D., Sheng, X., Zhao, B. and Chilkoti, A. (2006) Protein resistant polymer brushes on silicon oxide by surface initiated atom transfer radical polymerization. *Langmuir*, **22**, 3751–6.

13 Ma, H., Hyun, J., Stiller, P. and Chilkoti, A. (2004) "Non-fouling" oligo(ethylene glycol) polymer brushes synthesized by surface initiated atom transfer radical polymerization. *Advanced Materials*, **16**, 338–41.

14 Tsukagoshi, T., Kondo, Y. and Yoshino, N. (2007) Protein adsorption and stability of poly(ethylene oxide)-modified surfaces having hydrophobic layer between substrate and polymer. *Colloids and Surfaces B, Biointerfaces*, **54**, 82–7.

15 Dalsin, J.L., Lin, L., Tosatti, S., Voros, J., Textor, M. and Messersmith, P.B. (2005) Surface modification for protein resistance using a biomimetic approach. *Langmuir*, **21**, 640–6.

16 Fan, X., Lin, L. and Messersmith, P.B. (2006) Cell fouling resistance of polymer brushes grafted from Ti substrates by surface-initiated polymerization: effect of ethylene glycol side chain length. *Biomacromolecules*, **7**, 2443–8.

17 Lee, H., Dellatore, S.M., Miller, W.M. and Messersmith, P.B. (2007) Mussel-inspired surface chemistry for multifunctional coatings. *Science*, **318**, 426–30.

18 Zürcher, S., Wäckerlin, D., Bethuel, Y., Malisova, B., Textor, M., Tosatti, S. and Gademann, K. (2006) Biomimetic surface modifications based on the cyanobacterial iron chelator anachelin. *Journal of the American Chemical Society*, **128**, 1064–5.

19 Chi, Y.S., Lee, B.S., Kil, M., Jung, H.J., Oh, E. and Choi, I.S. (2009) Asymmetrically functionalized, four-armed, poly(ethylene glycol) compounds for construction of chemically functionalizable non-biofouling surfaces. *Chemistry: An Asian Journal*, **4**, 135–42.

20 Groll, J., Ameringer, T., Spatz, J.P. and Moeller, M. (2005) Ultrathin coatings from isocyanate-terminated star PEG prepolymers: layer formation and characterization. *Langmuir*, **21**, 1991–9.

21 Harbers, G.M., Emoto, K., Greef, C., Steven, W., Metzger, S.W., Woodward, H.N., Mascali, J.J., Grainger, D.W. and Lochhead, M.J. (2007) Functionalized poly(ethylene glycol)-based bioassay surface chemistry that facilitates bio-immobilization and inhibits non-specific protein, bacterial, and mammalian cell adhesion. *Chemistry of Materials*, **19**, 4405–14.

22 Hölzl, M., Tinazli, A., Leitner, C., Hahn, C.D., Lackner, B., Tampé, R. and Gruber, H.J. (2007) Protein-resistant self-assembled monolayers on gold with latent aldehyde functions. *Langmuir*, **23**, 5571–7.

23 Raynor, J.E., Petrie, T.A., García, A.J. and Collard, D.M. (2007) Controlling cell adhesion to titanium: functionalization of poly[oligo(ethylene glycol) methacrylate] brushes with cell-adhesive peptides. *Advanced Materials*, **19**, 1724–8.

24 Zhou, Y., Liedberg, B., Gorochovceva, N., Makuska, R., Dedinaite, A. and Claesson, P.M. (2007) Chitosan-*N*-poly(ethylene oxide) brush polymers for reduced non-specific protein adsorption. *Journal of Colloid and Interface Science*, **305**, 62–71.

25 Murata, H., Chang, B.-J., Prucker, O., Dahm, M. and Rühe, J. (2004) Polymeric coatings for biomedical devices. *Surface Science*, **570**, 111–18.

26 Deng, L., Mrksich, M. and Whitesides, G.M. (1996) Self-assembled monolayers of alkanethiolates presenting tri(propylene sulfoxide) groups resist the adsorption of protein. *Journal of the American Chemical Society*, **118**, 5136–7.

27 Krishnan, S., Weinman, C.J. and Ober, C.K. (2008) Advances in polymers for anti-biofouling surfaces. *Journal of Materials Chemistry*, **18**, 3405–13.

28 Teare, D.O.H., Schofield, W.C.E., Garrod, R.P. and Badyal, J.P.S. (2005) Poly(*N*-acryloylsarcosine methyl ester) protein-resistant surfaces. *Journal of Physical Chemistry B*, **109**, 20923–8.

29 Dilly, S.J., Beecham, M.P., Brown, S.P., Griffin, J.M., Clark, A.J., Griffin, C.D., Marshall, J., Napier, R.M., Taylor, P.C. and Marsh, A. (2006) Novel tertiary amine oxide surfaces that resist non-specific protein adsorption. *Langmuir*, **22**, 8144–50.

30 Yang, H., Lazos, D. and Ulbricht, M. (2005) Thin, highly crosslinked polymer layer synthesized via photoinitiated graft copolymerization on a self-assembled-monolayer-coated gold surface. *Journal of Applied Polymer Science*, **97**, 158–64.

31 Hayward, J.A. and Chapman, D. (1984) Biomembrane surfaces as models for polymer design: the potential for haemocompatibility. *Biomaterials*, **5**, 135–42.

32 Kudaibergenov, S., Jaeger, W. and Laschewsky, A. (2006) Polymeric betaines characterization, and application. *Advances in Polymer Science*, **201**, 157–224.

33 Chen, S., Zheng, J., Li, L. and Jiang, S. (2005) Strong resistance of phosphorylcholine self-assembled monolayers to protein adsorption: insights into nonfouling properties of zwitterionic materials. *Journal of the American Chemical Society*, **127**, 14473–8.

34 Cho, W.K., Kong, B. and Choi, I.S. (2007) Highly efficient non-biofouling coating of zwitterionic polymers: Poly((3-(methacryloylamino)propyl)-dimethyl(3-sulfopropyl)ammonium hydroxide). *Langmuir*, **23**, 5678–82.

35 Chang, Y., Chen, S., Zhang, Z. and Jiang, S. (2006) Highly protein-resistant coatings from well defined diblock copolymers containing sulfobetaines. *Langmuir*, **22**, 2222–6.

36 Yang, W., Chen, S., Cheng, G., Vaisocherová, H., Xue, H., Li, W., Zhang, J. and Jiang, S. (2008) Film thickness dependence of protein adsorption from blood serum and plasma onto poly(sulfobetaine)-grafted surfaces. *Langmuir*, **24**, 9211–14.

37 Ladd, J., Zhang, Z., Chen, S., Hower, J.C. and Jiang, S. (2008) Zwitterionic polymers exhibiting high resistance to non-specific protein adsorption from human serum and plasma. *Biomacromolecules*, **9**, 1357–61.

38 Chang, Y., Liao, S.C., Higuchi, A., Ruaan, R.C., Chu, C.W. and Chen, W.Y. (2008) A highly stable nonbiofouling surface with well-packed grafted zwitterionic polysulfobetaine for plasma protein repulsion. *Langmuir*, **24**, 5453–8.

39 Li, G., Xue, H., Cheng, G., Chen, S., Zhang, F. and Jiang, S. (2008) Ultralow fouling zwitterionic polymers grafted from surfaces covered with an initiator via an adhesive mussel mimetic linkage. *Journal of Physical Chemistry B*, **112**, 15269–74.

40 Chang, Y., Chen, S., Yu, Q., Zhang, Z., Bernards, M. and Jiang, S. (2007) Development of biocompatible interpenetrating polymer networks containing a sulfobetaine-based polymer and a segmented polyurethane for protein resistance. *Biomacromolecules*, **8**, 122–7.

41 Bolduc, O.R. and Masson, J.-F. (2008) Monolayers of 3-mercaptopropyl-amino acid to reduce the non-specific adsorption of serum proteins on the surface of biosensors. *Langmuir*, **24**, 12085–91.

42 Statz, A.R., Barron, A.E. and Messersmith, P.B. (2008) Protein, cell and bacterial fouling resistance of polypeptoid-modified surfaces: effect of side-chain chemistry. *Soft Matter*, **4**, 131–9.

43 Konradi, R., Pidhatika, B., Mühlebach, A. and Textor, M. (2008) Poly-2-methyl-2-oxazoline: a peptide-like polymer for protein-repellent surfaces. *Langmuir*, **24**, 613–16.

44 Holland, N.B., Qiu, Y., Ruegsegger, M. and Marchant, R.E. (1998) Biomimetic engineering of non-adhesive glycocalyx-like surfaces using oligosaccharide surfactant polymers. *Nature*, **392**, 799–801.

45 Ruegsegger, M.A. and Marchant, R.E. (2001) Reduced protein adsorption and platelet adhesion by controlled variation of oligomaltose surfactant polymer coatings. *Journal of Biomedical Materials Research*, **56**, 159–67.

46 Perrino, C., Lee, S., Choi, S.W., Maruyama, A. and Spencer, N.D. (2008) A biomimetic alternative to poly(ethylene glycol) as an antifouling coating: resistance to non-specific protein adsorption of poly(L-lysine)-*graft*-dextran. *Langmuir*, **24**, 8850–6.

47 Gan, D., Mueller, A. and Wooley, K.L. (2003) Amphiphilic and hydrophobic surface patterns generated from hyperbranched fluoropolymer/linear polymer networks: minimally adhesive coatings via the crosslinking of hyperbranched fluoropolymers. *Journal of Polymer Science Part A: Polymer Chemistry*, **41**, 3531–40.

48 Yeh, P.Y.J., Kainthan, R.K., Zou, Y., Chiao, M. and Kizhakkedathu, J.N. (2008) Self-assembled monothiol-terminated hyperbranched polyglycerols on a gold surface: a comparative study on the structure, morphology, and protein adsorption characteristics with linear poly(ethylene glycol)s. *Langmuir*, **24**, 4907–16.

49 Siegers, C., Biesalski, M. and Haag, R. (2004) Self-assembled monolayers of dendritic polyglycerol derivatives on gold that resist the adsorption of proteins. *Chemistry – A European Journal*, **10**, 2831–8.

50 Massia, S.P., Stark, J. and Letbetter, D.S. (2000) Surface-immobilized dextran limits cell adhesion and spreading. *Biomaterials*, **21**, 2253–61.

51 Jiang, H., Manolache, S., Wong, A.C. and Denes, F.S. (2006) Synthesis of dendrimer-type poly(ethylene glycol) structures from plasma-functionalized silicon rubber surfaces. *Journal of Applied Polymer Science*, **102**, 2324–37.

52 Chen, S., Yu, F., Yu, Q., He, Y. and Jiang, S. (2006) Strong resistance of a thin crystalline layer of balanced charged groups to protein adsorption. *Langmuir*, **22**, 8186–91.

53 Salloum, D.S. and Schlenoff, J.B. (2004) Protein adsorption modalities on polyelectrolyte multilayers. *Biomacromolecules*, **5**, 1089–96.

54 Ladam, G., Gergely, C., Senger, B., Decher, G., Voegel, J.-C., Schaaf, P. and Cuisinier, F.J.G. (2000) Protein interactions with polyelectrolyte multilayers: interactions between human serum albumin and polystyrene sulfonate/polyallylamine multilayers. *Biomacromolecules*, **1**, 674–87.

55 Ekblad, T., Andersson, O., Tai, F.-I., Ederth, T. and Liedberg, B. (2009) Lateral control of protein adsorption on charged polymer gradients. *Langmuir*, **25**, 3755–62.

56 Heuberger, R., Sukhorukov, G., Vörös, J., Textor, M. and Möhwald, H. (2005) Biofunctional polyelectrolyte multilayers and microcapsules: control of non-specific and bio-specific protein adsorption. *Advanced Functional Materials*, **15**, 357–66.

57 Kim, J., Wacker, B.K. and Elbert, D.L. (2007) Thin polymer layers formed using multiarm poly(ethylene glycol) vinylsulfone by a covalent layer-by-layer method. *Biomacromolecules*, **8**, 3682–6.

58 Reisch, A., Voegel, J.-C., Decher, G., Schaaf, P. and Mésini, P.J. (2007) Synthesis of polyelectrolytes bearing phosphorylcholine moieties,

Macromolecular Rapid Communications, **28**, 2217–23.

59 Reisch, A., Hémmerle, J., Voegel, J.C., Gonthier, E., Decher, G., Benkirane-Jessel, N., Chassepot, A., Mertz, D., Lavalle, P., Mésini, P. and Schaaf, P. (2008) Polyelectrolyte multilayer coatings that resist protein adsorption at rest and under stretching. *Journal of Materials Chemistry*, **18**, 4242–5.

60 Olenych, S.G., Moussallem, M.D., Salloum, D.S., Schlenoff, J.B. and Keller, T.C.S. (2005) Fibronectin and cell attachment to cell and protein resistant polyelectrolyte surfaces. *Biomacromolecules*, **6**, 3252–8.

61 Sperling, C., Schweiss, R.B., Streller, U. and Werner, C. (2005) In vitro hemocompatibility of self-assembled monolayers displaying various functional groups. *Biomaterials*, **26**, 6547–57.

62 Rodrigues, S.N., Gonçalves, I.C., Martins, M.C.L., Barbosa, M.A. and Ratner, B.D. (2006) Fibrinogen adsorption, platelet adhesion and activation on mixed hydroxyl-/methyl-terminated self-assembled monolayers. *Biomaterials*, **27**, 5357–67.

63 Chung, Y.C., Chiu, Y.H., Wu, Y.W. and Tao, Y.T. (2005) Self-assembled biomimetic monolayers using phospholipid-containing disulfides. *Biomaterials*, **26**, 2313–24.

64 Zhang, Z., Zhang, M., Chen, S., Horbett, T.A., Ratner, B.D. and Jiang, S. (2008) Blood compatibility of surfaces with superlow protein adsorption. *Biomaterials*, **29**, 4285–91.

65 Zhou, J., Yuan, J., Zang, X., Shen, J. and Lin, S. (2005) Platelet adhesion and protein adsorption on silicone rubber surface by ozone-induced grafted polymerization with carboxybetaine monomer. *Colloids and Surfaces B, Biointerfaces*, **41**, 55–62.

66 Ishii, T., Wada, A., Tsuzuki, S., Casolaro, M. and Ito, Y. (2007) Copolymers including L-histidine and hydrophobic moiety for preparation of nonbiofouling surface. *Biomacromolecules*, **8**, 3340–4.

67 Ayres, N., Holt, D.J., Jones, C.F., Corum, L.E. and Grainger, D.W. (2008) Polymer brushes containing sulfonated sugar repeat units: synthesis, characterization, and in vitro testing of blood coagulation activation. *Journal of Polymer Science Part A: Polymer Chemistry*, **46**, 7713–24.

68 Gupta, A.S., Wang, S., Link, E., Anderson, E.H., Hofmann, C., Lewandowski, J., Kottke-Marchant, K. and Marchant, R.E. (2006) Glycocalyx-mimetic dextran-modified poly(vinyl amine) surfactant coating reduces platelet adhesion on medical-grade polycarbonate surface. *Biomaterials*, **27**, 3084–95.

69 Sandberg, T., Carlsson, J. and Ott, M.K. (2009) Interactions between human neutrophils and mucin-coated surfaces. *Journal of Materials Science: Materials in Medicine*, **20**, 621–31.

70 Chen, H., Wang, L., Zhang, Y., Li, D., McClung, W.G., Brook, M.A., Sheardown, H. and Brash, J.L. (2008) Fibrinolytic poly(dimethyl siloxane) surfaces. *Macromolecular Bioscience*, **8**, 863–70.

71 McClung, W.G., Clapper, D.L., Hu, S.-P. and Brash, J.L. (2001) Lysine-derivatized polyurethane as a clot lysing surface: conversion of adsorbed plasminogen to plasmin and clot lysis in vitro. *Biomaterials*, **22**, 1919–24.

72 McClung, W.G., Clapper, D.L., Hu, S.-P. and Brash, J.L. (2000) Adsorption of plasminogen from human plasma to lysine-containing surfaces. *Journal of Biomedical Materials Research*, **49**, 409–14.

73 Asuri, P., Karajanagi, S.S., Kane, R.S. and Dordick, J.S. (2007) Polymer-nanotube-enzyme composites as active antifouling films. *Small*, **3**, 50–3.

74 Messersmith, P.B. and Textor, M. (2007) Enzymes on nanotubes thwart fouling. *Nature Nanotechnology*, **2**, 138–9.

75 Radomski, M.W., Palmer, R.M. and Moncada, S. (1987) Endogenous nitric oxide inhibits human platelet adhesion to vascular endothelium. *Lancet*, **2**, 1057–8.

76 Konishi, R., Shimizu, R., Firestone, L., Walters, F.R., Wagner, W.R., Federspiel, W.J., Konishi, H. and Hattler, B.G. (1996) Nitric oxide prevents human platelet adhesion to fiber membranes in whole blood. *American Society of Artificial Internal Organs Journal*, **42**, M850–3.

77 Yang, J., Welby, J.L. and Meyerhoff, M.E. (2008) Generic nitric oxide (NO) generating surface by immobilizing organoselenium species via layer-by-layer assembly. *Langmuir*, **24**, 10265–72.

78 Brynda, E. and Houska, M. (1996) Multiple alternating molecular layers of albumin and heparin on solid surfaces. *Journal of Colloid and Interface Science*, **183**, 18–25.

79 Brynda, E. and Houska, M. (1997) Interactions of proteins with polyelectrolytes at solid/liquid interfaces: sequential adsorption of albumin and heparin. *Journal of Colloid and Interface Science*, **188**, 243–50.

80 Sakaguchi, H., Serizawa, T. and Akashi, M. (2003) Layer-by-layer assembly on hydrogel surfaces and control of human whole blood coagulation. *Chemistry Letters*, **32**, 174–5.

81 Tan, Q., Ji, J., Barbosa, M.A., Fonseca, C. and Shen, J. (2003) Constructing thromboresistant surface on biomedical stainless steel via layer-by-layer deposition anticoagulant. *Biomaterials*, **24**, 4699–705.

82 Thierry, B., Winnik, F.M., Merhi, Y., Silver, J. and Tabrizian, M. (2003) Bioactive coatings of endovascular stents based on polyelectrolyte multilayers. *Biomacromolecules*, **4**, 1564–71.

83 Benkirane-Jessel, N., Schwinté, P., Falvey, P., Darcy, R., Haïkel, Y., Schaaf, P., Voegel, J.-C. and Ogier, J. (2004) Build-up of polypeptide multilayer coatings with anti-inflammatory properties based on the embedding of piroxicam-cyclodextrin complexes. *Advanced Functional Materials*, **14**, 174–82.

84 Ji, J., Tan, Q. and Shen, J. (2004) Construction of albumin multilayer coating onto plasma treated poly(vinyl chloride) via electrostatic self-assembly. *Polymers for Advanced Technologies*, **15**, 490–4.

85 Tan, Q., Ji, J., Zhao, F., Fan, D.-Z., Sun, F.-Y. and Shen, J.-C. (2005) Fabrication of thromboresistant multilayer thin film on plasma treated poly (vinyl chloride) surface. *Journal of Materials Science: Materials in Medicine*, **16**, 687–92.

86 Ji, J., Tan, Q., Fan, D.-Z., Sun, F.-Y., Barbosa, M.A. and Shen, J. (2004) Fabrication of alternating polycation and albumin multilayer coating onto stainless steel by electrostatic layer-by-layer adsorption. *Colloids and Surfaces B, Biointerfaces*, **34**, 185–90.

87 Fu, J., Ji, J., Yuan, W. and Shen, J. (2005) Construction of anti-adhesive and antibacterial multilayer films via layer-by-layer assembly of heparin and chitosan. *Biomaterials*, **26**, 6684–92.

88 Cai, K., Rechtenbach, A., Hao, J., Bossert, J. and Jandt, K.D. (2005) Polysaccharide-protein surface modification of titanium via a layer-by-layer technique: characterization and cell behaviour aspects. *Biomaterials*, **26**, 5960–71.

89 Yang, S.Y., Mendelsohn, J.D. and Rubner, M.F. (2003) New class of ultrathin, highly cell-adhesion-resistant polyelectrolyte multilayers with micropatterning capabilities. *Biomacromolecules*, **4**, 987–94.

90 Dyer, M.A., Ainslie, K.M. and Pishko, M.V. (2007) Protein adhesion on silicon-supported hyperbranched poly(ethylene glycol) and poly(allylamine) thin films. *Langmuir*, **23**, 7018–23.

91 Yu, D.-G., Jou, C.-H., Lin, W.-C. and Yang, M.C. (2007) Surface modification of poly(tetramethylene adipate-*co*-terephthalate) membrane via layer-by-layer assembly of chitosan and dextran sulfate polyelectrolyte multilayer. *Colloids and Surfaces B, Biointerfaces*, **54**, 222–9.

92 Thierry, B., Winnik, F.M., Merhi, Y. and Tabrizian, M. (2003) Nanocoatings onto arteries via layer-by-layer deposition: Toward the *in vivo* repair of damaged blood vessels. *Journal of the American Chemical Society*, **125**, 7494–5.

93 Xu, F., Nacker, J.C., Crone, W.C. and Masters, K.S. (2008) The haemocompatibility of polyurethane-hyaluronic acid copolymers. *Biomaterials*, **29**, 150–60.

94 Wang, D.-A., Gao, C.-Y., Yu, G.-H. and Feng, L.-X. (2001) Surface coating of stearyl poly(ethylene oxide) coupling-polymer on polyurethane guiding catheters with poly(ether urethane) film

building additive for biomedical applications. *Biomaterials*, **22**, 1549–62.

95 Elkasabi, Y., Yoshida, M., Nandivada, H., Chen, H.-Y. and Lahann, J. (2008) Towards multipotent coatings: chemical vapor deposition and biofunctionalization of carbonyl-substituted copolymers. *Macromolecular Rapid Communications*, **29**, 855–70.

96 Kenawy, E.-R., Worley, S.D. and Broughton, R. (2007) Chemistry and applications of antimicrobial polymers: a state-of-the-art review. *Biomacromolecules*, **8**, 1359–84.

97 Haldar, J., Weight, A.K. and Kilbanov, A.M. (2007) Preparation, application and testing of permanent antibacterial and antiviral coatings. *Nature Protocols*, **2**, 2412–17.

98 Mukherjee, K., Rivera, J.J. and Klibanov, A.M. (2008) Practical aspects of hydrophobic polycationic bactericidal "paints". *Applied Biochemistry and Microbiology*, **151**, 61–70.

99 Tiller, J.C., Liao, C.-J., Lewis, K. and Klibanov, A.M. (2001) Designing surfaces that kill bacterial on contact. *Proceedings of the National Academy of Sciences of the United States of America*, **98**, 5981–5.

100 Isquith, A.J., Abbott, E.A. and Walters, P.A. (1972) Surface-bonded antimicrobial activity of an organosilicon quaternary ammonium chloride. *Applied Microbiology*, **24**, 859–63.

101 Krishnan, S., Ward, R.J., Hexemer, A., Sohn, K.E., Lee, K.L., Angert, E.R., Fischer, D.A., Kramer, E.J. and Ober, C.K. (2006) Surfaces of fluorinated pyridinium block copolymers with enhanced antibacterial activity. *Langmuir*, **22**, 11255–66.

102 Madkour, A., Dabkowski, J.M., Nüsslein, K. and Tew, G.N. (2009) Fast disinfecting antimicrobial surfaces. *Langmuir*, **25**, 1060–7.

103 Huang, J., Koepsel, R.R., Murata, H., Wu, W., Lee, S.B., Kowalewski, T., Russell, A.J. and Matyjaszewski, K. (2008) Nonleaching antibacterial glass surfaces via "grafting onto": the effect of the number of quaternary ammonium groups on biocidal activity. *Langmuir*, **24**, 6785–95.

104 Makal, U., Wood, L., Ohman, D.E. and Wynne, K.J. (2006) Polyurethane biocidal polymeric surface modifiers. *Biomaterials*, **27**, 1316–26.

105 Grunzinger, S.J., Kurt, P., Brunson, K.M., Wood, L., Ohman, D.E. and Wynne, K.J. (2007) Biocidal activity of hydantoin-containing polyurethane polymeric surface modifiers. *Polymer*, **48**, 4653–62.

106 Kurt, P., Wood, L., Ohman, D.E. and Wynne, K.J. (2007) Highly effective contact antimicrobial surfaces via polymer surface modifiers. *Langmuir*, **23**, 4719–23.

107 Yao, C., Li, X., Neoh, K.G., Shi, Z. and Kang, E.T. (2008) Surface modification and antibacterial activity of electrospun polyurethane fibrous membranes with quaternary ammonium moieties. *Journal of Membrane Science*, **320**, 259–67.

108 Sellenet, P.H., Allison, B., Applegate, B.M. and Youngblood, J.P. (2007) Synergistic activity of hydrophilic modification in antibiotic polymers. *Biomacromolecules*, **8**, 19–23.

109 Aymonier, C., Schlotterbeck, U., Antonietti, L., Zacharias, P., Thomann, R., Tiller, J.C. and Mecking, S. (2002) Hybrids of silver nanoparticles with amphiphilic hyperbranched macromolecules exhibiting antimicrobial properties. *Chemical Communications*, **24**, 3018–19.

110 Malcher, M., Volodkin, D., Heurtault, B., André, P., Schaaf, P., Möhwald, H., Voegel, J.-C., Sokolowski, A., Ball, V., Boulmedais, F. and Frisch, B. (2008) Embedded silver ions-containing liposomes in polyelectrolyte multilayers: cargos films for antibacterial agents. *Langmuir*, **24**, 10209–15.

111 Lee, D., Cohen, R.E. and Rubner, M.F. (2005) Antibacterial properties of Ag nanoparticle loaded multilayers and formation of magnetically directed antibacterial microparticles. *Langmuir*, **21**, 9651–9.

112 Li, Z., Lee, D., Sheng, X., Cohen, R.E. and Rubner, M.F. (2006) Two-level antibacterial coating with both release-killing and contact-killing capabilities. *Langmuir*, **22**, 9820–3.

113 Sambhy, V., Peterson, B.R. and Sen, A. (2008) Multifunctional silane polymers for persistent surface derivatization and their antimicrobial properties. *Langmuir*, **24**, 7549–58.

114 Voccia, S., Ignatova, M., Jérôme, R. and Jérôme, C. (2006) Design of antibacterial surfaces by a combination of electrochemistry and controlled radical polymerization. *Langmuir*, **22**, 8607–13.

115 Jampala, S.N., Sarmadi, M., Somers, E.B., Wong, A.C.L. and Denes, F.S. (2008) Plasma-enhanced synthesis of bactericidal quaternary ammonium thin layers on stainless steel and cellulose surfaces. *Langmuir*, **24**, 8583–91.

116 Gopishetty, V., Roiter, Y., Tokarev, I. and Minko, S. (2008) Multiresponsive biopolyelectrolyte membrane. *Advanced Materials*, **20**, 4588–93.

117 Lee, H., Lee, Y., Statz, A.R., Rho, J., Park, T.G. and Messersmith, P.B. (2008) Substrate-independent layer-by-layer assembly by using mussel-adhesive-inspired polymers. *Advanced Materials*, **20**, 1619–23.

118 Girshevitz, O., Nitzan, Y. and Sukenik, C.N. (2008) Solution-deposited amorphous titanium dioxide on silicone rubber: a conformal, crack-free antibacterial coating. *Chemistry of Materials*, **20**, 1390–6.

119 Yuan, W., Ji, J., Fu, J. and Shen, J. (2007) A facile method to construct hybrid multilayered films as a strong and multifunctional antibacterial coating. *Journal of Biomedical Materials Research. Part B, Applied Biomaterials*, **85B**, 556–63.

120 Zhang, F., Shi, Z.L., Chua, P.H., Kang, E.T. and Neoh, K.G. (2007) Functionalization of titanium surfaces via controlled living radical polymerization: from antibacterial surface to surface for osteoblast adhesion. *Industrial and Engineering Chemistry Research*, **46**, 9077–86.

121 Aumsuwan, N., Heinhorst, S. and Urban, M.W. (2007) Antibacterial surfaces on expanded polytetrafluoroethylene; penicillin attachment. *Biomacromolecules*, **8**, 713–18.

122 Martin, T.P., Kooi, S.E., Chang, S.H., Sedransk, K.L. and Gleason, K.K. (2007) Initiated chemical vapor deposition of antimicrobial polymer coatings. *Biomaterials*, **28**, 909–15.

123 Martin, T.P., Lau, K.K.S., Chan, K., Mao, Y., Gupta, M., O'Shaughnessy, W.S. and Gleason, K.K. (2007) Initiated chemical vapor deposition (iCVD) of polymer nanocoating. *Surface and Coatings Technology*, **201**, 9400–5.

124 Gabriel, G.J., Som, A., Madkour, A.E., Eren, T. and Tew, G.N. (2007) Infectious disease: connecting innate immunity to biocidal polymers. *Materials Science and Engineering*, **57**, 28–64.

125 Etienne, O., Gasnier, C., Taddei, C., Voegel, J.-C., Aunis, D., Schaaf, P., Metz-Boutigue, M.H., Bolcato-Bellemin, A.L. and Egles, C. (2005) Antifungal coating by biofunctionalized polyelectrolyte multilayered films. *Biomaterials*, **26**, 6704–12.

126 Etienne, O., Picart, C., Taddei, C., Haikel, Y., Dimarcq, J.L., Schaaf, P., Voegel, J.-C., Ogier, J.A. and Egles, C. (2004) Multilayer polyelectrolyte films functionalized by insertion of defensin: a new approach to protection of implants from bacterial colonization. *Antimicrobial Agents and Chemotherapy*, **48**, 3662–9.

127 Glinel, K., Jonas, A.M., Jouenne, T., Leprince, J., Galas, L. and Huck, W.T.S. (2009) Antibacterial and antifouling polymer brushes incorporating antimicrobial peptide. *Bioconjugate Chemistry*, **20**, 71–77.

128 Vacheethasanee, K. and Marchant, R.E. (2000) Surfactant polymers designed to suppress bacterial (*Staphylococcus epidermidis*) adhesion on biomaterials. *Journal of Biomedical Materials Research*, **50**, 302–12.

129 Boulmedais, F., Frisch, B., Etienne, O., Lavalle, Ph., Picart, C., Ogier, J., Voegel, J.-C., Schaaf, P. and Egles, C. (2004) Polyelectrolyte multilayer films with pegylated polypeptides as a new type of anti-microbial protection for biomaterials. *Biomaterials*, **25**, 2003–11.

130 Maddikeri, R.R., Tosatti, S., Schuler, M., Chessari, S., Textor, M., Richards, R.G. and Harris, L.G. (2008) Reduced medical infection related bacterial strains

adhesion on bioactive RGD modified titanium surfaces: a first step toward cell selective surfaces. *Journal of Biomedical Materials Research*, **84A**, 425–35.

131 Cheng, G., Zhang, Z., Chen, S., Bryers, J.D. and Jiang, S. (2007) Inhibition of bacterial adhesion and biofilm formation on zwitterionic surfaces. *Biomaterials*, **28**, 4192–9.

132 Hou, S., Burton, E.A., Simon, K.A., Blodgett, D., Luk, Y.-Y. and Ren, D. (2007) Inhibition of *Escherichia coli* biofilm formation by self-assembled monolayers of functional alkanethiols on gold. *Applied and Environmental Microbiology*, **73**, 4300–7.

133 Richert, L., Lavalle, P., Payan, E., Shu, X.Z., Prestwich, G.D., Stoltz, J.-F., Schaaf, P., Voegel, J.-C. and Picart, C. (2004) Layer by layer buildup of polysaccharide films: physical chemistry and cellular adhesion aspects. *Langmuir*, **20**, 448–58.

134 Lichter, J.A., Thompson, M.T., Delgadillo, M., Nishikawa, T., Rubner, M.F. and Vliet, K.J.V. (2008) Substrata mechanical stiffness can regulate adhesion of viable bacteria. *Biomacromolecules*, **9**, 1571–8.

135 Chen, K.-S., Ku, Y.-A., Lin, H.-R., Yan, T.-R., Sheu, D.-C. and Chen, T.-M. (2006) Surface grafting polymerization of *N*-vinyl-2-pyrrolidone onto a poly(ethylene terephthalate) nonwoven by plasma pretreatment and its antibacterial activities. *Journal of Applied Polymer Science*, **100**, 803–9.

136 Zhang, F., Zhang, Z., Zhu, X., Kang, E.-T. and Neoh, K.-G. (2008) Silk-functionalized titanium surfaces for enhancing osteoblast functions and reducing bacterial adhesion. *Biomaterials*, **29**, 4751–9.

137 Puskas, J.E., Chen, Y., Dahman, Y. and Padavan, D. (2004) Polyisobutylene-based biomaterials. *Journal of Polymer Science Part A: Polymer Chemistry*, **42**, 3091–109.

138 Krishna, Y., Sheridan, C.M., Kent, D.L., Grierson, I. and Williams, R.L. (2007) Polydimethylsiloxane as a substrate for retinal pigment epithelial cell growth. *Journal of Biomedical Materials Research*, **80A**, 669–78.

139 Goddard, J.M. and Hotchkis, J.H. (2007) Polymer surface modification for the attachment of bioactive compounds. *Progress in Polymer Science*, **32**, 698–725.

140 Krsko, P. and Libera, M. (2005) Biointeractive hydrogels. *Materials Today*, **8**, 36–44.

141 Murphy, W.L., Mercurius, K.O., Koide, S. and Mrksich, M. (2004) Substrates for cell adhesion prepared via active site-directed immobilization of a protein domain. *Langmuir*, **20**, 1026–30.

142 Chawla, K., Lee, S., Lee, B.P., Dalsin, J.L., Messersmith, P.B. and Spencer, N.D. (2009) A novel low-friction surface for biomedical applications: modification of poly(dimethylsiloxane) (PDMS) with polyethylene glycol(PEG)-DOPA-lysine. *Journal of Biomedical Materials Research. Part A*, **90**, 742–9.

143 Zhang, Z., Chen, S. and Jiang, S. (2006) Dual-functional biomimetic materials: nonfouling poly(carboxybetaine) with active functional groups for protein immobilization. *Biomacromolecules*, **7**, 3311–15.

144 Rose, S.F., Lewis, A.L., Hanlon, G.W. and Lloyd, A.W. (2004) Biological responses to cationically charged phosphocholine-based materials *in vitro*. *Biomaterials*, **25**, 5125–35.

145 Muthukrishnan, S., Nitschke, M., Gramm, S., Özyürek, Z., Voit, B., Werner, C. and Müller, A.H.E. (2006) Immobilized hyperbranched glycoacrylate films as bioactive supports. *Macromolecular Bioscience*, **6**, 658–66.

146 Mei, Y., Kumar, A., Gao, W., Gross, R., Kennedy, S.B., Washburn, N.R., Amis, E.J. and Elliott, J.T. (2004) Biocompatibility of sorbitol-containing polyesters. Part I: Synthesis, surface analysis and cell response *in vitro*. *Biomaterials*, **25**, 4195–201.

147 Benhabbour, S.R., Sheardown, H. and Adronov, A. (2008) Cell adhesion and proliferation on hydrophilic dendritically modified surfaces. *Biomaterials*, **29**, 4177–86.

148 Tourniaire, G., Collins, J., Campbell, S., Mizomoto, H., Ogawa, S., Thaburet, J.-F. and Bradley, M. (2006) Polymer

microarrays for cellular adhesion. *Chemical Communications*, 2118–20.

149 Mant, A., Tourniaire, G., Diaz-Mochon, J.J., Elliott, T.J., Williams, A.P. and Bradley, M. (2006) Polymer microarrays: identification of substrates for phagocytosis assays. *Biomaterials*, **27**, 5299–306.

150 Tare, R.S., Khan, F., Tourniaire, G., Morgan, S.M., Bradley, M. and Oreffo, R.O.C. (2009) A microarray approach to the identification of polyurethanes for the isolation of human skeletal progenitor cells and augmentation of skeletal cell growth. *Biomaterials*, **30**, 1045–55.

151 Alperin, C., Zandstrab, P.W. and Woodhouse, K.A. (2005) Polyurethane films seeded with embryonic stem cell-derived cardiomyocytes for use in cardiac tissue engineering applications. *Biomaterials*, **26**, 7377–86.

152 Zhu, A.P., Zhao, F. and Fang, N. (2008) Regulation of vascular smooth muscle cells on poly(ethylene terephthalate) film by O-carboxymethylchitosan surface immobilization. *Journal of Biomedical Materials Research*, **86A**, 467–76.

153 Müller, R., Abke, J., Schnell, E., Macionczyk, F., Gbureck, U., Mehrl, R., Ruszczak, Z., Kujat, R., Englert, C., Nerlich, M. and Angele, P. (2005) Surface engineering of stainless steel materials by covalent collagen immobilization to improve implant biocompatibility. *Biomaterials*, **26**, 6962–72.

154 Jaber, J.A. and Schlenoff, J.B. (2006) Recent developments in the properties and applications of polyelectrolyte multilayers., *Current Opinion in Colloid and Interface Science*, **11**, 324–9.

155 Tang, Z., Wang, Y., Podsiadlo, P. and Kotov, N.A. (2006) Biomedical applications of layer-by-layer assembly: from biomimetics to tissue engineering. *Advanced Materials*, **18**, 3203–24.

156 Elbert, D.L., Herbert, C.B. and Hubbell, J.A. (1999) Thin polymer layers formed by polyelectrolyte multilayer techniques on biological surfaces. *Langmuir*, **15**, 5355–62.

157 Tryoen-Tóth, P., Vautier, D., Haikel, Y., Voegel, J.-C., Schaaf, P., Chluba, J. and Ogier, J. (2002) Viability, adhesion, and bone phenotype of osteoblast-like cells on polyelectrolyte multilayer thin films. *Journal of Biomedical Materials Research*, **60**, 657–67.

158 Zhu, H., Ji, J., Barbosa, M.A. and Shen, J. (2004) Protein electrostatic self-assembly on poly (DL-lactide) scaffold to promote osteoblast growth. *Journal of Biomedical Materials Research*, **71B**, 159–65.

159 Liao, S., Wang, W., Uo, M., Ohkawa, S., Akasaka, T., Tamura, K., Cui, F. and Watari, F. (2005) A three-layered nano-carbonated hydroxyapatite/collagen/PLGA composite membrane for guided tissue regeneration. *Biomaterials*, **26**, 7564–71.

160 Chluba, J., Voegel, J.-C., Decher, G., Erbacher, P., Schaaf, P. and Ogier, J. (2001) Peptide hormone covalently bound to polyelectrolytes and embedded into multilayer architectures conserving full biological activity. *Biomacromolecules*, **2**, 800–5.

161 Kuhl, P.R. and Griffith-Cima, L.G. (1996) Tethered epidermal growth factor as a paradigm of growth factor-induced stimulation from the solid phase. *Nature Medicine*, **2**, 1022–7.

162 Bochu, W., Yoshikoshi, A., Sakanishi, A. and Ito, Y. (1998) Tissue engineering by immobilized growth factors. *Materials Science & Engineering: C, Biomimetic and Supramolecular*, **6**, 267–74.

163 Kapur, T.A. and Shoichet, M.S. (2003) Chemically-bound nerve growth factor for neural tissue engineering applications. *Journal of Biomaterials Science, Polymer Edition*, **14**, 383–94.

164 Taguchi, T., Kishida, A., Akashi, M. and Maruyama, I. (2000) Immobilization of human vascular endothelial growth factor (VEGF165) onto biomaterials: an evaluation of the biological activity of immobilized VEGF165. *Journal of Bioactive and Compatible Polymers*, **15**, 309–20.

165 Backer, M.V., Patel, V., Jehning, B.T., Claffey, K.P. and Backer, J.M. (2006) Surface immobilization of active vascular endothelial growth factor via a cysteine-containing tag. *Biomaterials*, **27**, 5452–8.

166 Boura, C., Menu, P., Payan, E., Picart, C., Vogel, J.C., Muller, S. and Stoltz, J.F. (2003) Endothelial cells grown on thin polyelectrolyte mutlilayered films: an evaluation of a new versatile surface modification. *Biomaterials*, **24**, 3521–30.

167 Mendelsohn, J.D., Yang, S.Y., Hiller, J.A., Hochbaum, A.I. and Rubner, M.F. (2003) Rational design of cytophilic and cytophobic polyelectrolyte multilayer thin films. *Biomacromolecules*, **4**, 96–106.

168 Richert, L., Lavalle, P., Vautier, D., Senger, B., Stoltz, J.-F., Schaaf, P., Voegel, J.-C. and Picart, C. (2002) Cell interactions with polyelectrolyte multilayer films. *Biomacromolecules*, **3**, 1170–8.

169 Schneider, A., Bolcato-Bellemin, A.-L., Francius, G., Jedrzejwska, J., Schaaf, P., Voegel, J.-C., Frisch, B. and Picart, C. (2006) Glycated polyelectrolyte multilayer films: differential adhesion of primary versus tumor cells. *Biomacromolecules*, **7**, 2882–9.

170 Wittmer, C.R., Phelps, J.A., Saltzman, W.M. and Van Tassel, P.R. (2007) Fibronectin terminated multilayer films: protein adsorption and cell attachment studies. *Biomaterials*, **28**, 851–60.

171 Kerdjoudj, H., Boura, C., Moby, V., Montagne, K., Schaaf, P., Voegel, J.-C., Stoltz, J.-F. and Menu, P. (2007) Re-endothelialization of human umbilical arteries treated with polyelectrolyte multilayers: a tool for damaged vessel replacement. *Advanced Functional Materials*, **17**, 2667–73.

172 Berthelemy, N., Kerdjoudj, H., Gaucher, C., Schaaf, P., Stoltz, J.-F., Lacolley, P., Voegel, J.-C. and Menu, P. (2008) Polyelectrolyte films boost progenitor cell differentiation into endothelium-like monolayers. *Advanced Materials*, **20**, 2674–8.

173 Wittmer, R.C., Phelps, J.A., Lepus, C.M., Saltzman, W.M., Harding, M.J. and Van Tassel, P.R. (2008) Multilayer nanofilms as substrates for hepatocellular applications. *Biomaterials*, **29**, 4082–90.

174 Ren, K., Crouzier, T., Roy, C. and Picart, C. (2008) Polyelectrolyte multilayer films of controlled stiffness modulate myoblast cell differentiation. *Advanced Functional Materials*, **18**, 1378–89.

175 Salloum, D.S., Olenych, S.G., Keller, T.C.S. and Schlenoff, J.B. (2005) Vascular smooth muscle cells on polyelectrolyte multilayers: hydrophobicity-directed adhesion and growth. *Biomacromolecules*, **6**, 161–7.

176 Wu, Z.-R., Ma, J., Liu, B.-F., Xu, Q.-Y. and Cui, F.-Z. (2007) Layer-by-layer assembly of polyelectrolyte films improving cytocompatibility to neural cells. *Journal of Biomedical Materials Research*, **81A**, 355–62.

177 Nadiri, A., Kuchler-Bopp, S., Mjahed, H., Hu, B., Haikel, Y., Schaaf, P., Voegel, J.-C. and Benkirane-Jessel, N. (2007) Cell apoptosis control using BMP4 and noggin embedded in a polyelectrolyte multilayer film. *Small*, **3**, 1577–83.

178 Haynie, D.T., Zhang, L., Rudra, J.S., Zhao, W., Zhong, Y. and Palath, N. (2005) Polypeptide multilayer films. *Biomacromolecules*, **6**, 2895–913.

179 Moby, V., Boura, B., Kerdjoudj, H., Voegel, J.-C., Marchal, L., Dumas, D., Schaaf, P., Stoltz, J.-F. and Menu, P. (2007) Poly(styrenesulfonate)/Poly(allylamine) multilayers: a route to favor endothelial cell growth on expanded poly(tetrafluoroethylene) vascular grafts. *Biomacromolecules*, **8**, 2156–60.

180 He, W. and Bellamkonda, R.V. (2005) Nanoscale neuro-integrative coatings for neural implants. *Biomaterials*, **26**, 2983–90.

181 Gheith, M.K., Sinani, V.A., Wicksted, J.P., Matts, R.L. and Kotov, N.A. (2005) Single-walled carbon nanotube polyelectrolyte multilayers and freestanding films as a biocompatible platform for neuroprosthetic implants. *Advanced Materials*, **17**, 2663–70.

182 Lee, J., Shanbhag, S. and Kotov, N.A. (2006) Inverted colloidal crystals as three-dimensional microenvironments for cellular co-cultures. *Journal of Materials Chemistry*, **16**, 3558–64.

183 De Geest, B.G., Van Camp, W., Du Prez, F.E., De Smedt, S.C., Demeester, J. and Hennink, W.E. (2008) Degradable multilayer films and hollow capsules via a 'click' strategy. *Macromolecular Rapid Communications*, **29**, 1111–18.

184 da Silva, R.M.P., Mano, J.F. and Reis, R.L. (2007) Smart thermoresponsive coatings and surfaces for tissue engineering: switching cell-material boundaries. *Trends in Biotechnology*, **25**, 577–83.

185 Yamato, M. and Okano, T. (2004) Cell sheet engineering. *Materials Today*, **7**, 42–7.

186 Yang, J., Yamato, M., Kohno, C., Nishimoto, A., Sekine, H., Fukai, F. and Okano, T. (2005) Cell sheet engineering: recreating tissues without biodegradable scaffolds. *Biomaterials*, **26**, 6415–22.

187 Kushida, A., Yamato, M., Konno, C., Kikuchi, A., Sakurai, Y. and Okano, T. (1999) Decrease in culture temperature releases monolayer endothelial cell sheets together with deposited fibronectin matrix from temperature-responsive culture surfaces. *Journal of Biomedical Materials Research*, **45**, 355–62.

188 Okano, T., Yamada, N., Sakai, H. and Sakurai, Y. (1993) A novel recovery system for cultured cells using plasma-treated polystyrene dishes grafted with poly(N-isopropylacrylamide). *Journal of Biomedical Materials Research*, **27**, 1243–51.

189 Schmaljohann, D., Oswald, J., Jorgensen, B., Nitschke, M., Beyerlein, D. and Werner, C. (2003) Thermo-responsive PNiPAAm-g-PEG films for controlled cell detachment. *Biomacromolecules*, **4**, 1733–9.

190 Okano, T., Yamada, N., Okuhara, M., Sakai, H. and Sakurai, Y. (1995) Mechanism of cell detachment from temperature-modulated, hydrophilic-hydrophobic polymer surfaces. *Biomaterials*, **16**, 297–303.

191 Yamato, M., Okuhara, M., Karikusa, F., Kikuchi, A., Sakurai, Y. and Okano, T. (1999) Signal transduction and cytoskeletal reorganization are required for cell detachment from cell culture surfaces grafted with a temperature-responsive polymer. *Journal of Biomedical Materials Research*, **44**, 44–52.

192 Akiyama, Y., Kikuchi, A., Yamato, M. and Okano, T. (2004) Ultrathin poly(N-isopropylacrylamide) grafted layer on polystyrene surfaces for cell adhesion/detachment control. *Langmuir*, **20**, 5506–11.

193 Canavan, H.E., Cheng, X., Graham, D.J., Ratner, B.D. and Castner, D.G. (2005) Cell sheet detachment affects the extracellular matrix: a surface science study comparing thermal liftoff, enzymatic, and mechanical methods. *Journal of Biomedical Materials Research*, **75A**, 1–13.

194 Tsuda, Y., Kikuchi, A., Yamato, M., Nakao, A., Sakurai, Y., Umezu, M. and Okano, T. (2005) The use of patterned dual thermoresponsive surfaces for the collective recovery as co-cultured cell sheets. *Biomaterials*, **26**, 1885–93.

195 Harimoto, M., Yamato, M., Hirose, M., Takahashi, C., Isoi, Y.; Kikuchi, A. and Okano, T. (2002) Novel approach for achieving double-layered cell sheets co-culture: overlaying endothelial cell sheets onto monolayer hepatocytes utilizing temperature-responsive culture dishes. *Journal of Biomedical Materials Research*, **62**, 464–70.

196 Ide, T., Nishida, K., Yamato, M., Sumide, T., Utsumi, M., Nozaki, T., Kikuchi, A., Okano, T. and Tano, Y. (2006) Structural characterization of bioengineered human corneal endothelial cell sheets fabricated on temperature-responsive culture dishes. *Biomaterials*, **27**, 607–14.

197 Nishida, K., Yamato, M., Hayashida, Y., Watanabe, K., Maeda, N., Watanabe, H., Yamamoto, K., Nagai, S., Kikuchi, A., Tano, Y. and Okano, T. (2004) Functional bioengineered corneal epithelial sheet grafts from corneal stem cells expanded ex vivo on a temperature-responsive cell culture surface. *Transplantation*, **77**, 379–85.

198 Nishida, K., Yamato, M., Hayashida, Y., Watanabe, K., Yamamoto, K., Adachi, E., Nagai, S., Kikuchi, A., Maeda, N., Watanabe, H., Okano, T. and Tano, Y. (2004) Corneal reconstruction with tissue-engineered cell sheets composed of autologous oral mucosal epithelium. *New England Journal of Medicine*, **351**, 1187–96.

199 Nitschke, M., Gramm, S., Götze, T., Valtink, M., Drichel, J., Voit, B., Engelmann, K. and Werner, C. (2007)

Thermo-responsive poly(NiPAAm-*co*-DEGMA) substrates for gentle harvest of human corneal endothelial cell sheets. *Journal of Biomedical Materials Research*, **80A**, 1003–10.
200 Shimizu, T., Yamato, M., Kikuchi, A. and Okano, T. (2003) Cell sheet engineering for myocardial tissue reconstruction. *Biomaterials*, **24**, 2309–16.
201 Shimizu, T., Yamato, M., Isoi, Y., Akutsu, T., Setomaru, T., Abe, K., Kikuchi, A., Umezu, M. and Okano, T. (2002) Fabrication of pulsatile cardiac tissue grafts using a novel 3-dimensional cell sheet manipulation technique and temperature-responsive cell culture surfaces. *Circulation Research*, **90**, e40–8.
202 Zammaretti, P. and Jaconi, M. (2004) Cardiac tissue engineering: regeneration of the wounded heart. *Current Opinion in Biotechnology*, **15**, 430–4.
203 Yamato, M., Kwon, O.H., Hirose, M., Kikuchi, A. and Okano, T. (2001) Novel patterned cell coculture utilizing thermally responsive grafted polymer surfaces. *Journal of Biomedical Materials Research*, **55**, 137–40.
204 Yamato, M., Konno, C., Utsumi, M., Kikuchi, A. and Okano, T. (2002) Thermally responsive polymer-grafted surfaces facilitate patterned cell seeding and co-culture. *Biomaterials*, **23**, 561–7.
205 Okajima, S., Yamaguchi, T., Sakai, Y. and Nakao, S.-I. (2005) Regulation of cell adhesion using a signal-responsive membrane substrate. *Biotechnology and Bioengineering*, **91**, 237–43.
206 Okajima, S., Sakai, Y. and Yamaguchi, T. (2005) Development of a regenerable cell culture system that senses and releases dead cells. *Langmuir*, **21**, 4043–9.
207 Ebara, M., Yamato, M., Aoyagi, T., Kikuchi, A., Sakai, K. and Okano, T. (2004) Immobilization of cell-adhesive peptides to temperature-responsive surfaces facilitates both serum-free cell adhesion and noninvasive cell harvest. *Tissue Engineering*, **10**, 1125–35.
208 Ebara, M., Yamato, M., Aoyagi, T., Kikuchi, A., Sakai, K. and Okano, T. (2004) Temperature-responsive cell culture surfaces enable "on-off" affinity control between cell integrins and RGD ligands. *Biomacromolecules*, **5**, 505–10.
209 Yeo, W.-S. and Mrksich, M. (2006) Electroactive self-assembled monolayers that permit orthogonal control over the adhesion of cells to patterned substrates. *Langmuir*, **22**, 10816–20.
210 Mendes, P.M. (2008) Stimuli-responsive surfaces for bio-applications. *Chemical Society Reviews*, **37**, 2512–29.
211 Koberstein, J.T. (2004) Molecular design of functional polymer surfaces. *Journal of Polymer Science Part B: Polymer Physics*, **42**, 2942–56.
212 Krishnan, S., Ayothi, R., Hexemer, A., Finlay, J.A., Sohn, K.E., Perry, R., Ober, C.K., Kramer, E.J., Callow, M.E., Callow, J.A. and Fischer, D.A. (2006) Anti-biofouling properties of comblike block copolymers with amphiphilic side chains. *Langmuir*, **22**, 5075–86.
213 Ernsting, M., Labow, R.S. and Santerre, J.P. (2003) Surface modification of a polycarbonate-urethane using a vitamin E-derivatized fluoroalkyl surface modifier. *Journal of Biomaterials Science, Polymer Edition*, **14**, 1411–26.
214 Ernsting, M.J., Bonin, G.C., Yang, M., Labow, R.S. and Santerre, J.P. (2005) Generation of cell adhesive substrates using peptide fluoralkyl surface modifiers. *Biomaterials*, **26**, 6536–46.
215 O'Shaughnessy, W.S., Murthy, S.K., Edell, D.J. and Gleason, K.K. (2007) Stable biopassive insulation synthesized by initiated chemical vapor deposition of poly(1,3,5-trivinyltrimethylcyclotrisiloxane). *Biomacromolecules*, **8**, 2564–70.
216 Karp, J.M., Shoichet, M.S. and Davies, J.E. (2003) Bone formation on two-dimensional poly(DL-lactide-*co*-glycolide) (PLGA) films and three-dimensional PLGA tissue engineering scaffolds *in vitro*. *Journal of Biomedical Materials Research*, **64A**, 388–96.
217 Kim, D.-H., Richard-Burns, S.M., Hendricks, J.L., Sequera, C. and Martin, D.C. (2007) Effect of immobilized nerve growth factor on conductive polymers: electrical properties and cellular response. *Advanced Functional Materials*, **17**, 79–86.
218 Li, B., Ma, Y., Wang, S. and Moran, P.M. (2005) Influence of carboxyl group

density on neuron cell attachment and differentiation behavior: gradient-guided neurite outgrowth. *Biomaterials*, **26**, 4956–63.
219 Jhaveri, S.J., Hynd, M.R., Dowell-Mesfin, N., Turner, J.N., Shain, W. and Ober, C.K. (2009) Release of nerve growth factor from HEMA hydrogel-coated substrates and its effect on the differentiation of neural cells. *Biomacromolecules*, **10**, 174–83.
220 Bridges, A.W., Singh, N., Burns, K.L., Babensee, J.E., Lyon, L.A. and Garcia, A.J. (2008) Reduced acute inflammatory responses to microgel conformal coatings. *Biomaterials*, **29**, 4605–15.
221 Singh, N., Bridges, A.W., Garcia, A.J. and Lyon, L.A. (2007) Covalent tethering of functional microgel films onto poly(ethylene terephthalate) surfaces. *Biomacromolecules*, **8**, 3271–5.
222 Krol, S., Guerra, S.D., Grupillo, M., Diaspro, A., Gliozzi, A. and Marchetti, P. (2006) Multilayer nanoencapsulation. New approach for immune protection of human pancreatic islets. *Nano Letters*, **6**, 1933–9.
223 Wilson, J.T., Cui, W. and Chaikof, E.L. (2008) Layer-by-layer assembly of a conformal nanothin PEG coating for intraportal islet transplantation. *Nano Letters*, **8**, 1940–8.
224 Wyman, J.L., Kizilel, S., Skarbek, R., Zhao, X., Connors, M., Dillmore, W.S., Murphy, W.L., Mrksich, M., Nagel, S.R. and Garfinkel, M.R. (2007) Immunoisolating pancreatic islets by encapsulation with selective withdrawal. *Small*, **3**, 683–90.
225 Vernon, R.B., Gooden, M.D., Lara, S.L. and Wight, T.N. (2005) Microgrooved fibrillar collagen membranes as scaffolds for cell support and alignment. *Biomaterials*, **26**, 3131–40.
226 Draghi, L. and Cigada, A. (2007) Nanostructured surfaces for biomedical applications. Part I: Nanotopography. *Journal of Applied Biomaterials and Biomechanics*, **5**, 61–9.
227 Feinberg, A.W., Wilkerson, W.R., Seegert, C.A., Gibson, A.L., Hoipkemeier-Wilson, L. and Brennan, A.B. (2007) Systematic variation of microtopography, surface chemistry and elastic modulus and the state dependent effect on endothelial cell alignment. *Journal of Biomedical Materials Research*, **86A**, 522–34.
228 Senaratne, W., Sengupta, P., Jakubek, V., Holowka, D., Ober, C.K. and Baird, B. (2006) Functionalized surface arrays for spatial targeting of immune cell signaling. *Journal of the American Chemical Society*, **128**, 5594–5.
229 Tezcaner, A. and Hicks, D. (2008) In vitro characterization of micropatterned PLGA-PHBV8 blend films as temporary scaffolds for photoreceptor cells. *Journal of Biomedical Materials Research*, **86A**, 170–81.
230 Cheng, X., Wang, Y., Hanein, Y., Bohringer, K.F. and Ratner, B.D. (2004) Novel cell patterning using microheater-controlled thermoresponsive plasma films. *Journal of Biomedical Materials Research*, **70A**, 159–68.
231 Mills, C.A., Navarro, M., Engel, E., Martinez, E., Ginebra, M.P., Planell, J., Errachid, A. and Samitier, J. (2006) Transparent micro- and nano-patterned poly(lactic acid) for biomedical applications. *Journal of Biomedical Materials Research*, **76A**, 781–7.
232 Mills, C.A., Martinez, E., Errachid, A., Engel, E., Funes, M., Moormann, C., Wahlbrink, T., Gomilal, G., Plane, J. and Samitier, J. (2007) Nanoembossed polymer substrates for biomedical surface interaction studies. *Journal of Nanoscience and Nanotechnology*, **7**, 4588–94.
233 Fernandez, J.G., Mills, C.A., Martinez, E., Lopez-Bosque, M.J., Sisquella, X., Errachid, A. and Samitier, J. (2008) Micro- and nanostructuring of freestanding, biodegradable, thin sheets of chitosan via soft lithography. *Journal of Biomedical Materials Research*, **85A**, 242–7.
234 Li, J., McNally, H. and Shi, R. (2008) Enhanced neurite alignment on micro-patterned poly-L-lactic acid films. *Journal of Biomedical Materials Research*, **87A**, 392–404.
235 Li, J. and Shi, R. (2007) Fabrication of patterned multi-walled poly-L-lactic conduits for nerve regeneration. *Journal of Neuroscience Methods*, **165**, 257–64.

236 Krsko, P., McCann, T.E., Thach, T.-T., Laabs, T.L., Geller, H.M. and Libera, M.R. (2009) Length-scale mediated adhesion and directed growth of neural cells by surface-patterned poly(ethylene glycol) hydrogels. *Biomaterials*, **30**, 721–9.

237 Leguen, E., Chassepot, A., Decher, G., Schaaf, P., Voegel, J.-C. and Jessel, N. (2007) Bioactive coatings based on polyelectrolyte multilayer architectures functionalized by embedded proteins, peptides or drugs. *Biomolecular Engineering*, **24**, 33–41.

238 Wood, K.C., Boedicker, J.Q., Lynn, D.M. and Hammond, P.T. (2005) Tunable drug release from hydrolytically degradable layer-by-layer thin films. *Langmuir*, **21**, 1603–9.

239 Wood, K.C., Chuang, H.F., Batten, R.D., Lynn, D.M. and Hammond, P.T. (2006) Controlling interlayer diffusion to achieve sustained, multiagent delivery from layer-by-layer thin films. *Proceedings of the National Academy of Sciences of the United States of America*, **103**, 10207–12.

240 Macdonald, M., Rodriguez, N.M., Smith, R. and Hammond, P.T. (2008) Release of a model protein from biodegradable self assembled films for surface delivery applications. *Journal of Controlled Release*, **131**, 228–34.

241 Kim, B.-S., Park, S.W. and Hammond, P.T. (2008) Hydrogen-bonding layer-by-layer-assembled biodegradable polymeric micelles as drug delivery vehicles from surfaces. *ACS Nano*, **2**, 386–92.

242 Nguyen, P.M., Zacharia, N.S., Verploegen, E. and Hammond, P.T. (2007) Extended release antibacterial layer-by-layer films incorporating linear-dendritic block copolymer micelles. *Chemistry of Materials*, **19**, 5524–30.

243 Addison, T., Cayre, O.J., Biggs, S., Armes, S.P. and York, D. (2008) Incorporation of block copolymer micelles into multilayer films for use as nanodelivery systems. *Langmuir*, **24**, 13328–33.

244 Caruso, F., Caruso, R.A. and Möhwald, H. (1998) Nanoengineering of inorganic and hybrid hollow spheres by colloidal templating. *Science*, **282**, 1111–14.

245 Erel-Unal, I. and Sukhishvili, S.A. (2008) Hydrogen-bonded hybrid multilayers: film architecture controls release of macromolecules. *Macromolecules*, **41**, 8737–44.

246 De Geest, B.G., Déjugnat, C., Sukhorukov, G.B., Braeckmans, K., De Smedt, S.C. and Demeester, J. (2005) Self-rupturing microcapsules. *Advanced Materials*, **17**, 2357–61.

247 Chen, M.-C., Liang, H.-F., Chiu, Y.-L., Chang, Y., Wei, H.-J. and Sung, H.-W. (2005) A novel drug-eluting stent spray-coated with multi-layers of collagen and sirolimus. *Journal of Controlled Release*, **108**, 178–89.

248 Jewell, C.M., Fuchs, S.M., Flessner, R.M., Raines, R.T. and Lynn, D.M. (2007) Multilayered films fabricated from an oligoarginine-conjugated protein promote efficient surface-mediated protein transduction. *Biomacromolecules*, **8**, 857–63.

249 Lau, K.K.S. and Gleason, K.K. (2007) All-dry synthesis and coating of methacrylic acid copolymers for controlled release. *Macromolecular Bioscience*, **7**, 429–34.

250 Yoshida, M., Langer, R., Lendlein, A. and Lahann, J. (2006) From advanced biomedical coatings to multi-functionalized biomaterials. *Journal of Macromolecular Science: Part C: Polymer Reviews*, **46**, 347–75.

251 Ranade, S.V., Richard, R.E. and Helmus, M.N. (2005) Styrenic block copolymers for biomaterial and drug delivery applications. *Acta Biomaterialia*, **1**, 137–44.

252 Zhang, J., Chua, L.S. and Lynn, D.M. (2004) Multilayered thin films that sustain the release of functional DNA under physiological conditions. *Langmuir*, **20**, 8015–21.

253 Jewell, C.M., Zhang, J., Fredin, N.J. and Lynn, D.M. (2005) Multilayered polyelectrolyte films promote the direct and localized delivery of DNA to cells. *Journal of Controlled Release*, **106**, 214–23.

254 Jewell, C.M., Zhang, J., Fredin, N.J., Wolff, M.R., Hacker, T.A. and Lynn, D.M. (2006) Release of plasmid DNA from intravascular stents coated with

ultrathin multilayered polyelectrolyte films. *Biomacromolecules*, **7**, 2483–91.

255 Zhang, J., Fredin, N.J., Janz, J.F., Sun, B. and Lynn, D.M. (2006) Structure/property relationships in erodible multilayered films: influence of polycation structure on erosion profiles and the release of anionic polyelectrolytes. *Langmuir*, **22**, 239–45.

256 Chen, J., Huang, S.-W., Lin, W.-H. and Zhuo, R.-X. (2007) Tunable film degradation and sustained release of plasmid DNA from cleavable polycation/plasmid DNA multilayers under reductive conditions. *Small*, **3**, 636–43.

257 Lvov, Y., Decher, G. and Sukhorukov, G. (1993) Assembly of thin films by means of successive deposition of alternate layers of DNA and poly(allylamine). *Macromolecules*, **26**, 5396–9.

258 Wang, J., Zhang, P.-C., Mao, H.-Q. and Leong, K.W. (2002) Enhanced gene expression in mouse muscle by sustained release of plasmid DNA using PPE-EA as a carrier. *Gene Therapy*, **9**, 1254–61.

259 Lu, Z.-Z., Wu, J., Sun, T.-M., Ji, J., Yan, L.-F. and Wang, J. (2008) Biodegradable polycation and plasmid DNA multilayer film for prolonged gene delivery to mouse osteoblasts. *Biomaterials*, **29**, 733–41.

260 Liu, X., Zhang, J. and Lynn, D.M. (2008) Ultrathin multilayered films that promote the release of two DNA constructs with separate and distinct release profiles. *Advanced Materials*, **20**, 4148–53.

261 Liu, X., Zhang, J. and Lynn, D.M. (2008) Polyelectrolyte multilayers fabricated from charge-shifting anionic polymers: a new approach to controlled film disruption and the release of cationic agents from surfaces. *Soft Matter*, **4**, 1688–95.

262 Zhang, X., Sharma, K.K., Boeglin, M., Ogier, J., Mainard, D., Voegel, J.-C., Mely, Y. and Benkirane-Jessel, N. (2008) Transfection ability and intracellular DNA pathway of nanostructured gene-delivery systems. *Nano Letters*, **8**, 2432–6.

263 Wang, F., Wang, J., Zhai, Y., Li, G., Li, D. and Dong, S. (2008) Layer-by-layer assembly of biologically inert inorganic ions/DNA multilayer films for tunable DNA release by chelation. *Journal of Controlled Release*, **132**, 65–73.

264 Sakai, S., Yamada, Y., Yamaguchi, T., Ciach, T. and Kawakami, K. (2008) Surface immobilization of poly(ethyleneimine) and plasmid DNA on electrospun poly(L-lactic acid) fibrous mats using a layer-by-layer approach for gene delivery. *Journal of Biomedical Materials Research*, **88A**, 281–7.

2
Biofunctionalization of Polymeric Thin Films and Surfaces

Holger Schönherr

2.1
Introduction: The Case of Biofunctionalized Surfaces and Interfaces

The surface properties of both natural and artificial materials are among the main determinants for a wide range of important properties, including (but not limited to) biocompatibility, biointegration, and functionality [1, 2]. These properties are crucial not only for biomedical devices and implants, but also for (bio)sensors in various formats, among other applications [3, 4]. In addition, since interactions with the local environment play a decisive role in stem cell differentiation and in stem cell dedifferentiation [5], cell–surface interactions [6, 7] are also an important aspect of molecular biology, and especially in the area of research into cancer.

In general, whenever an object comes into contact with a biological system, such as cells, or with correspondingly derived fluids (e.g., blood or serum), the processes that occur at the newly formed interface determine whether that object will fulfill its function or will fail. These newly formed interfaces, where the initial contact between material and the biological system occurs, are frequently referred to as a *biointerfaces* [1, 8]. A substantial fraction of reactions and molecular interactions in biology – including molecular recognition processes – occur at natural interfaces. Although the role of biointerfaces was first appreciated many years ago, it is fair to say that biointerface science – as an identifiable discipline that attempts to address the complex phenomena involved in a quantitative and physico-chemically rigorous manner – has emerged only quite recently.

In the case of medical implants (as one class of illustrative example), the surface roughness and chemical composition may determine the success of, for instance, the integration of the implant with newly grown bone – that is, the osteoinduction and osseointegration of the implant material [9]. Likewise, improvements in the biocompatibility and integration of materials in biomedical applications can be enhanced by employing tailored polymeric coatings; an example is the surface-attached hydrogel coatings that become attached photochemically to heart valves [10].

Nanomaterials for the Life Sciences Vol.5: Nanostructured Thin Films and Surfaces.
Edited by Challa S. S. R. Kumar
Copyright © 2010 WILEY-VCH Verlag GmbH & Co. KGaA, Weinheim
ISBN: 978-3-527-32155-1

The same holds true for biosensors, whether catheter-based *in situ* sensors that can be used to quantify glucose levels in blood, or *ex situ* sensors that address the presence and abundance of particular antibodies. Here, the interaction between the analytes in solution and the receptors immobilized on a sensor surface can depend crucially on the orientation of the receptor. In this respect, the suppression of nonspecific protein adsorption is a key element [11, 12].

A third area, where these interfaces are important and where polymers play a pivotal role, relates to the interactions of cells with their immediate surroundings [6]. It is well known that cells interact with their external environment via transmembrane proteins, many of which are receptors that function not only as transmitters of information but also as transporters of molecules from the outside to the inside of the cell. As so-called "membrane-spanning proteins," these receptors are characterized by an extracellular ligand-binding domain and an intracellular signaling domain. The binding of a ligand with the receptor causes an intracellular cascade of enzyme-mediated reactions that in turn leads to the signal being amplified. This process is triggered either by a change in the receptor's conformation, or by a change in its affinity for other molecules. As this so-called "signal transduction" affects gene regulation, it should not be surprising that a variety of cell functions, including survival, proliferation, migration and differentiation, have been shown to be governed by integrated signals from many types of molecule.

Interestingly, these signaling cascades and cellular processes can also be controlled or triggered by more or less specific interactions with tailored surfaces (interfaces). In their early investigations, Whitesides, Ingber and colleagues patterned monolayers by microcontact printing [13, 14] with cell-adhesive and cell-repellent functionalities, and further investigated the effect of the feature dimensions of these patterns on cell behavior [15]. The extent of the cell-adhesive areas was shown to control cell behavior, such as proliferation or apoptosis (Figure 2.1).

In addition to surface chemistry [16], the topographical structures [17] and control of the substrate modulus are important in determining cell behavior on surfaces [18–20]. Apart from their low-cost manufacture, this combination of factors and determinants predestines *polymers* (i.e., macromolecular materials) for the fabrication of functional biointerfaces. Polymers can be structured conveniently on length scales that cross many orders of magnitude (from <10nm to several meters). Polymers also possess hierarchical structures, with many levels of different complexities. The primary chemical structure, and in many cases also the encoded information for higher-level hierarchical structures, are controlled by the synthetic process. Unless the functional groups in complex monomers are compatible with a direct polymerization, then a subsequent modification will be required. As will be discussed in this chapter, a plethora of very successful functionalization and modification strategies, which impart the required functionality to the polymer surface in question, has been developed during the past decades. However, more recently a variety of new approaches has been introduced, such that the level of control achieved and the level of complexity that can be addressed in the fabrication of advanced biointerfaces bridges from the molecular to macroscopic length scales. Finally, linear and crosslinked polymers offer the possibility

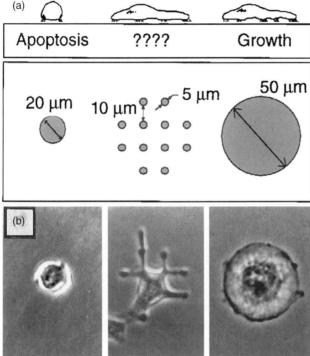

Figure 2.1 Cell–extracellular matrix (ECM) contact area versus cell spreading as a regulator of cell fate. (a) Diagram of substrates used to vary cell shape independently of the cell–ECM contact area. Substrates were patterned with small, closely spaced circular islands (center) so that cell spreading could be promoted as in cells on larger, single round islands, although the ECM contact area would be low as in cells on the small islands; (b) Phase-contrast micrographs of cells spread on single 20 μm or 50 μm-diameter circles or multiple 5 μm circles patterned as shown in (a). Reproduced with permission from Ref. [15]; © 1997, AAAS.

to tune mechanical properties, such as the elastic modulus, over a broad range. Depending on the choice of backbone, comonomers, side chains and crosslink density, the entire spectrum of required moduli can be realized.

This chapter will focus primarily on the *chemistry* and *surface engineering* involved in the fabrication of polymer-based biofunctional surfaces and biointerfaces. In addition to discussing the sophisticated and optimized surface chemistry to facilitate efficient and controllable bioconjugation, and to implement the important function of a suppressed nonspecific adsorption of proteins, both micropatterning and nanopatterning methodologies will be outlined [21]. The initial discussion covers the general requirements for biofunctionalized surfaces, after which strategies and selected examples are provided with regards to surface modifications using functional polymers and polymer-based approaches. The chemical surface modification of polymer surfaces and polymeric thin films, in order to fabricate advanced biointerfaces, is described, after which attention is focused on studies of cell–surface interactions and the routes compatible with patterning. Finally, the most recent advances in the field will be considered, together with future perspectives.

2.2
Polymer-Based Biointerfaces

Thin structured polymeric films currently form the focal point for the development of advanced materials and tailored interfaces for applications such as biosensors and biologically oriented research involving biointerfaces [1, 8, 22]. Patterned polymer films that incorporate reactive functional groups provide a platform that can be further modified by chemical reactions [23]. Prime examples include the fabrication of DNA arrays on glassy polymers, such as poly(methyl methacrylate) (PMMA) [24], the synthesis of star-shaped poly(ethylene glycol)s (PEGs) [25], and the immobilization of proteins and peptide sequences on polyurethanes [26]. In addition, new approaches, which include the fabrication of reactive surfaces by electrografting [27], using click chemistry [28, 29], or surface-initiated polymerizations (SIPs), have been reported [30, 31].

Apart from differences in their chemical composition, thin and ultrathin polymer films may possess different architectures, as shown schematically for passivating PEG films in Figure 2.2 (see also Section 2.2.1). The lower protein repellence properties of the grafted non-crosslinked star polymers has been attributed to the gaps between the stars exposing the underlying surface to small proteins. By contrast, linear PEGs and crosslinked star PEGs provide a much lower exposure of the underlying surface to the protein solution.

In addition to these architectural considerations, the altered behavior of polymers in thin films and at interfaces must be taken into account. As well as changes in the glass transition temperatures (T_g) of ultrathin substrate-supported polymer films [33], the confinement into sub-100 nm thin films has been reported to affect diffusion of the polymer molecules comprising the film [34], as well as diffusion of small, low-molar-mass tracer molecules, among other altered properties [35].

2.2.1
Requirements for Biofunctionalized Polymer Surfaces

The first target is to identify the key features and functions that must be introduced when fabricating macromolecular biofunctional surfaces and biointerfaces. The direct deposition of biologically relevant molecules and immobilization via non-specific physisorption onto surfaces is successful only in certain cases. While DNA and certain robust proteins do not denature, and hence may not lose functionality upon adsorption, many biological molecules are denatured. In addition, the requirement for certain applications (as noted above) require an enhanced control over, for example, protein position, coverage and, in particular, spatial orientation.

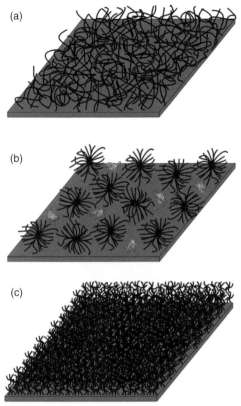

Figure 2.2 Schematic diagram of: (a) grafted linear PEG; (b) grafted star PEG; and (c) grafted and intermolecularly crosslinked star PEG on a surface. Reproduced with permission from Ref. [32]; © 2007, Royal Society of Chemistry.

DNA microarrays, as one typical example of an established biointerface application, have witnessed a steady improvement since the first pioneering reports of the 1990s [36]. In principle, single-stranded DNA (ssDNA) of known composition is immobilized in various spots on a biochip surface, and serves as a means to recognize and quantify solution-borne DNA. In addition, on-chip DNA synthesis using, for example, advanced photochemistry has been developed [37, 38]. More recently, DNA arrays which exploit protein–DNA constructs were shown to comprise a feasible route towards the fabrication of protein arrays [39]. The central requirements for any functional biointerface in the context of this chapter are demonstrated with this example:

- Any nonspecific interaction of solution-borne DNA with the biochip surface must be minimized or eliminated.

- Likewise, the active sensing moieties – the ssDNA – must be immobilized such that it is accessible to the target DNA in solution; this allows the ssDNA to discriminate the presence of base-pair mismatches.

Hence, the deposition or coupling of a passivating layer or coating is as important as the immobilization of a probe DNA. In this context, the reproducibility, variations in surface activity, DNA chip aging, and reliability of the coupling chemistry have been critically discussed [40, 41]. In addition to a sophisticated and optimized surface chemistry to facilitate an efficient and controllable bioconjugation, and to implement the function of suppressing the nonspecific adsorption of proteins, these requirements include compatibility with micropatterning and nanopatterning methodologies.

The requirements of sensor surface fabrication for label-free detection techniques, such as surface plasmon resonance (SPR) detection, may be even more stringent, according to a recent review [4]. These requirements include:

- A simple and rapid fabrication method.
- Intimate contact of the biomolecules with the transducer surface, favoring good and unhindered accessibility of the analyte.
- The immobilization of a large amount of biomolecules so as to ensure sufficient sensitivity.
- Long-term stability.
- Oriented immobilization of the biomolecules.
- The prevention of nonspecific adsorption to avoid erroneous or misleading measurements.

Many of these mandatory requirements are met successfully with different advanced surface engineering approaches (Figures 2.3 and 2.4). The prevention of nonspecific adsorption is often achieved (as alluded to above) by coating the surface with a PEG layer. Whilst PEG is characterized mainly by its hydrophilicity, it can also accommodate water and thereby eliminate hydrophobic interactions, as well as protein denaturation that would result in an entropic conformational gain for the protein. Finally, the presence of an adsorbed PEG layer acts to block sites for protein adsorption.

In general, protein adsorption can be prevented by eliminating the attractive interactions between the protein and the surface (*thermodynamic control*), or alternatively by increasing the activation energy barrier for adsorption (*kinetic control*). A kinetic control can be achieved by preventing the protein from sensing the surface, for example, via a coating with a hydrogel coating (that may be based on PEG).

The protein resistance of PEG has been attributed to a combination of many factors (Figure 2.3). Electrical double layer forces are eliminated, since the neutral PEG screens the substrate charge. The primary adsorption of proteins to the surface is prevented by the formation of dense PEG layers, although sufficiently thick PEG layers can also prevent additional (secondary) adsorption. An *entropic barrier* is also built up which renders the compression of chains by proteins unfavorable.

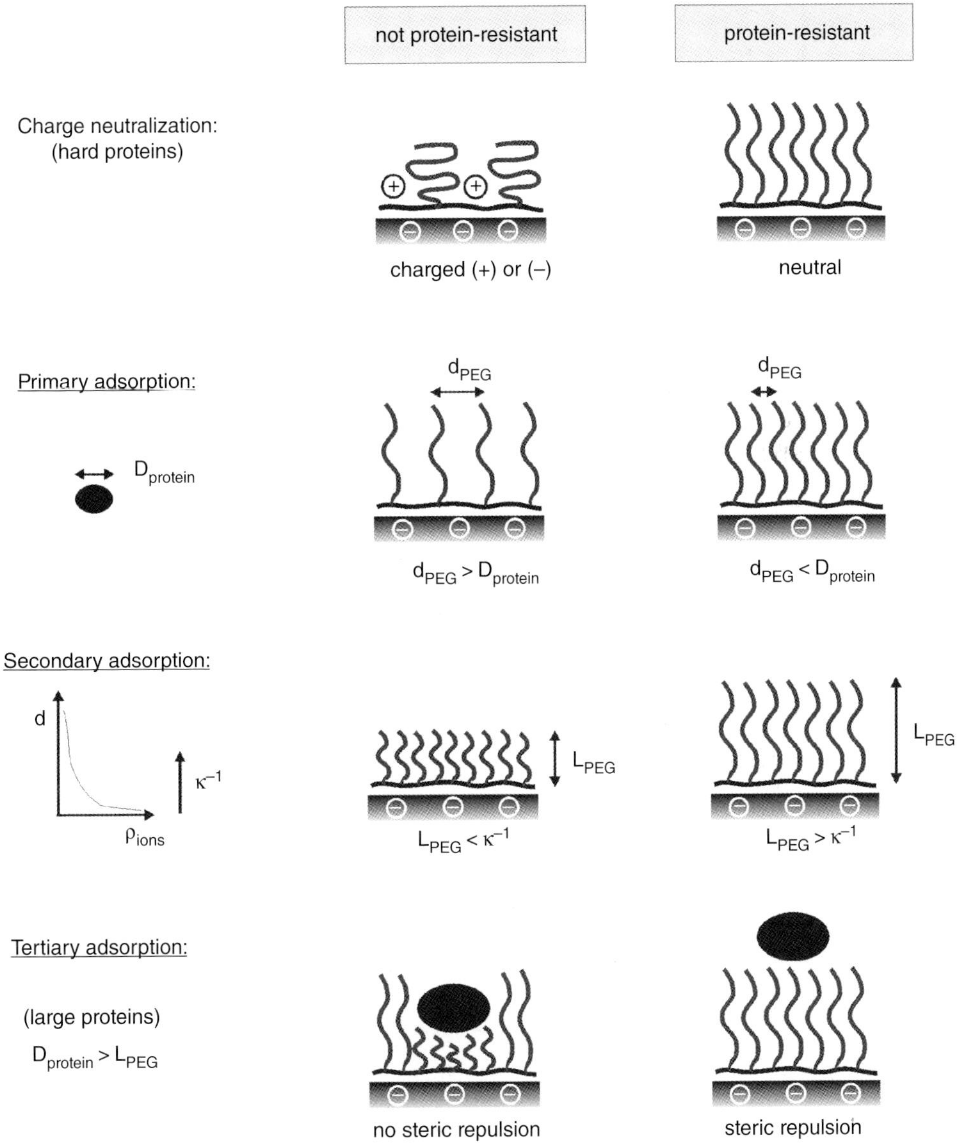

Figure 2.3 Design rules for PEG-based coatings in order to achieve protein resistance. Here, L denotes the layer thickness, κ^{-1} the Debye length, d the distance between the grafted PEG chains, and D the protein diameter. Reproduced with permission from Ref. [42].

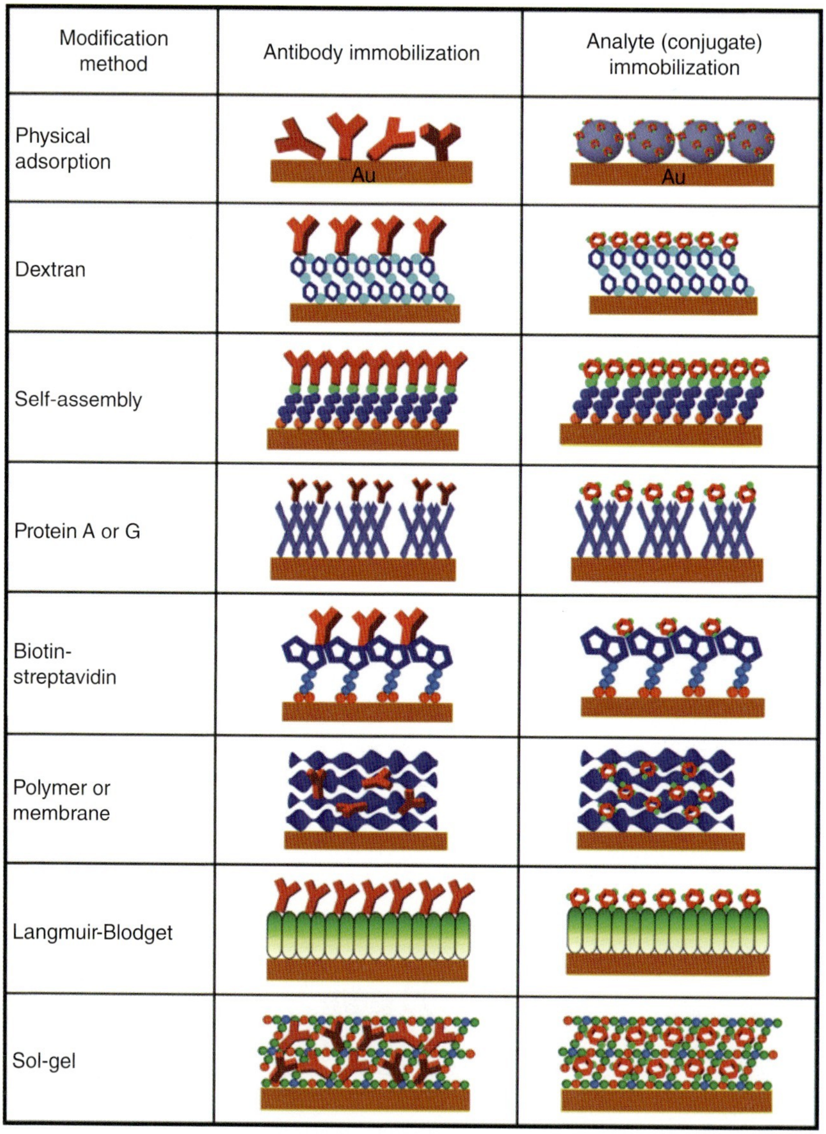

Modification method	Antibody immobilization	Analyte (conjugate) immobilization
Physical adsorption		
Dextran		
Self-assembly		
Protein A or G		
Biotin-streptavidin		
Polymer or membrane		
Langmuir-Blodget		
Sol-gel		

Figure 2.4 Schematic of different interfacial architectures for the immobilization of capture moieties. Reproduced with permission from Ref. [4]; © 2007, IOP Publishing.

Among the possible formats for the immobilization of antibodies or DNA for biosensing, the creation of thin organic films (referred to as self-assembled monolayers; SAMs) on solid supports, polymeric substrates or polymer thin films, and of tailored polymer brushes, have been reported [21]. Some of the established immobilization procedures for antibodies, and the reaction with the solution-borne analyte, are summarized in Figure 2.4.

2.2.2
Surface Modification Using Functional Polymers and Polymer-Based Approaches

It is possible to differentiate several methods of functionalizing surfaces with polymers to fabricate functional biointerfaces (Figure 2.5). Depending on the type of attachment and the process, this includes: (i) the attachment of preformed polymers to coupling moieties on surfaces or polymers; (ii) the growth of polymers from surface-immobilized initiators; (iii) the covalent immobilization or

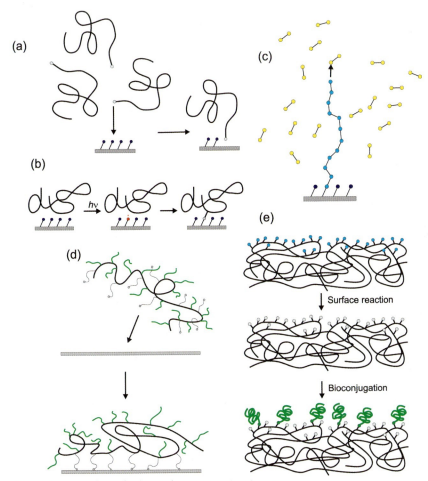

Figure 2.5 (a) Grafting of polymer chains to reactive sites on the surface; (b) Capture of polymer chains via photochemically activated reactive groups; (c) Surface-initiated polymerization; (d) Adsorption and coupling of multifunctional graft copolymer; (e) Surface modification and subsequent bioconjugation of the polymer surface.

physisorption of multifunctional copolymers or terpolymers; and (iv) gas-phase processes, including plasma and chemical vapor deposition (CVD) polymerizations. Alternative approaches, such as layer-by-layer (LBL) deposition are also important, but details of these are not included in this chapter.

All these approaches differ in terms of their complexity, the level of control, and also the functionality of the resultant biointerfaces. Simple surface engineering approaches are, of course, attractive if they provide the desired functional and stable biointerfaces. However, it is à priori impossible to decide which approach is appropriate to address a particular problem. The use of slightly more complex macromolecular architectures, such as end-functionalized star polymers [25, 32], may represent a route to these interfaces.

Several classic approaches comprise multistep procedures, including several subsequent chemical activation and functionalization reactions at the surfaces, while others consist of a single immobilization step. However, in the latter case the immobilized adsorbate may be the product of a multistep synthesis, such as multipotent copolymers that possess the necessary functional elements for surface attachment, passivation and function via active sites. Multipotent polymeric coatings may also be synthesized *in situ* (see Section 2.2.2.4).

2.2.2.1 Grafting of Polymers to Surfaces

Polymer chains can be covalently tethered to the surface using the so-called "grafting to" approach (Figure 2.6) [44]; this refers to the covalent linkage of an

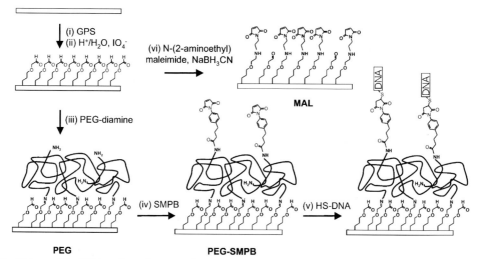

Figure 2.6 Schematic representation of the procedure used for preparing PEG-grafted and maleimide-terminated surfaces for the immobilization of thiolated DNA oligonucleotides. Reproduced with permission from Ref. [43]; © 2006, American Chemical Society.

a The thickness of the PEG layer is not drawn to scale.

end-functionalized polymer (or a protein) to a surface that presents complementary reactive groups. Functional groups on the polymer can also be further reacted in subsequent conjugation reactions. This approach results in lower grafting densities compared to grafting-from procedures (as discussed in Section 2.2.2.2), which are in the range of ~10–40 mg mm^{-2}, because the immobilized polymer chains sterically hinder the diffusion of polymer molecules from the solution to reactive sites at the surface.

Furthermore, depending upon the surface density of the polymer and the interaction forces between the polymer and underlying surface, the thickness and density of the polymer graft is difficult to control, due to the various conformations of the polymer at the surface. This limitation is especially important when fabricating a nonfouling surface using PEG that is grafted from solution, since the protein resistance of grafted PEG depends on both the surface density and length of the PEG chains [45].

The grafting of polymers equipped with a terminal functional group (grafting-to approach) allows the chains to be immobilized at any surface that exposes the reactive functional groups. Despite the concomitant disadvantage of intrinsically limited grafted film thicknesses, this approach is useful in many applications. The surface-immobilized species is not necessarily exposed at the surface of a monolayer; rather, it may also be localized as a pending functional group of a polymer (*vide infra*).

A related approach is the photochemically induced generation of highly reactive species, such as those formed by benzophenone [46]. Here, the radicals react with nearby C–H bonds (for example, of a polymer chain) and thus form covalent bonds between the surface-immobilized benzophenone and the polymer in question (Figure 2.7). This simple and effective way to covalently link thin polymeric layers to solid surfaces can be controlled by varying the illumination time and the molecular weight of the polymer. In the case of poly(styrene) (PS) and poly(ethyloxazoline), the thickness of the layer was found to increase linearly with the radius of gyration of the corresponding polymer.

Another such example is the covalent (and thus very robust) attachment of (bio)molecules to wet chemically activated polymer surfaces under well-controlled conditions in buffer (see Figure 2.8). In this particular example of block copolymer, the microphase-separated morphology of the block copolymer with PS cylinders in a matrix of the reactive *tert*-butyl ester acrylate (PtBA) affords an improved stability of the films. At the film surface, an approximately 8 nm-thick PtBA skin layer is exposed that can be converted to poly(acrylic acid) by a facile deprotection reaction using, for example, organic acids (trifluoroacetic acid, TFA) or heat. This deprotection is followed by an established activation using N-hydroxysuccinimide (NHS) chemistry, to yield active esters that are reactive towards primary amines in an aqueous medium (Figure 2.8a). The data for the immobilization of amino-end functionalized PEG with different molar masses are presented in Figure 2.8b. Based on ellipsometry and X-ray photoelectron spectroscopy (XPS) data, grafting densities of ~2.4, 0.8, and 0.5 PEG molecules per nm^2 were determined for PEG with different molar masses. Comparable SAMs of a NHS ester-terminated

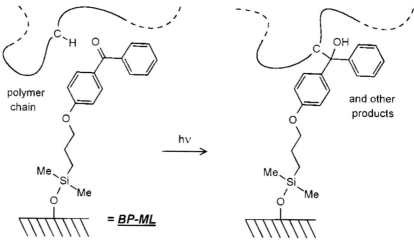

Figure 2.7 Schematic representation of the photochemical attachment of polymer chains to solid surfaces via illumination of polymer-covered monolayers of a benzophenone derivative (BP-ML). Reproduced with permission from Ref. [46]; © 1999, American Chemical Society.

disulfide on gold showed much reduced grafted thicknesses, which indicated an enhanced availability of NHS ester groups at and near the surface of the polymer film. Thus, the polymer film could be functionalized in a quasi-three-dimensional (3-D) manner. The improved coverages of PEG led to a better functionality of the PEG layer to suppress the nonspecific adsorption of proteins and, depending on the molar mass, to reduce and eliminate the interaction of cells with correspondingly functionalized surfaces (Figure 2.9).

The reliable conjugation of proteins, as well as PEG for the required surface passivation, on activated (PS-*b*-PtBA), by using the grafting-to approach, is shown in Figure 2.9. In particular, the blocking ability of PEG layers with different number average molar masses (M_n) was addressed. It has been well established that the molar mass/chain length and the grafting density of PEG molecules determine, to a large extent, the material's antifouling properties [47]. On homogeneous layers of PLL and fibronectin the cells were shown not only to adhere, but also to stretch out on the film (Figure 2.9a,b); this was indicative of the full adhesive capacity of PaTu8988T cells, and the unrestricted formation of focal adhesions. On the PEG layers with a high molar mass, only isolated cells were observed that exhibited a rounded appearance; these cells were unable to attach and died after prolonged contact with the surface. The observed differences in cell-surface coverage are shown, quantitatively, in the histogram of Figure 2.9e.

2.2.2.2 Polymer Brushes by Surface-Initiated Polymerization

A *polymer brush* is a system of polymer chains that are densely end-tethered or end-grafted onto a surface [48]. The attachment can be made either by end-grafting

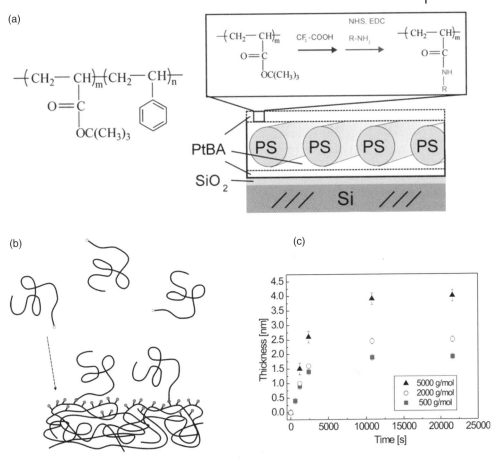

Figure 2.8 (a) Schematic of the block copolymer architecture and the surface chemistry employed for immobilization; (b) "Grafting-to" reaction employed for the NHS-activated polymer surface; (c) Thickness changes of activated PS-b-PtBA films after grafting PEG$_n$–NH$_2$ (n = 11, 45, 114) for various times as determined by ex situ ellipsometry. Reprinted with permission from Ref. [120]; © 2007, Elsevier.

of a homopolymer, or by the selective adsorption of one of the blocks of a diblock copolymer [49]. Grafting has been utilized as an important technique to modify the chemical and physical properties of polymers (see also Section 2.2.2.1). The major advantage of polymer brushes over other surface-modification methods (e.g., SAMs) is their mechanical and chemical robustness, coupled with a high degree of synthetic flexibility towards the introduction of a variety of functional groups (also via brush copolymerization; see below). There is also an increasing interest in the use of functional or diblock copolymer brushes for "smart" or responsive surfaces, which can change their physical properties upon stimulation [50].

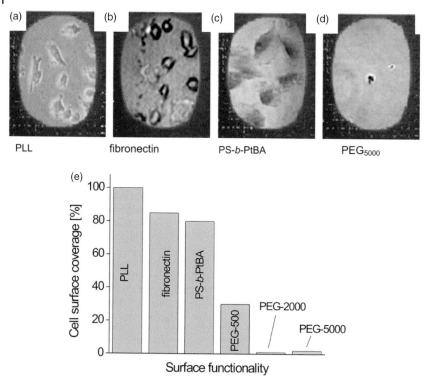

Figure 2.9 Representative wide-field optical micrograph of: (a) PLL-covered PS$_{690}$-b-PtBA$_{1210}$ films (scale: 90 μm × 160 μm); (b) Fibronectin-covered film (scale: 90 μm × 160 μm); (c) PS-b-PtBA-covered film (scale: 90 μm × 160 μm); (d) PEG5000–NH$_2$- covered film (scale: 90 μm × 160 μm); (e) Histogram of cell-surface coverage (normalized to coverage on PLL) versus polymer surface composition. Reprinted with permission from Ref. [120]; © 2007, Elsevier.

An alternative approach for grafting polymers to surfaces is that of "grafting from" (also known as SIP), as shown in Figure 2.10. In this approach, a surface is activated to present an initiator, and a polymer of interest is grown from the surface [51]. First, the substrate of choice is modified with initiator-bearing SAMs. These SAMs can be formed on almost any surface, as long as an appropriate functionality is chosen (e.g., thiols on gold, silanes on glass, Si/SiO$_2$ and plasma-oxidized polymers). The initiator surfaces are then exposed to solutions containing a catalyst and monomer (plus a solvent, if necessary). Ideally, the polymerization should be not only surface-initiated but also surface-confined; that is, no polymerization should occur in solution [52].

In order to achieve maximum control over the brush density, polydispersity, and composition, a controlled polymerization is highly desirable. During the past few years this field has rapidly evolved, such that today all of the major controlled polymerization strategies have been used to grow polymer brushes [52, 53]. Of these, controlled radical polymerizations have become the most popular routes,

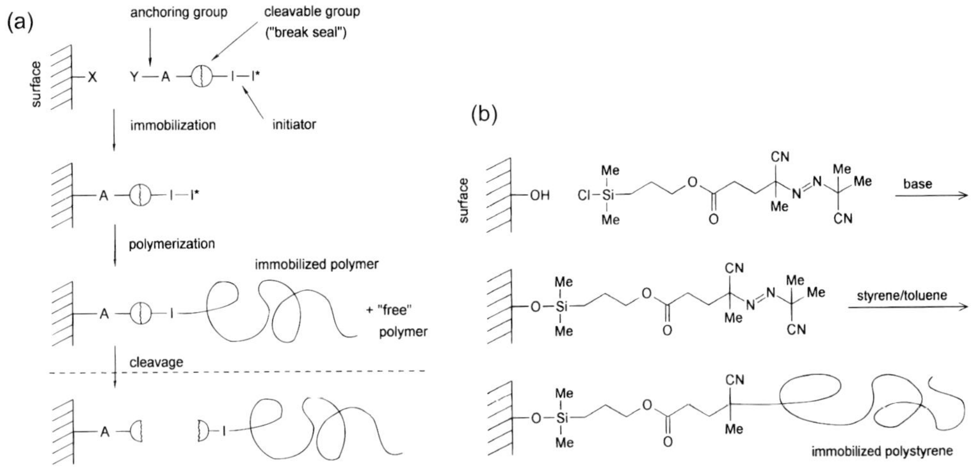

2.2 Polymer-Based Biointerfaces | 69

Figure 2.10 (a) Concept for the synthesis of covalently attached polymer monolayers via radical graft polymerization ("grafting-from" technique) using immobilized initiators; (b) Synthesis of monolayers of polystyrene terminally attached to SiO$_x$ surfaces by using a self-assembled monolayer of AIBN-like azo compound with a chlorosilane headgroup. Reproduced with permission from Ref. [51]; © 1998, American Chemical Society.

mainly because of their tolerance to a wide range of functional monomers and less stringent experimental conditions. Different controlled polymerizations for brush growth have been used, including living ring-opening polymerization (ROP), living anionic polymerization, living cationic polymerization, ring-opening metathesis polymerization (ROMP), nitroxide-mediated polymerization (NMP), reversible addition-fragmentation chain-transfer polymerization, and atom transfer radical polymerization (ATRP) [54].

In fact, ATRP has become probably the most widely employed technique for the formation of polymer brushes via SIP. ATRP is compatible with a variety of functionalized monomers, and the living/controlled character of the ATRP process yields polymers with a low polydispersity that are end-functionalized and so can be used as macroinitiator for the formation of di- and triblock copolymers. Equally important, surface-initiated ATRP is experimentally more accessible than, for example, the living anionic polymerization, which requires rigorously dry conditions.

The controlled nature of ATRP is due to the reversible activation–deactivation reaction between the growing polymer chain and a copper-ligand species [55]. One typical monomer for the formation of polymer brushes via surface-initiated ATRP (SI-ATRP) is oligoethylene glycol methacrylate (Figure 2.11). For example, Ma et al. synthesized an alkanethiol-functionalized with a terminal ATRP initiator, and subsequently prepared a SAM of this ATRP initiator-functionalized thiol on gold [56]. This initiator-functionalized SAM on gold was then used to carry out a SI-ATRP of poly(oligo(ethyleneglycol)methacrylate) (poly(OEGMA)) on gold (as

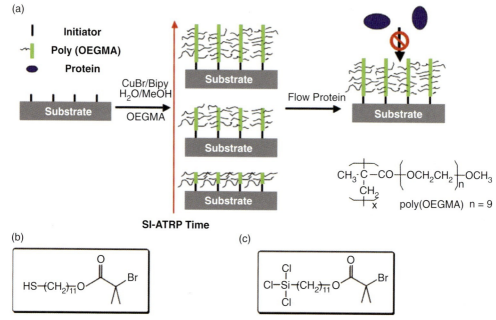

Figure 2.11 (a) Schematic illustration of SI-ATRP of OEGMA strategy to create a protein-resistant surface. Molecular structure of ATRP initiator; (b) Thiol-terminated initiator; (c) Silane-terminated initiator (11-(2-bromo-2-methyl)propionyloxy) undecyltrichloro silane. Adapted with permission from Ref. [31]; © 2006, American Chemical Society.

shown in Figure 2.11). Later, Chilkoti and coworkers exploited this SIP for the fabrication of bioinert surfaces.

Polystyrene-*block*-poly(*tert*-butyl acrylate) (PS-*b*-PtBA) brushes were also synthesized using ATRP. Here, PtBA can be regarded as a precursor for carboxylic acid functionalities, as shown by Mengel et al., who hydrolyzed PtBA in LB films [57] (see also Section 2.2.2.1).

The "grafting from" approach possesses several attractive features:

- A very high surface density of up to 85 mg mm^{-2} of a polymer can be obtained [58].

- A linear increase in the polymer thickness with polymer molecular mass is observed, because the high surface density of the polymer forces the polymer chains into a brush conformation so as to minimize steric constraints.

- The procedure is easy, because the synthesized polymer is solely localized at the interface; hence, the need to extract any loosely physisorbed polymer is eliminated [59].

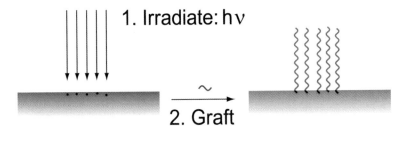

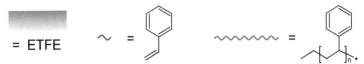

Figure 2.12 Scheme of the UV-initiated grafting process. Selective irradiation of an ethylene tetrafluoroethylene (ETFE) foil with extreme ultraviolet (EUV) radiation creates radicals in the polymer. Subsequent grafting of polystyrene leads to the growth of a 3-D structure. Reproduced with permission from Ref. [60]; © 2004, American Vacuum Society.

Graft polymerizations were also initiated on neat polymer surfaces, for example by intense UV radiation, which generates radicals. These radicals may subsequently polymerize monomers that are present close to the surface, via free-radical pathways (Figure 2.12) [60].

These exciting advances relating to the nanostructuring of polymer and biomacromolecular brushes were recently reviewed by the Zauscher group. The integration of various brush synthesis techniques with advanced nanolithographic and templating approaches can provide access to well-defined nanostructures based on polymers (Figure 2.13) [63]. Gradient structures may be fabricated by spatial variations in the irradiation dose for light-induced SIP. Responsive polymers, such as poly(N-isopropylacrylamide) (poly-NIPAAm) have also been synthesized in a spatially highly controlled manner.

One very new development in this area has been the adaptation of solid-phase peptide synthesis to graft peptide monolayers from solids for the modification of interfacial properties. By using a commercial microwave peptide synthesizer, the group of Ducker grew a 15-residue peptide from the surface of an aminosilanized silicon wafer [64]. This method of direct synthesis of peptides at a surface represents a novel means of introducing a peptide layer at surfaces. In this way, complex adsorption or grafting procedures, or the purchase of expensive peptides, are circumvented.

Previously (Figure 2.14), simple peptides, such as poly(gamma-benzyl-L-glutamate) (PBLG), were grafted in monolayers onto silicon with native oxide

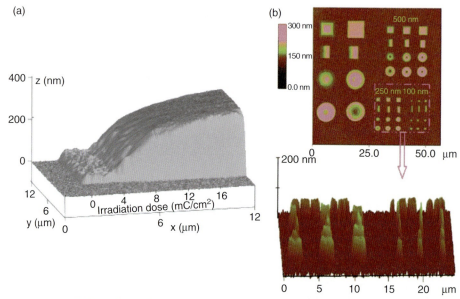

Figure 2.13 (a) Atomic force microscopy (AFM) height image of polymer brush gradient formed via a variation in irradiation dose during electron-beam chemical lithography; (b) AFM images of pNIPAAm brush patterns, showing gradual height variations within some of the patterned features. Adapted with permission from Refs [61, 62]; © 2007, Wiley-VCH.

surfaces via an initiation of the polymerization of the N-carboxy anhydride of gamma-benzyl-L-glutamate by a surface-attached amino-terminated silane [65].

2.2.2.3 Physisorbed Multifunctional Polymers

A very successful and rapid approach for the modification of oxidic and other surfaces with functional polymers has been developed by the groups of Hubbell and Textor [66–68]. By exploiting multipoint attachment of designed copolymers with grafted side chains for attachment, functional side chains and passivating chains dense layers can be fabricated by a simple dip-coating process. The adsorbed copolymer layer forms presumably a comb-like structure at the surface, with the anchor groups bound to the surface, while the hydrophilic and uncharged PEG side chains are exposed to the solution phase. The copolymer architecture was shown in various studies to represent an important factor in the resulting protein resistance. In particular, the packing-density and the radii of gyration of the PEG are seen to control protein repellence.

On oxidic surfaces poly-L-lysines are exploited, whilst for gold surfaces polysulfides were developed. The protonated primary amino groups of the lysine residues interact strongly with the negatively charged species on the oxidic surfaces, whereas the sulfides bind strongly to the gold surfaces. In addition to this anchor/immobilization function, these polymers comprise two further functions.

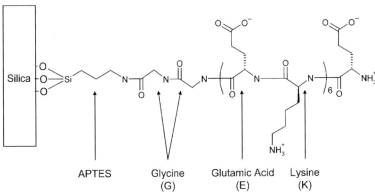

Figure 2.14 Schematic of the peptide synthesized on silica. The charge is a function of the pH. Reproduced with permission from Ref. [64]; © 2009, American Chemical Society.

First, the pending PEG chains in poly(L-lysine)-g-poly(ethylene glycol) (PLL-g-PEG) render the modified surfaces protein-resistant. Second, telechelic PEG units bearing terminal biotin or other active groups, such as nitrilotriacetic acid-equipped PEGs for the Ni-histidine-based complexation of proteins, render them functional (Figure 2.15 and Figure 2.16).

The adsorbed PLL-g-PEG layers were shown to be highly effective towards reducing the adsorption of proteins also from blood serum. The coatings were also found to be resistant towards fibrinogen, which plays an important role in the cascade of events leading to biomaterial-surface-induced blood coagulation and thrombosis. For example, Textor et al. reported adsorbed protein levels as low as <5 ng cm^{-2} for an optimized polymer architecture. The PLL-g-PEG layers were also reported to show good long-term stability, while retaining their resistance to protein adsorption [70].

In detailed studies addressing the interaction mechanism and strength of PLL-PEG layers, by using colloidal force microscopy, the steric component was found to be most prominent for high PEG grafting densities [71]. At low ionic strengths and high grafting densities, an additional repulsive electrostatic surface force was also observed. For lower grafting densities and lower ionic strengths, Pasche et al. observed a substantial attractive electrostatic contribution arising from an interaction of the electrical double layer of the protonated amine groups. Thus, for the PLL-g-PEG coatings the net interfacial force could be described as a superposition of electrostatic and steric-entropic contributions.

When investigating the specific immobilization of biomolecules, the biotin–streptavidin interaction has been successfully implemented. The specific binding of avidin, neutral avidin, or streptavidin, and the subsequent immobilization of biotinylated biomolecules (such as antibodies) has been demonstrated [67]. More recently, the advantages of the PLL-g-PEG-based layers were combined with the well-established NiII–His-tag chemistry (Figure 2.16) [72].

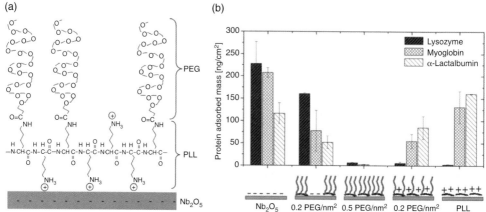

Figure 2.15 (a) Idealized scheme of the interfacial structure of a monolayer of PLL-g-PEG adsorbed onto a metal oxide substrate (Nb$_2$O$_5$) via electrostatic interactions between the negatively charged metal oxide substrate and positively charged amino-terminated PLL side chains (at neutral pH); (b) Protein-adsorbed mass measured in 10 mM HEPES buffer (pH 7.4) by optical waveguide lightmode spectroscopy (OWLS), on the five surfaces with various surface charges and PEG surface densities. From left to right the surface charge gradually changes from negative to positive; the PEG surface density passes through a maximum of 0.5 PEG nm^{-2} for the PLL(20)-g[3.5]-PEG(2) polymer layer shown in the center. Surfaces with a medium value for the average PEG surface density of 0.2 PEG nm^{-2} were obtained with PLL(20)-g[3.5]-PEG(2) (1/2; i.e., half monolayer coverage), charged negatively, and a monolayer of PLL(20)-g[10.1]-PEG(2), charged positively. Reproduced with permission from Ref. [69]; © 2005, American Chemical Society.

2.2.2.4 Multipotent Covalent Coatings

A similar principle can be exploited if polymers with functional groups are synthesized that can couple covalently to species exposed at a film surface. An early example of this was the concept of using reactive polymer interlayers as an immobilization platform. For instance, the alternating copolymer of vinyl isocyanate and maleic anhydride (poly[(1-methylvinyl isocyanate)-*alt*-(maleic anhydride)]) (IAP) [73], has been employed successfully by Beyer et al. to generate reactive polymer surfaces that expose anhydride and isocyanate groups [74]. These surfaces are reactive towards amines, among others. In addition, by using bifunctional spacers (e.g., 2-aminoethanol), multilayers with tunable thickness and functionality can easily be obtained (Figure 2.17).

A related approach has been reported by Park et al. [75] who, in a terpolymer with pending trimethoxy silanes as the anchors for silanol-terminated surfaces, combined PEG side chains to reduce protein adsorption and NHS esters for the subsequent conjugation of active species (including biomolecules). In this way the central functions of bioreactivity and antibiofouling could be incorporated together with anchors for stable surface attachment.

A similar result was obtained by exploiting cleverly chosen multicomponent formulations that comprised an α-NHS ester-ω-trimethoxy silane, an azidosulfo-

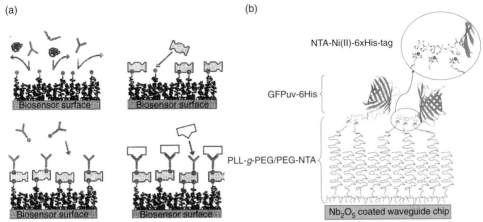

Figure 2.16 (a) Schematic of biointerface fabrication using multisite adsorption of poly-L-lysine-PEG conjugates by dip-coating. The charged lysine residues adhere via electrostatic interactions on the oxidic biosensor surface, the PEG brushes prevent nonspecific adsorption of proteins, while the incorporated terminated biotin can be exploited to immobilize tetrafunctional streptavidin. The remaining binding site of streptavidin serve as attachment sites for biotin-modified antibodies. Reproduced with permission from Ref. [67]; © 2002, American Chemical Society.; (b) The capture of His-tagged proteins in the correct orientation has been demonstrated using the modified PLL-g-PEG polymer. A schematic view of the molecular assembly monolayer of the polycationic polymer PLL-g-PEG/PEG-NTA on a negatively charged Nb_2O_5-coated waveguide chip. The GFPuv-6His is shown to be attached via the six adjacent histidine residues to three and two Ni–NTA complexes, with the NTA covalently bound to the end of the long PEG chains (molecular weight 3.4 kDa). The nonfunctionalized PEG chains have a molecular weight of 2 kDa. The magnified inset shows details of the NTA–Ni^{II}/6 × His-tag interaction. Reproduced with permission from Ref. [72]; © 2006, Wiley-VCH.

nylhexyl-triethoxy silane, and a polyethylene sorbitan tetraoleate in a polar solvent [76]. The films obtained by spin-coating and thermal curing possessed the necessary functions and were successfully used in studies of advanced biointerfaces. In particular, the bioactive species were successfully immobilized, while the adhesion of nonspecific protein, bacteria and cells was inhibited.

2.2.2.5 Plasma Polymerization and Chemical Vapor Deposition (CVD) Approaches

Plasma treatments (low-pressure glow discharge) represent versatile methods for the modification and functionalization of polymer surfaces [77]. In addition to a broad range of surface characteristics, which can be obtained by introducing different functional groups (e.g., by changing the type of gas, hydrophilic and hydrophobic surfaces can be achieved) [78], the deposition of polymers [79], the immobilization of (for example) surfactant molecules [80], and/or etching of the specimen surface [81] can each be utilized to tailor the surface properties by using this technique.

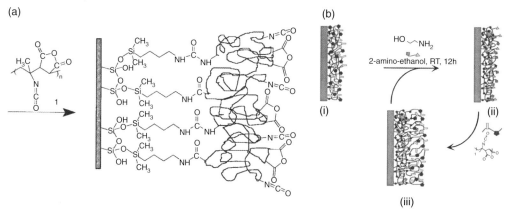

Figure 2.17 (a) Schematic representation of the assembly of the reactive polymer after reaction with the silanized substrate. Only the top layer of the functional groups is shown; (b) Schematic representation of the assembly of multilayers of IAP on silicon oxide surfaces: (i) reactive polymer layer assembled on a silanized surface; (ii) after reaction with 2-aminoethanol; (iii) after reaction with another layer of IAP. The pentagonal and rectangular shapes symbolize the reactive moieties of IAP; squares and triangular shapes symbolize the functional groups of 2-aminoethanol. Black circles denote amide, urea, and urethane functional groups formed by the reaction of 2-aminoethanol with IAP. Reproduced with permission from Ref. [74]; © 1996, American Chemical Society.

The formation of polymer films by plasma deposition is a very attractive approach to thin coatings prepared in a "clean" process. The precursor molecules are first introduced into a low-pressure glow discharge chamber by using a carrier gas, and are activated in the plasma (here, the term "activation" refers to dissociation and radical formation). Polymerization reactions initiated by these radicals can lead to a reaction with monomers and monomer fragments in the chamber, with such reactions taking place both in the plasma and on the substrate surface. If suitable monomers are chosen, then a more or less crosslinked film can be deposited. Plasma-polymerized films have been reported to show interesting characteristics, such as improved adhesion [82] or improved blood compatibility [83]; they have also been applied as coatings for DNA sensors [84].

In the particular case of polyallylamine layers obtained by pulsed plasma deposition, the plasma deposition parameters allow one to tune important properties such as swelling and loading. The adsorption of ssDNA onto pulsed plasma-polymerized allylamine films, as assessed by *in situ* SPR measurements, correlated with the chemical nature of the plasma films that can be tuned by varying the corresponding plasma on and off times (Figure 2.18).

The earlier-introduced electrostatic adsorption of PLL-g-PEG onto oxidic surfaces, suffers from the drawback that the deposited films are unstable for use in extreme pH values or in high ionic strength solutions. However, this limitation can be addressed by a combination of plasma modification and covalent attachment of the PLL-g-PEG copolymers. As reported by Blättler and coworkers, aldehyde plasma-modified substrates provide an excellent platform to couple, and thereby permanently immobilize, PLL-g-PEG [85]. Even after a 24 h exposure of

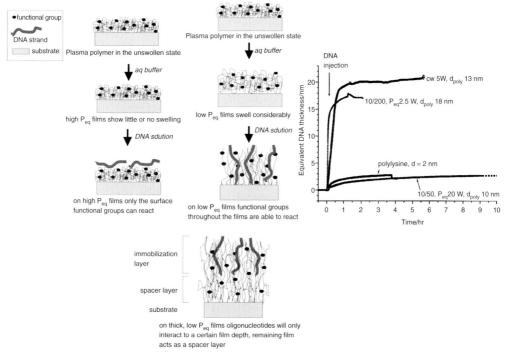

Figure 2.18 (a) Schematic representation of plasma polymer swelling and DNA binding to low and high duty cycle (DC) plasma polymer films (DC is defined via the plasma on and off times ($t_{on}/t_{on} + t_{off}$); (b) Surface plasmon resonance (SPR) kinetic measurements of DNA binding on a low DC (10/200), a 5 W continuous wave and a high DC (10/50) plasma polymerized allylamine compared to that on a polylysine film ($d \sim 2$ nm): c_{DNA} 100 nM, PBS buffer at pH 7.4. Reproduced with permission from Ref. [84]; © 2003, American Chemical Society.

covalently immobilized layers of PLL-g-PEG to high ionic strength buffer (2400 mM NaCl), no significant change in the protein resistance was observed (Figure 2.19). By contrast, simply electrostatically adsorbed PLL-g-PEG coatings would have lost their protein resistance under the same conditions. This combined approach also broadens the scope of PLL-g-PEG, since the requirement for charged surfaces is lifted.

For the surface modification of microfluidics devices made from poly(dimethyl siloxane) (PDMS), the gas-phase processes possess a variety of advantages. One very successful route which was introduced by Langer and colleagues [86, 87] is based on a CVD polymerization using poly-[para-xylylene carboxylic acid pentafluorophenolester-co-para-xylylene] (PPX-PPF) (Figure 2.20). In this case, the deposited reactive coating exposes reactive pentafluorophenol ester groups at the surface, and these can be utilized to immobilize amino end-functionalized biomolecules.

The deposition, which can be tuned such that no byproducts are formed [88], is carried out at room temperature and does not require any catalyst, solvent, or

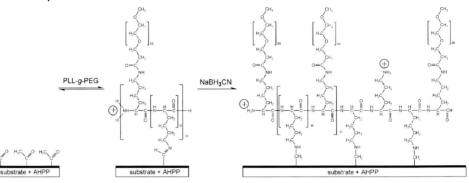

Figure 2.19 Schematic diagram of the covalent immobilization of PLL-g-PEG onto propanal plasma-modified substrates via reductive amination. Reproduced with permission from Ref. [85]; © 2006, American Chemical Society.

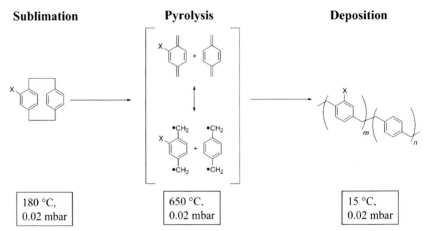

Figure 2.20 Chemical vapor deposition (CVD) polymerization of a reactive coating. Reproduced with permission from Ref. [86]; © 2001, American Chemical Society.

initiator. In addition to PPX-PPF, the use of poly-(p-xylylene-2,3-dicarboxylic acid anhydride) has also been reported [89]. For these coatings the functional groups are retained and the subsequent reaction with biological ligands or proteins does not require any further activation. The extension of this approach to different pending reactive groups allows the selective functionalization of confined microgeometries (Figure 2.21a) [90]. Moreover, the process has also been extended to orthogonal chemical functionalities which render surfaces reactive towards different ligands in a selective manner [91]. In this way, defined and stable surface properties are ensured, along with the capability to immobilize active biomolecules onto a surfaces of miniaturized biodevices.

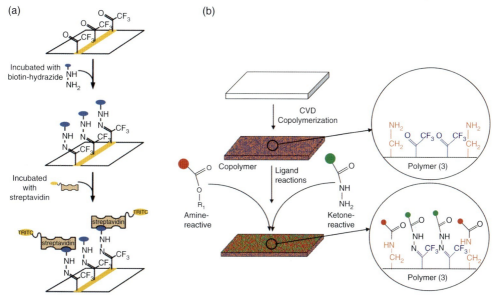

Figure 2.21 (a) Immobilization reactions used for the chemical immobilization of biotin and subsequent binding of streptavidin onto a reactive coating, which exposes a trifluoromethyl-substituted ketone. Reproduced with permission from Ref. [90]; © 2006, American Chemical Society; (b) Schematic outlining the selective reactivity of the multivalent surface. The activated ester only reacts with the aminomethyl group, while the hydrazide group shows selective reactivity toward ketones. Reproduced with permission from Ref. [91]; © 2006, Wiley-VCH.

2.2.3
Surface Modification of Polymer Surfaces, and Selected Examples

The modification of polymers by postsynthesis modification, and the concomitant formation of functional polymers, has been recently reviewed by Gauthier, Gibson and Klok [92]. In principle, all approaches developed as "polymer-analogous reactions" can also be considered at this point, and provide a "toolbox" for the modification of polymer surfaces and thin films [93].

2.2.3.1 Coupling and Bioconjugation Strategies

In general, surface reactions differ in several aspects from reactions carried out in solution. These differences are most pronounced in ordered environments, such as those encountered in monolayers, although they may also affect reactions at the surfaces of polymers. These effects can be categorized as "solvent" (e.g., altered surface $pK_{1/2}$ values), "steric," and "electronic/anchimeric" [94], among which the solvent and anchimeric effects do not necessarily possess analogues in solution. Further surface-specific peculiarities include the susceptibility to contamination, a limited mobility of surface-immobilized functional groups and spacers, a hindered access of the reactants to reactive sites in organized assemblies, crowded

2 Biofunctionalization of Polymeric Thin Films and Surfaces

transition states, the presence of surface forces, and problems related to reaction byproducts.

2.2.3.1.1 Covalent Coupling and Bioconjugation Strategies Covalent coupling reactions can be grouped into substitution and addition reactions, the difference between which relates to the presence of a leaving group in the former case. While a good leaving group may increase the reactivity – as in the case of reactive (active) esters versus conventional esters – the transition state may be more complex as the bulky leaving group contributes to the steric hindrance. A recent review by Jonkheijm et al. has summarized the most important classes of covalent attachment chemistries, as well as the different formats and approaches (Table 2.1) [95].

Table 2.1 Methods used for covalent protein immobilization. Reproduced with permission from Ref. [95]; © 2008, Wiley-VCH.

Surface functional group		Protein functional group		Product
NHS ester	[structure: surface-C(O)-O-N-succinimide]	H_2NR	[structure: surface-C(O)-N(H)-R]	amide
Aldehyde	[structure: surface-C(O)H]	H_2NR	[structure: surface-CH=N-R]	imine
Isothiocyanate	[structure: surface-N=C=S]	H_2NR	[structure: surface-NH-C(S)-NH-R]	thiourea
Epoxide	[structure: surface-epoxide]	H_2NR	[structure: surface-CH(OH)-CH2-NH-R]	aminoalcohol
Amine[a]	[structure: surface-NH$_2$]	$HO(O)CCH_2R$	[structure: surface-NH-C(O)-R]	amide

a With coupling reagent [e.g., carbonyldiimidazole (CDI)].

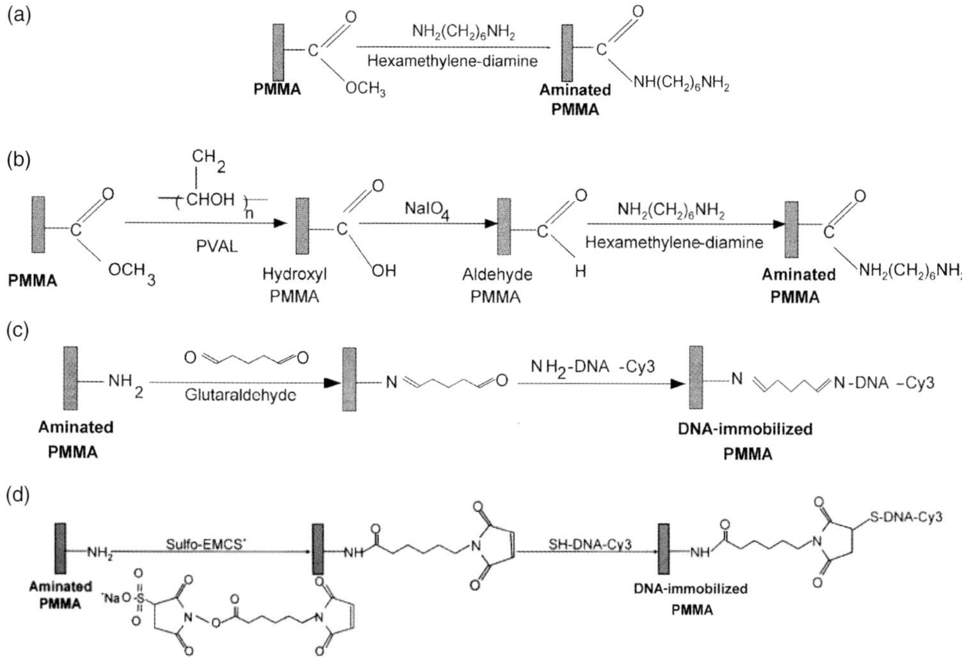

Figure 2.22 Chemistries used to modify the PMMA substrates and to covalently immobilize 5′-end-modified DNA oligonucleotides to amine-terminated surfaces. (a) The available methyl esters of the PMMA, under basic pH conditions, are reacted with an electron donor (N) present on the hexamethylenediamine, yielding primary amines on the surface; (b) Protocol described by Bulmus et al. [115]; (c) Attaching the aminated DNA to aminated PMMA surfaces. The aminated PMMA was treated with the homobifunctional crosslinker glutardialdehyde, having a terminal aldehyde group reacting to the primary amino groups of the PMMA, through an imine bond. The NH_2-DNA probes are subsequently reacted with the other aldehyde terminal of the crosslinker, establishing a covalent bond between the surface and the oligonucleotide probe; (d) Attaching thiolated DNA to aminated PMMA surfaces. The aminated PMMA is reacted with the NHS ester group of the heterobifunctional crosslinker sulfo-EMCS, and subsequently the maleimide portion of the sulfo-EMCS is reacted with a 5′-end-thiolated DNA probe, to achieve a covalent immobilized DNA. Reproduced with permission from Ref. [24]; © 2004, Oxford University Press.

The basis for the surface attachment of, for example DNA, can be the design of polymers or conventional polymers that are activated. Examples comprise typical photoresist materials, such as SU-8 [113], elastomeric materials that are used in soft lithography (including PDMS) [114], or typical mass polymers, such as PMMA [24]. PMMA can be functionalized as shown in Figure 2.22.

A well very established and widespread substitution reaction in aqueous media is the already many-fold mentioned substitution employing active esters, such as NHS esters (see Table 2.1) [109, 116, 117]. Under mild conditions, (bio)molecules with primary amino groups are efficiently coupled. For each type of protein the relevant reaction parameters, including pH, concentration, ionic strength and reaction time, must be optimized in order to achieve high coverages. This reaction is of particular value as many proteins contain a number of lysine residues that often are located on the outer periphery of the protein. However, it has also been shown that an abundance of lysines may often result in multipoint attachment. In this case, the resultant layers and biointerfaces comprise unwanted heterogeneity, while the immobilized protein may exhibit a restricted conformational flexibility.

In reactions on model SAMs, the yield of the activation step has been reported to be limited to 50% also in the case of sulfonated NHS esters [118]. This limit has been attributed to the steric packing of the NHS ester intermediate. By repeated carboxylic acid activation and reaction with NH_3, up to 80% all of the carboxylic acid-terminated molecules were converted to amides. In the case of polymers, the transition state would be expected to be less crowded [119], and hence higher yields should be observed. Examples with polymers include the NHS ester-based coupling reactions of, for example, DNA, proteins, and PEG [120]. If the pH of the buffer is well controlled, then the hydrolysis of the NHS ester is only a negligible side reaction (compare Figure 2.8). However, this side reaction will limit the shelf life of commercially available DNA array support slides that employ NHS chemistry [40]. In such films, the inefficient surface activation reaction is circumvented by using NHS ester-equipped polymers.

The Schiff base (imin) formation between a surface-immobilized aldehyde and a primary amine on the biomolecule may serve as an illustrative example of an addition reaction, an example being the immobilization of DNA [121]. The imin reaction product is formed either as the *cis* or *trans* isomer, and is in dynamic equilibrium with the free reactants; this equilibrium can be exploited for reversible and reusable biochips. However, if a robust attachment is sought, then a reduction of the imine with borane chemistry would typically be carried out to yield a more stable secondary amine.

One class of reaction, with a unique applicability in the context of biointerfaces and biofunctionalization of polymers, has been developed during the past decade. This so-called archetypal "click" reaction is based on the long-known 1,3-dipolar cycloaddition (Huisgen reaction), where the reaction between an azide and a terminal acetylene proceeds in the presence of a copper catalyst, with essentially quantitative yield under the mildest conditions, to yield a stable 1,2,3-triazole. The tolerance of this reaction towards other functional groups, coupled with the mild conditions, render this reaction highly valuable within the context of this chapter. The 1,3-dipolar cycloaddition, which was pioneered by Sharpless and others, has since found ample application in the field of bioconjugation and surface modification [28, 122, 123].

Here, we will briefly discuss a recent example that combines the above-introduced gas-phase CVD polymerization with click chemistry [28]. The CVD polymerization of a suitable monomer affords the *para*-phenylene-based polymer with pending acetylene groups. The reaction proceeds under mild conditions, with an azide-functionalized biotin derivative (Figure 2.23, left). The efficient capture

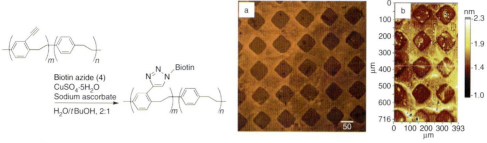

Figure 2.23 Left: Huisgen 1,3-dipolar cycloaddition between a biotin-based azide-ligand and acetylene-polymer. Right: (a) Fluorescence micrograph showing the binding of tetramethylrhodaminisothiocyanate-streptavidin to patterns of biotin azide; (b) Corresponding ellipsometric image for determining the layer thickness. Reproduced with permission from Ref. [28]; © 2006, Wiley-VCH.

of fluorescently labeled streptavidin can be monitored using fluorescence microscopy and imaging ellipsometry (Figure 2.23, right).

2.2.3.1.2 Noncovalent Coupling and Bioconjugation Strategies

The covalent coupling reactions discussed above are complemented by a series of noncovalent bioconjugation reactions that are frequently used as alternatives if an enhanced selectivity of the surface reactions is targeted. Two of the most prominent examples are briefly discussed here, based on their relevance and widespread use (see also the example *vide supra*; Figure 2.23).

The interaction of the proteins streptavidin or avidin on the one hand, and of biotin (vitamin H) on the other hand, is characterized by the largest free energies of association yet observed for the noncovalent binding of a protein and a small ligand in an aqueous medium. The strength of this interaction is characterized by an extraordinarily high binding constant of $\sim 10^{15} M^{-1}$ for the interaction between biotin and streptavidin (for avidin and biotin, $\sim 10^{13} M^{-1}$) [124]. In practice, such a bond can be regarded as permanent, even over a wide range of temperature and pH. Compared to chemical reactions that may require particular conditions (cf. NHS ester chemistry), this complexation is very specific. Furthermore, the positioning of two pairs of biotin-binding sites on opposite faces of the protein allow this system to be exploited as a molecular platform for the construction of designed interfacial architectures.

Many of the covalent reactions summarized in Table 2.1 also require biomolecules with defined pending functional groups. Other than in custom-synthesized DNA, where the chain end function is determined by design, the lysine and cysteine residues of naturally occurring proteins are rarely unique. Hence, the covalent attachment of proteins via their lysine residues does not provide any control over the correct orientation of the immobilized protein. This lack of regiospecificity may ultimately lead to the undesired blocking of a reactive site of a protein as a result of the immobilization reaction.

These shortcomings can be overcome by using the biotin–streptavidin system described above, although metal ion chelator systems have also been further developed and refined in this context. One well-established approach utilizes the N-nitrilotriacetic acid (NTA)/His$_6$-tag chelator system, which is also well known in the isolation and purification of gene products as a single step [125, 126]. The NTA is typically immobilized on a surface and, in the presence of nickel ions, will capture histidine-tagged biomolecules by exploiting the high affinity of histidine for Ni^{2+} ions. In this way, the His-tagged protein is bound to the surface [127]. As the His$_6$-tag is small and flexible, the function of the protein is preserved. In order to generate reusable biochip surfaces, the complexation can be reversed by adding imidazole or ethylene diamine tetraacetic acid (EDTA). More recently, protein binding by multivalent chelator surfaces was shown to be a strategy by which to overcome the known problems with the low affinity and rapid dissociation of the mono-NTA–His$_6$-tag interaction. As demonstrated by Tampe and coworkers, immobilized histidine-tagged proteins were uniformly oriented and retained their function [128]. These approaches should also be feasible for polymer-based platforms.

As noted above, the attachment chemistries applicable to protein immobilization do not differ fundamentally, if one compares the monolayer- and polymer-based approaches. However, some striking differences have been noted with, in many cases, the polymer-based approaches comparing in many ways very favorably with SAMs [129]. In addition to the inherent advantage of a higher loading per unit surface area (due to the increase in dimensionality from 2-D to quasi-3-D; this was alluded to in Section 2.2.2.2), it is mainly the improved robustness and stability, as well as the compatibility with micropatterning and nanopatterning methodologies, that can be proposed, among other factors.

In order to increase the molecular loading—that is, the number of molecules immobilized per unit area—hydrogels [130, 131], dendrimers [132, 133], hyperbranched polymers [134], CVD approaches [89], self-assembled polyelectrolyte multilayers [135], self-assembled polymer layers [136], plasma polymers [84, 112], and comb copolymers [137] have been investigated, among others [138].

The aspect of a high loading per unit surface area is illustrated in Figure 2.24 for a dendron-modified cellulose [139]. Dendrons, which can be considered as the building blocks for highly branched, well-defined dendrimer macromolecules, provide within their coupling layers many attachment points. Biofunctionalized surfaces based on dendronized cellulose were prepared, for example, by the heterogeneous functionalization of a deoxy-azido cellulose film, with the polyamidoamine (PAMAM) dendron exploiting the above-mentioned "click" chemistry. The amino groups provided by the solid supports were exploited to covalently couple the enzyme glucose oxidase after activation with glutardialdehyde. These biofunctionalized surfaces, when based on dendronized cellulose, were reported to provide excellent reproducibility and a good storage stability.

The same theme has been exploited commercially in hydrogel sensor chips. Most notably, crosslinked carboxymethyl (CM) dextran-based platforms have found widespread application since, as hydrophilic highly swollen hydrogels, they

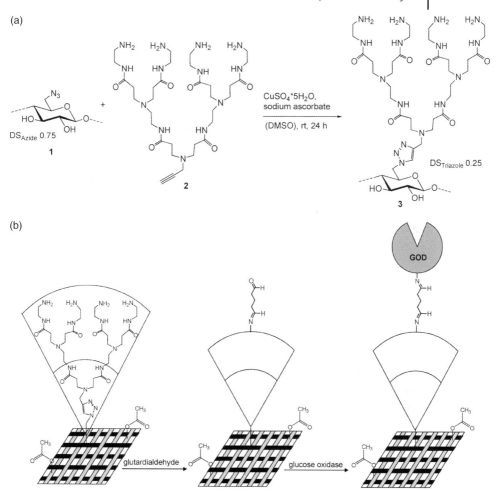

Figure 2.24 (a) Syntheses of 6-deoxy-6-(1,2,3-triazolo)-4-polyamidoamine (PAMAM) cellulose of 2.5th generation obtained by copper-catalyzed Huisgen reaction (idealized structure); (b) Schematic preparation of a blend of 6-deoxy-6-(1,2,3-triazolo)-4-polyamidoamine cellulose (degree of substitution (DS) = 0.25) and cellulose acetate (DS = 2.50) and subsequent surface activation with glutardialdehyde for covalent immobilization of glucose oxidase (GOD; light gray bars), cellulose acetate (black bars), and 6-deoxy-6-(1,2,3-triazolo)-4-PAMAM cellulose. Reproduced with permission from Ref. [139]; © 2009, American Chemical Society.

provide a large surface area which enables high-loading biomolecule immobilization [140]. Such polymeric hydrogels afford up to 100-fold higher capacity of immobilization than is found for planar surfaces. In order to realize stable microarray chips, antibodies or other recognition units can be coupled covalently to the pending carboxyl groups in the dextran matrix. The hydrogel nature of the matrix also afford low levels of nonspecific protein adsorption.

As an alternative approach, natural, synthetic and biohybrid hydrophilic polymers have each been investigated for the attachment or entrapment of biomolecules [96]. Polymer networks are readily accessible via photochemically or thermally initiated crosslinking and, due to the wide variety of possible chemical structures of the monomers used, and the precise control of the molecular architecture and morphology, a variety of polymer supports for protein immobilization have been developed. In addition to hydrophilic polymers such as PEG and poly(vinyl alcohol) (PVA), hydrophobic polymers – including nitrocellulose or polyacrylamide or acrylate gels – have been widely used for the immobilization of biomolecules. Both, agarose and acrylamide were photopolymerized onto surfaces which had been functionalized with acryl groups; subsequently, amine groups were generated on the polymer surface by reaction with hydrazine or ethylenediamine. Other examples of biopolymer-derived gels applied to the immobilization of proteins include polysaccharides. One such example is *chitosan*, an amine-modified, natural, nontoxic polysaccharide that is pH-responsive and can (depending on the state of charge) adhere via electrostatic interaction with both glass surfaces and solution-borne proteins.

2.2.3.2 Interaction with Cells

Investigations into the interactions of cells with designed surfaces is one area where polymers have made a considerable impact. As noted by several groups, the microenvironment determines cell adhesion and other important processes [97]. Yet, in addition to the required biochemical patterns, polymers can easily be structured and tuned in terms of their mechanical properties.

Shi *et al.* reported recently on the development of materials with optimized biointerfaces for the modulation of spreading, proliferation, and differentiation of human mesenchymal stem cells on gelatin-immobilized poly(L-lactide-*co*-ε-caprolactone) (PLCL) substrates [98]. The fabrication of the substrates involves a classical radiation-induced modification of a PLCL film surface, followed by classic activation by ethyl-3-(3-dimethyl aminopropyl) carbodiimide hydrochloride/*N*-hydroxy succinimide (EDC/NHS) chemistry. Finally, gelatin was immobilized covalently (Figure 2.25).

Tailored surfaces on the basis of designed polymer surfaces also the interaction of cells and selected proteins to be interrogated, with unprecedented resolution. In the atomic force microscopy (AFM) images shown in Figure 2.26a, a *PaTu8988T* cancer cell interacted with fibronectin that was covalently immobilized on the film. (Note: these AFM experiments were carried out with fixed cells, although live cell imaging is also possible.) The length scale of the filopodia and of the microphase separation of the block copolymer was comparable; thus, nanometer-scale control may ultimately be achieved if the block copolymer domains can be selectively functionalized with ligands. These fundamental interaction studies can be expanded to a microarray format, as shown in Figure 2.26b. Here, Hook *et al.* demonstrated the suitability of the robotic approach for chip-based functional genomics and high-density cell assays that may become relevant within the context of toxicology and drug screening [99].

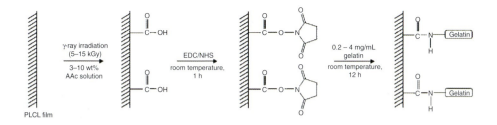

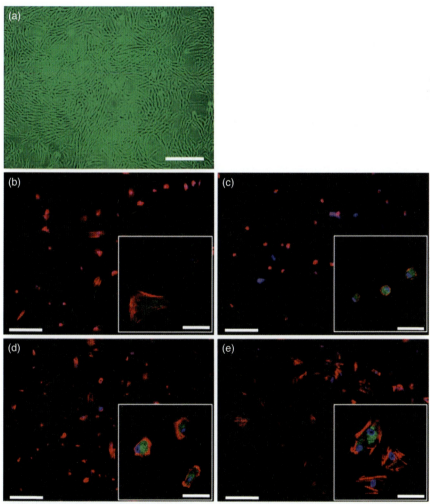

Figure 2.25 Upper panel: Schematic diagram of the biofuctionalization of PLCL films using γ-ray irradiation, activation of the carboxylic acid groups formed during the irradiation, and the subsequent grafting of gelatin. Lower panel: Morphologies of human mesenchymal stem cells (hMSC) cultured on various substrates. (a) Phase-contrast light microscopy image of the adherent hMSC on the tissue cell culture plate (scale bar = 300 μm) and immunofluorescent staining of the adherent cells on the films; (b) Fibronectin (FN)-coated glass; (c) PLCL; (d) AAc-PLCL; (e) Gelatin-AAc-PLCL. Larger images indicate the population of the cells on each film, obtained using fluorescence microscopy at ×100 magnification (scale bar = 200 μm). The inserted square images show the detailed morphology of the adherent cells captured using laser scanning confocal microscopy (×1000 magnification; scale bar = 50 μm). Reproduced with permission from Ref. [98]; © 2008, American Chemical Society.

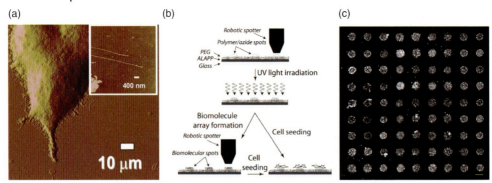

Figure 2.26 (a) Tapping mode AFM amplitude image of *PaTu8988T* cell on fibronectin-functionalized polystyrene-block-poly(*tert*-butyl acrylate) (PS_{690}-*b*-$PtBA_{1210}$). Inset: tapping mode AFM height image. Reproduced with permission from Ref. [100]; © 2009, Wiley-VCH; (b) Schematic of the formation of a chemically patterned surface for cell microarray applications. A polymer functionalized with a photoreactive azide group is arrayed onto a PEG surface using a robotic spotter. Subsequent irradiation with UV light covalently crosslinks the polymer to the PEG surface, resulting in the formation of a patterned surface, which can subsequently be used as a base substrate for the additional formation of a DNA, protein, or small molecule array formed by the same robotic spotter that was utilized for the polymer array formation. When cells are seeded to these patterned surfaces, cell attachment follows the crosslinked polymer pattern, while attachment on the PEG surface between the printed spots is prevented. (Note: the schematic is not drawn to scale.); (c) Fluorescence microscopy image of Hoechst 33342-stained SK-N-SH neuroblastoma cell microarray (×4 magnification; scale bar = 500 μm). Reproduced with permission from Ref. [101]; © 2009, American Chemical Society.

Simultaneous biochemical and topographical patterning has been reported (for example, by the Textor group) in the investigation of isolated cells and their interaction with the microenvironment. The first successful experiments on single cells that were enclosed individually into well-like structures demonstrated the feasibility of investigating the relationship between the shape and function of single cells (or clusters of cells) in a 3-D microenvironment (Figure 2.27) [102].

These structures were fabricated using a two-step replication process. First, the polymer (polystyrene) was hot-embossed to provide the topographic profile. The plateau surface between the microwells was then decorated selectively with a passivating PLL-g-PEG layer by inverted microcontact printing. Finally, the microwells were functionalized by their adsorbing proteins or functionalized PLL-g-PEG. Whilst this outstanding example demonstrated the feasibility of fabricating well-defined microwell structures, the process is beyond the scope of currently available surface and interface engineering techniques to implement patterning and structures on all relevant length scales. In this case, one length scale is set by the size of the cells themselves, while the other length scale is controlled by the clustering processes of receptors in the cell membrane. As shown by Spatz and coworkers, the distances between individual RGD-(Arginine-Glycine-Aspartate)-peptides [103]

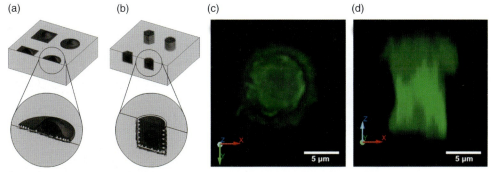

Figure 2.27 Scheme of the concept of (a) conventional 2-D patterning of cells and (b) micro-3-D culturing of single cells. The surface of the microwells exhibits cell-binding properties, while the plateau surface inhibits adsorption of proteins and attachment of cells. (c,d) 3-D reconstruction of a confocal laser scanning microscopy (CLSM z-stack of a single epithelial cell (MDCK II) attached inside a microwell (10 μm diameter, 11 μm depth); 3 h culture time in fetal calf serum (FCS) containing medium; fixed in PFA and filamentous actin stained with FITC-phalloidin. (a) View from below; (b) View from the side. Reprinted with permission from Ref. [102]; © 2005, Elsevier.

of ≤58 nm and ≥73 nm caused the presence or absence, respectively, of focal adhesions. The corresponding length scale was attributed to the characteristic length scale for integrin-receptor clustering [104–106]. Hence, there is a need to control surface chemical functionality and structures on more than one length scale, starting from the scale of cells down to the molecular level [107].

2.2.3.3 Patterned Polymeric Thin Films in Biosensor Applications

In addition to cell–surface interaction studies, patterned polymer-based biointerfaces are important in applications related to biochips. Typical commercial biochip-based sensors possess spot sizes with dimensions on the order of several micrometers to 100 μm [108]. A novel approach to prepare a stable, crosslinked, bioreactive polymer film is shown in Figure 2.28 [110] where, by exploiting star polymers, glass substrates have been modified as a support material for oligonucleotide microarrays.

A decrease in spot dimensions to the nanometer scale may offer several added advantages. As reported by Colpo and coworkers [111], the immunoreaction efficiency of uniformly functionalized polyacrylic acid (PAA) surfaces and chemically nanopatterned PAA surfaces differ significantly for antigen–antibody interactions. By using a combination of colloidal lithography and plasma-enhanced chemical vapor deposition (PE-CVD) (Figure 2.29), Colpo and colleagues were able to fabricate nanopatterned surfaces that exposed COOH-functionalized nanoareas in a passivating poly(ethylene oxide) (PEO)-like matrix. Moreover, the immunoreaction efficiency was found to be enhanced on the nanopatterned film in the absence of protein A (pA) as the orienting protein.

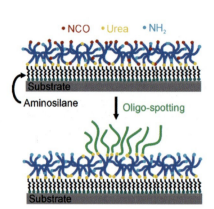
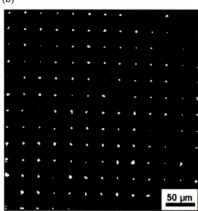

Figure 2.28 (a) Scheme of the immobilization strategy of amino-terminal oligonucleotides (green). Freshly prepared star PEG layers contain isocyanate groups (red) that covalently bind the printed or spotted oligonucleotides. After hydrolysis and crosslinking, the reactive isocyanate groups have been removed, and the nonfunctionalized parts of the star PEG coating are nonadhesive; (b) Fluorescence microscopy image of oligonucleotide stamped onto a star PEG layer prepared from a solution in tetrahydrofuran (THF) (17.7 nm thickness) by microcontact printing. Reproduced with permission from Ref. [110]; © 2005, American Chemical Society.

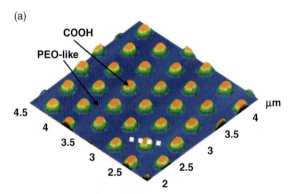
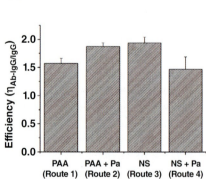

Figure 2.29 (a) AFM image of the nanopatterned surface; (b) Efficiency for the immunoreaction IgG/Ab-IgG.
NS = nanostructure. Routes 1 and 2 refer to unstructured PAA surfaces. In the first and third routes, human IgG was immobilized by direct adsorption onto, respectively, homogeneously PAA-functionalized and nanopatterned PAA/PEO surfaces. In the second and fourth routes, the human IgG immobilization onto, respectively, homogeneously COOH-functionalized and nanopatterned COOH/PEO surfaces, was protein A-assisted. Reproduced with permission from Ref. [111]; © 2008, American Chemical Society.

2.3
Summary and Future Perspectives

In this chapter, the fabrication of polymer-based biofunctional surfaces and biointerfaces was treated from the point of view of *chemistry* and *surface engineering*. Established and new methods for altering the surface chemistry to ensure efficient and controllable bioconjugation, and to implement the important function of suppressed nonspecific adsorption of proteins, were introduced. Emphasis was placed on polymeric materials as substrates or molecules for modification because of the central role of macromolecules in this context. Such a role stems from the unique combination of factors and determinants which render polymers ideal in many respects. In polymers, the structuring, control of mechanical properties, and control of chemical composition are paired with low cost and optimized processing.

The topic was developed based on a discussion of the general requirements for biofunctionalized surfaces and the polymer-specific features of surface modification. Starting from the chemical surface modification of polymer surfaces and polymeric thin films for the fabrication of advanced biointerfaces, the focus was shifted toward cell–surface interactions studies and routes that are compatible with patterning approaches.

The surface properties of medical devices, sensor platforms and other biointerfaces can be tuned, as outlined in this chapter, over a broad range. The powerful arsenal of methods already available to researchers is constantly complemented by new approaches; hence, the level of control of both chemical composition and structure at these biointerfaces is improved continually and significantly. The relevance of structured and biofunctionalized polymer architectures is not limited to the application areas of biosensing and controlled cell–surface interactions. Indeed, with more recent developments in the suppression of any unwanted nonspecific adsorption of proteins and efficient bioconjugation/surface decoration, numerous new areas for potential applications and uses have materialized. It is however, also important to point out that the suppression of unwanted nonspecific adsorption of biomolecules is both material- and biosystem-dependent. Thus, it is questionable as to whether one single solution exists that would satisfy the requirements of all applications. This lack of a generic solution must be acknowledged, notably in relation to microbial adhesion and the bacterial colonization of surfaces.

Within the specific context of biointerfaces, the developments outlined in this chapter promise to provide new incentives across the traditional boundaries of the disciplines involved. Tailored microenvironments, as demonstrated by the groups of Textor and others, already allow the behavior of single cells to be interrogated. In view of the challenges encountered in toxicology screening and drug testing, cell microarray technology might also be expected to play an important role in the future. Of equal importance are the microenvironments and niches for stem cell differentiation. The biofunctionalization and structuring approaches developed so far offer control in terms of dimensionality and sophistication. By following the pathways via designed/engineered surfaces and structures, based on the approaches detailed in this chapter, routes to systematic biomedical studies have been opened. It is expected that research groups in both biology and biomedicine in general will expand the pioneering studies referred to here to include a wide range of different cell types, as well as more complex microenvironments. Likewise, the creation of low-cost, high-efficiency biosensors – an example being personalized point-of-care applications – remains a clear target for future activities.

References

1 Kasemo, B. (2002) Biological surface science. *Surface Science*, **500**, 656–77.
2 Kumar, C.S.S.R. (ed.) (2006) *Tissue, Cell and Organ Engineering*, Wiley-VCH Verlag GmbH.
3 Schena, M. (2003) *Microarray Analysis*, Wiley-Liss, John Wiley & Sons, Inc., NJ.
4 Shankaran, D.R. and Miura, N. (2007) Trends in interfacial design for surface plasmon resonance based immunoassays. *Journal of Physics D: Applied Physics*, **40**, 7187–200 and references therein.
5 (a) Mitsiadis, T.A., Barrandon, O., Rochat, A., Barrandon, Y. and De Bari, C. (2007) Stem cell niches in mammals. *Experimental Cell Research*, **313**, 3377–85.
(b) Jones, D.L. and Wagers, A.J. (2008) No place like home: anatomy and function of the stem cell niche. *Nature Reviews. Molecular Cell Biology*, **9**, 11–21.
(c) Moore, K.A. and Lemischka, I.R. (2006) Stem cells and their niches. *Science*, **311**, 1880–5.
6 Spatz, J.P. (2004) Cell-nanostructure interactions, in *Nanobiotechnology* (eds C.M. Niemeyer and C.A. Mirkin), Wiley-VCH Verlag GmbH.
7 Chen, C.S., Jiang, X. and Whitesides, G.M. (2005) Microengineering the environment of mammalian cells in culture. *MRS Bulletin*, **30**, 194–201.
8 Castner, D.G. and Ratner, B.D. (2002) Biomedical surface science: foundations to frontiers. *Surface Science*, **500**, 28–60.
9 Habibovic, P., Yuan, H., van der Valk, C.M., Meijer, G., van Blitterswijk, C.A. and de Groot, K. (2005) 3D microenvironment as essential element for osteoinduction by biomaterials. *Biomaterials*, **26**, 3565–75.
10 Murata, H., Chang, B.-J., Prucker, O., Dahm, M. and Rühe, J. (2004) Polymeric coatings for biomedical devices. *Surface Science*, **570**, 111–18.
11 Morra, M. (2000) On the molecular basis of fouling resistance. *Journal of Biomaterials Science, Polymer Edition*, **11**, 547–69.
12 (a) Bergstrom, K., Holmberg, K., Safranj, A., Hoffman, A.S., Edgell, M.J., Kozlowski, A., Hovanes, B.A. and Harris, J.M. (1992) Reduction of fibrinogen adsorption on PEG-coated polystyrene surfaces. *Journal of Biomedical Materials Research*, **26**, 779–90.
(b) Harris, M.J. (1992) *Poly(Ethylene Glycol) Chemistry Biotechnical and Biomedical Applications*, Plenum Press, New York.
(c) Prime, K.L. and Whitesides, G.M. (1993) Adsorption of proteins onto surfaces containing end-attached oligo(ethylene oxide) – a model system using self-assembled monolayers. *Journal of the American Chemical Society*, **115**, 10714–21.
(d) Kingshott, P. and Griesser, H.J. (1999) Surfaces that resist bioadhesion. *Current Opinion in Solid State and Materials Science*, **4**, 403–12.
(e) Kingshott, P., McArthur, S., Thissen, H., Castner, D.G. and Griesser, H.J. (2002) Ultrasensitive probing of the protein resistance of PEG surfaces by secondary ion mass spectrometry. *Biomaterials*, **23**, 4775–85.

(f) Ostuni, E., Chapman, R.G., Holmlin, R.E., Takayama, S. and Whitesides, G.M. (2001) A survey of structure-property relationships of surfaces that resist the adsorption of protein. *Langmuir*, **17**, 5605–20.

(g) Glasmastar, K., Larsson, C., Hook, F. and Kasemo, B.J. (2002) Protein adsorption on supported phospholipid bilayers. *Colloid and Interface Science*, **246**, 40–7.

13 Kumar, A. and Whitesides, G.M. (1993) Features of gold having micrometer to centimeter dimensions can be formed through a combination of stamping with an elastomeric stamp and an alkanethiol "ink" followed by chemical etching. *Applied Physics Letters*, **63**, 2002–4.

14 (a) Bernard, A., Renault, J.P., Michel, B., Bosshard, H.R. and Delamarche, E. (2000) Microcontact printing of proteins. *Advanced Materials*, **12**, 1067–70.

(b) Falconnet, D., Koenig, A., Assi, T. and Textor, M. (2004) A combined photolithographic and molecular-assembly approach to produce functional micropatterns for applications in the biosciences. *Advanced Functional Materials*, **14**, 749–56.

15 Chen, C.S., Mrksich, M., Huang, S., Whitesides, G.M. and Ingber, D.E. (1997) Geometric control of cell life and death. *Science*, **276**, 1425–8.

16 Goddard, J.M. and Hotchkiss, J.H. (2007) Polymer surface modification for the attachment of bioactive compounds. *Progress in Polymer Science*, **32**, 698–725.

17 Britland, S., Morgan, H., Wojciak-Stothard, B., Riehle, M., Curtis, A. and Wilkinson, C. (1996) Synergistic and hierarchical adhesive and topographic guidance of BHK cells. *Experimental Cell Research*, **228**, 313–25.

18 (a) Ingber, D.E. (2003) Tensegrity I. Cell structure and hierarchical systems biology. *Journal of Cell Science*, **116**, 1157–73.

(b) Ingber, D.E. (2003) Tensegrity II. How structural networks influence cellular information processing networks. *Journal of Cell Science*, **116**, 1397–408.

19 (a) Ito, Y. (1999) Surface micropatterning to regulate cell functions. *Biomaterials*, **20**, 2333–42.

(b) Falconnet, D., Csucs, G., Grandin, H.M. and Textor, M. (2006) Surface engineering approaches to micropattern surfaces for cell-based assays. *Biomaterials*, **27**, 3044–63.

(c) Curtis, A. and Wilkinson, C. (1997) Topographical control of cells. *Biomaterials*, **18**, 1573–83.

(d) Pelham, R.J. and Wang, Y.-L. (1997) locomotion and focal adhesions are regulated by substrate flexibility. *Proceedings of the National Academy of Sciences of the United States of America*, **94**, 13661–5.

(e) Curtis, A.S., Casey, B., Gallagher, J.O., Pasqui, D., Wood, M.A. and Wilkinson, C.D. (2001) Substratum nanotopography and the adhesion of biological cells. Are symmetry or regularity of nanotopography important? *Biophysical Chemistry*, **94**, 275–83.

(f) Curtis, A.S., Gadegaard, N., Dalby, M.J., Riehle, M.O., Wilkinson, C.D. and Aitchison, G. (2004) Cells react to nanoscale order and symmetry in their surroundings. *IEEE Transactions on Nanobioscience*, **3**, 61–5.

(g) Dalby, M.J., Silvio, L.D., Harper, E.J. and Bonfield, W. (2002) In vitro adhesion and biocompatibility of osteoblast-like cells to poly(methylmethacrylate) and poly(ethylmethacrylate) bone cements. *Journal of Materials Science: Materials in Medicine*, **13**, 311–14.

20 Wong, J.Y., Leach, J.B. and Brown, X.Q. (2004) Balance of chemistry, topography, and mechanics at the cell-biomaterial interface: Issues and challenges for assessing the role of substrate mechanics on cell response. *Surface Science*, **570**, 119–33.

21 Förch, R., Schönherr, H. and Jenkins, A.T.A. (eds) (2009) *Surface Design: Applications in Bioscience and Nanotechnology*, Wiley-VCH Verlag GmbH.

22 (a) Lahann, J. (2006) Reactive polymer coatings for biomimetic surface engineering. *Chemical Engineering Communications*, **193**, 1457–68.

(b) Yoshida, M., Langer, R., Lendlein, A. and Lahann, J. (2006) From advanced biomedical coatings to multi-functionalized biomaterials. *Polymer Reviews*, **46**, 347–75.

23 (a) Bruening, M.L., Zhou, Y., Aguilar, G., Agee, R., Bergbreiter, D.E. and Crooks, R.M. (1997) Synthesis and characterization of surface-grafted, hyperbranched polymer films containing fluorescent, hydrophobic, ion-binding, biocompatible, and electroactive groups. *Langmuir*, **13**, 770–8.
(b) Chance, J.J. and Purdy, W.C. (1997) Fabrication of carboxylic acid-terminated thin films using poly(ethyleneimine) on a gold surface. *Langmuir*, **13**, 4487–9.

24 Fixe, F., Dufva, M., Telleman, P. and Chistensen, C.B.V. (2004) Functionalization of poly(methyl methacrylate) (PMMA) as a substrate for DNA microarrays. *Nucleic Acids Research*, **32**, e9.

25 Amirgoulova, E.V., Groll, J., Heyes, C.D., Ameringer, T., Rocker, C., Möller, M. and Nienhaus, G.U. (2004) Biofunctionalized polymer surfaces exhibiting minimal interaction towards immobilized proteins. *ChemPhysChem*, **5**, 552–5.

26 Wang, D.-A., Ji, J., Sun, Y.-H., Shen, J.-C., Feng, L.-X. and Elisseeff, J.H. (2002) In situ immobilization of proteins and RGD peptide on polyurethane surfaces via poly(ethylene oxide) coupling polymers for human endothelial cell growth. *Biomacromolecules*, **3**, 1286–95.

27 Jerome, C., Gabriel, S., Voccia, S., Detrembleur, C., Ignatova, M., Gouttebaron, R. and Jerome, R. (2003) Preparation of reactive surfaces by electrografting. *Chemical Communications*, 2500–1.

28 Nandivada, H., Chen, H.-Y., Bondarenko, L. and Lahann, J. (2006) Reactive polymer coatings that click. *Angewandte Chemie, International Edition*, **45**, 3360–3.

29 Cummins, D., Duxbury, C.J., Quaedflieg, P.J.L.M., Magusin, P.C.M.M., Koning, C.E. and Heise, A. (2009) Click chemistry as a means to functionalize macroporous PolyHIPE. *Soft Matter*, **5**, 804–11.

30 Dai, J., Bao, Z., Sun, L., Hong, S.U., Baker, G.L. and Bruening, M.L. (2006) High-capacity binding of proteins by derivatized poly(acrylic acid) brushes. *Langmuir*, **22**, 4274–81.

31 Ma, H., Li, D., Sheng, X., Zhao, B. and Chilkoti, A. (2006) Protein-resistant polymer coatings on silicon oxide by surface-initiated atom transfer radical polymerization. *Langmuir*, **22**, 3751–6.

32 Heyes, C.D., Groll, J., Möller, M. and Nienhaus, G.U. (2007) Synthesis, patterning and applications of star-shaped poly(ethylene glycol) biofunctionalized surfaces. *Molecular BioSystems*, **3**, 419–30.

33 (a) Keddie, J.L., Jones, R.A.L. and Cory, R.A. (1994) Size-dependent depression of the glass transition temperature in polymer films. *Europhysics Letters*, **27**, 59–64.
(b) Keddie, J.L., Jones, R.A.L. and Cory, R.A. (1994) Interface and surface effects on the glass-transition temperature in thin polymer films. *Faraday Discussions*, **98**, 219–30.
(c) Forrest, J.A. and Mattson, J. (2000) Reductions of the glass transition temperature in thin polymer films: probing the length scale of cooperative dynamics. *Physical Review E*, **61**, R53–6 and references cited therein.
(d) Prucker, O., Christian, S., Bock, H., Rühe, J., Frank, C.W. and Knoll, W. (1998) On the glass transition in ultrathin polymer films of different molecular architecture. *Macromolecular Chemistry and Physics*, **199**, 1435–44.

34 (a) Frank, B., Gast, A.P., Russell, T.P., Brown, H.R. and Hawker, C. (1996) Polymer mobility in thin films. *Macromolecules*, **29**, 6531–4.
(b) Zheng, X., Rafailovich, M.H., Sokolov, J., Strzhemechny, Y., Schwarz, S.A., Sauer, B.B. and Rubinstein, M. (1997) Long-range effects on polymer diffusion induced by a bounding interface. *Physical Review Letters*, **79**, 241–4.
(c) Zheng, X., Sauer, B.B., van Alsten, J.G., Schwarz, S.A., Rafailovich, M.H., Sokolov, J. and Rubinstein, M. (1995) Reptation dynamics of a polymer melt

near an attractive solid interface. *Physical Review Letters*, **74**, 407–10.
(d) Russell, T.P. and Kumar, S.K. (1997) The one that got away. *Nature*, **386**, 771–2.
(e) Lin, E.K., Kolb, R., Satija, S.K. and Wu, W.-L. (1999) Reduced polymer mobility near the polymer solid interface as measured by neutron reflectivity. *Macromolecules*, **32**, 3753–7.

35 (a) Tseng, K.C., Turro, N.J. and Durning, C.J. (2000) Tracer diffusion in thin polystyrene films. *Polymer*, **41**, 4751–5.
(b) Hall, D.B. and Torkelson, J.M. (1998) Small molecule probe diffusion in thin and ultrathin supported polymer films. *Macromolecules*, **31**, 8817–25.

36 Schena, M., Shalon, D., Davis, R.W. and Brown, P.O. (1995) Quantitative monitoring of gene expression patterns with complementary DNA microarray. *Science*, **270**, 467–70.

37 Fodor, S.P.A., Read, J.L., Pirrung, M.C., Stryer, L., Lu, A.T. and Solas, D. (1991) Light-directed, spatially addressable parallel chemical synthesis. *Science*, **251**, 767–73.

38 (a) Chee, M., Yang, R., Hubbell, E., Berno, A., Huang, X.C., Stern, D., Winkler, J., Lockhart, D.J., Morris, M.S. and Fodor, S.P.A. (1996) Accessing genetic information with high-density DNA arrays. *Science*, **274**, 610–14.
(b) Pirrung, M.C. (2002) How to make a DNA chip, *Angewandte Chemie, International Edition*, **41**, 1277–89.

39 He, M., Stoevesandt, O., Palmer, E.A., Khan, F., Ericsson, O. and Taussig, M.J. (2008) Printing protein arrays from DNA arrays. *Nature Methods*, **5**, 175–7.

40 (a) Gong, P. and Grainger, D.W. (2004) Comparison of DNA immobilization efficiency on new and regenerated commercial amine-reactive polymer microarray surfaces. *Surface Science*, **570**, 67–77.
(b) Gong, P., Harbers, G.M. and Grainger, D.W. (2006) Multi-technique comparison of immobilized and hybridized oligonucleotide surface density on commercial amine-reactive microarray slides. *Analytical Chemistry*, **78**, 2342–51.

41 Canales, R.D., Luo, Y.L., Willey, J.C. *et al.* (2006) Evaluation of DNA microarray results with quantitative gene expression platforms. *Nature Biotechnology*, **24**, 1115–22.

42 Pasche, S. (2004) Mechanisms of protein resistance of adsorbed PEG-graft copolymers. PhD thesis. ETH Zurich, Dissertation no. 15712.

43 Schlapak, R., Pammer, P., Armitage, D., Zhu, R., Hinterdorfer, P., Vaupel, M., Fruhwirth, T. and Howorka, S. (2006) Glass surfaces grafted with high-density poly(ethylene glycol) as substrates for DNA oligonucleotide microarrays. *Langmuir*, **22**, 277–85.

44 Currie, E.P.K., Norde, W. and Stuart, M.A.C. (2003) Tethered polymer chains: surface chemistry and their impact on colloidal and surface properties. *Advances in Colloid and Interface Science*, **100**, 205–65.

45 Feng, C.L., Zhang, Z., Förch, R., Knoll, W., Vancso, G.J. and Schönherr, H. (2005) Reactive thin polymer films as platforms for the immobilization of biomolecules. *Biomacromolecules*, **6**, 3243–51.

46 Prucker, O., Naumann, C.A., Rühe, J., Knoll, W. and Frank, C.W. (1999) Photochemical attachment of polymer films to solid surfaces via monolayers of benzophenone derivatives. *Journal of the American Chemical Society*, **121**, 8766–70.

47 Deible, C.R., Petrosko, P., Johnson, P.C., Beckman, E.J., Russell, A.J. and Wagner, W.R. (1998) Molecular barriers to biomaterial thrombosis by modification of surface proteins with polyethylene glycol. *Biomaterials*, **19**, 1885–93.

48 Advincula, R.C., Brittain, W.J., Caster, K.C. and Rühe, J. (eds) (2005) *Polymer Brushes*, Wiley-VCH Verlag GmbH.

49 Whitmore, M.D. and Noolandi, J. (1990) Theory of adsorbed block copolymers. *Macromolecules*, **23**, 3321–39.

50 Konradi, R. and Rühe, J. (2005) Binding of oppositely charged surfactants to poly(methacrylic acid) brushes. *Macromolecules*, **38**, 6140–51.

51 Prucker, O. and Rühe, J. (1998) Polymer layers through self-assembled

monolayers of initiators. *Langmuir*, **14**, 6893–8.
52 Edmondson, S., Osborne, V.L. and Huck, W.T.S. (2004) Polymer brushes via surface-initiated polymerizations. *Chemical Society Reviews*, **33**, 14–22.
53 (a) Senaratne, W., Andruzzi, L. and Ober, C.K. (2005) Self-assembled monolayers and polymer brushes in biotechnology: current applications and future perspectives. *Biomacromolecules*, **6**, 2427–48.
(b) Zhao, B. and Brittain, W.J. (2000) Polymer brushes: surface-immobilized macromolecules. *Progress in Polymer Science*, **25**, 677–710.
54 (a) Jaworek, T., Neher, D., Wegner, G., Wieringa, R.H. and Schouten, A.J. (1998) Electromechanical properties of an ultrathin layer of directionally aligned helical polypeptides. *Science*, **279**, 57–60.
(b) Jordan, R., Ulman, A., Rafailovich, M.H. and Sokolov, J. (1999) Surface-initiated anionic polymerization of styrene by means of self-assembled monolayers. *Journal of the American Chemical Society*, **121**, 1016–22.
(c) Zhao, B. and Brittain, W.J. (2000) Synthesis of polystyrene brushes on silicate substrates via carbocationic polymerization from self-assembled monolayers. *Macromolecules*, **33**, 342–8.
(d) Kim, N.Y., Jeon, N.L., Choi, I.S., Takami, S., Harada, Y.K., Finnie, R., Girolami, G.S., Nuzzo, R.G., Whitesides, G.M. and Laibinis, P.E. (2000) Surface initiated ring-opening metathesis polymerization on Si/SiO$_2$. *Macromolecules*, **33**, 2793–5.
(e) Blomberg, S., Ostberg, S., Harth, E., Bosman, A.W., Van Horn, B. and Hawker, C.J. (2002) Production of crosslinked, hollow nanoparticles by surface-initiated living free-radical polymerization. *Journal of Polymer Science Part A: Polymer Chemistry*, **40**, 1309–20.
(f) Baum, M. and Brittain, W.J. (2002) Synthesis of polymer brushes on silicate substrates via reversible addition fragmentation chain transfer technique. *Macromolecules*, **35**, 610–15.
(g) Matyjaszewski, K., Miller, P.J., Shukla, N., Immaraporn, B., Gelman, A., Luokala, B.B., Siclovan, T.M., Kickelbick, G., Vallant, T., Hoffmann, H. and Pakula, T. (1999) Polymers at interfaces: using atom transfer radical polymerization in the controlled growth of homopolymers and block copolymers from silicon surfaces in the absence of untethered sacrificial initiator. *Macromolecules*, **32**, 8716–24.
55 Ejaz, M., Yamamoto, S., Ohno, K., Tsujii, Y. and Fukuda, T. (1998) Controlled graft polymerization of methyl methacrylate on silicon substrate by the combined use of the Langmuir-Blodgett and atom transfer radical polymerization techniques. *Macromolecules*, **31**, 5934–6.
56 Ma, H., Hyun, J., Stiller, P. and Chilkoti, A. (2004) Non-fouling oligo(ethylene glycol)- functionalized polymer brushes synthesized by surface-initiated atom transfer radical. *Advanced Materials*, **16**, 338–41.
57 Mengel, C., Esker, A.R., Meyer, W.H. and Wegner, G. (2002) Preparation and modification of poly(methacrylic acid) and poly(acrylic acid) multilayers. *Langmuir*, **18**, 6365–72.
58 Prucker, O. and Rühe, J. (1998) Mechanism of radical chain polymerizations initiated by azo compounds covalently bound to the surface of spherical particles. *Macromolecules*, **31**, 602–13.
59 Nath, N., Hyun, J., Ma, H. and Chilkoti, A. (2004) Surface engineering strategies for control of protein and cell interactions. *Surface Science*, **570**, 98–110.
60 Padeste, C., Solak, H.H., Brack, H.P., Slaski, M., Gürsel, S.A. and Scherer, G.G. (2004) Patterned grafting of polymer brushes onto flexible polymer substrates. *Journal of Vacuum Science Technology B*, **22**, 3191–5.
61 Steenackers, M., Kuller, A., Ballav, N., Zharnikov, M., Grunze, M. and Jordan, R. (2007) Morphology control of structured polymer brushes. *Small*, **3**, 1764–73.
62 He, Q., Kuller, A., Schilp, S., Leisten, F., Kolb, H.-A., Grunze, M. and Li, J. (2007)

Fabrication of controlled thermosensitive polymer nanopatterns with one-pot polymerization through chemical lithography. *Small*, **3**, 1860–5.

63 Ducker, R., Garcia, A., Zhang, J., Chen, T. and Zauscher, S. (2008) Polymeric and biomacromolecular brush nanostructures: progress in synthesis, patterning and characterization. *Soft Matter*, **4**, 1774–86.

64 Mosse, W.K.J., Koppens, M.L., Gengenbach, T.R., Scanlon, D.B., Gras, S.L. and Ducker, W.A. (2009) Peptides grafted from solids for the control of interfacial properties. *Langmuir*, **25**, 1488–94.

65 Chang, Y.C. and Frank, C.W. (1996) Grafting of poly(gamma-benzyl-L-glutamate) on chemically modified silicon oxide surfaces. *Langmuir*, **12**, 5824–9.

66 Kenausis, G.L., Voros, J., Elbert, D.L., Huang, N., Hofer, R., Ruiz-Taylor, L., Textor, M., Hubbell, J.A. and Spencer, N.D. (2000) Poly(L-lysine)-g-poly(ethylene glycol) layers on metal oxide surfaces: attachment mechanism and effects of polymer architecture on resistance to protein adsorption. *Journal of Physical Chemistry B*, **104**, 3298–309.

67 Huang, N.-P., Voros, J., De Paul, S.M., Textor, M. and Spencer, N.D. (2002) Biotin-derivatized poly(L-lysine)-g-poly(ethylene glycol): a novel polymeric interface for bioaffinity sensing. *Langmuir*, **18**, 220–30.

68 Pasche, S., De Paul, S.M., Voros, J., Spencer, N.D. and Textor, M. (2003) Poly(L-lysine)-graft-poly(ethylene glycol) assembled monolayers on niobium oxide surfaces: a quantitative study of the influence of polymer interfacial architecture on resistance to protein adsorption by ToF-SIMS and in situ OWLS. *Langmuir*, **19**, 9216–25.

69 Pasche, S., Voros, J., Griesser, H.J., Spencer, N.D. and Textor, M. (2005) Effects of ionic strength and surface charge on protein adsorption at PEGylated surfaces. *Journal of Physical Chemistry B*, **109**, 17545–52.

70 Faraasen, S., Vörös, J., Csucs, G., Textor, M., Merkle, H.P. and Walter, E. (2003) Ligand-specific targeting of microspheres to phagocytes by surface modification with poly(L-lysine)-grafted poly(ethylene glycol) conjugate. *Pharmaceutical Research*, **20**, 237–43.

71 Pasche, S., Textor, M., Meagher, L., Spencer, N.D. and Griesser, H.J. (2005) Relationship between interfacial forces measured by colloid-probe atomic force microscopy and protein resistance of poly(ethylene glycol)-grafted poly(L-lysine) adlayers on niobia surfaces. *Langmuir*, **21**, 6508–20.

72 Zhen, G., Falconnet, D., Kuennemann, E., Vörös, J., Spencer, N.D., Textor, M. and Zürcher, S. (2006) Nitrilotriacetic acid functionalized graft copolymers: a polymeric interface for selective and reversible binding of histidine-tagged proteins. *Advanced Functional Materials*, **16**, 243–51.

73 Mormann, W. and Schmalz, K. (1994) Polymers from multifunctional isocyanates. 9. Alternating copolymers from 2-propenyl isocyanate and maleic anhydride. *Macromolecules*, **27**, 7115–20.

74 Beyer, D., Bohanon, T.M., Knoll, W., Ringsdorf, H., Elender, G. and Sackmann, E. (1996) Surface modification via reactive polymer interlayers. *Langmuir*, **12**, 2514–18.

75 Park, S., Lee, K.-B., Choi, I.S., Langer, R. and Jon, S. (2007) Dual functional, polymeric self-assembled monolayers as a facile platform for construction of patterns of biomolecules. *Langmuir*, **23**, 10902–5.

76 Harbers, G.M., Emoto, K., Greef, C., Metzger, S.W., Woodward, H.N., Mascali, J.J., Grainger, D.W. and Lochhead, M.J. (2007) Functionalized poly(ethylene glycol)-based bioassay surface chemistry that facilitates bio-immobilization and inhibits nonspecific protein, bacterial, and mammalian cell adhesion. *Chemistry of Materials*, **19**, 4405–14.

77 d'Agostino, R. (ed.) (1990) *Plasma Deposition, Treatment, and Etching of Polymers*, Academic Press, Boston.

78 Iriyama, Y., Yasuda, T., Cho, D.I. and Yasuda, H. (1990) Plasma-surface treatment on nylon fabrics by

fluorocarbon compounds. *Journal of Applied Polymer Science*, **39**, 249–64.

79 (a) For a general reference on plasma polymerization, see Yasuda, H. (1995) *Plasma Polymerization*, Academic Press, Orlando.
(b) Hsieh, M.C., Farris, R.J. and McCarthy, T.J. (1997) Surface "priming" for layer-by-layer deposition: polyelectrolyte multilayer formation on allylamine plasma-modified poly(tetrafluoroethylene). *Macromolecules*, **30**, 8453–8.
(c) Chatelier, R.C., Drummond, C.J., Chan, D.Y.C., Vasic, Z.R., Gengenbach, T.R. and Griesser, T.J. (1995) Theory of contact angles and the free energy of formation of ionizable surfaces–application to heptylamine radiofrequency plasma-deposited films. *Langmuir*, **11**, 4122–8.

80 Terlingen, J.G.A., Feijen, J. and Hoffman, A.S. (1993) Immobilization of surface active compounds on polymer supports using glow-discharge processes. 1. Dodecylsulfate on polypropylene. *Journal of Colloid and Interface Science*, **155**, 55–65.

81 Manos, D.M. and Flamm, D.L. (1989) *Plasma Etching, An Introduction*, Academic Press, Boston.

82 (a) Wertheimer, M.R. and Schreiber, H.P. (1981) Surface-property modification of aromatic polyamides by microwave plasmas. *Journal of Applied Polymer Science*, **26**, 2087–96.
(b) Moshonov, A. and Avny, Y. (1980) The use of acetylene glow-discharge for improving adhesive bonding of polymeric films. *Journal of Applied Polymer Science*, **25**, 771–81.

83 (a) Ertel, S.I., Ratner, B.D. and Horbett, T.A. (1990) Radiofrequency plasma deposition of oxygen-containing films on polystyrene and poly(ethylene-terephthalate) substrates improves endothelial cell growth. *Journal of Biomedical Materials Research*, **24**, 1637–59.
(b) Yeh, Y.-S., Iriyama, Y., Matsuzawa, Y., Hanson, S.R. and Yasuda, H. (1988) Blood compatibility of surfaces modified by plasma polymerization. *Journal of Biomedical Materials Research*, **22**, 795–818.

84 Zhang, Z., Chen, Q., Knoll, W., Forch, R., Holcomb, R. and Roitman, D. (2003) Plasma polymer film structure and DNA probe immobilization. *Macromolecules*, **36**, 7689–94.

85 Blättler, T.M., Pasche, S., Textor, M. and Griesser, H.J. (2006) High salt stability and protein resistance of poly(L-lysine)-g-poly(ethylene glycol) copolymers covalently immobilized via aldehyde plasma polymer interlayers on inorganic and polymeric substrates. *Langmuir*, **22**, 5760–9.

86 Lahann, J., Balcells, M., Lu, H., Rodon, T., Jensen, K.F. and Langer, R. (2003) Reactive polymer coatings: a first step toward surface engineering of microfluidic devices. *Analytical Chemistry*, **75**, 2117–22.

87 Lahann, J., Choi, I.S., Lee, J., Jensen, K.F. and Langer, R. (2001) A new method toward microengineered surfaces based on reactive coating. *Angewandte Chemie, International Edition*, **40**, 3166–9.

88 Greiner, A. (1997) Polymer films by chemical vapour deposition. *Trends in Polymer Science*, **5**, 12–16.

89 Lahann, J., Balcells, M., Rodon, T., Lee, J., Choi, I.S., Jensen, K.F. and Langer, R. (2002) Reactive polymer coatings: a platform for patterning proteins and mammalian cells onto a broad range of materials. *Langmuir*, **18**, 3632–8.

90 Chen, H.-Y., Elkasabi, Y. and Lahann, J. (2006) Surface modification of confined microgeometries via vapor-deposited polymer coatings. *Journal of the American Chemical Society*, **128**, 374–80.

91 Elkasabi, Y., Chen, H.-Y. and Lahann, J. (2006) Multipotent polymer coatings based on chemical vapor deposition copolymerization. *Advanced Materials*, **18**, 1521–6.

92 Gauthier, M.A., Gibson, M.I. and Klok, H.-A. (2009) Synthesis of functional polymers by post-polymerization modification. *Angewandte Chemie, International Edition*, **48**, 48–58.

93 Plate, N.A., Litmanovich, A.D. and Noah, O.V. (1995) *Macromolecular Reactions:*

Peculiarities, Theory and Experimental Approaches, John Wiley & Sons, Ltd, Chichester.

94 Chechik, V., Crooks, R.M. and Stirling, C.J.M. (2000) Reactions and reactivity in self-assembled monolayers. *Advanced Materials*, **12**, 1161–71.

95 Jonkheijm, P., Weinrich, D., Schroder, H., Niemeyer, C.M. and Waldmann, H. (2008) Chemical strategies for generating protein biochips. *Angewandte Chemie, International Edition*, **47**, 9618–47.

96 (a) Phillips, K.S., Han, J.H., Martinez, M., Wang, Z., Carter, D. and Cheng, Q. (2006) Nanoscale glassification of gold substrates for surface plasmon resonance analysis of protein toxins with supported lipid membranes. *Analytical Chemistry*, **78**, 596–603.
(b) Bossi, A., Bonini, F., Turner, A.P.F. and Piletsky, S.A. (2007) Molecularly imprinted polymers for the recognition of proteins: the state of the art. *Biosensors and Bioelectronics*, **22**, 1131–7.
(c) Cooper, M.A., Try, A.C., Carroll, J., Ellar, D.J. and Williams, D.H. (1998) Surface plasmon resonance analysis at a supported lipid monolayer. *Biochimica et Biophysica Acta*, **1373**, 101–11.
(d) Lotierzo, M., Henry, O.Y.F., Piletsky, S., Tothill, I., Cullen, D., Kania, M., Hock, B. and Turner, A.P.F. (2004) Surface plasmon resonance sensor for domoic acid based on grafted imprinted polymer. *Biosensors and Bioelectronics*, **20**, 145–52.
(e) Besenicar, M., Macek, P., Lakey, J.H. and Anderluh, G. (2006) Surface plasmon resonance in protein–membrane interactions. *Chemistry and Physics of Lipids*, **141**, 169–78.
(f) Matsui, J., Akamatsu, K., Hara, N., Miyoshi, D., Nawafune, H., Tamaki, K. and Sugimoto, N. (2005) SPR sensor chip for detection of small molecules using molecularly imprinted polymer with embedded gold nanoparticles. *Analytical Chemistry*, **77**, 4282–5.
(g) Toyama, S., Shoji, A., Yoshida, Y., Yamauchi, S. and Ikariyama, Y. (1998) Surface design of SPR-based immunosensor for the effective binding of antigen or antibody in the evanescent field using mixed polymer matrix. *Sensors and Actuators B*, **52**, 65–71.

97 Houseman, B.T. and Mrksich, M. (2001) The microenvironment of immobilized Arg-Gly-Asp peptides is an important determinant of cell adhesion. *Biomaterials*, **222**, 943–55.

98 Shin, Y.M., Kim, K.-S., Lim, Y.M., Nho, Y.C. and Shin, H. (2008) Modulation of spreading, proliferation, and differentiation of human mesenchymal stem cells on gelatin-immobilized poly(L-lactide-co-ε-caprolactone) substrates. *Biomacromolecules*, **9**, 1772–81.

99 Tourniaire, G., Collins, J., Campbell, S., Mizomoto, H., Ogawa, S., Thaburet, J.F. and Bradley, M. (2006) Polymer microarrays for cellular adhesion. *Chemical Communications*, **20**, 2118–20.

100 Embrechts, A., Feng, C.L., Bredebusch, I., Rommel, C.E., Schenkenburger, J., Vansco, G.J. and Schönherr, H. (2009) *Interaction of Structured and Functionalized Polymers with Cancer Cells* in Förch, R., Schönherr, H. and Jenkins, A.T.A. (eds) *Surface design: Applications in Bioscience and Nanotechnology*, Wiley-VCH Verlag GmbH, Chapter 3.4, 233–250.

101 Hook, A.L., Thissen, H. and Voelcker, N.H. (2009) Advanced substrate fabrication for cell microarrays. *Biomacromolecules*, **10**, 573–9.

102 Dusseiller, M.R., Schlaepfer, D., Koch, M., Kroschewski, R. and Textor, M. (2005) An inverted microcontact printing method on topographically structured polystyrene chips for arrayed micro-3-D culturing of single cells. *Biomaterials*, **26**, 5917–25.

103 Ruoslahti, E. (1996) RGD and other recognition sequences for integrins. *Annual Review of Cell and Developmental Biology*, **12**, 697–715.

104 Cavalcanti-Adam, E., Micoulet, A., Blümmel, J., Auernheimer, J., Kessler, H. and Spatz, J.P. (2006) Lateral spacing of integrin ligands influences cell spreading and focal adhesion assembly. *European Journal of Cell Biology*, **85**, 219–24.

105 (a) Maheshwari, G., Brown, G., Lauffenburger, D.A., Wells, A. and Griffith, L.G. (2000) Cell adhesion and

motility depend on nanoscale RGD clustering. *Journal of Cell Science*, **113**, 1677–86.
(b) Koo, L.Y., Irvine, D.J., Mayes, A.M., Lauffenburger, D.A. and Griffith, L.G. (2002) Coregulation of cell adhesion by nanoscale RGD organization and mechanical stimulus. *Journal of Cell Science*, **115**, 1423–33.

106 (a) Horbett, T.A. (1994) The role of adsorbed proteins in animal cell adhesion. *Colloids and Surfaces B, Biointerfaces*, **2**, 225–40.
(b) Siebers, M.C., ter Brugge, P.J., Walboomers, X.F. and Jansen, J.A. (2005) Integrins as linker proteins between osteoblasts and bone replacing materials. A critical review. *Biomaterials*, **26**, 137–46.

107 Blättler, T., Huwiler, C., Ochsner, M., Stadler, B., Solak, H., Voros, J. and Grandin, H.M. (2006) Nanopatterns with biological functions. *Journal of Nanoscience and Nanotechnology*, **6**, 2237–64.

108 Seurynck-Servoss, S.L., White, A.M., Baird, C.L., Rodland, K.D. and Zangar, R.C. (2007) Evaluation of surface chemistries for antibody microarrays. *Analytical Biochemistry*, **371**, 105–15.

109 Staros, J.V., Wright, R.W. and Swingle, D.M. (1986) Enhancement by N-hydroxysulfosuccinimide of water-soluble carbodiimide mediated coupling reactions. *Analytical Biochemistry*, **156**, 220–2.

110 Ameringer, T., Hinz, M., Mourran, C., Seliger, H., Groll, J. and Möller, M. (2005) Ultrathin functional star PEG coatings for DNA microarrays. *Biomacromolecules*, **6**, 1819–23.

111 Valsesia, A., Colpo, P., Mannelli, I., Mornet, S., Bretagnol, F., Ceccone, G. and Rossi, F. (2008) Use of nanopatterned surfaces to enhance immunoreaction efficiency. *Analytical Chemistry*, **80**, 1418–24.

112 Zhang, Z., Knoll, W., Forch, R., Holcomb, R. and Roitman, D. (2005) DNA hybridization on plasma-polymerized allylamine. *Macromolecules*, **38**, 1271–6.

113 Tao, S.L., Popat, K.C., Norman, J.J. and Desai, T.A. (2008) Surface modification of SU-8 for enhanced biofunctionality and non-fouling properties. *Langmuir*, **24**, 2631–6.

114 Delamarche, E., Donzel, C., Kamounah, F.S., Wolf, H., Geissler, M., Stutz, R., Schmidt-Winkel, P., Michel, B., Mathieu, H.J. and Schaumburg, K. (2003) Microcontact printing using poly(dimethylsiloxane) stamps hydrophilized by poly(ethylene oxide) silanes. *Langmuir*, **19**, 8749–58.

115 Bulmus, V., Ayhan, H. and Piskin, E. (1997) Modified PMMA monosize microbeads for glucose oxidase immobilization. *Chemical Engineering Journal*, **65**, 71–6.

116 Lahiri, J., Isaacs, L., Tien, J. and Whitesides, G.M. (1999) A Strategy for the generation of surfaces presenting ligands for studies of binding based on an active ester as a common reactive intermediate: a surface plasmon resonance study. *Analytical Chemistry*, **71**, 777–90.

117 (a) Yang, M., Teeuwen, R.L.M., Giesbers, M., Baggerman, J., Arafat, A., de Wolf, F.A., van Hest, J.C.M. and Zuilhof, H. (2008) One-step photochemical attachment of NHS-terminated monolayers onto silicon surfaces and subsequent functionalization. *Langmuir*, **24**, 7931–8.
(b) Wojtyk, J.T.C., Morin, K.A., Boukherroub, R. and Wayner, D.D.M. (2002) Modification of porous silicon surfaces with activated ester monolayers. *Langmuir*, **18**, 6081–7.
(c) Duhachek, S.D., Kenseth, J.R., Casale, G.P., Small, G.J., Porter, M.D. and Jankowiak, R. (2000) Monoclonal antibody–gold biosensor chips for detection of depurinating carcinogen–DNA adducts by fluorescence line-narrowing spectroscopy. *Analytical Chemistry*, **72**, 3709–16.

118 Frey, B.L. and Corn, R.M. (1996) Covalent attachment and derivatization of poly(L-lysine) monolayers on gold surfaces as characterized by polarization-modulation FT-IR spectroscopy. *Analytical Chemistry*, **68**, 3187–93.

119 Schönherr, H., Feng, C.L. and Shovsky, A. (2003) Interfacial reactions in confinement: kinetics and temperature dependence of reactions in self-

assembled monolayers compared to ultrathin polymer films. *Langmuir*, **19**, 10843–51.

120 Feng, C.L., Embrechts, A., Bredebusch, I., Bouma, A., Schnekenburger, J., García-Parajó, M., Domschke, W., Vancso, G.J. and Schönherr, H. (2007) Tailored interfaces for biosensors and cell-surface interaction studies via activation and derivatization of polystyrene-block-poly(*tert*-butyl acrylate) thin films. *European Polymer Journal*, **43**, 2177–90.

121 (a) Peelen, D. and Smith, L.M. (2005) Immobilization of amine-modified oligonucleotides on aldehyde-terminated alkanethiol monolayers on gold. *Langmuir*, **21**, 266–71.
(b) Choi, H.J., Kim, N.H., Chung, B.H. and Seong, G.H. (2005) Micropatterning of biomolecules on glass surfaces modified with various functional groups using photoactivatable biotin. *Analytical Biochemistry*, **347**, 60–6.
(c) Betancor, L., Lopez-Gallego, F., Hidalgo, A., Alonso-Morales, N., Mateo, C., Fernandez-Lafuente, R. and Guisan, J.M. (2006) Different mechanisms of protein immobilization on glutaraldehyde activated supports: Effect of support activation and immobilization conditions. *Enzyme and Microbial Technology*, **39**, 877–82.
(d) MacBeath, G. and Schreiber, S.L. (2000) Printing proteins as microarrays for high-throughput function determination. *Science*, **289**, 1760–3.

122 (a) Duckworth, B.P., Xu, J., Taton, T.A., Guo, A. and Distefano, M.D. (2006) Site-specific, covalent attachment of proteins to a solid surface. *Bioconjugate Chemistry*, **17**, 967–74.
(b) Collman, J.P., Devaraj, N.K., Eberspacher, T.P.A. and Chidsey, C.E.D. (2006) Mixed azide-terminated monolayers: a platform for modifying electrode surfaces. *Langmuir*, **22**, 2457–64.
(c) Lummerstorfer, T. and Hoffmann, H. (2004) Click chemistry on surfaces: 1,3-dipolar cycloaddition reactions of azide-terminated monolayers on silica. *Journal of Physical Chemistry B*, **108**, 3963–6.

123 Devaraj, N.K., Miller, G.P., Ebina, W., Kakaradov, B., Collman, J.P., Kool, E.T. and Chidsey, C.E.D. (2005) Chemoselective covalent coupling of oligonucleotide probes to self-assembled monolayers. *Journal of the American Chemical Society*, **127**, 8600–1.

124 Wilchek, M. and Bayer, E. (1990) *Avidin-Biotin Technology Methods in Enzymology*, Vol. **184**, Academic Press, San Diego, CA.

125 Arnold, F.H. (1992) *Metal-Affinity Protein Separations*, Academic Press, San Diego.

126 (a) Hochuli, E. (1990) Purification of recombinant proteins with metal chelate adsorbent. *Genetic Engineering*, **12**, 87–98.
(b) Hochuli, E., Bannwarth, W., Dçbeli, H., Gentz, R. and Stuber, D. (1988) Genetic approach to facilitate purification of recombinant proteins with a novel metal chelate adsorbent. *Bio/Technology*, **6**, 1321–5.

127 Nieba, L., Nieba-Axmann, S.E., Persson, A., Hamalainen, M., Edebratt, F., Hansson, A., Lidholm, J., Magnusson, K., Karlsson, A.F. and Pluckthun, A. (1997) BIACORE analysis of histidine-tagged proteins using a chelating NTA sensor chip. *Analytical Biochemistry*, **252**, 217–28.

128 Tinazli, A., Tang, J., Valiokas, R., Picuric, S., Lata, S., Piehler, J., Liedberg, B. and Tampe, R. (2005) High-affinity chelator thiols for switchable and oriented immobilization of histidine-tagged proteins: a generic platform for protein chip technologies. *Chemistry: A European Journal*, **11**, 5249–59.

129 Love, J.C., Estroff, L.A., Kriebel, J.K., Nuzzo, R.G. and Whitesides, G.M. (2005) Self-assembled monolayers of thiolates on metals as a form of nanotechnology. *Chemical Reviews*, **105**, 1103–70.

130 Malmqvist, M. and Karlsson, R. (1997) Biomolecular interaction analysis: affinity biosensor technologies for functional analysis of proteins. *Current Opinion in Chemical Biology*, **1**, 378–83.

131 (a) Sigal, G.B., Bamdad, C., Barberis, A., Strominger, J. and Whitesides, G.M. (1996) A self-assembled monolayer for the binding and study of histidine-

tagged proteins by surface plasmon resonance. *Analytical Chemistry*, **68**, 490–7.
(b) Gehrke, S.H., Vaid, N.R. and McBride, J.F. (1998) Protein sorption and recovery by hydrogels using principles of aqueous two-phase extraction. *Biotechnology and Bioengineering*, **58**, 416–27.

132 (a) Niemeyer, C.M. and Blohm, D. (1999) DNA microarrays. *Angewandte Chemie, International Edition*, **38**, 2865–9.
(b) Schulze, A. and Downward, J. (2001) Gene expression using microarrays – a technology review, *Nature Cell Biology*, **3**, E190–5.

133 Benters, R., Niemeyer, C.M., Drutschmann, D., Blohm, D. and Wöhrle, D. (2002) DNA microarrays with PAMAM dendritic linker systems. *Nucleic Acids Research*, **30**, e10.

134 Rowan, B., Wheeler, M.A. and Crooks, R.M. (2002) Patterning bacteria within hyperbranched polymer film templates. *Langmuir*, **18**, 9914–17.

135 Zhou, X., Wu, L. and Zhou, J. (2004) Fabrication of DNA microarrays on nanoengineered polymeric ultrathin film prepared by self-assembly of polyelectrolyte multilayers. *Langmuir*, **20**, 8877–85.

136 Park, S.J., Lee, K.B., Choi, I.S., Langer, R. and Jon, S.Y. (2007) Dual functional, polymeric self-assembled monolayers as a facile platform for construction of patterns of biomolecules. *Langmuir*, **23**, 10902–5.

137 Kuhlman, W., Taniguchi, I., Griffith, L.G. and Mayes, A.M. (2007) Interplay between PEO tether length and ligand spacing governs cell spreading on RGD-modified PMMA-g-PEO comb copolymers. *Biomacromolecules*, **8**, 8206–13.

138 Kumar, N., Parajuli, O., Dorfman, A., Kipp, D. and Hahm, J.I. (2007) Activity study of self-assembled proteins on nanoscale diblock copolymer templates. *Langmuir*, **23**, 7416–22.

139 Pohl, M., Michaelis, N., Meister, F. and Heinze, T. (2009) Biofunctional surfaces based on dendronized cellulose. *Biomacromolecules*, **10**, 382–9.

140 (a) Touil, F., Pratt, S., Mutter, R. and Chen, B. (2006) Screening a library of potential prion therapeutics against cellular prion proteins and insights into their mode of biological activities by surface plasmon resonance. *Journal of Pharmaceutical and Biomedical Analysis*, **40**, 822–32.
(b) Ferracci, G., Seagar, M., Joel, C., Miquelis, R. and Leveque, C. (2004) Real time analysis of intact organelles using surface plasmon resonance. *Analytical Biochemistry*, **334**, 367–75.
(c) Rich, R.L. and Myszka, D.G. (2005) Survey of the year 2004 commercial optical biosensor literature. *Journal of Molecular Recognition*, **18**, 431–78.

3
Stimuli-Responsive Polymer Nanocoatings
Ana L. Cordeiro

3.1
Introduction

Polymers that are capable of responding to changes in the surrounding environment are often referred to as being "stimuli-responsive," "smart," "intelligent" or "environmentally sensitive." Such polymers are of widespread interest in nanotechnology, pharmacy, medicine, biomaterials, materials sciences, and engineering, mainly because they mimic biological systems, responding to environmental changes by changing their conformation, solubility, hydrophilic/hydrophobic balance, reaction rate, and molecular recognition. In general, these responses are reversible, and this type of "switching" may occur repeatedly. One key parameter when defining the behavior of smart polymers is their nonlinear response to external stimuli. In other words, a system may remain stable over a wide variation of an external parameter to which it is sensitive (such as temperature or pH), but will then undergo a drastic conformational change when that parameter is varied over a very narrow range that is close to a critical value. Polymers can be classified according to the stimuli towards which they are sensitive; some typical stimuli are summarized in Table 3.1. Polymers that respond to a combination of two or more external stimuli have also been developed.

Stimuli-responsive polymers can also be classified according to their physical form [19]. Responsive polymers have found applications as free chains dissolved in aqueous solutions (e.g., for the recovery of target molecules) [1], as hydrogels, whether covalently or noncovalently crosslinked [2, 20] (e.g., for drug delivery), and immobilized onto solid surfaces, whether adsorbed or surface-grafted [21, 22] (e.g., for cell sheet engineering). In recent years, stimuli-responsive polymer surfaces have attracted substantial research interest, and the advances in their applications have been particularly outstanding [21, 22]. The grafting of stimuli-responsive polymers onto substrates, for example silica or polymeric surfaces, has led to the generation of smart surfaces with stimuli-modulated properties that allow inclusively the production of "on-off" systems by varying either the surface hydrophilicity/hydrophobicity or the pore/channel dimensions. Both, the design

Table 3.1 Environmental stimuli.

Stimulus	Reference(s)
Physical	
Temperature	[2, 3]
Ionic strength	[4]
Solvents	[5]
Radiation (ultraviolet, visible, ultrasonic)	[6, 7]
Electric and magnetic fields	[8, 9]
Mechanical stress and strain	[10]
Chemical	
pH	[11, 12]
Specific ions	[13, 14]
Chemical agents	
Biochemical	
Enzyme substrates	[14]
Affinity ligands	[15–17]
Other biochemical agents	[18]

Reprinted in part with permission from Ref. [1].

and development of stimuli-responsive or smart surfaces are defined by the applications for which those materials are intended. For example, surfaces may be designed to modulate cell adhesion (which may be achieved by controlling surface hydrophilicity), or to modulate diffusion profiles (achievable by controlling the pore dimensions in porous substrates). The dynamic control of surface properties using stimuli-responsive polymers is of major interest for the control of biomolecular adsorption, cell adhesion and function, and also biocompatibility [23–25]. The development of dynamic surfaces capable of regulating biological functions in response to applied stimuli is very important for a variety of medical applications, including tissue engineering and regenerative medicine, and offers new possibilities for fundamental cellular studies. One excellent example of this is to provide the capability to understand cell–extracellular matrix (ECM) interactions.

In recent years, the field of stimuli-responsive materials has attracted intense research efforts, with several reviews having summarized progress in areas such as biomedical applications (tissue engineering, cell culture, drug delivery) [2–4, 8, 19, 22, 26–49], separation [19, 50–53], bioconjugates [54–58], and microfluidics [21, 59]. Despite many reports having been produced, only selected examples are included in this chapter. In general, attention will be focused on those polymers that are responsive to temperature and pH, as these have been recognized for their efficiency and convenience in many applications. The conjugation of responsive polymers with biological molecules, such as proteins, oligopeptides and polysaccharides, is of particular interest as it enables these materials to respond to physical, chemical and biological stimuli, thus expanding their potential. As an understanding of structure–property relationships is essential for the design and development of new, "smart" materials, brief details are provided of the preparation of polymer films and the characterization of their interfacial properties. Finally, an overview of the major advances and applications of stimuli-responsive surfaces is provided, and future perspectives are discussed.

3.2
Stimuli-Responsive Polymers

3.2.1
Polymers Responsive to Temperature

Temperature-responsive polymer systems are possibly the most widely studied class of stimuli-responsive polymers [28, 58]. Thermosensitive polymers are especially important in biomedicine, as they can be prepared to gel *in situ* by simply increasing the temperature from ambient to physiological [2]. They can also be used to prepare materials capable of releasing drugs when the local temperature is changed, perhaps as the result of fever, infection, or disease [28].

The most widely studied thermoresponsive polymer is poly(N-isopropylacrylamide) (PNIPAAm) and its derived copolymers [58, 60]. These polymers are characterized by a lower critical solution temperature (LCST), defined as the critical temperature above which a polymer solution undergoes a phase transition from soluble to insoluble [58]. Below the LCST, the strong hydrogen bonding between the polymer hydrophilic (i.e., amide) groups and water exceeds the unfavorable free energy associated with the exposure of hydrophobic groups to water, leading to good polymer solubility. Above the LCST, the interactions between the hydrophobic side groups are dominant, which leads to a release of the structured water and to polymer collapse. PNIPAAm is the most investigated thermoresponsive polymer because it undergoes a reversible sharp coil-to-globule transition at 32 °C in water [60–63]. The unique properties of PNIPAAm and derived copolymers have been widely used in the fabrication of drug delivery systems [28, 42, 43, 45, 51], to modulate cell adhesion and protein adsorption [24, 64–70], in cell sheet engineering [71–73], and in a variety of separation processes [50, 51, 74].

Many efforts have been made towards the modification of the swelling/deswelling behavior of thermoresponsive hydrogels, to control the swelling degree, transition dynamics, and transition temperature to generate materials that fit specific applications [75–81]. The transition temperature of PNIPAAm can be tuned by various means; an example is by the copolymerization of NIPAAm (Figure 3.1a) with hydrophilic or hydrophobic comonomers. In general, the LCST is lowered by the introduction of hydrophobic comonomers, but increased by the introduction of hydrophilic comonomers [43, 76, 78, 83–86]. Changes in the LCST triggered by the incorporation of comonomers with different properties are believed to be due to changes in the overall polymer hydrophilicity, and not to the direct influence on the structuring of water around the hydrophobic groups. The introduction of

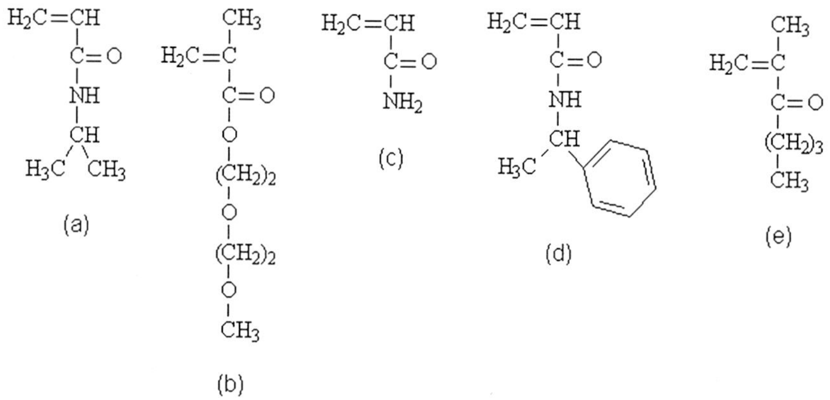

Figure 3.1 Chemical structures of: (a) NIPAAm. Examples of (b), (c) more hydrophilic and (d), (e) more hydrophobic comonomers copolymerized with NIPAAm to tune its LCST; (b) Diethyleneglycol methacrylate [71]; (c) Acrylamide; (d) N-(1-phenylethyl) acrylamide) [83]; (e) Butyl methacrylate [24].

hydrophobic comonomers decreases the content of the hydrophilic groups and, as a consequence, the polymer hydrophobicity increases due to the strong interactions between the polymer hydrophobic side groups [78]. Depending on the application purpose, many comonomers have been used to tune the LCST of PNIPAAm. A recent review on advances in the synthesis, structural phenomena and properties of copolymers of NIPAAm with various types of comonomers has been provided [56]. The structural formula of NIPAAm and examples of comonomers introduced to tune the LCST of PNIPAAm are presented in Figure 3.1.

In addition to copolymerization with comonomers of variable hydrophilicity, the LCST can be further tuned simply by the addition of surfactants [87] or salts [62] to the aqueous solution. The effect of adding salt on the transition temperature and the degree of swelling is especially important in many practical applications, and has been widely investigated [60, 62, 87–93]. The effect of salt addition on the LCST was attributed to the promotion or disruption of the ordered water molecules surrounding the polymer [89, 91]. The LCST may be either increased ("salting-in") or decreased ("salting-out") by the addition of salt, with the effect being determined by the nature of the salt that is added [87, 89].

Control of the kinetic and thermodynamic aspects of thermoresponsive materials is fundamental in practical applications. The main strategies to improve the response dynamics of PNIPAAm-based hydrogels have been summarized by Zhang and coworkers [81]. These strategies can be classified as: *physical*, if changes in the microstructure of the hydrogel are introduced (e.g., by generating a porous structure or phase-separated structures); or *chemical*, if a chemical modification of the hydrogel at the molecular level is involved (e.g., by introducing residual amounts of hydrophilic moieties) [81].

The immobilization of PNIPAAm-containing layers on surfaces has been achieved using a variety of different techniques, including plasma polymerization [94–96], electron beam irradiation [61, 97, 98], photografting [99], gamma irradiation [100], low-pressure plasma immobilization [76], and *in-situ* polymerization on activated surfaces [101]. The resultant surface-immobilized hydrogels preserved their bulk properties, responding to changes in environmental conditions, and were therefore susceptible for use in practical applications. The method used for surface immobilization plays a critical role in determining the final temperature-dependent properties of the immobilized hydrogel. The graft architecture and grafting density were shown strongly to influence temperature-dependent changes in wetting due to the different motilities of the grafted polymer chains [102]. Terminally grafted PNIPAAm chains are characterized by an enhanced chain mobility, as compared with multipoint grafted chains; this is reflected in the larger changes in water contact angles with temperature variation [102]. Additionally, by altering the thickness (and crosslinking degree) of the thermoresponsive layer, it is possible to tune the swelling/deswelling degree of the immobilized hydrogel [103]. The properties of the surface used to graft the thermoresponsive layer also play a role in the temperature-dependent properties of the grafted polymer. For example, the grafting of PNIPAAm onto nanotextured surfaces greatly influences surface wetting. The difference between the water contact angle above and below the LCST was shown to increase with increasing pore size, which can be explained by the topographical changes associated with the expansion and collapse of the grafted polymer layer with the transition temperature [104].

3.2.2
Polymers Responsive to pH

Polymers sensitive to variations in environmental pH contain pendent acidic (as for example carboxylic) or basic (e.g., amino) groups that either accept or release protons in response to changes in the environmental pH. Polymers that contain a large number of ionizable groups are designated by polyelectrolytes [32]. The change in net charge of pendent groups causes a change in the polymer chain hydrodynamic volume. Hydrogels constituted by polymer chains with ionizable groups are therefore characterized by improved swelling characteristics, as compared to uncharged polymers due to electrostatic repulsion between charges in the polymer chain. Consequently, environmental conditions capable of influencing electrostatic repulsion – such as pH, ionic strength and the type of counterion – strongly influence the swelling/deswelling degree of charged hydrogels [12, 28, 105, 106]. Ionic gels containing weakly acidic pendent groups are characterized by an increasing degree of swelling with increasing solution pH, whereas gels containing weakly basic pendent groups are characterized by an increasing degree of swelling with decreasing pH. The electrostatic repulsion between charges in the polymer chains drives the precipitation or solubilization of a polymer in solution, to the swelling or deswelling of hydrogels, and to the hydrophilic or hydrophobic properties of a surface [58]. The most widely investigated pH-responsive polyacids

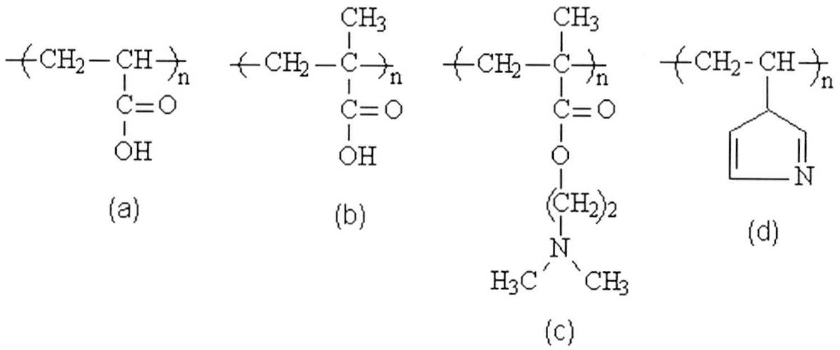

Figure 3.2 Chemical structure of examples of pH-responsive polyacids. (a) Poly(acrylic acid); (b) Poly(methacrylic acid)) and polybases; (c) Poly(N,N'-dimethyl aminoethylmethacrylate); (d) Poly(vinyl imidazole).

are poly(acrylic acid) (PAAc) and poly(methacrylic acid) (PMAAc) (Figure 3.2). Poly(N,N'-dimethyl aminoethylmethacrylate) and poly(vinyl imidazole) are examples of pH-responsive polybases (Figure 3.2). Hydrogels based on weakly ionizable polysaccharides, such as alginate and chitosan, and some polypetides consisting of amino acids with ionizable pendent groups, show pH-responsive phase transition [58, 107]. A recent comprehensive review has been produced on the synthesis and properties of pH-responsive polymers [12].

The pH-sensitivity and swelling properties of pH-responsive polymers can be tuned by introducing neutral comonomers [105, 108]. The introduction of hydrophobic moieties was shown to have the effect of shifting the transition pH to higher values in polyacid-based hydrogels, and to lower values in polybasic-based hydrogels [58]. The applicability of pH-responsive hydrogels can be mostly found in drug delivery systems (as oral and implantable drug delivery agents) [28, 32]. In that respect, pH-responsive materials are particularly suitable, as significant variations in pH exist in the different organs and tissues. The unique properties of pH-responsive polymers have also been used in separation processes [109], in DNA and gene delivery [110], and in biosensing.

3.2.3
Dual Responsive/Multiresponsive Polymers

Efforts have been directed towards the preparation of polymers that can respond to more than one stimulus. For example, PNIPAAm copolymers that can respond to other stimuli in addition to temperature have been successfully prepared. By adding monomers such as acrylic acid (AAc), methacrylic acid, maleic acid, and N-isopropyl-maleamic acid, it was possible to generate polymers that were sensitive to both temperature and pH [11, 89, 111–116], whilst when using photodimerizable chromophores (e.g., acridizinium) it was possible to obtain polymers that responded to both temperature and light [117]. Using a different approach, it has been shown that by grafting PNIPAAm onto the surface of a pH-responsive alginate, a porous structure with a fast swelling/deswelling response to both changes in pH and temperature can be obtained [118]. Polymers with distinct temperature transitions (i.e., doubly thermoresponsive polymers) can be obtained by combining copolymers with different thermosensitivities. A comprehensive description on the combination of thermoresponsive properties with other types of sensitivity (pH, light, solvent, magnetic field) has been provided in a recent review by Dimitrov et al. [119].

3.2.4
Intelligent Bioconjugates

The conjugation of biomolecules with stimuli-responsive polymers results in polymer–biomolecule systems of increased versatility, as these are responsive not only to chemical and/or physical but also biological stimuli. For this reason, responsive polymer–biopolymer conjugates have been the target of many investigations, notably by the Hoffman group [1, 54, 120]. The reversible LCST behavior of thermoresponsive polymers was conjugated to various biomacromolecules (e.g., antibodies, proteins and enzymes) to create dually responsive or multiresponsive polymer–biomolecule systems. In order to facilitate the chemical incorporation of biomacromolecules, a variety of polyfunctional thermoresponsive copolymers containing reactive functional groups have been designed and synthesized by various routes (for details, see Ref. [56] and references therein). The conjugation of proteins with responsive polymers has been achieved both by randomly using the protein reactive sites (usually lysine groups) [121] and by site-specific conjugation to genetically engineered specific amino acids (e.g., cysteine) into selected sites of the amino acid sequence of the protein [17, 122]. Random conjugation may result in a decrease in the protein's bioactivity as a consequence of possible interference with the active site, or changes in the microenvironment. Site-specific conjugation can only be used when the amino acid sequence of the protein is known. However, site-specific conjugation offers the advantage that the binding site may be selected to be located either near or within the active site (in order to control ligand–protein processes and protein bioactivity), or far away from it (so as to avoid influencing bioactivity) [17]. Further details on the approaches for the synthesis of polymer bioconjugates can be found in recent reviews [54, 55, 123].

Conjugates of stimuli-responsive polymers and biomolecules have found applications in biotechnology, medicine and nanotechnology, being inclusively used for affinity separation, as molecular sensors for lab-on-a-chip devices, and as switches for the control of protein and enzyme activities [1, 6, 16, 17, 54, 57, 58, 120, 124, 125]. The activity of thermoresponsive polymer–biomolecule conjugates depends on the hydrophilic/hydrophobic changes of the responsive polymer chains as a result of changes in environmental temperature. Target proteins can be successfully separated from multicomponent aqueous solutions through affinity binding to thermoresponsive polymers to which a specific ligand to the target protein is coupled. After binding of the target protein to the coupled ligand, the protein may be easily recovered by triggering hydrophilicity/hydrophobicity changes by a variation of temperature [1, 58, 120]. For example, it was possible to control the reversible binding and release of biotin to streptavidin site-specifically conjugated with PNIPAAm, simply by varying the ambient temperature [17, 122]. Stimuli-responsive polymer–enzyme conjugates are of special interest, as the possibility of regulating enzyme activity through changes in the environmental conditions (e.g., temperature and light) opens new perspectives for the development of diagnostic assays, and for bioprocessing [6, 124].

Parameters such as the conjugation site, length of polymer chain, molecular weight and type of stimuli-responsive polymer, should be considered when developing bioconjugates with optimal activity [58]. Although thermoresponsive bioconjugates have been mainly used in separation applications as free chains in solution, they have also been grafted onto surfaces for their use for the separation of target proteins from flowing aqueous solutions [126, 127]. The target proteins are bound to the surface-immobilized affinity ligands above the LCST, and can then be recovered by lowering the temperature below the transition temperature of the responsive polymer, as a result of polymer dehydration and collapse.

3.2.5
Responsive Biopolymers

Biomimetic approaches have been used to produce polypeptides with characteristics similar to those of synthetic responsive polymers. One of the most widely studied responsive peptide-based biopolymers was inspired by the mammalian protein elastin. Elastin is an abundant ECM protein that is responsible for providing elasticity to tissues, due to its unique mechanical properties that allow repeated extensibility followed by elastic recoil [128–131]. *Tropoelastin*, the soluble precursor of elastin, is secreted by cells during elastogenesis, and is composed of alternating hydrophobic and hydrophilic domains containing lysine crosslinking sites [129]. Elastin-like polypeptides (ELPs) are based on the repeating motif of tropoelastin. The most studied EPLs consist of the pentapeptide sequence Val-Pro-Gly-X-Gly, where the residue X is any amino acid, with exception of Pro [132–134]. ELPs are thermally responsive polypeptides that undergo phase transition in aqueous solution upon changes in the environmental temperature. In the literature related to elastin, the LCST behavior of ELPs is also designated by inverse transition behavior, and the LCST is referred to as *inverse temperature transition* (T_t) [130, 134]. Below the LCST, ELPs are highly solvated and therefore soluble in water. However, when the temperature is increased above the LCST the polypeptides undergo a phase transition that is characterized by desolvation and aggregation. This transition is completely reversible, and can be triggered not only by changes in ambient temperature but also by changes in pH or ionic strength [134]. The LCST is determined by the choice of the guest residue X [132–137]. The effect of each guest residue on the transition temperature on EPLs containing a single guest residue was investigated by Urry and coworkers [132, 134]. The LCST of the ELPs decreases as the hydrophobicity of the guest residues is increased [134], an effect which depends on the molar content of specific guest residues [132]. The dependence of temperature transition on chain length allows further adjustments of the LCST over a wide range [134, 136, 138]. Additional functionalities and environmental sensitivities, without disrupting the transition, can be added by introducing specific guest residues. Residues with chemically reactive side chains (e.g., lysine) can be added to allow for crosslinking or for the incorporation of further functionalities. The introduction of guest residues with ionizable side chains enables a triggering of the transition not only by changing the temperature but also by varying the pH [130, 139, 140].

Although ELPs have been chemically synthesized [136], recent syntheses have been mainly based on genetically encoded ELPs [130, 135, 140, 141]. Genetic polymerization is highly advantageous as it allows the control of important macromolecular properties ensuring monodispersity, exact stereochemistry and molecular weight, while permitting a high yield. Additionally, the type, number and location of reactive sites in the polypeptide can be specified with precision, which in turn makes possible the controlled grafting of ELPs to surfaces [135]. One of the main advantages associated with ELPs, as compared with other polymer systems, is that they can be synthesized with precisely tuned LCSTs between 0 °C and 100 °C; making the design of responsive materials to fit specific applications possible. The LCST behavior of ELPs has the additional advantage of permitting their easy purification based simply on their responsive properties.

Biopolymers, such as peptide-based materials, have become increasingly important, and have found a variety of potential applications in biomedicine, an example being in regenerative medicine. In particular, ELPs have shown good *in vivo* biocompatibility, with a controllable degradation rate, and minimal cytotoxicity, immune and inflammation responses [130]. The potential uses of ELPs have been demonstrated over a very wide range of applications, including controlled drug delivery [45, 46, 142, 143], protein purification [133], tissue engineering and regenerative medicine [142–149], and biosensing [150–152]. Further details on the medical and biotechnological applications of ELPs can be found in recent reviews and references therein [46, 48, 130, 140, 153].

ELPs have been successfully immobilized onto a variety of surfaces, while preserving responsive properties, and undergoing a reversible phase transition as a result of changes in the environmental conditions. The use of environmentally sensitive surfaces based on ELPs opens up new possibilities as compared with conventional smart surfaces, due to a precise control of the structure and function of the biopolymers, the possibility of incorporating useful ligands, and an improved biocompatibility and biodegradability [130]. The phase transition and stimuli-dependent properties of surface-immobilized ELPs have been used to dynamically control the immobilization of proteins and other biomolecules [135, 144, 150–152, 154–156], as well controlling cell adhesion [157–160]. This opened a new route to create regenerable protein arrays for immunoassays and drug screening, and for the development of novel methods for cell sheet engineering.

3.3
Polymer Films and Interfacial Analysis

Stimuli-responsive polymer films may be prepared using a wide variety of deposition techniques (e.g., by spin-coating or plasma deposition). The choice of technique used depends on the characteristics of the substrate to be coated, of the polymer to be deposited, and of the requirements of the final film. Covalent attachment is usually preferred as this results in polymer films with enhanced stability. The covalent grafting of the polymer to the surface may be achieved by using either "grafting-to" procedures (i.e., by the reaction of an end-functionalized polymer in solution with an activated surface) or "grafting-from" approaches (i.e., by the polymerization of monomers from surface-anchored initiators). Although the "grafting-to" method makes possible a predetermination of the properties of the polymer to be surface-grafted, the "grafting-from" method has the advantage of permitting larger grafting densities, as "grafting-to" is restricted by the steric hindrance of surrounding bonded chains [19, 22]. Among "grafting-from" methods, atom transfer free radical polymerization (ATRP) and reversible addition-fragmentation chain transfer (RAFT) polymerization have recently received increasing attention, and have been used to generate responsive polymer films with a high control over film thickness, grafting density, composition, and chain architecture [74, 161–163].

Correlations between the surface structure and properties of polymer films are fundamental, as these determine whether a surface will have the determined required properties. An understanding of structure–properties relationships is therefore fundamental to the development and rational design of novel smart surfaces. The structure of stimuli-responsive polymer films has been analyzed using multiple surface-sensitive techniques, including atomic force microscopy (AFM) [83, 94, 164–167], time-of-flight-secondary ion mass spectrometry (ToF-SIMS) [94], contact angle measurements using several different techniques [83, 101, 102, 165, 168–170], ellipsometry [76, 83, 164, 171], surface plasmon resonance (SPR) [170], laser scanning confocal microscopy [172], sum frequency generation (SFG) vibrational spectroscopy [94], neutron reflectivity measurements [80, 173], Raman spectroscopy [174], and electrokinetic measurements [83, 175].

Among stimuli-responsive surfaces, the PNIPAAm-based surfaces have been the most widely investigated. Surface analyses using a wide range of techniques have enabled a detailed characterization of the temperature dependent properties of surface-immobilized thermoresponsive films and to reveale their structural features, providing information for the design of responsive surfaces with improved characteristics. The temperature-dependent swelling/deswelling of PNIPAAm-based films in different media (pure water, phosphate-buffered saline solution, artificial sea water) has been investigated using *in situ* ellipsometry measurements [76, 83, 176]. The results revealed the impact of temperature and the presence of electrolytes on the swelling degree of the immobilized films, and also provided an insight into the effects of heating/cooling rates on swelling/deswelling kinetics [76, 83, 176]. AFM studies, aimed at probing the mechanical properties of surface-immobilized NIPAAm films, have shown that the collapsed film is stiffer than the expanded film, with moduli below and above the transition temperature (depending on the degree of crosslinking) in the range of kilopascals to a few megapascals, respectively [94, 167].

Wetting properties are considered to be the most important temperature-dependent interfacial properties of thermoresponsive films. The ability to modulate surface wettability by varying temperature has found broad applicability, namely in the design of surfaces for the controlled release of proteins [64, 66, 67, 69, 70] and cells [36, 37, 72, 113, 177]. A wide range of water contact angles for PNIPAAm-immobilized films has been reported [94, 101, 165, 168–170]. These different angles can be explained by considering the variations in film preparation (which result in different surface coverages, film thicknesses and compositions) and the various experimental procedures adopted for the measurements. Recently, the captive bubble method, in combination with a contact angle technique known as axisymmetric drop shape analysis (ADSA), was used to determine the advancing and receding contact angle of fully hydrated thermoresponsive surfaces in contact with aqueous solution, and its dependence on temperature [83]. The measurements provided a novel understanding of the characteristics and environmentally dependent properties of the responsive surfaces in aqueous media. In addition, an analysis of the hysteresis between advancing and receding contact angles allowed an inference to be made of the extent of changes that occur in surface properties during wetting and dewetting which, again, were seen to depend on the temperature. Inverse receding contact angle measurements of poly(NIPAAm-co-N-(1-phenylethyl)acrylamide) films supported the intrinsic properties of the thermoresponsive film, as the surface hydrophilicity was shown to decrease with increasing temperature [83]. Couriously, the advancing water contact angle decreased with increasing temperature; this was explained by the different molecular mobilities at different temperatures, which either allowed or hampered the reorientation of hydrophobic segments at the solid–liquid and solid–fluid interfaces [83]. The rearrangement of surface groups of plasma-polymerized PNIPAAm films with temperature was recently investigated using SFG vibrational spectroscopy [94]. The results indicated that, above the transition temperature, the surface organizes to resemble the dehydrated film so as to free any bound water and allow hydrogen bonding below the surface. As a consequence, the hydrophobic isopropyl side chains organize towards the aqueous environment. Below the transition temperature the hydrophobic isopropyl groups are disordered, either away from the surface, towards the bulk, or in the surface plane [94]. These observations support the existing proposed mechanisms for the LCST behavior of thermoresponsive polymers [178]. Below the transition temperature, the well-hydrated polymer chains adopt a random conformation, with the amide groups forming hydrogen bonds with water. When the temperature is increased the polymer adopts a compact conformation with enhanced hydrophobic interactions, accompanied by dehydration.

As will be shown below, the magnitude of the conformational changes that occur in grafted polymer chains with temperature, determines the temperature-dependent surface properties and plays a critical role in defining whether a particular polymer film would be suitable for a certain application, or not. Recent investigations have clearly shown the importance of addressing interdependent variables such as the swelling ratio, molecular mobility, wettability, chain density, and the concentration of hydrophobic groups when designing and developing responsive surfaces. An understanding of how to design surfaces while controlling any temperature-dependent conformational changes would enable not only the modulation of surface interactions with proteins, cells and analytes, but also the production of surfaces to fit specific applications.

3.4
Applications

3.4.1
Release Matrices

Surface-immobilized stimuli-responsive hydrogels have been used for the active release of drugs to prevent the formation of biofilms on the surfaces of medical implants. Bacterial infection is a major limitation associated with the use of medical implants (such as catheters, artificial prostheses and subcutaneous sensors), and represents a serious problem in biomedicine [179, 180]. Infections associated with medical implants result from the attachment of bacteria to the surface of the medical device, and the subsequent formation of a *biofilm*. As the bacterial cells in a biofilm are characterized by an increased resistance to antimicrobial agents [181, 182], the incorporation of an antimicrobial agent (e.g., an antibiotic) into the surface of the implanted device might represent an effective means of preventing bacterial colonization and infection, avoiding the problems associated with a high-doseage of antibiotics. Although recently much progress has been made in the incorporation of antimicrobial compounds into medical devices, strategies for the adequate loading and controlled release of agents remain a challenge [180, 183].

PNIPAAm and related copolymers have demonstrated their applicability to be used as immobilized hydrogels for drug delivery [184–186]. Recently, Jones and coworkers developed surface-immobilized PNIPAAm-based hydrogels with improved mechanical and swelling properties which permitted their use as coatings for medical devices [185]. The expanded hydrogels (i.e., when maintained below the LCST) enabled the loading/entrapment of aqueous soluble drugs, which could be released in pulsatile mode by increasing the temperature above the LCST, so that hydrogel collapse occurred. In particular, poly(NIPAAm-*co*-2-(hydroxyethyl)methacrylate)) hydrogels showed high loading and release capabilities for the drug chlorhexidine diacetate, which in turn resulted in a very encouraging decrease in the viability of *Staphylococcus epidermidis* [185]. The *in-situ* replenishment of the surface-immobilized hydrogels was made possible simply by providing a contact with the drug solution and a decreasing temperature. This possibility of *in-situ* drug reloading of surfaces is especially attractive as it enhances not only the lifetime but also the clinical efficacy of the coated device. In another example, polypropylene surfaces were grafted with interpenetrating networks of PNIPAAm and AAc [186]. The introduction of AAc had the advantage of decreasing the friction coefficient and increasing drug loading capacity due to an enhanced hydrophilicity and the presence of ionizable groups. The grafted interpenetrating networks allowed both the drug loading and drug release rate to be tuned, in turn reducing the formation of biofilms of *Staphylococcus aureus* [186]. These results clearly demonstrated the potential benefits of using responsive coatings to prevent biofilm formation on implanted surfaces.

3.4.2
Cell Sheet Engineering

One of the most innovative applications of thermoresponsive surfaces in tissue engineering for regenerative medicine was introduced by Okano and coworkers during the early 1990s [61, 187], and consists of cell sheet engineering [36, 37, 72, 73].

In tissue engineering, the fundamental approach is to grow cells at a surface and subsequently to harvest them without damage. Classical methods of cell culture involve growing cells to confluency on polystyrene (PS) tissue culture dishes, and subsequently detaching them, either by mechanical dissociation or by using chemical agents (e.g., EDTA) or biochemical reactions (with proteolytic enzymes, such as trypsin). Unfortunately, these methods have certain disadvantages as they damage the cell membrane, the cellular junctions, and the ECM [95, 188]. Such disruption of the formed tissue-like structures is clearly a drawback from the tissue-engineering perspective.

The cell sheet engineering approach to harvest cultured cells is based on the observation that (in general) cells preferentially attach to and grow on hydrophobic rather than on hydrophilic surfaces. Since the hydrophilicity of PNIPAAm-grafted surfaces increases with decreasing temperature, it was hypothesized that cells cultured on these surfaces above the LCST might become detached simply by lowering the temperature–that is, by increasing the surface hydrophilicity [61]. This hypothesis was first verified through the successive subculture of bovine hepatocytes onto PNIPAAm-grafted PS dishes. At 37 °C, the slightly hydrophobic PNIPAAm surface allowed for cell attachment and proliferation (similar to PS surfaces). However, on lowering the temperature below the polymer LCST (32 °C), the surface became hydrophilic and swelled due to the phase transition of the polymer, which in turn caused the adhered cells to detach [61]. To date, the concept of using of NIPAAm-based thermoresponsive surfaces as cell culture carriers for cell sheet harvesting has been demonstrated for a very wide variety of cell types, including fibroblasts [176], vascular endothelial cells [189], aortic endothelial cells [24, 190, 191], cardiomyocytes [192], microglial cells [193], and, corneal endothelial [71] and epithelial cells [194, 195]. As different cells have cellular metabolisms with different temperature sensitivities, the optimal temperature for cell recovery from a thermoresponsive surface will depend on the cell type. For example, the optimal temperature for recovery of endothelial cells from PNIPAAm-grafted PS surfaces was found to be 20 °C, whereas for the recovery of hepatocytes this was 10 °C [98].

Comparative studies between classical methods of cell detachment (mechanical dissociation and addition of trypsin) and the use of thermoresponsive surfaces revealed that the latter approach resulted in less damage to the cells and to the ECM [187, 188, 190, 191]. Cells released from PNIPAAm-based surfaces were harvested as closed monolayers (preserving tight cell junctions) and with an almost intact ECM [82, 95, 97, 190, 191, 196]. Whilst the use of conventional methods disrupt cell–cell junctions and the attachment of the ECM to the surface, the use of a thermoresponsive surface will disrupt only interactions between the cell-adhesive proteins at the surface and the polymer surface [36, 95, 191]. It has also been shown that cell detachment using thermoresponsive surfaces enables a higher cell recovery efficiency, and will maintain cell properties as compared to conventional methods (which may cause damage to the cell receptors and alter gene expression) [193]. Although harvested cell sheets using thermoresponsive surfaces are accompanied by the majority of the ECM components, a detailed analysis of the surfaces following cell lift-off, by using immunoassay, X-ray photoelectron spectroscopy (XPS) and ToF-SIMS, revealed that some extracellular proteins (as well as some collagen) had remained at the surface [190]. Experiments performed by seeding cells onto surfaces from which cells had already been lifted revealed a higher cell adhesion than with cells seeded onto surfaces from which cells had been released using conventional methods [190]. These observations further support the proposal that the detachment of cells from surfaces based on PNIPAAm causes less damage to the ECM proteins than classic cell culture methods.

The detachment of cell sheets from thermoresponsive surfaces by varying the temperature is driven by changes in surface properties such as hydrophobicity/hydrophilicity, swelling degree and exposure/nonexposure of functionalities. Investigations into the mechanisms of cell detachment on temperature change from PNIPAAm surfaces revealed that cell detachment mediated by active energy-consuming metabolic processes, including intracellular signaling and cytoskeletal organization [98, 189]. Confluent cell sheets can be detached from PNIPAAm-grafted surfaces together with a deposited fibronectin matrix by decreasing temperature [191]; indicating that, below the LCST, the interactions between the deposited ECM and the cell receptors were maintained. Interestingly, fibronectin could not be released from the surface in the absence of cells upon surface hydration alone, which indicated that cellular activity is necessary for fibronectin release [97]. Initially, cell adhesion is controlled by the physicochemical interactions between the cell, the ECM and the surface; a second step involves cellular metabolic processes such as the development of focal adhesion and cytoskeletal reorganization [97, 98, 197]. Above the LCST, the physicochemical properties of the collapsed polymer surface enable cell adhesion and proliferation. However, when the temperature is lowered below the LCST, hydration of the polymer surface causes a decrease in the anchorage strength of the deposited ECM to the surface. This loss of cell–ECM–surface equilibrium, despite cellular activity (including cytoskeleton dynamics) being retained, causes the cells to become rounded and to detach from the surface. As the cell–ECM binding is preserved, any ECM that is deposited beneath the cells will become detached together with the cells [97]. The small fraction of proteins that remains at the thermoresponsive surface after cell detachment [190] may be due to the fact that cells mainly interact with the ECM at focal points, and consequently the ECM proteins do not interact strongly with the cells retained on the surface [95]. Since each specific cell phenotype creates different ECM structures, it is likely that the deposited basal structure will vary with the cell type. It is therefore also expected that a different response to the temperature-triggered detachment will occur, depending on the cell source [30].

Thermoresponsive surfaces based on PNIPAAm for cell sheet engineering applications have been fabricated using a wide variety of methods, including electron beam [61, 98, 103, 187, 193, 198] and plasma polymerization of NIPAAm [70, 94, 95, 190]. It was, however, found that some of the surfaces produced based on PNIPAAm do not support cell adhesion (even above the LCST), and are therefore unsuitable for use as culture substrates. The most frequently reported method for fabricating thermoresponsive surfaces for cell sheet engineering is electron beam polymerization of NIPAAm onto PS substrates. Investigations into the influence of the properties of electron beam-grafted PNIPAAm layers on cellular behavior revealed that the layer thickness played a clear role in cell adhesion and detachment [103]. Although, changes in surface wettability were observed for all tested grafting densities between 20 °C and 37 °C, temperature-dependent cell adhesion and detachment occurred only on PNIPAAm layers with a thickness of 15–20 nm (grafting density ca. 1.4 μg cm^{-2}). No cell adhesion occurred on PNIPAAm layers thicker than 30 nm (grafting density ca. 2.9 μg cm^{-2}) above the LCST [103], while no cell detachment upon temperature decrease occurred on layers thinner than 15 nm. Kikuchi and Okano suggested that this dependence of cell behavior on the thickness of the PNIPAAm layer is due to an enhanced or limited mobility of the grafted polymer chains, which in turn depended on the grafting density, hydration, and temperature [37]. The grafting density of the PNIPAAm layer was shown to influence chain mobility and to have an effect on the temperature-dependent surface wetting [102, 168]. PNIPAAm layers grafted onto PS, and of thickness 20–30 nm, were divided into two regions: at the PS interface, the PNIPAm chains are highly hydrophobic and aggregated, whilst at the top-most surface the PNIPAAm chains have a restricted mobility and limited hydration – which possibly enable a temperature-based regulation of cell attachment and detachment [37, 72]. In contrast, PNIPAAm layers of thickness >30 nm are constituted by the regions as described above, but with an additional region at the top-most surface of relatively hydrated and less restricted PNIPAAm chains, which do not permit strong cell–surface interactions, and so influence cell adhesion [37].

A remarkable contrast was observed when PNIPAAm was grafted onto glass substrates instead of PS, with no cell attachment being observed above the LCST for grafting densities of 1.28 µg cm^{-2} [199]. For the same grafting density of 1.4 µg cm^{-2} the grafted layer thickness was larger on PS (15.5 nm) than on glass (8.8 nm), which indicated that the PNIPAAm grafted onto glass was extremely compact [103, 199]. The fact that no cell attachment was observed for lower grafting densities onto glass substrates than onto PS, indicated that the surface properties of the underlying substrate were playing a role in the properties of the grafted layer. In the case of glass substrates, less dehydration at the surface would be expected as compared with PS, due to the presence of silanol groups; consequently, a thinner, more dense layer with a restricted molecular motion would provide temperature-dependent cell adhesiveness.

Temperature-triggered cell attachment and release was also found to depend on the layer thickness and amount of PNIPAAm when it was grafted onto PS surfaces by surface-initiated ATRP; in this case, the number of adhered cells increased with decreasing PNIPAAm layer thickness [200]. An opposite tendency was observed, however, when PNIPAAm was grafted by ATRP onto silicon substrates (i.e., the number of attached cells increased with increasing layer thickness). These different tendencies may, again, be attributable to the different properties of the underlying substrates [201]. In analogy to the situation observed for the electron beam grafting of PNIPAAm onto PS surfaces [103], the cell attachment/detachment behavior of PNIPAAm that had been grafted by ATRP was seen to depend not only on the amount of PNIPAAm that had been grafted, but also on the layer thickness [200].

As noted in Section 2.1, the copolymerization of NIPAAm with hydrophobic or hydrophilic comonomers allows the generation of thermoresponsive copolymers with transition temperatures that are tunable over a broad range [56]. A variety of different monomers have been copolymerized with NIPAAm and used in cell sheet engineering. For example, the addition of poly(ethylene glycol) (PEG) permitted an adjustment of the LCST to biologically relevant temperatures. Graft copolymers of PNIPAAm or poly(N,N-diethylacrylamide) (PDEAAm) as components of the polymer backbone, and oligo(ethylene glycol) or PEG as side chain, were synthesized by free radical copolymerization [75, 202] with an optimized composition for a desirable hydrophilic/hydrophobic balance and an adjustment of the transition temperature to 37 °C. When the copolymers were successfully immobilized using a low-pressure plasma treatment [76], the resultant films maintained their temperature-dependent properties [76] and were utilized as fast responsive surfaces for temperature-triggered cell detachment [71, 176]. The incorporation of PEG was shown to allow the surface to maintain its gel properties in the collapsed state, as it did not allow for complete water extraction above the LCST. The gels produced were nontoxic, and permitted cell attachment, spreading and proliferation below the LCST, indicating that the PEG content (ca. 19 wt%) did not impair cell adhesion when the hydrogel was in the collapsed state [176]. When the temperature was decreased by only a few degrees (from 37 °C to 34 °C) – and long before the hydrogel was completely swollen – the cells became

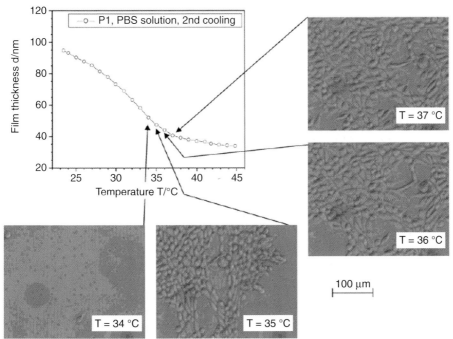

Figure 3.3 Micrographs of mouse fibroblast during cell detachment when decreasing the temperature: correlation to the film thickness of the hydrogel. Cells were cultured overnight and then placed in a Zeiss Cell Observer; the temperature was reduced at a rate of ca. 0.1 K min^{-1}. The temperature-dependence of the film thickness was measured using *in situ* spectroscopic ellipsometry. Reproduced with permission from Ref. [176]; © 2003, American Chemical Society.

detached from the surface within a few minutes (Figure 3.3). This immediate response of the cells towards temperature variation was attributed to the increased mobility of the hydrophilic PEG contained in the hydrogel diffusing towards the surface.

The acceleration of cell detachment from NIPAAm-containing surfaces by incorporating ionic groups or PEG was also explored by Okano and coworkers [204]. Interestingly however, in those experiments the incorporation of amounts of PEG as low as 0.5 wt% had a dramatic effect on cell adhesion [204]. This difference can be explained by the different cell types used in both studies, which may behave differently, and also to the possibly differing graft architecture of the plasma-immobilized films exhibiting a strong interaction of the PEG unit with the PNIPAAm (or PDEAAm) above the transition temperature [176].

Research has also been directed toward lowering the LCST, by incorporating hydrophobic monomers (e.g., *n*-butyl methacrylate; BMA). The temperature transition of the generated poly(NIPAAm-*co*-BMA) films, and the magnitude of hydrophilic to hydrophobic changes, were each decreased with increasing the content of the hydrophobic monomer [24]. These effects were attributed to a suppression of the cooperative hydration of NIPAAm in the vicinity of BMA, and to the hydrophobic interactions produced by the BMA domains leading to aggregation of the BMA component [24]. Cell detachment studies showed that the time required for cell sheet detachment increased with increasing BMA content, although this may have been accelerated by decreasing the treatment temperature [24].

As lengthy treatments at low temperature may have negative effects on cell function, a rapid detachment and recovery of cell sheets is required to maintain the cell phenotype and biological functions. Much research has been directed towards developing strategies to improve the response rate of PNIPAAm hydrogels (for a review, see Ref. [81]). In order to accelerate cell sheet detachment, microporous culture membranes were grafted with PNIPAAm by electron beam irradiation [205]. Subsequently, cells cultured on PNIPAAm-grafted membranes detached more rapidly than from PNIPAAm-grafted PS culture dishes. The kinetics of cell detachment was enhanced when using porous membranes, as water for hydration of the grafted PNIPAAm was supplied not only by diffusion from the periphery of the cell sheet, but also from the pores beneath adherent cells. The rapid access of water molecules through the pores beneath the cells facilitated a rapid hydration of the grafted PNIPAAm, thus accelerating cell sheet detachment [205]. The introduction of a limited amount of PEG chains on the microporous culture surfaces caused a dramatic enhancement of the diffusion of water molecules, decreasing the time necessary for cell detachment as compared with grafted membranes without PEG moieties [198].

The incorporation of matrix polymers and growth factors into the surface-immobilized hydrogel enhanced the functionalities of these coatings [206–211]. For example, the introduction of reactive carboxylate functional groups by using the comonomers 2-carboxyisopropylacrylamide or 3-carboxy-n-propylacerylamide [204, 210] enabled the immobilization of biomacromolecules via amide bonding. The immobilization of RGD cell adhesion peptides (example: Arg-Gly-Asp-Ser) onto reactive thermoresponsive surfaces facilitated cell spreading under culture conditions in the absence of serum, at physiological temperature [212]. The decrease in temperature resulted in the spontaneous detachment of the cells as the hydrated grafted polymer chains dissociated the immobilized RGDS from the cell-surface integrins. The coimmobilization of RGDS, and of the cell growth factor insulin (INS), onto thermoresponsive surfaces resulted in both a facilitated initial cell adhesion and an induction of cell proliferation (shortening the culture time) above the LCST [211]. When decreasing the temperature, the polymer conformational changes weakened the interactions of the surface-immobilized molecules and respective receptors on the cell membrane, thus inducing cell detachment [211].

Recently, cell sheet engineering based on a novel genetically engineered ECM protein containing ELP was developed [159]. The genetic engineering of proteins enables the modification of a constructed thermoresponsive ECM with functional domains. Investigations using biocompatible coatings fabricated by the layer-by-layer (LBL) assembly of ELP-polyethyleneimine (PEI) and ELP-PAAc [160] showed that cell proliferation, focal adhesion and cytoskeletal organization each depended on the number of bilayers constituting the coating. The advantages of ELP-based coatings fabricated by LBL technology offer new possibilities for investigating the influence of layer thickness and the mechanical properties of responsive coatings on cellular responses.

Heterotypic cell interactions are fundamental to achieve and maintain specific functions in tissues and organs, as these modulate cell growth, migration, and/or differentiation; they are, therefore, fundamental to regenerative medicine [213]. The fabrication of patterned grafted PNIPAAM surfaces generated by electron beam radiation using mask patterns enabled the coculture of heterotypic cells and the recovery of cocultured cell sheets [214, 215]. The combination of controlled micropatterning of different cells and three-dimensional (3-D) cell sheet tissue engineering allowed the generation of multilayered tissue constructs, without the need for scaffolds or acellular materials [216]. Controlled cell adhesion, growth and thermally triggered detachment were further achieved using advanced patterned functionalized thermoresponsive surfaces [217]. The site-selective biofunctionalization of patterned carboxyl-functional thermoresponsive polymers with RGDS and/or cell growth factors (e.g., insulin) induced site-selective cell adhesion and growth, along with patterned biofunctional domains. The biomolecular patterned thermoresponsive surface allowed the fabrication of either contiguous cell monolayers or mesh-like monolayer tissues, thus revealing a multifunctional potential in cell culture technology [217].

In order to precisely control the structural organization of cell sheets, microtextured thermoresponsive substrates were fabricated by grafting PNIPAAm (via electron beam radiation) onto microtextured PS produced by hot embossing [218]. This combination of surface texturing providing guidance cues for a precise control of cellular organization and temperature sensitivity, allowing for the detachment and harvest of cultured cell sheets. This made possible the generation of cell sheets with a defined organization; indeed, the cell sheets maintained their defined tissue organization when transferred from patterned to nonpatterned surfaces [218].

Harvesting of the ECM deposited during cell culture, together with the released cell sheets from the thermoresponsive surfaces, enables the ready adhesion of harvested cells to various surfaces, as well as to other cell sheets and host tissues. This facilitated adhesion results in a very high efficiency in cell delivery to host tissues by using cell sheet engineering transplantation [36]. Cell sheets fabricated through cell sheet engineering can be directly transplanted into host tissues, without the need to use scaffolds or carriers; alternatively, they can be layered to create 3-D tissue-like structures [33, 34, 38]. By using a direct transplantation process, corneal epithelium [194], bladder urethelium, and periodental ligaments have each been reconstructed. Likewise, 3-D structures can be created by the layering of homotypic cell sheets; an example is the re-creation of cardiac muscle [192, 219, 220]. The heterotypic layering of cell sheets allows the engineering of higher (organ-like) structures [216], an example being kidney glomeruli.

3.4.3
Biofilm Control

Natural and artificial surfaces are rapidly colonized by microorganisms when immersed in natural aqueous environments [221–223]. The formation of biofilms is highly advantageous for the species involved, as it allows survival under hostile environments (e.g., protection against dehydration, predation and antimicrobial agents) and provides optimal conditions in terms of available nutrients (due to the relative higher concentration of nutrients at the solid/liquid interface). Biofilms are formed in a wide variety of surfaces, including medical devices (e.g., implants) and industrial systems (e.g., cooling systems). Unfortunately, biofilms may have a severe negative impact, as they can reduce the lifetime and operation of devices, increase the operating and maintenance costs and, in the case of biomedical surfaces, cause infection and pain; in severe cases they may endanger human life.

The potential of stimuli-responsive polymer coatings as bacterial anti-biofouling agents was first investigated by the group of G. Lopez during the late 1990s [101, 177, 224]. In these studies, PNIPAAm was surface-grafted onto PS surfaces, after which the effect of surface hydration upon temperature decrease on the attachment of two bacterial species which differed in the degree to which they attach to hydrophilic and hydrophobic surfaces (*Halomonas marina* and *Staphylococcus epidermis*) [225], was investigated [224]. Interestingly, the marine bacterium *H. marina* (Gram-negative), which had been observed to more readily attach to hydrophobic surfaces than to hydrophilic surfaces [225], attached onto the collapsed PNIPAAm-grafted surface at 37 °C and was released upon lowering the temperature by rinsing at 4 °C (i.e., upon hydration of the surface). The opposite effect was observed for *S. epidermis*, which attached readily onto the hydrated PNIPAAm-grafted surface and detached upon increasing the surface temperature (i.e., on hydrogel collapse). The attachment of *S. epidermis* to the surfaces was seen to correlate directly with the hydrophilicity of the surface, with a preference for hydrophilic surfaces [225, 226]. A similar preference was observed for the food-borne pathogen *Listeria monocytogenes*, the attachment of which was shown to be affected by the phase transition of PNIPAAm copolymer surfaces, with a higher bacterial attachment on surfaces below the LCST (i.e., to more hydrophilic surfaces) [227]. Recently, carboxyl-terminated thermoresponsive polymers with LCST-values of 20 °C, 32 °C, and 42 °C, all of which were prepared by free-radical polymerization of NIPAAm, were grafted onto aminofunctionalized surfaces and used to investigate the effect of the phase transition of the polymers on the attachment of the pathogens *Salmonella typhimurium* and *Bacillus cereus*. The results showed a higher attachment of both bacterial strains above the LCST, revealing a preference of the strains tested for hydrophobic surfaces [66, 113]. A short-term reversibility of bacterial attachment to the thermoresponsive surfaces was also demonstrated. Although the long-term effects remain to be investigated, these are expected to require surfaces capable of responding to adaptive adhesion mechanisms.

The use of stimuli-responsive surfaces in investigations with microorganisms is especially important, since such surfaces can be used as probes of biological adhesion mechanisms. Indeed, possible applications in biotechnology may be envisaged, for example to reduce biofouling in industrial applications. These dynamic surfaces may enable the removal of formed biofilms, simply by changing the temperature of the washing solutions, without any need to use antimicrobial agents or surfactants. Clearly, this would represent an environmentally friendly approach to classical methods of biofilm control.

3.4.4
Cell Sorting

The sorting of specific cells from multicellular tissues or aggregates is essential for the development of engineered tissues. Existing cell-sorting techniques include density gradient isolation via centrifugation and fluorescence-activated cell sorting (FACS) [228]. Recently, a novel method based on surfaces containing a mixture of the cell adhesive PNIPAAm-gelatin and nonadhesive PNIPAAm was introduced by Matsuda and coworkers [229]. These authors were able to sort two bovine vascular cell types (endothelial cells and smooth muscle cells) of different adhesiveness to well-defined, mixed coatings. This was achieved by seeding and culturing the cell mixture at physiological temperature, with subsequent release of endothelial cells by lowering the temperature to below the LCST (i.e., to room temperature). Sequential procedures of culturing the collected cells at 37 °C and the release of target cells at room temperature resulted in highly pure cultures of target cells. The mechanisms by which both cell types adhered to coatings containing variable mixing ratios of cell-adhesive and non-cell-adhesive materials remains unclear, but it is believed to be related to the different amounts/densities of cell membrane adhesion receptors (integrins) [229].

In a different approach, thermoresponsive membranes constituted of PNIPAAm grafted onto polypropylene (PP) membranes by plasma-induced graft polymerization [230] were used to adsorb monoclonal antibodies to allow for the capture of specific cell types from a cell suspension, thus enabling cell separation and/or enrichment [231, 232]. The simple adsorption of antibodies onto the PNIPAAm-g-PP membrane effectively enabled the binding of target cells, which could easily be recovered by lowering the temperature below the LCST, as a result of the hydration of PNIPAAm and increase in surface hydrophilicity [232].

3.4.5
Stimuli-Modulated Membranes

Modification of the surface of porous membranes with stimuli-responsive polymers allows the membrane pore size to be controlled, due to the coil-to-globule transition of the responsive polymer. Stimuli-responsive polymers have been widely used as responsive valves to control diffusion and permeation [19, 51, 100, 233–235]. (For a recent review, see Ref. [19].) The grafting of PNIPAAm onto porous membranes allows control of the solute diffusion rate, with higher diffusion at temperatures above the LCST as a result of polymer collapse and opening of membrane pores [19, 51]. For example, a composite material consisting of PNIPAAm grafted onto sintered glass filters was developed to function as a membrane of controllable permeation [184]. The on-off control of permeation was possible by temperature variation. Below the LCST, the expansion of PNIPAAm resulted in a closing of the pores of the sintered glass filter, and consequently in a decrease of the permeation rate. An increase in temperature above the LCST resulted in chain collapse, an opening of pores, and an increase in permeation rate [184]. The opposite effect was observed with drug delivery capsules, which are coated with PNIPAAm, since drug release from the capsule interior was hampered with increasing temperature as result of polymer aggregation above the LCST; closing the pores on the capsule surface and blocking solute diffusion [236]. Porous polypropylene membranes were grafted with PNIPAAm for the separation of macromolecules from solutions containing hydrophilic and hydrophobic solutes [237]. Above the LCST, the surface of the membrane pores were hydrophobic, which resulted in an adsorption of the hydrophobic solutes onto the pore surface and a diffusion of hydrophilic solutes. However, by decreasing the temperature below the LCST, the hydrophilicity of the surface of the pores increased the desorption of adsorbed hydrophobic solutes, which condensed in the permeate side [237]. The grafting of membranes with responsive polymers has revealed certain drawbacks, including modifications of the pore size, pore size distribution and nonhomogeneous grafting densities between the pore interior and exterior. Consequently, recent investigations have been directed towards the fabrication of responsive membranes (e.g., by phase inversion) [58, 238, 239]. One alternative approach involved the fabrication of polymeric composite membranes; for example, nanoparticles of poly(NIPAAm-co-MAAc) were embedded into a matrix of a hydrophobic polymer. Then, in response to external stimuli (e.g., temperature, pH), the variation in particle size caused opening/closing of the matrix channels, controlling the permeation of various proteins and peptides [240].

3.4.6
Chromatography

Temperature- and pH-responsive polymers are the most widely used stimuli-responsive polymers in applications related to the recovery of target molecules (e.g., proteins and drugs) [50]. Hence, particular attention has been focused on their potentialities in chromatographic separation [50, 51, 53, 74, 93, 126, 163, 241–249].

The temperature-dependent hydrophilicity/hydrophobicity of surface-immobilized PNIPAAm has been used to fabricate high-performance liquid chromatography (HPLC) packing materials to separate a wide variety of molecules [53, 93, 247, 250, 251]. Below the LCST the polymer chains are hydrated, and therefore the analytes will interact weakly with the polymer surface; this will result in short retention times and poor resolution. However, when the temperature is raised above the LCST the polymer collapses, increasing the hydrophobic interactions between the analytes and the thermoresponsive stationary phase. This results in increased retention times allowing the resolution of the components of the solution. The retention times are increased as the hydrophobicity of the components is increased; in other words, the elution order of the components reflects their hydrophobic properties. The applicability of PNIPAAm-based coated surfaces as a stationary phase for chromatographic separation was demonstrated for different types of samples, including steroids, peptides, proteins, enzymes, and environmentally relevant samples [52, 53, 203]. In the case of complex samples consisting of components of polarity which vary over a wide range, separation with the required resolution within a reasonable time frame, can be achieved by using temperature gradients. For example, a complex mixture of steroids was successfully resolved by separating the components of lower hydrophobicity at high temperatures. The subsequent temperature decrease weakened the hydrophobic interactions of the remaining components with the surface of the stationary phase, thereby decreasing their retention times [53].

The use of temperature-controlled stationary phases for chromatographic separation has the advantage of enabling the separation of biomolecules in an aqueous environment and under isocratic conditions, avoiding the denaturation of proteins and peptides. The separation of bioactive molecules (e.g., enzymes) without any loss of bioactivity has been successfully achieved. Additionally, thermoresponsive chromatography columns can be easily cleaned by washing the column with water at a low temperature, as the increasing surface hydrophilicity will allow for the release of any remaining adsorbed biomolecules. Moreover, as there is no need to use organic solvents, the method is environmentally advantageous.

The selection of the polymer grafting method determines the molecular architecture of the grafted thermoresponsive layers, and is critically important for defining the separation capacity of the grafted column. As mentioned above, different grafting methods lead to different grafted polymer configurations, which in turn greatly influences the temperature-dependent physico-chemical properties (such as wettability) of the surface. Consequently, the grafting architecture of the grafted polymer influences solute elution [51, 102]. For example, the introduction of freely mobile PNIPAAm-grafted chains onto multipoint immobilized PNIPAAm showed improved temperature-dependent changes in wettability and thickness [102]. As a result, longer retention times were observed on the surface following the inclusion of freely mobile chains onto the multipoint immobilized PNIPAAm. In the case of thin hydrogel layers, a 3-D crosslinked structure with limited mobility resulted in a peak broadening that promoted the partitioning of analytes [51, 102].

The modification of chromatography column materials with PNIPAAm-based polymers has been mainly achieved by using "grafting-to" procedures [203, 248]. The grafting of pure PNIPAAm may not achieve acceptable levels of biomolecular separation, possibly due to an insufficient dehydration of the PNIPAAm chains below the LCST. Hydrophobic comonomers were therefore introduced to improve the hydrophobicity of NIPAAm. The introduction of hydrophobic comonomers will increase the hydrophobic interactions of the surface with analytes and, as a result, improve the separation. For example, polymers composed of BMA and NIPAAm [93, 203], or poly(N-(N'-alkylcarbamidepropyl methacrylamide) analogues [93, 203] showed promising potential for use as matrices for the chromatographic separation of several biomolecules. It should be noted however that, as discussed in Section 2.1, the introduction of hydrophobic comonomers has the additional advantage of allowing the LCST to be tuned to lower temperatures.

Recent investigations have been directed towards the fabrication of thermoresponsive surfaces for separation by using a "grafting-from" approach, for example via ATRP [74, 163] or RAFT polymerization [162]. The "grafting-from" method has the advantage of permitting larger grafting densities, with amounts of PNIPAAm grafted onto silica beads by ATRP approximately 10-fold higher than when prepared using conventional methods [163]. The resultant enhanced hydrophobicity of the densely grafted surfaces above the LCST enabled the separation of peptides [163], without any need to incorporate hydrophobic comonomers, as was necessary when using other grafting methods [203]. ATRP polymerization allows the regulation of grafting density by varying the concentration of the ATRP initiator on the surface [74], and of the grafted chain length by varying the polymerization time [163]. For example, a variation in chain length of the grafted PNIPAAm allows an adjustment of the temperature-dependent interactions with analytes. In this case, short chains would not allow sufficient dehydration and aggregation such that the separation of steroids could not be achieved, whereas long, densely packed chains allowed such separation with high resolution [163].

As discussed in Section 3.4.2, the temperature-dependent changes in the surface properties of PNIPAAm grafted surfaces appear to also depend on the properties of the underlying substrate, and this is reflected in cell adhesion and detachment. Thus, it was hypothesized that a possible influence of substrate polarity on the phase transition behavior of grafted PNIPAAm might be reflected on the separation efficacy of biomolecules. Nagase and coworkers used PNIPAAm brushes grafted onto mixed silane self-assembled monolayers (SAMs) to investigate the effect of substrate properties on the retention of steroids [252]. The results showed the retention time of hydrophobic steroids to be increased with decreasing interface polarity, due to an enhanced hydration of the PNIPAAm brushes when grafted onto polar surfaces, when compared to apolar surfaces. As a consequence of this greater temperature-dependent hydration/dehydration of the PNIPAAm layers, wider variations in retention factor related to temperature were observed on the polar grafted interface. Clearly, the polarity of the grafted interface plays an important role in the temperature-dependent wetting properties of the grafted PNIPAAm, and consequently on temperature-modulated separation [252].

The simultaneous modulation of electrostatic and hydrophobic interactions with temperature is of great interest in the separation of biomolecules. The introduction of weakly charged comonomers into PNIPAAm, allows to create surfaces that

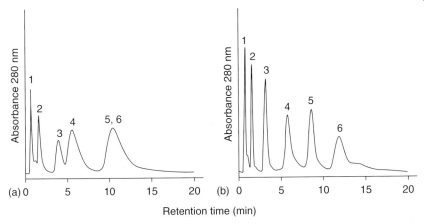

Figure 3.4 Chromatograms of a mixture of phenylthiohydantoin (PTH)-amino acids at: (a) 10 °C; (b) 50 °C on a P(NIPAAm-co-tBAAm-co-AAc) gel column. Peaks: 1 = Asp; 2 = Asn; 3 = His; 4 = Met; 5 = Arg; 6 = Phe. Flow-rate 1.0 ml min^{-1}; Eluent PBS (pH 6.0, I = 0.1); Detection UV 280 nm. Reproduced with permission from Ref. [245]; © Elsevier.

respond to temperature variations with changes in hydrophilicity/hydrophobicity, as well as with changes in surface charge density. Bioactive ionic compounds have been successfully separated using copolymer surfaces by introducing either anionic monomers (e.g., acrylic acid) or cationic monomers [2-(dimethyl-amino) ethyl methacrylate or N,N-(dimethyl-amino)propyl acrylamide] into PNIPAAm [126, 244, 249, 250, 253–256]. Below the LCST, the polymer chains are hydrated, which allows electrostatic interactions to occur between the charged groups on the immobilized polymer chains and the charged analytes. By raising the temperature above the LCST, the ion-exchange groups on the polymer chain surface become hidden due to chain collapse, and this in turn changes the surface from charged to uncharged [52, 245, 256]. By changing the temperature in this way it is possible to trigger changes in both the charge and hydrophobicity of the pH- and temperature-responsive polymers. Thermoresponsive surfaces containing ionic-charged monomers can be used to separate components of similar hydrophobicities, but with differing ion-exchange groups. For example, poly(NIPAAm-co-tBAAm-co-Aac)-immobilized layers allowed the separation of a mixture of phenylthiohydantoin (PTH)-amino acids [245]. Whilst the retention time of apolar amino acids was increased with increasing temperature, the inverse was observed for basic amino acids (Figure 3.4). The results indicated that the elution behavior of nonpolar amino acids might be attributable to hydrophobic interactions, whereas the elution of polar amino acids was due to electrostatic interactions.

Attempts have also been made to develop size-selective separation media based on thermoresponsive polymers [50, 51, 257–260]. For example, the use of silica beads grafted with PNIPAAm-based polymers as a stationary phase for high-performance size-exclusion chromatography showed temperature-responsive elution changes for low-molecular-weight proteins and polysaccharides [259]. However, the elution was found to be temperature-independent for high-molecular-weight substances. A higher resolution was observed below the LCST, possibly due to the hydrophilic properties of the surface beads, which improved porosity [50, 259]. For some proteins, the elution time was extended when the temperature was increased, possibly due to hydrophobic interactions between the protein and the collapsed polymer surface [51]. In another example, hydroxypropylcellulose beads exhibiting temperature-dependent porosity were shown to be promising for chromatographic applications since, by raising the temperature a decrease in pore size due to shrinkage of the beads allowed the separation of proteins [258].

Bioconjugates, as proteins conjugated with PNIPAAm, were revealed as showing promise for the affinity separation of biomolecules in immunoassays and enzyme recovery [54, 121, 231]. Affinity separation consists basically of the separation of target biomolecules from a complex mixture via their specific interaction with immobilized affinity moieties. The formation of conjugates of biomolecules and thermoresponsive polymers allows the one-step separation of biomolecules such as proteins and antibodies, simply by cyclical heating and cooling [1, 17, 54, 127]. In general, bioconjugates can function in solution and be recovered by precipitation as a result of temperature changes, although they may also be used whilst immobilized on surfaces [261, 262]. For example, the possibility of using temperature-controllable molecular recognition based on masking and forced-release effects was investigated by Okano and coworkers, who independently grafted PNIPAAm and the ligand Cibacron Blue F3G-A (CB) onto a matrix surface [261]. Under optimal immobilization conditions (i.e., density, spacer length, and polymer) the PNIPAAm collapsed chains below the LCST, which permitted an easy access of the target molecules (albumin) in solution to the immobilized ligands. On decreasing the temperature, the PNIPAAm chains extended, both sheltering the immobilized ligand (which made access of the target molecules to the ligand very difficult) and forcing out any previously bound target molecules (forced-release) [261].

For applications of thermoresponsive polymers in separation applications, an enhanced kinetics of swelling/deswelling (i.e., thermosensitivity) is usually desirable, in order to improve resolution and selectivity [50]. The thermosensitivity of PNIPAAm polymer systems can be improved by copolymerization with hydrophilic monomers (such as acrylic acid) or by preparing comb-like structures which create hydrophobic regions that facilitate water extraction upon polymer collapse [263].

3.4.7
Microfluidics and Laboratory-on-a-Chip

A temperature-dependent microfluidic chromatographic matrix consisting of PNIPAAm-coated latex beads has proven capable of separating biomolecules from a flow stream, revealing a high potential for use in diagnostics [241, 242]. PNIPAAm-coated beads were functionalized with an affinity moiety (biotin [242]) and flowed through a microfluidic channel. On raising the temperature above the LCST the beads aggregated adhering to the channel walls. These beads were shown of binding target molecules (streptavidin [242]) which subsequently was flowed through the channel. The coated beads and captured biomolecules were easily eluted by lowering the ambient temperature to below the LCST. This easy removal of the matrix was advantageous, as it not only allowed the beads to be reused but also added flexibility to the microfluidic devices. A single device can be used to separate different target molecules, depending on the packed matrix [242]. The smart microfluidic system was further improved by developing a complementary switchable surface trap [264] by grafting PNIPAAm onto the walls of the poly(dimethyl siloxane) (PDMS) channels. Above the LCST, a more uniform bead deposition was observed on the PNIPAAm-grafted channel walls as compared to the nongrafted walls. The PNIPAAm-grafted walls allowed a faster and more complete bead release when the temperature was decreased below the transition temperature [264].

Huber and coworkers have developed a microfluidic device that allows the programmed adsorption and release of proteins [25]. For this, a 4 nm-thick coating was generated by *in situ* polymerization of NIPAAm on functionalized SAMs, and integrated onto a microfluidic hot plate device. The device allowed the adsorption of proteins, their retention with negligible denaturation, and their release upon thermal switching. The rapid response characteristics of the device (<1 s) can be manipulated for proteomic functions, including the preconcentration and separation of soluble proteins on an integrated fluidic chip [25].

The potential for using ELP nanostructures as reversible smart biomolecular switches for the "on-chip" capture and release of target proteins from complex mixtures in bioanalytical devices, has been investigated by fabricating a nanoscale array of ELPs (Figure 3.5) [154]. The phase-transition behavior of the ELPs was used to reversibly immobilize a thioredoxin-ELP (Trx-ELP) fusion protein onto the ELP nanopattern above the LCST. The immobilized protein remained bioactive, which allowed the binding of an antithioredoxin (anti-Trx) antibody. This concept was further used to reversibly bind proteins directly from cell lysates [155]. The incubation of a patterned ELP surface into a solution of cell lysate containing an expressed ELP fusion protein resulted, upon temperature transition, in the capture of the ELP fusion protein onto the immobilized ELP. The captured proteins allowed the binding of its target from solution. The bound complex could be reversibly dissociated below the LCST [135, 154, 155], revealing a great potential for the use of immobilized responsive polypeptides in the fabrication of regenerable protein arrays for bioanalytic devices.

Stimuli-responsive polymer coatings have been also used for fluid flow control, acting as both sensors and actuators. For example, by coating the interior of capillaries with PNIPAAm, fluid flow could be controlled by varying the temperature, due to the temperature-dependent wetting of the grafted capillary surface [265, 266]. Using a different approach, Beebe and coworkers have incorporated

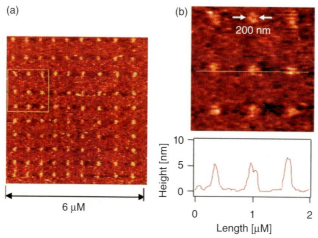

Figure 3.5 (a) Atomic force microscopy (AFM) tapping mode height image of a 10 × 9 ELP dot array in PBS buffer at room temperature; (b) Enlarged view of the area indicated in panel (a) and a representative cross-section, showing a typical feature height of 5–6 nm and a lateral feature size of approximately 200 nm. Reproduced with permission from Ref. [145]; © 2004, American Chemical Society.

responsive hydrogels into a microfluidic device by direct *in situ* photopatterning, and shown that the immobilized hydrogel can be used as an actuatable valve for fluid flow control [267]. The potentialities of controlling fluid flow in microchannels (including the control of flow rate and flow direction) as a result of the stimuli-dependent expansion and contraction of hydrogel-coated structures were nicely shown by these authors, using different designs [267]. The coating of nanopores with stimuli-responsive polypeptide chains to provide a novel mechanism for fluid flow control via the helix-coil transition of the grafted chains was also recently demonstrated [268]. Temperature- and pH-responsive hydrogels have also been used in the fabrication of autonomous micropumps and micromixers, by exploring the volume changes experienced by the responsive hydrogels under stimuli [269]. Recently, PNIPAAm was used to fabricate bidirectional actuating flaps to control a microchannel concentrator [270]. Some further developments of autonomous microfluidics incorporating stimuli-responsive hydrogels are highlighted in a recent review [59].

3.5
Summary and Future Perspectives

During the past few years, much effort has been expended towards the generation of smart switchable surfaces that mimic biological systems and enable the modulation of complex systems. The unique dynamic and controllable properties of stimuli-responsive materials have not only made possible the detailed investigation and control of cell–cell, cell–protein and protein–surface interactions, but have also found diverse applications that include controlled drug release, cell culture, chromatography, biosensing, and microfluidics. Progress in the field of cell sheet engineering for regenerative therapies has been particularly impressive. The use of surface-immobilized responsive layers in cell culture technologies is highly beneficial due to their advantageous mechanical properties that allow cellular matrix reorganization while retaining cell anchorage, to the facilitated diffusion and delivery of nutrients and growth factors, and their ability to release cell sheets without damage and without altering the cells' properties. The possibility of incorporating different bioactive molecules on the surface of immobilized layers will open new perspectives for the control of specific cell–protein interactions and the regulation of functions between cell-surface proteins and the corresponding ligands. The recent development of patterned coculture methods, using thermoresponsive polymers, will allow the further investigation of cell–cell communication and angiogenesis, the control of cell migration, and the fabrication of complex tissue grafts for regenerative medicine. New, elegant concepts for the design of intelligent materials and surfaces are continuously being reported to improve the properties (as higher biocompatibility or enhanced mechanical properties) and to extend the applicability of these materials. It is expected that surfaces of increased complexity will continue to be generated, aimed at a close resemblance to biological systems. Today, exciting new perspectives lie ahead, notably to use smart surfaces for the fabrication of self-controlled biosensors, for the controlled release of molecules in medicine and biotechnology, and for the modulated fabrication of complex tissues for regenerative medicine. The combined efforts of scientists from a wide variety of disciplines, including chemistry, materials science, engineering, biology and medicine, will enable the exploration and expansion of the already wide potential of this ongoing field.

Acknowledgments

Financial support from the AMBIO project (NMP4-CT-2005-011827), funded by the 6th framework program of the European Community, is gratefully acknowledged.

References

1 Hoffman, A.S. (2000) Bioconjugates of intelligent polymers and recognition proteins for use in diagnostics and affinity separations. *Clinical Chemistry*, **46** (9), 1478–86.
2 Klouda, L. and Mikos, A.G. (2008) Thermoresponsive hydrogels in biomedical applications. *European Journal of Pharmaceutics and Biopharmaceutics*, **68** (1), 34–45.
3 Bromberg, L.E. and Ron, E.S. (1998) Temperature-responsive gels and thermogelling polymer matrices for protein and peptide delivery. *Advanced Drug Delivery Reviews*, **31** (3), 197–221.

4 Jeong, B. and Gutowska, A. (2002) Lessons from nature: stimuli-responsive polymers and their biomedical applications. *Trends in Biotechnology*, **20** (7), 305–11.

5 Minko, S. (2006) Responsive polymer brushes. *Polymer Reviews*, **46** (4), 397–420.

6 Shimoboji, T., Larenas, E., Fowler, T., Kulkarni, S., Hoffman, A.S. and Stayton, P.S. (2002) Photoresponsive polymer-enzyme switches. *Proceedings of the National Academy of Sciences of the United States of America*, **99** (26), 16592–6.

7 Lendlein, A., Jiang, H., Junger, O. and Langer, R. (2005) Light-induced shape-memory polymers. *Nature*, **434** (7035), 879–82.

8 Kulkarni, R.V. and Biswanath, S. (2007) Electrically responsive smart hydrogels in drug delivery: a review. *Journal of Applied Biomaterials and Biomechanics*, **5** (3), 125–39.

9 Lai, J.J., Hoffman, J.M., Ebara, M., Hoffman, A.S., Estournes, C., Wattiaux, A. and Stayton, P.S. (2007) Dual magnetic-/temperature-responsive nanoparticles for microfluidic separations and assays. *Langmuir*, **23** (13), 7385–91.

10 Zhu, X.Y., Mills, K.L., Peters, P.R., Bahng, J.H., Liu, E.H., Shim, J., Naruse, K., Csete, M.E., Thouless, M.D. and Takayama, S. (2005) Fabrication of reconfigurable protein matrices by cracking. *Nature Materials*, **4** (5), 403–6.

11 Tasdelen, B., Kayaman-Apohan, N., Güven, O. and Baysal, B.M. (2004) pH-thermoreversible hydrogels. I. Synthesis and characterization of poly(N-isopropylacrylamide/maleic acid) copolymeric hydrogels. *Radiation Physics and Chemistry*, **69** (4), 303–10.

12 Sheng, D., Ravi, P. and Tam, K.C. (2008) pH-Responsive polymers: synthesis, properties and applications. *Soft Matter*, **4**, 435–49.

13 Mi, P., Chu, L.Y., Ju, X.J. and Niu, C.H. (2008) A smart polymer with ion-induced negative shift of the lower critical solution temperature for phase transition. *Macromolecular Rapid Communications*, **29** (1), 27–32.

14 Holtz, J.H. and Asher, S.A. (1997) Polymerized colloidal crystal hydrogel films as intelligent chemical sensing materials. *Nature*, **389** (6653), 829–32.

15 Miyata, T., Asami, N. and Uragami, T. (1999) A reversibly antigen-responsive hydrogel. *Nature*, **399** (6738), 766–9.

16 Ding, Z., Fong, R.B., Long, C.J., Stayton, P.S. and Hoffman, A.S. (2001) Size-dependent control of the binding of biotinylated proteins to streptavidin using a polymer shield. *Nature*, **411**, 59–62.

17 Stayton, P.S., Shimiboji, T., Long, C., Chilkoti, A., Ghen, G., Harris, J.M. and Hoffman, A.S. (1995) Control of protein-ligand recognition using a stimuli-responsive polymer. *Nature*, **378**, 472–4.

18 Hoare, T. and Pelton, R. (2008) Charge-switching, amphoteric glucose-responsive microgels with physiological swelling activity. *Biomacromolecules*, **9** (2), 733–40.

19 Kumar, A., Srivastava, A., Galaev, I.Y. and Mattiasson, B. (2007) Smart polymers: physical forms and bioengineering applications. *Progress in Polymer Science*, **32** (10), 1205–37.

20 Ahn, S.K., Kasi, R.M., Kim, S.C., Sharma, N. and Zhou, Y.X. (2008) Stimuli-responsive polymer gels. *Soft Matter*, **4** (6), 1151–7.

21 Liu, Y., Mu, L., Liu, B.H. and Kong, J.L. (2005) Controlled switchable surface. *Chemistry – A European Journal*, **11** (9), 2622–31.

22 Mendes, P.M. (2008) Stimuli-responsive surfaces for bio-applications. *Chemical Society Reviews*, **37**, 2512–29.

23 Hoffman, A.S. (2002) Hydrogels for biomedical applications. *Advanced Drug Delivery Reviews*, **54** (1), 3–12.

24 Tsuda, Y., Kikuchi, A., Yamato, M., Sakurai, Y., Umezu, M. and Okano, T. (2004) Control of cell adhesion and detachment using temperature and thermoresponsive copolymer grafted culture surfaces. *Journal of Biomedical Materials Research Part A*, **69A** (1), 70–8.

25 Huber, D.L., Manginell, R.P., Samara, M.A., Kim, B. and Bunker, B.C. (2003) Programmed adsorption and release of proteins in a microfluidic devices. *Science*, 301, 352–4.

26 Mano, J.F. (2008) Stimuli-responsive polymeric systems for biomedical applications. *Advanced Engineering Materials*, 10 (6), 515–27.

27 Chaterji, S., Kwon, I.K. and Park, K. (2007) Smart polymeric gels: redefining the limits of biomedical devices. *Progress in Polymer Science*, 32 (8-9), 1083–122.

28 Schmaljohann, D. (2006) Thermo- and pH-responsive polymers in drug delivery. *Advanced Drug Delivery Reviews*, 58, 1655–70.

29 de las Heras Alarcon, C., Pennadam, S. and Alexander, C. (2005) Stimuli responsive polymers for biomedical applications. *Chemical Society Reviews*, 34, 276–85.

30 da Silva, R.M.P., Mano, J.O.F. and Reis, R.L. (2007) Smart thermoresponsive coatings and surfaces for tissue engineering: switching cell-material boundaries. *Trends in Biotechnology*, 25 (12), 577–83.

31 Hoffman, A.S. (1987) Applications of thermally reversible polymers and hydrogels in therapeutics and diagnostics. *Journal of Controlled Release*, 6 (1), 297–305.

32 Qiu, Y. and Park, K. (2001) Environment-sensitive hydrogels for drug delivery. *Advanced Drug Delivery Reviews*, 53 (3), 321–39.

33 Yang, J., Yamato, M., Nishida, K., Ohki, T., Kanzaki, M., Sekine, H., Shimizu, T. and Okano, T. (2006) Cell delivery in regenerative medicine: the cell sheet engineering approach. *Journal of Controlled Release*, 116 (2), 193–203.

34 Yang, J., Yamato, M., Shimizu, T., Sekine, H., Ohashi, K., Kanzaki, M., Ohki, T., Nishida, K. and Okano, T. (2007) Reconstruction of functional tissues with cell sheet engineering. *Biomaterials*, 28 (34), 5033–43.

35 Voit, B., Schmaljohann, D., Gramm, S., Nitschke, M. and Werner, C. (2007) Stimuli-responsive polymer layers for advanced cell culture technologies. *International Journal of Materials Research*, 98, 646–50.

36 Yamato, M., Akiyama, Y., Kobayashi, J., Yang, J., Kikuchi, A. and Okano, T. (2007) Temperature-responsive cell culture surfaces for regenerative medicine with cell sheet engineering. *Progress in Polymer Science*, 32 (8-9), 1123–33.

37 Kikuchi, A. and Okano, T. (2005) Nanostructured designs of biomedical materials: applications of cell sheet engineering to functional regenerative tissues and organs. *Journal of Controlled Release*, 101 (1-3), 69–84.

38 Yang, J., Yamato, M., Kohno, C., Nishimoto, A., Sekine, H., Fukai, F. and Okano, T. (2005) Cell sheet engineering: recreating tissues without biodegradable scaffolds. *Biomaterials*, 26 (33), 6415–22.

39 Yang, J., Yamato, M. and Okano, T. (2005) Cell-sheet engineering using intelligent surfaces. *MRS Bulletin*, 30, 189–93.

40 Alexander, C. and Shakesheff, K.M. (2006) Responsive polymers at the biology/materials science interface. *Advanced Materials*, 18 (24), 3321–8.

41 Roy, I. and Gupta, M.N. (2003) Smart polymeric materials: emerging biochemical applications. *Chemistry and Biology*, 10 (12), 1161–71.

42 Okano, T., Bae, Y.H., Jacobs, H. and Kim, S.W. (1990) Thermally on-off switching polymers for drug permeation and release. *Journal of Controlled Release*, 11 (1-3), 255–65.

43 Kost, J. and Langer, R. (2001) Responsive polymeric delivery systems. *Advanced Drug Delivery Reviews*, 46 (1-3), 125–48.

44 Kikuchi, A. and Okano, T. (2002) Pulsatile drug release control using hydrogels. *Advanced Drug Delivery Reviews*, 54 (1), 53–77.

45 Chilkoti, A., Dreher, M.R., Meyer, D.E. and Raucher, D. (2002) Targeted drug delivery by thermally responsive polymers. *Advanced Drug Delivery Reviews*, 54 (5), 613–30.

46 Haider, M., Megeed, Z. and Ghandehari, H. (2004) Genetically engineered polymers: status and prospects for

controlled release. *Journal of Controlled Release*, **95** (1), 1–26.
47 Singh, S., Webster, D.C. and Singh, J. (2007) Thermosensitive polymers: synthesis, characterization, and delivery of proteins. *International Journal of Pharmaceutics*, **341** (1-2), 68–77.
48 Chilkoti, A., Christensen, T. and MacKay, J.A. (2006) Stimulus responsive elastin biopolymers: applications in medicine and biotechnology. *Current Opinion in Chemical Biology*, **10** (6), 652–7.
49 Galaev, I.Y. and Mattiasson, B. (1999) "Smart" polymers and what they could do in biotechnology and medicine. *Trends in Biotechnology*, **17** (8), 335–40.
50 Maharjan, P., Woonton, B.W., Bennett, L.E., Smithers, G.W., DeSilva, K. and Hearn, M.T.W. (2008) Novel chromatographic separation – the potential of smart polymers. *Innovative Food Science & Emerging Technologies*, **9** (2), 232–42.
51 Kikuchi, A. and Okano, T. (2002) Intelligent thermoresponsive polymeric stationary phases for aqueous chromatography of biological compounds. *Progress in Polymer Science*, **27**, 1165–93.
52 Kanazawa, H. (2007) Thermally responsive chromatographic materials using functional polymers. *Journal of Separation Science*, **30** (11), 1646–56.
53 Eri, A. and Kanazawa, H. (2006) Aqueous chromatography system using temperature-responsive polymer-modified stationary phases. *Journal of Separation Science*, **29** (6), 738–49.
54 Hoffman, A.S. and Stayton, P.S. (2007) Conjugates of stimuli-responsive polymers and proteins. *Progress in Polymer Science*, **32** (8-9), 922–32.
55 Lutz, J.-F. and Börner, H.G. (2008) Modern trends in polymer bioconjugates design. *Progress in Polymer Science*, **33** (1), 1–39.
56 Rzaev, Z.M.O., Dincer, S. and Piskin, E. (2007) Functional copolymers of N-isopropylacrylamide for bioengineering applications. *Progress in Polymer Science*, **32** (5), 534–95.

57 Allan, S. and Hoffman, P.S.S. (2004) Bioconjugates of smart polymers and proteins: synthesis and applications. *Macromolecular Symposia*, **207** (1), 139–52.
58 Gil, E.S. and Hudson, S.M. (2004) Stimuli-responsive polymers and their bioconjugates. *Progress in Polymer Science*, **29** (12), 1173–222.
59 Dong, L. and Jiang, H. (2007) Autonomous microfluidics with stimuli-responsive hydrogels. *Soft Matter*, **3** (10), 1223–30.
60 Schild, H.G. (1992) Poly (N-isopropylacrylamide): experiment, theory and application. *Progress in Polymer Science*, **17**, 163.
61 Yamada, N., Okano, T., Sakai, H., Karikusaa, F., Sawasakia, Y. and Sakurai, Y. (1990) Thermo-responsive polymeric surfaces, control of attachment and detachment of cultured cells. *Makromolekulare Chemie – Rapid Communications*, **11**, 571–6.
62 Schild, H.G. and Tirrell, D.A. (1990) Microcalorimetric detection of lower critical solution temperatures in aqueous polymer solutions. *Journal of Physical Chemistry*, **94**, 4352–6.
63 Otake, K., Inomata, H., Konno, M. and Saito, S. (1990) Thermal analysis of the volume phase transition with N-isopropylacrylamide gels. *Macromolecules*, **23** (1), 283–9.
64 Ivanov, A.E., Ekeroth, J., Nilsson, L., Mattiasson, B., Bergenståhl, B. and Galaev, I.Y. (2006) Variations of wettability and protein adsorption on solid siliceous carriers grafted with poly(N-isopropylacrylamide). *Journal of Colloid and Interface Science*, **296**, 538–44.
65 Chen, G., Imanishi, Y. and Ito, Y. (1998) Effect of protein and cell behavior on pattern-grafted thermoresponsive polymer. *Journal of Biomedical Materials Research*, **42**, 38–44.
66 Cunliffe, D., de las Heras Alarcon, C., Peters, V., Smith, J.R. and Alexander, C. (2003) Thermoresponsive surface-grafted poly(N-isopropylacrylamide) copolymers: effect of phase transitions on protein and bacterial attachment. *Langmuir*, **19**, 2888–99.

67 Heinz, P., Bretagnol, F., Mannelli, I., Sirghi, L., Valsesia, A., Ceccone, G., Gilliland, D., Landfester, K., Rauscher, H. and Rossi, F. (2008) Poly(N-isopropylacrylamide) grafted on plasma-activated poly(ethylene oxide): thermal response and interaction with proteins. *Langmuir*, **24** (12), 6166–75.

68 Grabstain, V. and Bianco-Peled, H. (2003) Mechanisms controlling the temperature-dependent binding of proteins to poly(N-isopropylacrylamide) microgels. *Biotechnology Progress*, **19**, 1728–33.

69 Wu, J.-Y., Liu, S.-Q., Heng, P.W.-S. and Yang, Y.-Y. (2005) Evaluating proteins release from, and their interactions with, thermosensitive poly (N-isopropylacrylamide) hydrogels. *Journal of Controlled Release*, **102** (2), 361–72.

70 Cheng, X., Canavan, H.E., Graham, D.J., Castner, D.G. and Ratner, B.D. (2006) Temperature dependent activity and structure of adsorbed proteins on plasma polymerized N-isopropyl acrylamide. *Biointerphases*, **1** (1), 61–72.

71 Nitschke, M., Gramm, S., Götze, T., Valtink, M., Drichel, J., Voit, B., Engelmann, K. and Werner, C. (2007) Thermo-responsive poly(NiPAAm-co-DEGMA) substrates for gentle harvest of human corneal endothelial cell sheets. *Journal of Biomedical Materials Research Part A*, **80A** (4), 1003–10.

72 Matsuda, N., Shimizu, T., Yamato, M. and Okano, T. (2007) Tissue engineering based on cell sheet technology. *Advanced Materials*, **19**, 3089–99.

73 Yamato, M. and Okano, T. (2004) Cell sheet engineering. *Materials Today*, **7** (5), 42–7.

74 Nagase, K., Kobayashi, J., Kikuchi, A., Akiyama, Y., Kanazawa, H. and Okano, T. (2008) Effects of graft densities and chain lengths on separation of bioactive compounds by nanolayered thermoresponsive polymer brush surfaces. *Langmuir*, **24**, 511–17.

75 Schmaljohann, D. and Gramm, S. (2002) Novel polymers and hydrogels based on N-alkyl acrylamides and poly(ethyleneglycol). *Polymer Preprints*, **43**, 758–9.

76 Schmaljohann, D., Beyerlein, D., Nitschke, M. and Werner, C. (2004) Thermo-reversible swelling of thin hydrogel films immobilized by low-pressure plasma. *Langmuir*, **20**, 10107–14.

77 Ringsdorf, H., Simon, J. and Winnik, F.M. (1992) Hydrophobically modified poly(N-isopropylacrylamides) in water: a look by fluorescence techniques at the heat-induced phase transition. *Macromolecules*, **25** (26), 7306–12.

78 Feil, H., Bae, Y.H., Feijen, J. and Kim, S.W. (1993) Effect of comonomer hydrophilicity and ionization on the lower critical solution temperature of N-isopropylacrylamide copolymers. *Macromolecules*, **26**, 2496–500.

79 Pareek, P., Adler, H.-J.P. and Kuckling, D. (2006) Tuning the swelling behavior of chemisorbed thin PNIPAAm hydrogel layers by N,N-dimethyl acrylamide content. *Progress in Colloid and Polymer Science*, **132**, 145–51.

80 Yim, H., Kent, M.S., Mendez, S., Lopez, G.P., Satija, S. and Seo, Y. (2006) Effects of grafting density and molecular weight on the temperature-dependent conformational change of poly(N-isopropylacrylamide) grafted chains in water. *Macromolecules*, **39**, 3420–6.

81 Zhang, X.-Z., Xu, X.-D., Cheng, S.-X. and Zhuo, R.-X. (2008) Strategies to improve the response rate of thermosensitive PNIPAAm hydrogels. *Soft Matter*, **4**, 385–91.

82 Nitschke, M., Götze, T., Gramm, S. and Werner, C. (2007) Detachment of human endothelial cell sheets from thermo-responsive poly(NiPAAm-co-DEGMA) carriers. *eXPRESS Polymer Letters*, **1**, 660–6.

83 Cordeiro, A.L., Zimmermann, R., Gramm, S., Nitschke, M., Janke, A., Schaefer, N., Grundke, K. and Werner, C. (2009) Temperature dependent physicochemical properties of poly(N-isopropylacrylamide-co-N-(1-phenylethyl) acrylamide) thin films. *Soft Matter*, **5** (7), 1367–77.

84 Taylor, L.D. and Cerankowski, L.D. (1975) Preparation of films exhibiting a balanced temperature dependence to

permeation by aqueous solutions – a study of lower consolute behavior. *Journal of Polymer Science, Polymer Chemistry Edition*, **13**, 2551–70.

85 Bae, Y.H., Okano, T. and Kim, S.W. (1990) Temperature dependence of swelling of crosslinked poly(N,N-alkyl substituted acrylamides) in water. *Journal of Polymer Science Part B: Polymer Physics*, **28**, 923–36.

86 Xue, W. and Hamley, I.W. (2002) Thermoreversible swelling behaviour of hydrogels based on N-isopropylacrylamide with a hydrophobic comonomer. *Polymer*, **43** (10), 3069–77.

87 Eeckman, F., Amighi, K. and Moees, A.J. (2001) Effect of some physiological and non-physiological compounds on the phase transition temperature of thermoresponsive polymers intended for oral controlled-drug delivery. *International Journal of Pharmaceutics*, **222**, 259–70.

88 Park, T.G. and Hoffman, A.S. (1993) Sodium chloride-induced phase transition in nonionic poly(N-isopropylacrylamide) gel. *Macromolecules*, **26**, 5045–8.

89 Liu, X.-M., Wang, L.S.L.-S., Wang, L., Huang, J. and He, C. (2004) The effect of salt and pH on the phase-transition behaviors of temperature-sensitive copolymers based on N-isopropylacrylamide. *Biomaterials*, **25** (25), 5659–66.

90 Zhang, Y., Furyk, S., Bergbreiter, D.E. and Cremer, P.S. (2005) Specific ion effects on the water solubility of macromolecules: PNIPAM and the Hofmeister series. *Journal of the American Chemical Society*, **127**, 14505–10.

91 Jhon, Y.K., Bhat, R.R., Jeong, C., Rojas, O.J., Szleifer, I. and Genzer, J. (2006) Salt-induced depression of lower critical solution temperature in a surface-grafted neutral thermoresponsive polymer. *Macromolecular Rapid Communications*, **27**, 697–701.

92 Geever, L.M., Nugent, M.J.D. and Higginbotham, C.L. (2007) The effect of salts and pH buffered solutions on the phase transition temperature and swelling of thermoresponsive pseudogels based on N-isopropylacrylamide. *Journal of Materials Science*, **42** (23), 9845–54.

93 Kanazawa, H., Yamamoto, K., Kashiwase, Y., Matsushima, Y., Takai, N., Kikuchi, A., Sakurai, Y. and Okano, T. (1997) Analysis of peptides and proteins by temperature-responsive chromatographic system using N-isopropylacrylamide polymer-modified columns. *Journal of Pharmaceutical and Biomedical Analysis*, **15** (9-10), 1545–50.

94 Cheng, X., Canavan, H.E., Stein, M.J., Hull, J.R., Kweskin, S.J., Wagner, M.S., Somorjai, G.A., Castner, D.G. and Ratner, B.D. (2005) Surface chemical and mechanical properties of plasma-polymerized N-isopropylacrylamide. *Langmuir*, **21**, 7833–41.

95 Canavan, H.E., Cheng, X.-H., Graham, D.J., Ratner, B.D. and Castner, D.G. (2005) Cell sheet detachment affects the extracellular matrix: a surface science study comparing thermal liftoff, enzymatic, and mechanical methods. *Journal of Biomedical Materials Research, Part A*, **75A** (1), 1–13.

96 Pan, Y.V., Wesley, R.A., Luginbuhl, R., Denton, D.D. and Ratner, B.D. (2001) Plasma polymerized N-isopropylacrylamide: synthesis and characterization of a smart thermally responsive coating. *Biomacromolecules*, **2** (1), 32–6.

97 Yamato, M., Konno, C., Kushida, A., Hirose, M., Mika, U., Kikuchi, A. and Okano, T. (2000) Release of adsorbed fibronectin from temperature-responsive culture surfaces requires cellular activity. *Biomaterials*, **21** (10), 981–6.

98 Okano, T., Yamada, N., Okuhara, M., Sakai, H. and Sakurai, Y. (1995) Mechanism of cell detachment from temperature-modulated, hydrophilic-hydrophobic polymer surfaces. *Biomaterials*, **16** (4), 297–303.

99 Curti, P.S., de Moura, M.R., Veiga, W., Radovanovic, E., Rubira, A.F. and Muniz, E.C. (2005) Characterization of PNIPAAm photografted on PET and PS surfaces. *Applied Surface Science*, **245** (1-4), 223–33.

100 Lin, Z., Xu, T. and Zhang, L. (2006) Radiation-induced grafting of N-isopropylacrylamide onto the brominated poly(2,6-dimethyl-1,4-phenylene oxide) membranes. *Radiation Physics and Chemistry*, **75** (4), 532–40.

101 Ista, L.K., Mendez, S., Perez-Luna, V.H. and Lopez, G.P. (2001) Synthesis of poly(N-isopropylacrylamide) on initiator-modified self-assembled monolayers. *Langmuir*, **17**, 2552–5.

102 Yakushiji, T., Sakai, K., Kikuchi, A., Aoyagi, T., Sakurai, Y. and Okano, T. (1998) Graft architectural effects on thermoresponsive wettability changes of poly(N-isopropylacrylamide)-modified surfaces. *Langmuir*, **14** (16), 4657–62.

103 Akiyama, Y., Kikuchi, A., Yamato, M. and Okano, T. (2004) Ultrathin poly(N-isopropylacrylamide) grafted layer on polystyrene surfaces for cell adhesion/detachment control. *Langmuir*, **20** (13), 5506–11.

104 Fu, Q., Rao, G.V.R., Basame, S.B., Keller, D.J., Artyushkova, K., Fulghum, J.E. and Lopez, G.P. (2004) Reversible control of free energy and topography of nanostructured surfaces. *Journal of the American Chemical Society*, **126**, 8904–5.

105 Kairali Podual, N.A.P. (2005) Relaxational behavior and swelling-pH master curves of poly[(diethylaminoethyl methacrylate)-*graft*-(ethylene glycol)] hydrogels. *Polymer International*, **54** (3), 581–93.

106 Peppas, N.A., Hilt, J.Z., Khademhosseini, A. and Langer, R. (2006) Hydrogels in biology and medicine: from molecular principles to bionanotechnology. *Advanced Materials*, **18** (11), 1345–60.

107 Markland, P., Zhang, Y.H., Amidon, G.L. and Yang, V.C. (1999) A pH- and ionic strength-responsive polypeptide hydrogel: synthesis, characterization, and preliminary protein release studies. *Journal of Biomedical Materials Research*, **47** (4), 595–602.

108 Brannon-Peppas, L. and Peppas, N.A. (1990) Dynamic and equilibrium swelling behaviour of pH-sensitive hydrogels containing 2-hydroxyethyl methacrylate. *Biomaterials*, **11** (9), 635–44.

109 Ma, Q., Chen, M., Yin, H.-R., Shi, Z.-G. and Feng, Y.-Q. (2008) Preparation of pH-responsive stationary phase for reversed-phase liquid chromatography and hydrophilic interaction chromatography. *Journal of Chromatography A*, **1212** (1-2), 61–7.

110 Godbey, W.T. and Mikos, A.G. (2001) Recent progress in gene delivery using non-viral transfer complexes. *Journal of Controlled Release*, **72** (1-3), 115–25.

111 Lee, W.-F. and Shieh, C.-H. (1999) pH-thermoreversible hydrogels. II. Synthesis and swelling behaviors of N-isopropylacrylamide-*co*-acrylic acid-*co*-sodium acrylate hydrogels. *Journal of Applied Polymer Science*, **73** (10), 1955–67.

112 Xu, X.-D., Zhang, X.-Z., Cheng, S.-X., Zhuo, R.-X. and Kennedy, J.F. (2007) A strategy to introduce the pH sensitivity to temperature sensitive PNIPAAm hydrogels without weakening the thermosensitivity. *Carbohydrate Polymers*, **68**, 416–23.

113 de las Heras Alarcon, C., Twaites, B., Cunliffe, D., Smith, J.R. and Alexander, C. (2005) Grafted thermo- and pH responsive co-polymers: surface-properties and bacterial adsorption. *International Journal of Pharmaceutics*, **295**, 77–91.

114 Benrebouh, A., Avoce, D. and Zhu, X.X. (2001) Thermo- and pH-sensitive polymers containing cholic acid derivatives. *Polymer*, **42** (9), 4031–8.

115 Kuckling, D., Adler, H.J.P., Arndt, K.F., Ling, L. and Habicher, W.D. (2000) Temperature and pH dependent solubility of novel poly(N-isopropylacrylamide) copolymers. *Macromolecular Chemistry and Physics*, **201** (2), 273–80.

116 Chen, G.-H. and Hoffman, A.S. (1995) Graft copolymers that exhibit temperature-induced phase transitions over a wide range of pH. *Nature*, **373**, 49–52.

117 Kuckling, D., Ivanova, I.G., Adler, H.-J.P. and Wolff, T. (2002) Photochemical switching of hydrogel film properties. *Polymer*, **43** (6), 1813–20.

118 Kim, J.H., Lee, S.B., Kim, S.J. and Lee, Y.M. (2002) Rapid temperature/pH response of porous alginate-g-poly(N-isopropylacrylamide) hydrogels. *Polymer*, **43** (26), 7549–58.

119 Dimitrov, I., Trzebicka, B., Müller, A.H.E., Dworak, A. and Tsvetanov, C.B. (2007) Thermosensitive water-soluble copolymers with doubly responsive reversibly interacting entities. *Progress in Polymer Science*, **32** (11), 1275–343.

120 Hoffman, A.S., Stayton, P.S., Bulmus, V., Chen, G.H., Chen, J.P., Cheung, C., Chilkoti, A., Ding, Z.L., Dong, L.C., Fong, R., Lackey, C.A., Long, C.J., Miura, M., Morris, J.E., Murthy, N., Nabeshima, Y., Park, T.G., Press, O.W., Shimoboji, T., Shoemaker, S., Yang, H.J., Monji, N., Nowinski, R.C., Cole, C.A., Priest, J.H., Harris, J.M., Nakamae, K., Nishino, T. and Miyata, T. (2000) Really smart bioconjugates of smart polymers and receptor proteins. *Journal of Biomedical Materials Research*, **52** (4), 577–86.

121 Chen, J.P. and Huffman, A.S. (1990) Polymer-protein conjugates: II. Affinity precipitation separation of human immunogammaglobulin by a poly(N-isopropylacrylamide)-protein A conjugate. *Biomaterials*, **11** (9), 631–4.

122 Ding, Z., Long, C.J., Hayashi, Y., Bulmus, E.V., Hoffman, A.S. and Stayton, P.S. (1999) Temperature control of biotin binding and release with A Streptavidin-poly(N-isopropylacrylamide) site-specific conjugate. *Bioconjugate Chemistry*, **10** (3), 395–400.

123 Heredia, K.L. and Maynard, H.D. (2007) Synthesis of protein-polymer conjugates. *Organic and Biomolecular Chemistry*, **5**, 45–53.

124 Shimoboji, T., Larenas, E., Fowler, T., Hoffman, A.S. and Stayton, P.S. (2003) Temperature-induced switching of enzyme activity with smart polymer-enzyme conjugates. *Bioconjugate Chemistry*, **14**, 517–25.

125 Hoffman, A.S., Stayton, P.S., Bulmus, V., Chen, G., Chen, J., Cheung, C., Chilkoti, A., Ding, Z., Dong, L., Fong, R., Lackey, C.L., Long, C.J., Miura, M., Morris, J.E., Murthy, N., Nabeshima, Y., Park, T.G., Press, O.W., Shimoboji, T., Shoemaker, S., Yang, H.J., Monji, N., Nowinski, R.C., Cole, C.A., Priest, J.H., Harris, J.M., Nakame, K., Nishino, T. and Miyata, T. (2000) Really smart bioconjugates of smart polymers and receptor proteins. *Journal of Biomedical Materials Research*, **52**, 577–86.

126 Kobayashi, J., Kikuchi, A., Sakai, K. and Okano, T. (2002) Aqueous chromatography utilizing hydrophobicity-modified anionic temperature-responsive hydrogel for stationary phases. *Journal of Chromatography A*, **958** (1-2), 109–19.

127 Anastase-Ravion, S., Ding, Z., Pellé, A., Hoffman, A.S. and Letourneur, D. (2001) New antibody purification procedure using a thermally responsive poly(N-isopropylacrylamide)-dextran derivative conjugate. *Journal of Chromatography B: Biomedical Sciences and Applications*, **761** (2), 247–54.

128 Vrhovski, B., Jensen, S. and Weiss, A.S. (1997) Coacervation characteristics of recombinant human tropoelastin. *European Journal of Biochemistry*, **250** (1), 92–8.

129 Vrhovski, B. and Weiss, A.S. (1998) Biochemistry of tropoelastin. *European Journal of Biochemistry*, **258** (1), 1–18.

130 Chow, D., Nunalee, M.L., Lim, D.W., Simnick, A.J. and Chilkoti, A. (2008) Peptide-based biopolymers in biomedicine and biotechnology. *Materials Science and Engineering: R: Reports*, **62** (4), 125–55.

131 Urry, D.W. and Parker, T.M. (2002) Mechanics of elastin: molecular mechanism of biological elasticity and its relationship to contraction. *Journal of Muscle Research and Cell Motility*, **23**, 543–59.

132 Urry, D.W., Luan, C.H., Parker, T.M., Gowda, D.C., Prasad, K.U., Reid, M.C. and Safavy, A. (1991) Temperature of polypeptide inverse temperature transition depends on mean residue hydrophobicity. *Journal of the American Chemical Society*, **113** (11), 4346–8.

133 Meyer, D.E. and Chilkoti, A. (1999) Purification of recombinant proteins by fusion with thermally-responsive

polypeptides. *Nature Biotechnology*, **17** (11), 1112–15.

134 Urry, D.W. (1997) Physical chemistry of biological free energy transduction as demonstrated by elastic protein-based polymers. *Journal of Physical Chemistry B*, **101**, 11007.

135 Nath, N. and Chilkoti, A. (2002) Creating "smart" surfaces using stimuli responsive polymers. *Advanced Materials*, **14**, 1243–7.

136 Nuhn, H. and Klok, H.-A. (2008) Secondary structure formation and LCST behavior of short elastin-like peptides. *Biomacromolecules*, **9**, 2755–63.

137 Miao, M., Bellingham, C.M., Stahl, R.J., Sitarz, E.E., Lane, C.J. and Keeley, F.W. (2003) Sequence and structure determinants for the self-aggregation of recombinant polypeptides modeled after human elastin. *Journal of Biological Chemistry*, **278** (49), 48553–62.

138 Meyer, D.E. and Chilkoti, A. (2004) Quantification of the effects of chain length and concentration on the thermal behavior of elastin-like polypeptides. *Biomacromolecules*, **5** (3), 846–51.

139 Yamaoka, T., Tamura, T., Seto, Y., Tada, T., Kunugi, S. and Tirrell, D.A. (2003) Mechanism for the phase transition of a genetically engineered elastin model peptide (VPGIG)(40) in aqueous solution. *Biomacromolecules*, **4** (6), 1680–5.

140 Simnick, A.J., Lim, D.W., Chow, D. and Chilkoti, A. (2007) Biomedical and biotechnological applications of elastin-like polypeptides. *Polymer Reviews*, **47** (1), 121–54.

141 Banta, S., Megeed, Z., Casali, M., Rege, K. and Yarmush, M.L. (2007) Engineering protein and peptide building blocks for nanotechnology. *Journal of Nanoscience and Nanotechnology*, **7** (2), 387–401.

142 Furgeson, D.Y., Dreher, M.R. and Chilkoti, A. (2006) Structural optimization of a "smart" doxorubicin-polypeptide conjugate for thermally targeted delivery to solid tumors. *Journal of Controlled Release*, **110** (2), 362–9.

143 Dreher, M.R., Raucher, D., Balu, N., Michael Colvin, O., Ludeman, S.M. and Chilkoti, A. (2003) Evaluation of an elastin-like polypeptide-doxorubicin conjugate for cancer therapy. *Journal of Controlled Release*, **91** (1-2), 31–43.

144 Nath, N., Hyun, J., Ma, H. and Chilkoti, A. (2004) Surface engineering strategies for control of protein and cell interactions. *Surface Science*, **570** (1-2), 98–110.

145 Urry, D.W. (1999) Elastic molecular machines in metabolism and soft-tissue restoration. *Trends in Biotechnology*, **17** (6), 249–57.

146 Betre, H., Ong, S.R., Guilak, F., Chilkoti, A., Fermor, B. and Setton, L.A. (2006) Chondrocytic differentiation of human adipose-derived adult stem cells in elastin-like polypeptide. *Biomaterials*, **27** (1), 91–9.

147 Betre, H., Setton, L.A., Meyer, D.E. and Chilkoti, A. (2002) Characterization of a genetically engineered elastin-like polypeptide for cartilaginous tissue repair. *Biomacromolecules*, **3** (5), 910–16.

148 Woodhouse, K.A., Klement, P., Chen, V., Gorbet, M.B., Keeley, F.W., Stahl, R., Fromstein, J.D. and Bellingham, C.M.C.M. (2004) Investigation of recombinant human elastin polypeptides as non-thrombogenic coatings. *Biomaterials*, **25** (19), 4543–53.

149 Zhang, H., Iwama, M., Akaike, T., Urry, D.W., Pattanaik, A., Parker, T.M., Konishi, I. and Nikaido, T. (2006) Human amniotic cell sheet harvest using a novel temperature-responsive culture surface coated with protein-based polymer. *Tissue Engineering*, **12** (2), 391–401.

150 Frey, W., Meyer, D.E. and Chilkoti, A. (2003) Thermodynamically reversible addressing of a stimuli responsive fusion protein onto a patterned surface template. *Langmuir*, **19** (5), 1641–53.

151 Gao, D., McBean, N., Schultz, J.S., Yan, Y., Mulchandani, A. and Chen, W. (2006) Fabrication of antibody arrays using thermally responsive elastin fusion proteins. *Journal of the American Chemical Society*, **128** (3), 676–7.

152 Zhang, K., Diehl, M.R. and Tirrell, D.A. (2005) Artificial polypeptide scaffold for protein immobilization. *Journal of the*

American Chemical Society, **127** (29), 10136–7.
153 Rodriguez-Cabello, J.C., Prieto, S., Arias, F.J., Reguera, J. and Ribeiro, A. (2006) Nanobiotechnological approach to engineered biomaterial design: the example of elastin-like polymers. *Nanomedicine*, **1** (3), 267–80.
154 Hyun, J., Lee, W.-K., Nath, N., Chilkoti, A. and Zauscher, S. (2004) Capture and release of proteins on the nanoscale by stimuli-responsive elastin-like polypeptide "switches". *Journal of the American Chemical Society*, **126** (23), 7330–5.
155 Nath, N. and Chilkoti, A. (2003) Fabrication of a reversible protein array directly from cell lysate using a stimuli-responsive polypeptide. *Analytical Chemistry*, **75** (4), 709–15.
156 Frey, W., Meyer, D.E. and Chilkoti, A. (2003) Dynamic addressing of a surface pattern by a stimuli-responsive fusion protein. *Advanced Materials*, **15** (3), 248–51.
157 Na, K., Jung, J., Kim, O., Lee, J., Lee, T.G., Park, Y.H. and Hyun, J. (2008) "Smart" biopolymer for a reversible stimuli-responsive platform in cell-based biochips. *Langmuir*, **24** (9), 4917–23.
158 Janorkar, A.V., Rajagopalan, P., Yarmush, M.L. and Megeed, Z. (2008) The use of elastin-like polypeptide-polyelectrolyte complexes to control hepatocyte morphology and function in vitro. *Biomaterials*, **29** (6), 625–32.
159 Masayasu, M., Yasunori, M. and Kobatake, E. (2008) Novel extracellular matrix for cell sheet recovery using genetically engineered elastin-like protein. *Journal of Biomedical Materials Research. Part B: Applied Biomaterials*, **86B**, 283–90.
160 Swierczewska, M., Hajicharalambous, C.S., Janorkar, A.V., Megeed, Z., Yarmush, M.L. and Rajagopalan, P. (2008) Cellular response to nanoscale elastin-like polypeptide polyelectrolyte multilayers. *Acta Biomaterialia*, **4** (4), 827–37.
161 Roohi, F. and Titirici, M.M. (2008) Thin thermo-responsive polymer films onto the pore system of chromatographic beads via reversible addition-fragmentation chain transfer polymerization. *New Journal of Chemistry*, **32** (8), 1409–14.
162 Roohi, F., Antonietti, M. and Titirici, M.-M. (2008) Thermo-responsive monolithic materials. *Journal of Chromatography A*, **1203** (2), 160–7.
163 Nagase, K., Kobayashi, J., Kikuchi, A., Akiyama, Y., Kanazawa, H. and Okano, T. (2007) Interfacial property modulation of thermoresponsive polymer brush surfaces and their interaction with biomolecules. *Langmuir*, **23** (18), 9409–15.
164 Schmidt, S., Motschmann, H., Hellweg, T. and von Klitzing, R. (2008) Thermoresponsive surfaces by spin-coating of PNIPAM-*co*-PAA microgels: a combined AFM and ellipsometry study. *Polymer*, **49** (3), 749–56.
165 Schmitt, F.J., Park, C., Simon, J., Ringsdorf, H. and Israelachvili, J. (1998) Direct surface force and contact angle measurements of an adsorbed polymer with a lower critical solution temperature. *Langmuir*, **14** (10), 2838–45.
166 Kidoaki, S., Ohya, S., Nakayama, Y. and Matsuda, T. (2001) Thermoresponsive structural change of a poly(*N*-isopropylacrylamide) graft layer measured with an atomic force microscope. *Langmuir*, **17**, 2402–7.
167 Matzelle, T.R., Ivanov, D.A., Landwehr, D., Heinrich, L.A., Herkt-Bruns, C., Reichelt, R. and Kruse, N. (2002) Micromechanical properties of "smart" gels: studies by scanning force and scanning electron microscopy of PNIPAAm. *Journal of Physical Chemistry B*, **106** (11), 2861–6.
168 Takei, Y.G., Aoki, T., Sanui, K., Ogata, N., Sakurai, Y. and Okano, T. (1994) Dynamic contact angle measurement of temperature-responsive surface properties for poly(*N*-isopropylacrylamide) grafted surfaces. *Macromolecules*, **27**, 6163–6.
169 Gilcreest, V.P., Carroll, W.M., Rochev, Y.A., Blute, I., Dawson, K.A. and Gorelov, A.V. (2004) Thermoresponsive poly(*N*-isopropylacrylamide) copolymers:

contact angles and surface energies of polymer films. *Langmuir*, **20**, 10138–45.
170 Balamurugan, S., Mendez, S., Balamurugan, S.S., O'Brien, M.J. and Lopez, G.P. (2003) Thermal response of poly(*N*-isopropylacrylamide) brushes probed by surface plasmon resonance. *Langmuir*, **19** (7), 2545–9.
171 Schmaljohann, D., Nitschke, M., Schulze, R., Eing, A., Werner, C. and Eichhorn, K.-J. (2005) In situ study of the thermoresponsive behavior of micropatterned hydrogel films by imaging ellipsometry. *Langmuir*, **21** (6), 2317–22.
172 Hirokawa, Y., Jinnai, H., Nishikawa, Y., Okamoto, T. and Hashimoto, T. (1999) Direct observation of internal structures in poly(*N*-isopropylacrylamide) chemical gels. *Macromolecules*, **32** (21), 7093–9.
173 Zhang, J.-M., Nylander, T., Campbell, R.A., Rennie, A.R., Zauscher, S. and Linse, P. (2008) Novel evaluation method of neutron reflectivity data applied to stimulus-responsive polymer brushes. *Soft Matter*, **4**, 500–9.
174 Lynch, I. et al. (2005) Correlation of the adhesive properties of cells to *N*-isopropylacrylamide/*N*-tert-butylacrylamide copolymer surfaces with changes in surface structure using contact angle measurements, molecular simulations and Raman spectroscopy. *Chemistry of Materials*, **17**, 3889–98.
175 Duval, J.F.L., Zimmermann, R., Cordeiro, A.L., Rein, N. and Werner, C. (2009) Electrokinetics of diffuse soft interfaces. IV. Analysis of streaming current measurements at thermo-responsive thin films. *Langmuir* DOI: 10.1021/la 9011907.
176 Schmaljohann, D., Oswald, J., Jørgensen, B., Nitschke, M., Beyerleni, D. and Werner, C. (2003) Thermo-responsive PNiPAAm-g-PEG films for controlled cell detachment. *Biomacromolecules*, **4**, 1733–9.
177 Ista, L.K. and Lopez, G.P. (1998) Lower critical solubility temperature materials as biofouling release agents. *Journal of Industrial Microbiology and Biotechnology*, **20**, 121–5.

178 Lin, S.-Y., Chen, K.-S. and Run-Chu, L. (1999) Thermal micro ATR/FT-IR spectroscopic system for quantitative study of the molecular structure of poly(*N*-isopropylacrylamide) in water. *Polymer*, **40** (10), 2619–24.
179 Mack, D., Rohde, H., Harris, L.G., Davies, A.P., Horstkotte, M.A. and Knobloch, J.K.M. (2006) Biofilm formation in medical device-related infection. *International Journal of Artificial Organs*, **29** (4), 343–59.
180 Hetrick, E.M. and Schoenfisch, M.H. (2006) Reducing implant-related infections: active release strategies. *Chemical Society Reviews*, **35** (9), 780–9.
181 Costerton, J.W., Cheng, K.J., Geesey, G.G., Ladd, T.I., Nickel, J.C., Dasgupta, M. and Marrie, T.J. (1987) Bacterial biofilms in nature and disease. *Annual Review of Microbiology*, **41**, 435–64.
182 Mah, T.-F.C. and O'Toole, G.A. (2001) Mechanisms of biofilm resistance to antimicrobial agents. *Trends in Microbiology*, **9** (1), 34–9.
183 Montanaro, L., Campoccia, D. and Arciola, C.R. (2007) Advancements in molecular epidemiology of implant infections and future perspectives. *Biomaterials*, **28** (34), 5155–68.
184 Li, S.K. and D'Emanuele, A. (2001) On-off transport through a thermoresponsive hydrogel composite membrane. *Journal of Controlled Release*, **75** (1-2), 55–67.
185 Jones, D.S., Lorimer, C.P., McCoy, C.P. and Gorman, S.P. (2008) Characterization of the physicochemical, antimicrobial, and drug release properties of thermoresponsive hydrogel copolymers designed for medical device applications. *Journal of Biomedical Materials Research Part B: Applied Biomaterials*, **85B** (2), 417–26.
186 Ruiz, J.-C., Alvarez-Lorenzo, C., Taboada, P., Burillo, G., Bucio, E., De Prijck, K., Nelis, H.J., Coenye, T. and Concheiro, A. (2008) Polypropylene grafted with smart polymers (PNIPAAm/PAAc) for loading and controlled release of vancomycin. *European Journal of Pharmaceutics and Biopharmaceutics*, **70** (2), 467–77.

187 Okano, T., Yamada, N., Sakai, H. and Sakurai, Y. (1993) A novel recovery system for cultured cells using plasma-treated polystyrene dishes grafted with poly(N-isopropylacrylamide). *Journal of Biomedical Materials Research*, **27** (10), 1243–51.

188 Canavan, H.E., Cheng, X.H., Graham, D.J., Ratner, B.D. and Castner, D.G. (2006) A plasma-deposited surface for cell sheet engineering: advantages over mechanical dissociation of cells. *Plasma Processes and Polymers*, **3** (6-7), 516–23.

189 Yamato, M., Okuhara, M., Karikusa, F., Kikuchi, A., Sakurai, Y. and Okano, T. (1999) Signal transduction and cytoskeletal reorganization are required for cell detachment from cell culture surfaces grafted with a temperature-responsive polymer. *Journal of Biomedical Materials Research*, **44**, 44.

190 Canavan, H.E., Cheng, X., Graham, D.J., Ratner, B.D. and Castner, D.G. (2005) Surface characterization of the extracellular matrix remaining after cell detachment from a thermoresponsive polymer. *Langmuir*, **21**, 1949–55.

191 Kushida, A., Yamato, M., Konno, C., Kikuchi, A., Sakurai, Y. and Okano, T. (1999) Decrease in culture temperature releases monolayer endothelial cell sheets together with deposited fibronectin matrix from temperature-responsive culture surfaces. *Journal of Biomedical Materials Research*, **45** (4), 355–62.

192 Shimizu, T., Yamato, M., Isoi, Y., Akutsu, T., Setomaru, T., Abe, K., Kikuchi, A., Umezu, M. and Okano, T. (2002) Fabrication of pulsatile cardiac tissue grafts using a novel 3-dimensional cell sheet manipulation technique and temperature-responsive cell culture surfaces. *Circulation Research*, **90** (3), E40–8.

193 Nakajima, K., Honda, S., Nakamura, Y., López-Redondo, F., Kohsaka, S., Yamato, M., Kikuchi, A. and Okano, T. (2001) Intact microglia are cultured and non-invasively harvested without pathological activation using a novel cultured cell recovery method. *Biomaterials*, **22** (11), 1213–23.

194 Nishida, K., Yamato, M., Hayashida, Y., Watanabe, K., Maeda, N., Watanabe, H., Yamamoto, K., Nagai, S., Kikuchi, A., Tano, Y. and Okano, T. (2004) Functional bioengineered corneal epithelial sheet grafts from corneal stem cells expanded ex vivo on a temperature-responsive cell culture surface. *Transplantation*, **77** (3), 379–85.

195 Yang, J., Yamato, M., Nishida, K., Hayashida, Y., Shimizu, T., Kikuchi, A., Tano, Y. and Okano, T. (2006) Corneal epithelial stem cell delivery using cell sheet engineering: not lost in transplantation. *Journal of Drug Targeting*, **14** (7), 471–582.

196 Ide, T., Nishida, K., Yamato, M., Sumide, T., Utsumi, M., Nozaki, T., Kikuchi, A., Okano, T. and Tano, Y. (2006) Structural characterization of bioengineered human corneal endothelial cell sheets fabricated on temperature-responsive culture dishes. *Biomaterials*, **27** (4), 607–14.

197 Miyamoto, S., Teramoto, H., Coso, O.A., Gutkind, J.S., Burbelo, P.D., Akiyama, S.K. and Yamada, K.M. (1995) Integrin function–molecular hierarchies of cytoskeletal and signaling molecules. *Journal of Cell Biology*, **131** (3), 791–805.

198 Hyeong Kwon, O., Kikuchi, A., Yamato, M. and Okano, T. (2003) Accelerated cell sheet recovery by co-grafting of PEG with PIPAAm onto porous cell culture membranes. *Biomaterials*, **24** (7), 1223–32.

199 Fukumori, K., Akiyama, Y., Yamato, M., Kobayashi, J., Sakai, K. and Okano, T. (2009) Temperature-responsive glass coverslips with an ultrathin poly(N-isopropylacrylamide) layer. *Acta Biomaterialia*, **5** (1), 470–6.

200 Mizutani, A., Kikuchi, A., Yamato, M., Kanazawa, H. and Okano, T. (2008) Preparation of thermoresponsive polymer brush surfaces and their interaction with cells. *Biomaterials*, **29** (13), 2073–81.

201 Xu, F.J., Zhong, S.P., Yung, L.Y.L., Kang, E.T. and Neoh, K.G. (2004) Surface-active and stimuli-responsive polymer-Si(100) hybrids from surface-initiated atom transfer radical polymerization for control of cell adhesion. *Biomacromolecules*, **5** (6), 2392–403.

202 Gramm, S., Komber, H., and Schmaljohann, D. (2005) Copolymerization kinetics of N-isopropylacrylamide and diethylene glycol monomethylether monomethacrylate determined by online NMR spectroscopy. *Journal of Polymer Science Part A: Polymer Chemistry*, **43**, 142–48.

203 Kanazawa, H., Kashiwase, Y., Yamamoto, K., Matsushima, Y., Kikuchi, A., Sakurai, Y. and Okano, T. (1997) Temperature-responsive liquid chromatography. 2. Effects of hydrophobic groups in N-isopropylacrylamide copolymer-modified silica. *Analytical Chemistry*, **69** (5), 823–30.

204 Ebara, M., Yamato, M., Hirose, M., Aoyagi, T., Kikuchi, A., Sakai, K. and Okano, T. (2003) Copolymerization of 2-carboxyisopropylacrylamide with N-isopropylacrylamide accelerates cell detachment from grafted surfaces by reducing temperature. *Biomacromolecules*, **4** (2), 344–9.

205 Kwon, O.H., Kikuchi, A., Yamato, M., Sakurai, Y. and Okano, T. (2000) Rapid cell sheet detachment from poly(N-isopropylacrylamide)-grafted porous cell culture membranes. *Journal of Biomedical Materials Research*, **50** (1), 82–9.

206 von Recum, H., Okano, T. and Wan Kim, S. (1998) Growth factor release from thermally reversible tissue culture substrates. *Journal of Controlled Release*, **55** (2-3), 121–30.

207 Recum, H.V., Kikuchi, A., Yamato, M., Sakurai, Y., Okano, T. and Kim, S.W. (1999) Growth factor and matrix molecules preserve cell function on thermally responsive culture surfaces. *Tissue Engineering*, **5** (3), 251–65.

208 Moran, M.T., Carroll, W.M., Gorelov, A. and Rochev, Y. (2007) Intact endothelial cell sheet harvesting from thermoresponsive surfaces coated with cell adhesion promoters. *Journal of the Royal Society Interface*, **4** (17), 1151–7.

209 Moran, M.T., Carroll, W.M., Selezneva, I., Gorelov, A. and Rochev, Y. (2007) Cell growth and detachment from protein-coated PNIPAAm-based copolymers. *Journal Biomedical Materials Research A*, **81**, 870–6.

210 Ebara, M., Yamato, M., Nagai, S., Aoyagi, T., Kikuchi, A., Sakai, K. and Okano, T. (2004) Incorporation of new carboxylate functionalized co-monomers to temperature-responsive polymer-grafted cell culture surfaces. *Surface Science*, **570** (1-2), 134–41.

211 Hatakeyama, H., Kikuchi, A., Yamato, M. and Okano, T. (2006) Bio-functionalized thermoresponsive interfaces facilitating cell adhesion and proliferation. *Biomaterials*, **27** (29), 5069–78.

212 Ebara, M., Yamato, M., Aoyagi, T., Kikuchi, A., Sakai, K. and Okano, T. (2004) Immobilization of cell-adhesive peptides to temperature-responsive surfaces facilitates both serum-free cell adhesion and noninvasive cell harvest. *Tissue Engineering*, **10** (7-8), 1125–35.

213 Bhatia, S.N., Balis, U.J., Yarmush, M.L. and Toner, M. (1999) Effect of cell-cell interactions in preservation of cellular phenotype: cocultivation of hepatocytes and nonparenchymal cells. *The FASEB Journal*, **13** (14), 1883–900.

214 Tsuda, Y., Kikuchi, A., Yamato, M., Nakao, A., Sakurai, Y., Umezu, M. and Okano, T. (2005) The use of patterned dual thermoresponsive surfaces for the collective recovery as co-cultured cell sheets. *Biomaterials*, **26** (14), 1885–93.

215 Yamato, M., Konno, C., Utsumi, M., Kikuchi, A. and Okano, T. (2002) Thermally responsive polymer-grafted surfaces facilitate patterned cell seeding and co-culture. *Biomaterials*, **23** (2), 561–7.

216 Tsuda, Y., Shimizu, T., Yamato, M., Kikuchi, A., Sasagawa, T., Sekiya, S., Kobayashi, J., Chen, G. and Okano, T. (2007) Cellular control of tissue architectures using a three-dimensional tissue fabrication technique. *Biomaterials*, **28** (33), 4939–46.

217 Hatakeyama, H., Kikuchi, A., Yamato, M. and Okano, T. (2007) Patterned

biofunctional designs of thermoresponsive surfaces for spatiotemporally controlled cell adhesion, growth, and thermally induced detachment. *Biomaterials*, **28** (25), 3632–43.
218 Isenberg, B.C., Tsuda, Y., Williams, C., Shimizu, T., Yamato, M., Okano, T. and Wong, J.Y. (2008) A thermoresponsive, microtextured substrate for cell sheet engineering with defined structural organization. *Biomaterials*, **29** (17), 2565–72.
219 Shimizu, T., Yamato, M., Kikuchi, A. and Okano, T. (2003) Cell sheet engineering for myocardial tissue reconstruction. *Biomaterials*, **24** (13), 2309–16.
220 Kubo, H., Shimizu, T., Yamato, M., Fujimoto, T. and Okano, T. (2007) Creation of myocardial tubes using cardiomyocyte sheets and an in vitro cell sheet-wrapping device. *Biomaterials*, **28** (24), 3508–16.
221 Cooksey, K.E. and Wigglesworth-Cooksey, B. (1995) Adhesion of bacteria and diatoms to surfaces in the sea: a review. *Aquatic Microbial Ecology*, **9**, 87–96.
222 Prakash, B., Veeregowda, B.M. and Krishnappa, G. (2003) Biofilms: a survival strategy of bacteria. *Current Science*, **85** (9), 1299–307.
223 Costerton, J.W., Stewart, P.S. and Greenberg, E.P. (1999) Bacterial biofilms: a common cause of persistent infections. *Science*, **284** (5418), 1318–22.
224 Ista, L.K., Perez-Luna, V.H. and Lopez, G.P. (1999) Surface-grafted, environmentally sensitive polymers for biofilm release. *Applied and Environmental Microbiology*, **65**, 1603–9.
225 Ista, L.K., Fan, H., Baca, O. and López, G.P. (1996) Attachment of bacteria to model solid surfaces: oligo(ethylene glycol) surfaces inhibit bacterial attachment. *FEMS Microbiology Letters*, **142** (1), 59–63.
226 Ferreirós, C.M., Carballo, J., Criado, M.T., Sáinz, V. and del Río, M.C. (1989) Surface free energy and interaction of Staphylococcus epidermidis with biomaterials. *FEMS Microbiology Letters*, **60** (1), 89–94.
227 Cunliffe, D., Smart, C.A., Tsibouklis, J., Young, S., Alexander, C. and Vulfson, E.N. (2000) Bacterial adsorption to thermoresponsive polymer surfaces. *Biotechnology Letters*, **22**, 141–5.
228 Wognum, A.W., Eaves, A.C. and Thomas, T.E. (2003) Identification and isolation of hematopoietic stem cells. *Archives of Medical Research*, **34** (6), 461–75.
229 Matsuda, T., Saito, Y. and Shoda, K. (2007) Cell sorting technique based on thermoresponsive differential cell adhesiveness. *Biomacromolecules*, **8**, 2345–9.
230 Kim, S.Y., Kanamori, T. and Shinbo, T. (2002) Preparation of thermal-responsive poly(propylene) membranes grafted with N-isopropylacrylamide by plasma-induced polymerization and their permeation. *Journal of Applied Polymer Science*, **84** (6), 1168–77.
231 Okamura, A., Itayagoshi, M., Hagiwara, T., Yamaguchi, M., Kanamori, T., Shinbo, T. and Wang, P.-C. (2005) Poly(N-isopropylacrylamide)-graft-polypropylene membranes containing adsorbed antibody for cell separation. *Biomaterials*, **26** (11), 1287–92.
232 Okamura, A., Hagiwara, T., Yamagami, S., Yamaguchi, M., Shinbo, T., Kanamori, T., Kondo, S., Miwa, K. and Itagaki, L. (2008) Effective cell separation utilizing poly(N-isopropylacrylamide)-grafted polypropylene membrane containing adsorbed antibody. *Journal of Bioscience and Bioengineering*, **105** (3), 221–5.
233 Ying, L., Kang, E.T., Neoh, K.G., Kato, K. and Iwata, H. (2003) Novel poly(N-isopropylacrylamide)-graft-poly(vinylidene fluoride) copolymers for temperature-sensitive microfiltration membranes. *Macromolecular Materials and Engineering*, **288**, 11–16.
234 Zhou, J., Liu, J., Wang, G., Lu, X., Wen, Z. and Li, J. (2007) Poly(N-isopropylacrylamide) interfaces with dissimilar thermo-responsive behavior for controlling ion permeation and immobilization. *Advanced Functional Materials*, **17** (16), 3377–82.

235 Rao, G.V.R., Balamurugan, S., Meyer, D.E., Chilkoti, A. and Lopez, G.P. (2002) Hybrid bioinorganic smart membranes that incorporate protein-based molecular switches. *Langmuir*, **18** (5), 1819–24.

236 Okahata, Y., Noguchi, H. and Seki, T. (1986) Thermoselective permeation from a polymer-grafted capsule membrane. *Macromolecules*, **19** (2), 493–4.

237 Choi, Y.-J., Yamaguchi, T. and Nakao, S.-i. (2000) A novel separation system using porous thermosensitive membranes. *Industrial and Engineering Chemistry Research*, **39** (7), 2491–5.

238 Ying, L., Kang, E.T. and Neoh, K.G. (2002) Synthesis and characterization of poly(N-isopropylacrylamide)-graft-poly(vinylidene fluoride) copolymers and temperature-sensitive membranes. *Langmuir*, **18** (16), 6416–23.

239 Ying, L., Wang, P., Kang, E.T. and Neoh, K.G. (2002) Synthesis and characterization of poly(acrylic acid)-graft-poly(vinylidene fluoride) copolymers and pH-sensitive membranes. *Macromolecules*, **35** (3), 673–9.

240 Zhang, K. and Wu, X.Y. (2004) Temperature and pH-responsive polymeric composite membranes for controlled delivery of proteins and peptides. *Biomaterials*, **25** (22), 5281–91.

241 Malmstadt, N., Hoffman, A.S. and Stayton, P.S. (2004) "Smart" mobile affinity matrix for microfluidic immunoassays. *Lab on a Chip*, **4** (4), 412–15.

242 Malmstadt, N., Yager, P., Hoffman, A.S. and Stayton, P.S. (2003) A smart microfluidic affinity chromatography matrix composed of poly(N-isopropylacrylamide)-coated beads. *Analytical Chemistry*, **75**, 2943–9.

243 Piskin, E. (2004) Molecularly designed water soluble, intelligent, nanosize polymeric carriers. *International Journal of Pharmaceutics*, **277** (1-2), 105–18.

244 Ayano, E., Nambu, K., Sakamoto, C., Kanazawa, H., Kikuchi, A. and Okano, T. (2006) Aqueous chromatography system using pH- and temperature-responsive stationary phase with ion-exchange groups. *Journal of Chromatography A*, **1119** (1-2), 58–65.

245 Sakamoto, C., Okada, Y., Kanazawa, H., Ayano, E., Nishimura, T., Ando, M., Kikuchi, A. and Okano, T. (2004) Temperature- and pH-responsive aminopropyl-silica ion-exchange columns grafted with copolymers of N-isopropylacrylamide. *Journal of Chromatography*, **1030** (1-2), 247–53.

246 Jungbauer, A. (2005) Chromatographic media for bioseparation. *Journal of Chromatography A*, **1065** (1), 3–12.

247 Kanazawa, H., Sunamoto, T., Matsushima, Y., Kikuchi, A. and Okano, T. (2000) Temperature-responsive chromatographic separation of amino acid phenylthiohydantoins using aqueous media as the mobile phase. *Analytical Chemistry*, **72** (24), 5961–6.

248 Kanazawa, H., Yamamoto, K., Matsushima, Y., Takai, N., Kikuchi, A., Sakurai, Y. and Okano, T. (1996) Temperature-responsive chromatography using poly(N-isopropylacrylamide)-modified silica. *Analytical Chemistry*, **68** (1), 100–5.

249 Kobayashi, J., Kikuchi, A., Sakai, K. and Okano, T. (2001) Aqueous chromatography utilizing pH-/temperature-responsive polymer stationary phases to separate ionic bioactive compounds. *Analytical Chemistry*, **73** (9), 2027–33.

250 Kanazawa, H., Nishikawa, M., Mizutani, A., Sakamoto, C., Morita-Murase, Y., Nagata, Y., Kikuchi, A. and Okano, T. (2008) Aqueous chromatographic system for separation of biomolecules using thermoresponsive polymer modified stationary phase. *Journal of Chromatography A*, **1191** (1-2), 157–61.

251 Zhu, Y., Ni, C.H., Shao, D. and Jiang, X. (2008) The preparation of composites of poly(N-isopropylacrylamide) with silica and its application in HPLC for separating naphthalene derivatives. *Polymer Composites*, **29** (4), 415–20.

252 Nagase, K., Kobayashi, J., Kikuchi, A., Akiyama, Y., Annaka, M., Kanazawa, H. and Okano, T. (2008) Influence of graft interface polarity on hydration/dehydration of grafted thermoresponsive

polymer brushes and steroid separation using all-aqueous chromatography. *Langmuir*, **24** (19), 10981–7.
253 Kikuchi, A., Kobayashi, J., Okano, T., Iwasa, T. and Sakai, K. (2007) Temperature-modulated interaction changes with adenosine nucleotides on intelligent cationic, thermoresponsive surfaces. *Journal of Bioactive and Compatible Polymers*, **22** (6), 575–88.
254 Ayano, E., Suzuki, Y., Kanezawa, M., Sakamoto, C., Morita-Murase, Y., Nagata, Y., Kanazawa, H., Kikuchi, A. and Okano, T. (2007) Analysis of melatonin using a pH- and temperature-responsive aqueous chromatography system. *Journal of Chromatography A*, **1156** (1-2), 213–19.
255 Nagase, K., Kobayashi, J., Kikuchi, A., Akiyama, Y., Kanazawa, H. and Okano, T. (2008) Preparation of thermoresponsive cationic copolymer brush surfaces and application of the surface to separation of biomolecules. *Biomacromolecules*, **9** (4), 1340–7.
256 Ayano, E., Sakamoto, C., Kanazawa, H., Kikuchi, A. and Okano, T. (2006) Separation of nucleotides with an aqueous mobile phase using pH- and temperature-responsive polymer modified packing materials. *Analytical Sciences*, **22** (4), 539–43.
257 Hosoya, K., Kimata, K., Araki, T., Tanaka, N. and Frechet, J.M.J. (1995) Temperature-controlled high-performance liquid chromatography using a uniformly sized temperature-responsive polymer-based packing material. *Analytical Chemistry*, **67** (11), 1907–11.
258 Adrados, B.P., Galaev, I.Y., Nilsson, K. and Mattiasson, B. (2001) Size exclusion behavior of hydroxypropylcellulose beads with temperature-dependent porosity. *Journal of Chromatography A*, **930** (1-2), 73–8.
259 Lakhiari, H., Okano, T., Nurdin, N., Luthi, C., Descouts, P., Muller, D. and Jozefonvicz, J. (1998) Temperature-responsive size-exclusion chromatography using poly(N-isopropylacrylamide) grafted silica. *Biochimica et Biophysica Acta: General Subjects*, **1379** (3), 303–13.
260 Hosoya, K., Sawada, E., Kimata, K., Araki, T., Tanaka, N. and Frechet, J.M.J. (1994) In situ surface-selective modification of uniform size macroporous polymer particles with temperature-responsive poly-N-isopropylacrylamide. *Macromolecules*, **27** (14), 3973–6.
261 Yoshizako, K., Akiyama, Y., Yamanaka, H., Shinohara, Y., Hasegawa, Y., Carredano, E., Kikuchi, A. and Okano, T. (2002) Regulation of protein binding toward a ligand on chromatographic matrices by masking and forced-releasing effects using thermoresponsive polymer. *Analytical Chemistry*, **74** (16), 4160–6.
262 Yamanaka, H., Yoshizako, K., Akiyama, Y., Sota, H., Hasegawa, Y., Shinohara, Y., Kikuchi, A. and Okano, T. (2003) Affinity chromatography with collapsibly tethered ligands. *Analytical Chemistry*, **75** (7), 1658–63.
263 Xu, F.J., Zhong, S.P., Yung, L.Y.L., Tong, Y.W., Kang, E.-T. and Neoh, K.G. (2006) Thermoresponsive comb-shaped copolymer-Si(1 0 0) hybrids for accelerated temperature-dependent cell detachment. *Biomaterials*, **27**, 1236–45.
264 Ebara, M., Hoffman, J.M., Hoffman, A.S. and Stayton, P.S. (2006) Switchable surface traps for injectable bead-based chromatography in PDMS microfluidic channels. *Lab on a Chip*, **6**, 843–8.
265 Saitoh, T., Suzuki, Y. and Hiraide, M. (2002) Preparation of poly(N-isopropylacrylamide)-modified glass surface for flow control in microfluidics. *Analytical Sciences*, **18**, 203.
266 Idota, N., Kikuchi, A., Kobayashi, J., Sakai, K. and Okano, T. (2005) Microfluidic valves comprising nanolayered thermoresponsive polymer-grafted capillaries. *Advanced Materials*, **17** (22), 2723–7.
267 Beebe, D.J., Moore, J.S., Bauer, J.M., Yu, Q., Liu, R.H., Devadoss, C. and Jo, B.-H. (2000) Functional hydrogel structures for autonomous flow control inside microfluidic channels. *Nature*, **404**, 588.

268 Adiga, S.P. and Brenner, D.W. (2007) Toward designing smart nanovalves: modeling of flow control through nanopores via the helix-coil transition of grafted polypeptide chains. *Macromolecules*, **40** (4), 1342–8.

269 Agarwal, A.K., Sridharamurthy, S.S., Beebe, D.J. and Jiang, H.R. (2005) Programmable autonomous micromixers and micropumps. *Journal of Microelectromechanical Systems*, **14** (6), 1409–21.

270 Zhang, Y., Kato, S. and Anazawa, T. (2008) A microchannel concentrator controlled by integral thermoresponsive valves. *Sensors and Actuators B: Chemical*, **129** (1), 481–6.

4
Ceramic Nanocoatings and Their Applications in the Life Sciences
Eng San Thian

4.1
Introduction

During the past three decades, major advances have been made in the development of biomedical materials for use as bone replacements. The first such biomaterials to be developed were tolerated within the physiological environment, but were generally metallic in nature and lacked any ability to encourage direct bone apposition [1]. As a consequence of this, Hulbert and coworkers introduced a stable oxide layer onto the metallic surface in order to activate bone growth [2]. However, although this concept was shown to elicit a favorable tissue reaction, a much longer time scale was required before osteointegration could occur. Today, the emphasis has shifted towards the use of bioactive ceramic materials, in particularly calcium phosphates (CaPs), due to their similarity with the mineral phase that is found in natural bone [3].

Calcium phosphates – and hydroxyapatite (HA) specifically – are often used in the form of coatings on metallic implants, to confer the biological and mechanical advantages of both types of material. However, in comparison with bioactive glasses and glass-ceramics, HA unfortunately has a relatively low bone-bonding rate [4]. The main approach towards improving the osteointegration rate of HA has been to perform a chemical doping of the ceramic material, using small amounts of those elements which are commonly found in human bone [5], since the HA structure is very tolerant of ionic substitution. The ultimate aim of this procedure was to provide a rapid return to structural stability so that patients could live a normal life as quickly as possible.

The idea of incorporating silicon (Si) into HA is largely based on the beneficial role of Si^{4+} ions in bone, and the excellent bioactivity of silica-based glasses and glass-ceramics. Previously, Carlisle [6–8] had reported that chicks receiving dietary silicon had shown an enhanced bone growth and development. Likewise, a significant upregulation of bone cell proliferation and gene expression were observed for bioactive glasses, due to the release of low levels of Si^{4+} ions into the physiological environment [9–11]. These findings led to the development of

Nanomaterials for the Life Sciences Vol.5: Nanostructured Thin Films and Surfaces.
Edited by Challa S. S. R. Kumar
Copyright © 2010 WILEY-VCH Verlag GmbH & Co. KGaA, Weinheim
ISBN: 978-3-527-32155-1

silicon-substituted HA (SiHA) as a potential bone graft material, with several studies having demonstrated a significant increase in the amount of bone apposition and quality of bone repair, compared to the use of HA ceramics [12–14].

Based on the success of the SiHA bone grafts, many investigations have been conducted recently to explore alternative techniques for the deposition of SiHA as coatings on metallic surfaces. In this chapter we will discuss the fabrication of SiHA nanostructured coatings, and provide details of comprehensive studies of the physical, chemical, and biological properties of these materials.

4.2
Magnetron Sputtering

The production of nanostructured SiHA coatings can be conducted via a magnetron sputtering process at room temperature, using a combination of phase-pure HA and Si targets. The application of this technique leads to the formation of uniform, well-adhered thin coatings, with the additional possibility of incorporating controlled amounts of beneficial elements, in this case Si.

Thian et al. [15, 16] reported a systematic approach for the deposition of SiHA coatings, using a sputter deposition system; a typical set-up to produce such coatings is shown schematically in Figure 4.1. Here, a constant flow of argon gas at a working pressure of 0.6 Pa was supplied to the chamber at each deposition run. The composition of SiHA coating could be controlled by the relative discharge power density supplied to each target. Generally, a radiofrequency (RF) of 13.56 MHz, with a power density of $3.1\,W\,cm^{-2}$ was supplied to the HA target, while three different direct current (dc) power densities of 0.2, 0.5 and $0.8\,W\,cm^{-2}$ were supplied to the Si target to achieve coatings with three different Si contents.

Following the successful deposition of SiHA coatings by sputtering, a number of recent reports have recently emerged describing the preparation of SiHA coatings. For example, Solla et al. [17, 18] reported that SiHA coatings could be obtained using a pulsed laser deposition technique, with analyses having revealed that the Si had been incorporated into the HA structure. In a separate study, the same group found that osteoblasts could proliferate and differentiate well on these SiHA coatings [19]. Later, Hijon et al. [20] produced SiHA coatings with Si contents of up to 2.1 wt% using a sol–gel technique, and showed that these coatings could support cell proliferation *in vitro* [21]. SiHA coatings have also been produced by electrophoretic deposition [22] and biomimetic deposition [23, 24] techniques, with the bioactive properties of the coatings being enhanced by the presence of Si.

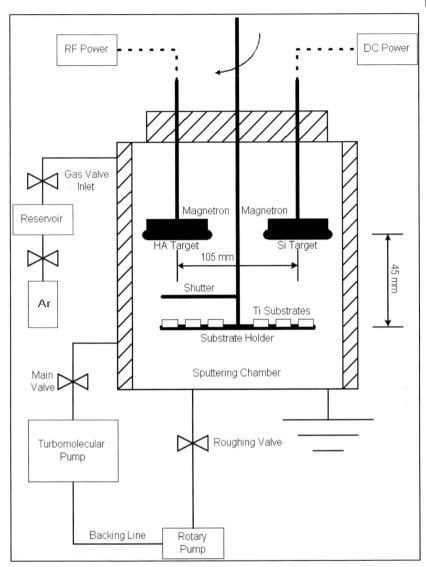

Figure 4.1 Schematic of the magnetron sputtering system set-up [16].

4.3 Physical and Chemical Properties of SiHA Coatings

When nanostructured SiHA coatings of thickness 0.7 μm were produced, the Si composition in the coating (0.8, 2.2, and 4.9 wt%) was shown to increase in relation to the direct current power density on the Si target [16].

Following heat-treatment at 600 °C, the amorphous coatings were transformed into phase-pure, polycrystalline coatings, with major peaks matching that of HA

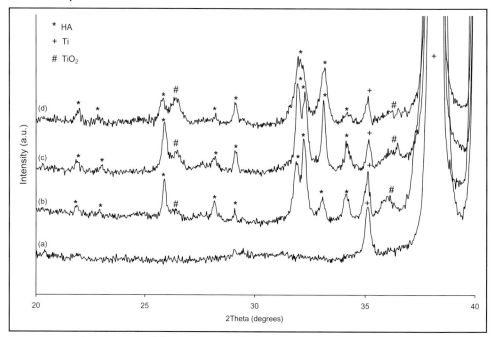

Figure 4.2 X-ray diffraction patterns of SiHA coatings of varying Si composition. (a) As-deposited; (b) Heat-treated, 0.8 wt% Si; (c) Heat-treated, 2.2 wt% Si; (d) Heat-treated, 4.9 wt% Si [16].

[Joint Committee on Powder Diffraction Standards (JCPDS) #9-432], as indicated by X-ray diffraction (XRD) scans (see Figure 4.2). The lattice plane reflections were also shown to become broader and less intense with the increasing Si level.

Heat-treatment led to several changes in the infrared (IR) spectra, with four typical absorption bands for HA being observed: (i) at $3571\,cm^{-1}$, which was assigned to the stretching mode of hydroxyl (O–H); (ii) at $1089\,cm^{-1}$, which was assigned to the v_3 phosphate (P–O) stretching mode, and became sharper and more intense; (iii) at $1038\,cm^{-1}$, which was also assigned to the v_3 P–O stretching mode; and (iv) at $962\,cm^{-1}$, which was assigned to the v_1 P–O stretching mode (Figure 4.3). Although all if these IR spectra exhibited absorption bands associated with HA, the band intensities corresponding to O–H and P–O (most noticeably at $962\,cm^{-1}$) decreased with the increasing Si level.

All of these coatings signified the transformation of an amorphous coating into a well-crystallized, phase-pure HA coating after heat treatment. Discussions have been on-going, however, concerning the performance of amorphous and crystalline coatings. Since crystalline HA has a considerably lower dissolution rate than amorphous HA [25, 26], it may be beneficial for long-term implant performance as it may lead to a reduced incidence of osteolysis [27, 28]. In contrast, amorphous HA was observed to enhance early bone apposition *in vivo* [29]. Phase-purity is

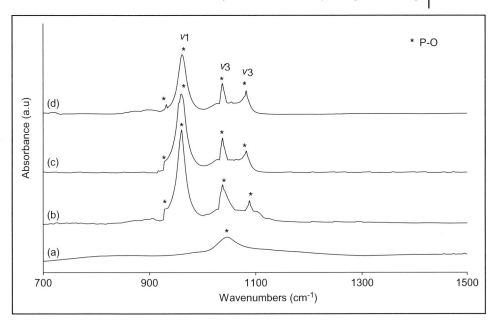

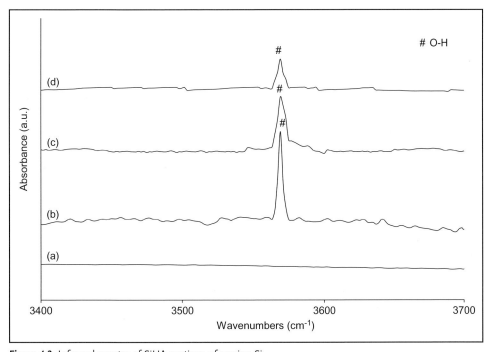

Figure 4.3 Infra-red spectra of SiHA coatings of varying Si composition. (a) As-deposited; (b) Heat-treated, 0.8 wt% Si; (c) Heat-treated, 2.2 wt% Si; (d) Heat-treated, 4.9 wt% Si [16].

also a critical issue for coatings of <1 μm thickness, as secondary calcium phosphate phases, such as tricalcium phosphate and tetracalcium phosphate, will dissolve more rapidly than HA [30]. This effect would cause the coating to dissolve completely in the physiological environment, even before it is able to trigger the promotion of a direct attachment at the bone/implant interface.

Changes in the crystallite size and in the unit cell parameters of the heat-treated coatings were noted with the substitution of Si into the HA structure [16]. With a Si addition from 0.8 to 4.9 wt%, the crystallite size of SiHA was seen to decrease from 50 to 19 nm. These results were in agreement with the findings of both Gibson et al. [31] and Arcos et al. [32], who showed that the addition of Si inhibited grain growth, with the effect being more significant as the Si level was increased.

The unit cell parameter's axes (a, from 0.9436 to 0.9466 nm; c, from 0.6915 to 0.6956 nm) were increased with Si substitution, following the trend as described by Kim et al. [33]. Although Si addition increased and decreased the cell parameters and crystallite size, respectively, the XRD scans did not highlight any significant changes in the relative peak intensities – as would normally be expected when there is an atomic substitution. This occurred because Si and phosphorus (P) are adjacent to each other in the Periodic Table (they differ by only one atomic number), and the level of Si substitution is relatively low (<5 wt%).

Contact angle measurements demonstrated that an uncoated titanium (Ti) substrate (67.9 ± 2.1°) possessed the highest contact angle value among all samples. As for the coated-Ti substrates, SiHA-coated Ti exhibited a lower contact angle when compared to the HA-coated Ti (50.1 ± 3.5°), with 0.8 wt% SiHA (34.0 ± 3.4°) showing the lowest value.

From both physical and chemical viewpoints, it appears that Si^{4+} ions are structurally incorporated into the HA lattice in solid solution, and are not segregated as a second phase. The phosphate tetrahedrals were basically replaced by silicate tetrahedrals, this being reflected in the decrease in IR P–O intensity with increasing Si. It also appeared that a substituted Si level as high as 4.9 wt% could be incorporated into the HA structure while maintaining phase purity. Clearly, the magnetron sputtering technique combined with heat treatment allows the production of nanostructured SiHA coatings with predictable properties.

4.4
Biological Properties of SiHA Coatings

4.4.1
In Vitro Acellular Testing

During the 1990s, a method for the rapid ranking of the bioactivity of materials was developed by Kokubo and coworkers [34], using a so-called "simulated body fluid" (SBF). The concentrations of ions present in this solution (which is known as SBF-K9) match the ionic concentrations of human blood plasma. Consequently, the rate at which a material is able to induce surface calcium phosphate precipitation from the SBF-K9 solution will provide a measure of its level of bioactivity, and this effect can be correlated with the likely activity *in vivo*.

Immersion tests using SBF-K9 solution have demonstrated that heat-treated SiHA coatings possessed a better bioactivity under physiological conditions, compared to as-deposited SiHA coatings or even as-deposited HA coatings and uncoated titanium (Ti) substrates [35].

Dissolution of the as-deposited SiHA coating, as evidenced by the appearance of submicrometer-sized pits on the surface, was observed after a one-day immersion in SBF-K9 solution, while pit formation was abundant by day 4 (Figure 4.4a). However, newly precipitated circular patches, comprising freshly nucleated crystallites, appeared to form on the surface by day 7 (Figure 4.4b), and a homogeneous, porous layer was observed after a 14-day period of immersion (Figure 4.4c).

Surface cracks on the heat-treated SiHA coating appeared to be exacerbated after immersion in SBF-K9 solution for 1 day, although this effect may be due to a preferential dissolution along the crack regions. By day 4, freshly nucleated crystallites were observed which covered up to 85% of the coating surface (Figure 4.5a). A homogeneous, porous layer was then observed after day 7 (Figure 4.5b), which had transformed into a dense layer by day 14 (Figure 4.5c).

In contrast, a porous layer consisting of short interconnecting rods (similar to Figure 4.4c) and loosely-packed globules (Figure 4.6) formed on the surface of as-deposited HA coating and uncoated Ti substrate after 14 days of immersion, respectively.

A detailed characterization of the newly formed layer confirmed that it was rich in calcium and phosphorus, along with the incorporation of sodium and magnesium. Broad and diffuse XRD peaks ascribed to HA were observed (Figure 4.7), indicating that the calcium phosphate (CaP) layer was apatitic in nature. With prolonged immersion in SBF, the peak intensity corresponding to lattice plane reflection (002) increased, which implied that this CaP layer was aligned preferentially along the *c*-axis. The appearance of carbonate (C–O) bands at 867 and 1416 cm^{-1} in the IR spectra (Figure 4.8) confirmed the presence of a carbonate-containing apatite layer. It was also noted that the hydroxyl (O–H) band intensity decreased, while the intensities of the C–O bands became more pronounced; this suggested a substitution of CO_3^{2-} ions for OH^- ions in the HA structure [36].

In the case of the as-deposited SiHA coating, the effect of precipitation process was attributed to a lack of coating crystallinity. Dissolution is known to occur during the initial stage when Ca^{2+} and P^{5+} ions are released into the SBF. These ions will increase the degree of supersaturation of SBF solution, thereby facilitating the initial formation of CaP nuclei on the coating surface, by consuming the Ca^{2+} and P^{5+} ions in the SBF solution. CaP crystallite growth may then proceed along with the continued formation of additional nuclei.

As for the heat-treated SiHA coating, the apatite nucleation process could be related to the structural-scale factor, which explains why heat-treated SiHA is more bioactive than as-deposited SiHA. It has been proposed that the nanocrystalline structure creates an increased grain boundary area, thereby providing an

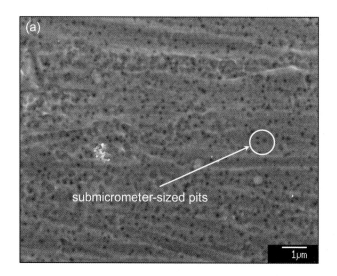
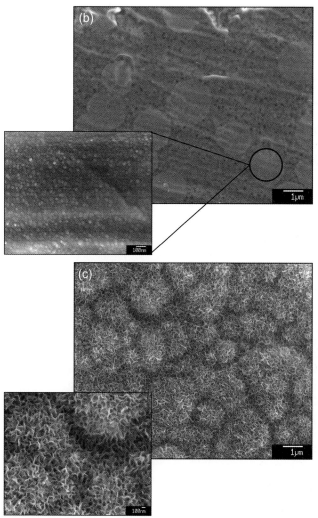

Figure 4.4 Micrographs of as-deposited 0.8 wt% SiHA coating after immersion in SBF solution for: (a) 4 days; (b) 7 days; and (c) 14 days [35].

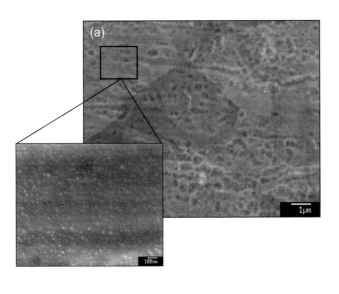
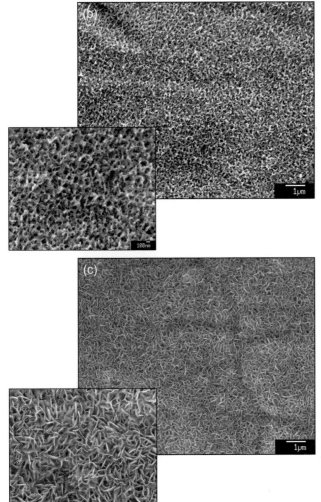

Figure 4.5 Micrographs of heat-treated 0.8 wt% SiHA coating after immersion in SBF solution for: (a) 4 days; (b) 7 days; and (c) 14 days [35].

158 4 Ceramic Nanocoatings and Their Applications in the Life Sciences

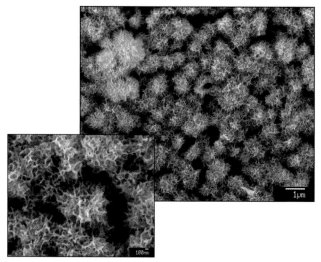

Figure 4.6 Micrograph of uncoated Ti substrate after immersion in SBF solution for 14 days.

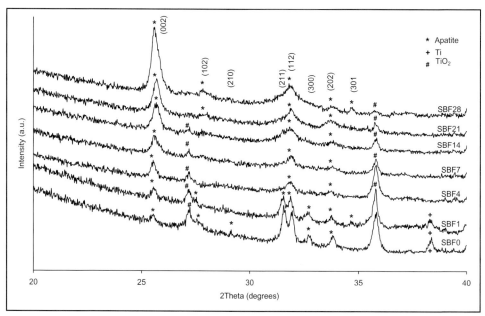

Figure 4.7 X-ray diffraction patterns of 0.8 wt% SiHA coatings before and after immersion in SBF solution [35].

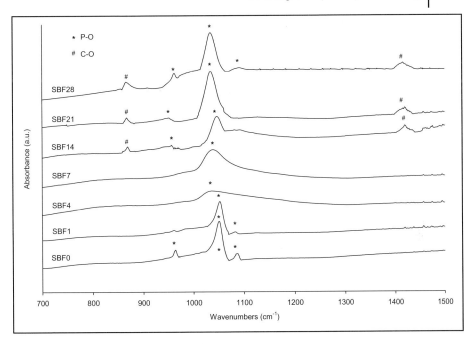

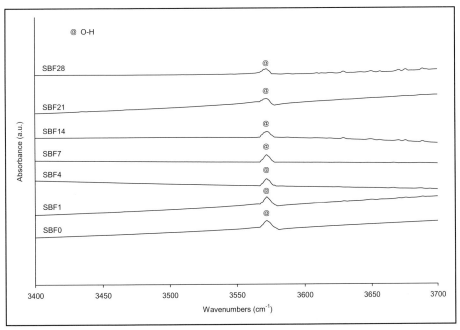

Figure 4.8 Infra-red spectra of 0.8 wt% SiHA coatings before and after immersion in SBF solution [35].

abundance of nucleation sites for CaP crystallites to form, since crystallite formation usually commences at areas of low surface energy. Once formed, these crystallites grew spontaneously and transformed into an apatite layer on the coating.

When comparing SiHA and HA coatings, a dense carbonate-containing apatite layer was formed on SiHA by day 14. However, this was not the case for HA, which suggested that SiHA was more bioactive than HA. As dissolution–precipitation is a surface-mediated process, the presence of surface silicate anions (due to the released Si^{4+} ions from SiHA) might have altered the surface properties of the material, thus attributing to the high bioactivity of SiHA. Botelho et al. [37] demonstrated that, in the presence of silicate anions, the surface charge of SiHA was decreased, inducing favorable sites for nucleation and crystallization of the apatite.

4.4.2
In Vitro Cellular Testing

Although the bioactivity of SiHA coatings has been established using acellular SBF, the conclusions drawn from such tests are often limited as the procedure does not take into consideration the effects of the cellular response to material surfaces. Hence, in this section we will detail the biological response of human osteoblast-like (HOB) cells to SiHA coatings, taking into account the effect of varying the Si substitution on the coating stability and its bioactivity.

The results of the AlamarBlue™ assay suggested that an increase in HOB cell growth with culture time was generally observed for all samples (Figure 4.9a) [38]. Although HOB cells growing on heat-treated 0.8 wt% SiHA (S2) samples showed the greatest growth level at all time points, significant differences were seen at days 3, 7, and 14 when compared to as-deposited 0.8 wt% SiHA (S1) and S2 samples, or to uncoated titanium (Ti) substrate and S2 samples.

Thian et al. [39] also showed that the metabolic activity of HOB cells varied with the different levels of Si substitution (Figure 4.9b). In general, all samples were capable of supporting cell growth, with the heat-treated 2.2 wt% SiHA (S3) sample exhibiting the highest growth level throughout the culture period. A statistically significant increase in cell growth was observed at day 14 on heat-treated 4.9 wt% SiHA (S4) as compared to S2 samples. Furthermore, a significant increase was noted for S3 when compared to S2 at days 4 and 7.

HOB cells attaching to the S1 and S2 samples displayed numerous well-developed actin filaments (green staining in Figure 4.10a,b) and distinct vinculin focal adhesion plaques (red staining) throughout the cell membranes. In contrast, very few adhesion plaques were expressed on the uncoated Ti substrate, and the cytoskeleton organization was poorly developed (Figure 4.10c) [38].

The cytoskeleton organization revealed distinctive differences in its quality among SiHA coatings with varying Si levels [39]. Well-developed actin filaments which aligned along the long axis of the cells were apparent within HOB cells on both the S3 and S4 samples (Figure 4.10d,e), but were less distinct and diffuse within HOB cells on the S2 sample.

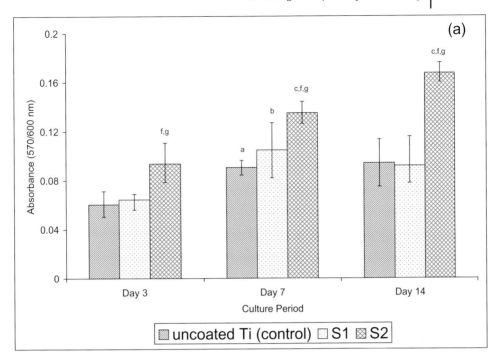

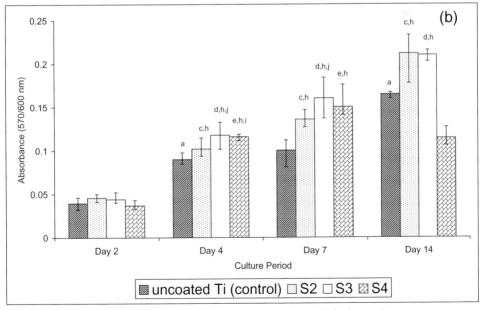

Figure 4.9 (a,b) Growth of HOB cells with culturing time. Key to samples: S1 = as-deposited 0.8 wt% SiHA; S2 = heat-treated 0.8 wt% SiHA; S3 = heat-treated 2.2 wt% SiHA; S4 = heat-treated 4.9 wt% SiHA. $^a p < 0.05$, higher on uncoated Ti substrate (control) between groups; $^b p < 0.05$, higher on S1 between groups; $^c p < 0.05$, higher on S2 between groups; $^d p < 0.05$, higher on S3 between groups; $^e p < 0.05$, higher on S4 between groups; $^f p < 0.05$, higher on S2 than control within groups; $^g p < 0.05$, higher on S2 than S1 within groups; $^h p < 0.05$, higher on coated (S2, S3 or S4) than control within groups, $^i p < 0.05$, higher on S4 than S2 within groups; $^j p < 0.05$, higher on S3 than S2 within groups [38, 39].

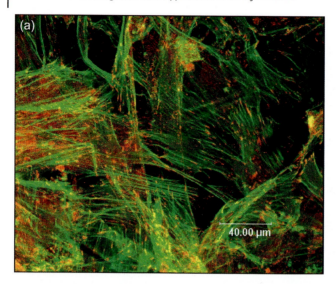

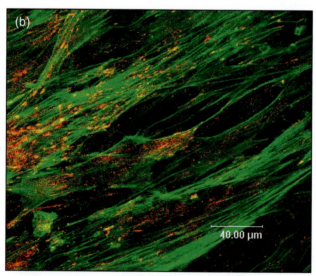

Figure 4.10 Confocal fluorescence microscopy of the nuclear DNA (blue), actin cytoskeleton (green) and vinculin plaque (red) in an HOB cell, as revealed with multiple labeling using TOTO-3, fluoroscein isothiocyanate (FITC)-conjugated phalloidin and Texas red-conjugated streptavidin on: (a) S1; (b) S2; (c) uncoated Ti substrate (control); (d) S3; and (e) S4 [38, 39].

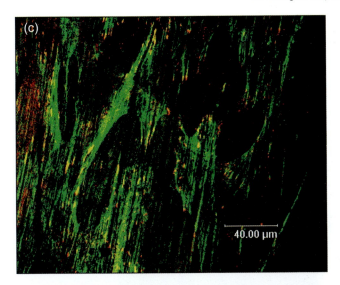

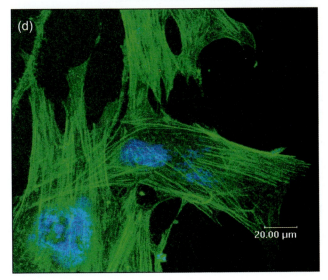

Figure 4.10 *Continued*

The cells appeared to attach, spread, and grow on all samples, with filopodia extending from the cell edges [39]. Cell spreading was less pronounced, with underdeveloped lamellipodia displayed, on the uncoated substrate (Figure 4.11a). In contrast, well-flattened cells were seen to cover the surfaces completely on all SiHA coatings (Figure 4.11b). Beyond day 42, signs of mineralization were observed on the surfaces of all samples. The formation of CaP mineral deposits on the uncoated Ti substrate was sparse (Figure 4.11c), whereas numerous and

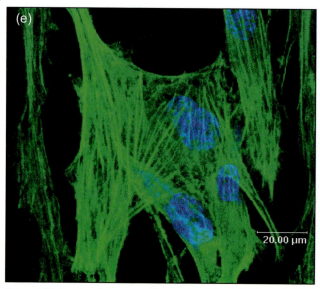

Figure 4.10 *Continued*

large mineral deposits were seen on all SiHA coatings, the most significant being for S4 (Figure 4.11d–f).

An apparent difference in the coating behavior was revealed after subjecting samples to a cell culture medium [38, 39]. For the S1 sample, the surface was covered by a lacey CaP network structure, with CaP spherulites seen precipitating and coalescing on the dissolving coating surface (Figure 4.12a). Beyond day 56, the CaP spherulite precipitates was detached completely from the S1 sample.

Nanocrystallites began to nucleate on the surfaces of the S2, S3, and S4 samples (Figure 4.12b–d). Furthermore, surface dissolution was observed on S4, as evidenced by the appearance of dissolution pits. A porous CaP structure was obtained for the S2 sample, while a dense structure was achieved for both the S3 and S4 samples. Beyond day 42, the CaP precipitate layer on all samples appeared similar, as it grew more thickly and more dense (Figure 4.12e). In contrast, a CaP layer was formed on the uncoated substrate after only 42 days in culture (Figure 4.12f).

The enhanced bioactivity of SiHA-coated substrate was attributed to the following effects: (i) the presence of HA promoting cell adhesion, growth and mineralization; and (ii) the presence of Si stimulating bone cell metabolism.

Si has been shown to inhibit crystal growth, and this effect was more significant with increasing Si levels [16, 31, 32]. As such, it might be expected that with increased Si content, the HA crystals would tend to dissolve faster, thereby facilitating the rapid precipitation of a carbonated-apatite layer which provides favorable sites at which the cells can attach and grow [40]. The released Si^{4+} ions might also have a positive protein adsorption. The formed silicate network structure was shown to be capable of adsorbing proteins [41], and consequently SiHA coatings would tend to contain a higher protein concentration than would either uncoated

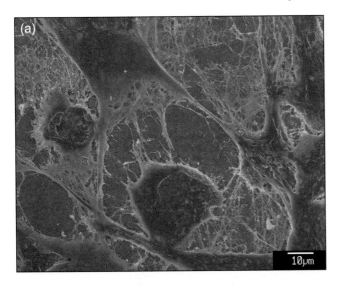

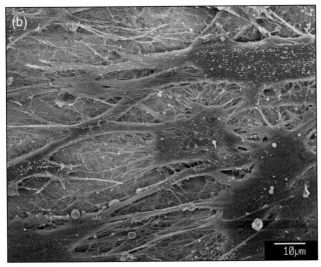

Figure 4.11 Cell morphology at different culture times.
(a) Uncoated Ti substrate (control) at day 2; (b) S2 at day 2;
(c) Control at day 42; (d) S2 at day 42; (e) S3 at day 42; (f) S4
at day 42 [16, 39].

Ti substrates or HA-coated substrates. These adsorbed proteins are likely to trigger a specific mRNA for gene expression, thereby carrying the coded information to ribosomal sites of protein synthesis in the cell, and stimulating osteoblast outgrowth. The results of studies conducted by Keeting *et al.* [42] and by Reffitt *et al.* [43] supported these hypotheses, as they showed that soluble Si simulated the proliferation and differentiation of HOB cells *in vitro*.

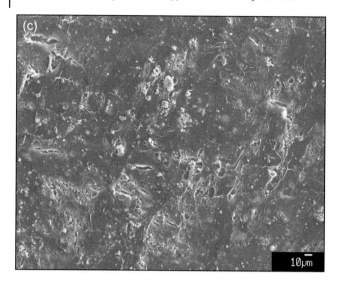

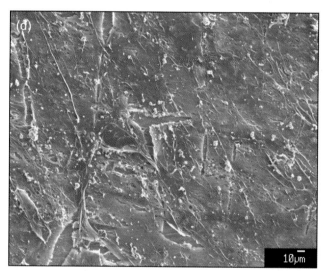

Figure 4.11 Continued

A correlation also existed between surface wettability and cell response. The results obtained from contact angle measurements suggested that the uncoated and coated Ti were hydrophobic and hydrophilic in nature, respectively. As such, it could be deduced that, by rendering a surface more hydrophilic, it would be possible to promote an initial cell adhesion and spreading, and thereby govern later cell proliferation and differentiation. This result is in accordance with the findings of Redey *et al.* [44] and Spriano *et al.* [45], who indicated that cells tended to spread better on a more wettable substratum.

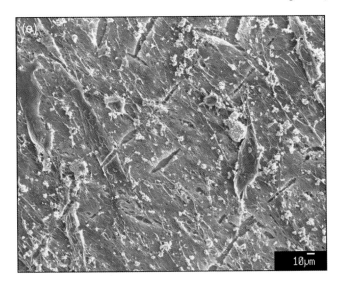

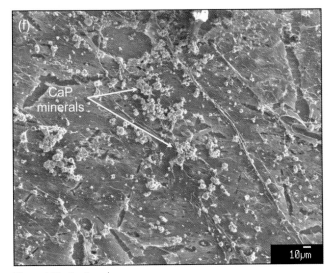

Figure 4.11 *Continued*

Cell growth was retarded at early culture period for coatings with a high Si level (4.9 wt%), due to a rapid dissolution of the coatings owing to their small crystallite size. However, when the surface was surface was modified by the growth of a new carbonated-apatite layer the cells began to migrate, adhere, and grow rapidly.

Taken together, all of these findings have indicated that Si tends to control the dissolution rate of the coating and, at the same time, plays an important role in the mineralization process. As such, of the three compositions studied, 2.2 wt% SiHA would be the preferred optimum, and merits further investigation.

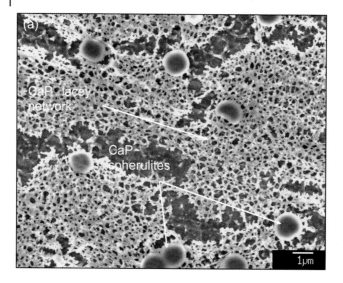

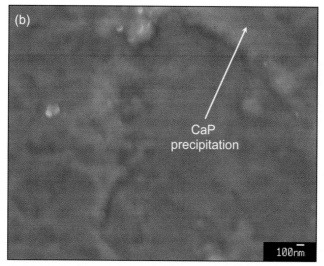

Figure 4.12 Coating morphology at different culture times.
(a) S1 at day 4; (b) S2 at day 2; (c) S3 at day 2; (d) S4 at day 2; (e) S2 at day 42; (f) Uncoated Ti substrate (control) at day 42 [39].

4.5
Future Perspectives

Whilst the recent introduction of SiHA as a nanostructured coating for biomedical applications continues to show great promise, there is still potential for major advances to be made in this area. These include:

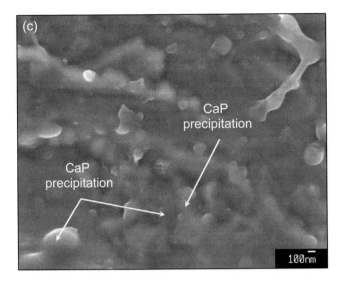

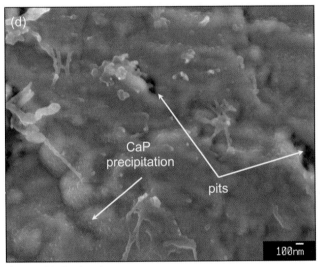

Figure 4.12 *Continued*

- An evaluation of the mechanical performance of existing deposited coatings.
- An evaluation of the bonding mechanism between bone and existing coatings, using an animal model *in vivo*.
- An improvement of existing coatings, with the ability to incorporate biological agents.
- An investigation of the precise mechanism by which Si activates cell activity.

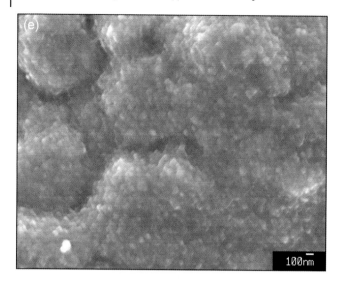

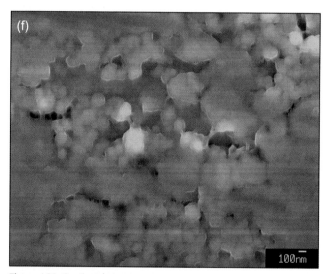

Figure 4.12 Continued

4.6
Conclusions

Novel, nanostructured SiHA coatings of varying Si levels have been deposited on metallic substrates using a magnetron sputtering technique. The results of *in vitro* studies indicate that these coatings exhibit a bioactivity which is clearly superior to that of uncoated and HA-coated substrates. Whilst a high Si level resulted in an enhancement of the biomineralization process, it also led to a rapid coating dissolution rate. In summary, the choice of coating depends on the final application; however, these SiHA coatings represent an excellent substitute for plasma-sprayed coatings and HA thin coatings.

References

1 Plenk, H. (1998) Prosthesis-bone interface. *Journal of Biomedical Materials Research*, **43**, 350–5.
2 Hulbert, S.F., Cooke, F. and Klawitter, J.J. (1973) Attachment of prostheses to the musculoskeletal system by tissue ingrowth and mechanical interlocking. *Journal of Biomedical Materials Research Symposium*, **4**, 1–23.
3 Aoki, H., Kato, K., Ogiso, M. and Tabata, T. (1977) Studies on applications of apatite dental materials. *Journal of Dental Engineering*, **18**, 86–9.
4 Oonishi, H., Hench, L.L., Wilson, J., Suginara, F., Tsuji, E., Kushitani, S. and Iwaki, H. (1999) Comparative bone growth behaviour in granules of bioceramic materials of various sizes. *Journal of Biomedical Materials Research*, **44**, 31–43.
5 Driessens, F.C.M. (1980) The mineral in bone, dentin and tooth enamel. *Bulletin des Sociétés Chimiques Belges*, **89**, 663–89.
6 Carlisle, E.M. (1970) Silicon: a possible factor in bone calcification. *Science*, **167**, 279–80.
7 Carlisle, E.M. (1972) Silicon: an essential element for the chick. *Science*, **178**, 619–21.
8 Carlisle, E.M. (1980) Biochemical and morphological changes associated with long bone abnormalities in silicon deficiency. *Journal of Nutrition*, **110**, 1046–55.
9 Xynos, I.D., Hukkanen, M.V., Batten, J.J., Buttery, L.D., Hench, L.L. and Polak, J.M. (2000) Bioglass 45S5® stimulates osteoblast turnover and enhances bone formation *in vitro*: implications and applications for bone tissue engineering. *Calcified Tissue International*, **67**, 321–9.
10 Xynos, I.D., Edgar, A.J., Buttery, L.D., Hench, L.L. and Polak, J.M. (2001) Gene-expression profiling of human osteoblasts following treatment with the ionic products of Bioglass® 45S5 dissolution. *Journal of Biomedical Materials Research*, **55**, 151–7.
11 Gough, J.E., Jones, J.R. and Hench, L.L. (2004) Nodule formation and mineralisation of human primary osteoblasts cultured on a porous bioactive glass scaffold. *Biomaterials*, **25**, 2039–46.
12 Patel, N., Best, S.M., Bonfield, W., Gibson, I.R., Hing, K.A., Damien, E. and Revell, P.A. (2002) A comparative study on the *in vivo* behaviour of hydroxyapatite and silicon substituted hydroxyapatite granules. *Journal of Materials Science: Materials in Medicine*, **13**, 1199–206.
13 Porter, A.E., Patel, N., Skepper, J.N., Best, S.M. and Bonfield, W. (2004) Effect of sintered silicate-substituted hydroxyapatite on remodeling processes at the bone-implant interface. *Biomaterials*, **25**, 3303–14.
14 Patel, N., Brooks, R.A., Clarke, M.T., Lee, P.M.T., Rushton, N., Gibson, I.R., Best, S.M. and Bonfield, W. (2005) *In vivo* assessment of hydroxyapatite and silicate-substituted hydroxyapatite granules using an ovine defect model. *Journal of Materials Science: Materials in Medicine*, **16**, 429–40.
15 Thian, E.S., Huang, J., Best, S.M., Barber, Z.H. and Bonfield, W. (2005) A novel way of incorporating silicon in hydroxyapatite (Si-HA) thin films. *Journal of Materials Science: Materials in Medicine*, **16**, 411–15.
16 Thian, E.S., Huang, J., Vickers, M.E., Best, S.M., Barber, Z.H. and Bonfield, W. (2006) Silicon-substituted hydroxyapatite (Si-HA): a novel calcium phosphate coating for biomedical applications. *Journal of Materials Science*, **41**, 709–17.
17 Solla, E.L., Gonzalez, P., Serra, J., Chiussi, S., Leon, B. and Lopez, J.G. (2007) Pulsed laser deposition of silicon substituted hydroxyapatite coatings from

synthetical and biological sources. *Applied Surface Science*, **254**, 1189–93.

18 Solla, E.L., Borrajo, J.P., Gonzalez, P., Serra, J., Chiussi, S., Serra, C., Leon, B. and Perez-Amor, M. (2008) Pulsed laser deposition of silicon-substituted hydroxyapatite coatings. *Vacuum*, **82**, 1383–5.

19 Lopez-Alvarez, M., Solla, E.L., Gonzalez, P., Serra, J., Leon, B., Marques, A.P. and Reis, R.L. (2009) Silicon-hydroxyapatite bioactive coatings (Si-HA) from diatomaceous earth and silica. Study of adhesion and proliferation of osteoblast-like cells. *Journal of Materials Science: Materials in Medicine*, **20**, 1131–6.

20 Hijon, N., Cabanas, M.V., Pena, J. and Vallet-Regi, M. (2006) Dip-coated silicon-substituted hydroxyapatite films. *Acta Biomaterialia*, **2**, 567–74.

21 Balamurugan, A., Rebelo, A.H.S., Lemos, A.F., Rocha, J.H.G., Ventura, J.M.G. and Ferreira, J.M.F. (2008) Suitability evaluation of sol-gel derived Si-substituted hydroxyapatite for dental and maxillofacial applications through in vitro osteoblast response. *Dental Materials*, **24**, 1374–80.

22 Xiao, X.F., Liu, R.F. and Tang, X.L. (2008) Electrophoretic deposition of silicon substituted hydroxyapatite coatings from n-butanol-chloroform mixture. *Journal of Materials Science: Materials in Medicine*, **19**, 175–82.

23 Zhang, E., Zou, C. and Yu, G. (2009) Surface microstructure and cell biocompatibility of silicon-substituted hydroxyapatite coating on titanium substrate prepared by a biomimetic process. *Materials Science and Engineering C: Biomimetic and Supramolecular*, **29**, 298–305.

24 Zhang, E., Zou, C. and Zeng, S. (2009) Preparation and characterization of silicon-substituted hydroxyapatite coating by a biomimetic process on titanium substrate. *Surface and Coatings Technology*, **203**, 1075–80.

25 Wolke, J.G.C., van Dijk, K., Schaeken, H.G., de Groot, K. and Jansen, J.A. (1994) Study of the surface characteristics of magnetron-sputter calcium phosphate coatings. *Journal of Biomedical Materials Research*, **28**, 1477–84.

26 Klein, C.P.A.T., Wolke, J.G.C., de Blieck-Hogervorst, J.M.A. and de Groot, K. (1994) Features of calcium phosphate coatings: an *in vitro* study. *Journal of Biomedical Materials Research*, **28**, 961–7.

27 Bloebaum, R.D. and Dupont, J.A. (1993) Osteolysis from a press-fit hydroxyapatite-coated implant. A case study. *Journal of Arthroplasty*, **8**, 195–202.

28 Morscher, E.W., Hefti, A. and Aebi, U. (1998) Severe osteolysis after third-body wear due to hydroxyapatite particles from acetubular cup coating. *Journal of Bone and Joint Surgery*, **80B**, 267–72.

29 de Brujin, J.D., Bovell, Y.P. and van Blitterswijk, C.A. (1994) Structural arrangements at the interface between plasma sprayed calcium phosphates and bone. *Biomaterials*, **15**, 543–50.

30 Driessens, F.C. (1988) Physiology of hard tissues in comparison with the solubility of synthetic calcium phosphates. *Annals of the New York Academy of Sciences*, **523**, 131–6.

31 Gibson, I.R., Best, S.M. and Bonfield, W. (2002) Effect of silicon substitution on the sintering and microstructure of hydroxyapatite. *Journal of the American Ceramic Society*, **85**, 2771–7.

32 Arcos, D., Rodriguez-Carvajal, J. and Vallet-Regi, M. (2004) Silicon incorporation in hydroxylapatite obtained by controlled crystallization. *Chemistry of Materials*, **16**, 2300–8.

33 Kim, S.R., Lee, J.H., Kim, Y.T., Riu, D.H., Jung, S.J., Lee, Y.J., Chung, S.C. and Kim, Y.H. (2003) Synthesis of Si,Mg substituted hydroxyapatite and their sintering behaviours. *Biomaterials*, **24**, 1389–98.

34 Kokubo, T., Kushitani, H., Sakka, S., Kitsugi, T. and Yamamuro, T. (1990) Solution able to reproduce *in vivo* surface-structure changes in bioactive glass-ceramic A-W. *Journal of Biomedical Materials Research*, **24**, 721–34.

35 Thian, E.S., Huang, J., Best, S.M., Barber, Z.H. and Bonfield, W. (2006) Novel silicon-doped hydroxyapatite (Si-HA) for biomedical coatings: an *in vitro* study using acellular simulated body fluid. *Journal of Biomedical Materials Research*, **76B**, 326–33.

36 Rehman, I. and Bonfield, W. (1997) Characterisation of hydroxyapatite and carbonate apatite by photo acoustic FTIR spectroscopy. *Journal of Materials Science: Materials in Medicine*, 8, 1–4.

37 Botelho, C.M., Lopes, M.A., Gibson, I.R., Best, S.M. and Santos, J.D. (2002) Structural analysis of Si-substituted hydroxyapatite: zeta potential and X-ray photoelectron spectroscopy. *Journal of Materials Science: Materials in Medicine*, 13, 1123–7.

38 Thian, E.S., Huang, J., Best, S.M., Barber, Z.H. and Bonfield, W. (2005) Magnetron co-sputtered silicon-containing hydroxyapatite thin films—an *in vitro* study. *Biomaterials*, 26, 2947–56.

39 Thian, E.S., Huang, J., Best, S.M., Barber, Z.H., Brooks, R.A., Rushton, N. and Bonfield, W. (2006) The response of osteoblasts to nanocrystalline silicon-substituted hydroxyapatite thin films. *Biomaterials*, 27, 2692–8.

40 Neo, M., Nakamura, T., Ohtsuki, C., Kokubo, T. and Yamamuro, T. (1993) Apatite formation on three kinds of bioactive materials at early stage *in vivo*: a comparative study by transmission electron microscopy. *Journal of Biomedical Materials Research*, 27, 999–1006.

41 Schwarz, K. (1974) Recent dietary trace element research, exemplified by tin, fluorine and silicon. *Federal Proceedings*, 33, 1748–57.

42 Keeting, P.E., Oursler, M.J., Wiegand, K.E., Bonde, S.K., Spelsberg, T.C. and Riggs, B.L. (1992) Zeolite A increases proliferation, differentiation, and transforming growth factor β production in normal adult human osteoblast-like cells *in vitro*. *Journal of Bone and Mineral Research*, 7, 1281–9.

43 Reffitt, D.M., Ogston, N., Jugdaoshsingh, R., Cheung, H.F.J., Evans, B.A.J., Thompson, R.P.H., Powell, J.J. and Hampson, G.N. (2003) Orthosilicic acid stimulates collagen type 1 synthesis and osteoblastic differentiation in human osteoblast-like cells *in vitro*. *Bone*, 32, 127–35.

44 Redey, S.A., Nardin, M., Bernache-Assolant, D., Rey, C., Delannoy, P., Sedel, L. and Marie, P.J. (2000) Behavior of human osteoblastic cells on stoichiometric hydroxyapatite and type A carbonate apatite: role of surface energy. *Journal of Biomedical Materials Research*, 50, 353–64.

45 Spriano, S., Bosetti, M., Bronzoni, M., Verne, E., Maina, G., Bergo, V. and Cannas, M. (2005) Surface properties and cell response of low metal ion release Ti-6Al-7Nb alloy after multi-step chemical and thermal treatments. *Biomaterials*, 26, 1219–29.

5
Gold Nanofilms: Synthesis, Characterization, and Potential Biomedical Applications

Shiho Tokonami, Hiroshi Shiigi, and Tsutomu Nagaoka

5.1
Introduction

In the past, metal nanoparticles have played important roles in many remarkable developments in nanotechnology. Gold nanoparticles (AuNPs), in particular, are recognized as the most stable metal nanoparticles. Such stability, coupled with their simple synthesis procedures and surface functionalization, have helped both scientists and engineers expand the use of AuNPs over a wide range of applications, including cytological staining, sensors, and electronics. When in aqueous solution, AuNPs exhibit a purplish red color in the dispersion state; in fact, colloidal gold has long been used as a pigment to produce ruby-colored glass and stained glass. Some examples of this date back to the fourth century BC, and the Lycurgus Cup [1], now a possession of the British Museum, is probably the most famous example. Although reports dating from the seventeenth century refer to the use of colloidal gold sols in medicine, Faraday's treatise of 1857 on the preparation of colloidal gold by the reduction of an aqueous solution of chloroaurate ($AuCl_4^-$), is recognized as the first such academic report [2]. The AuNPs that Faraday prepared are now housed at The Royal Institution of Great Britain, and still exhibit a deep red color despite their age. Faraday described the red suspended solution as a "divided gold metal" and, as a consequence, first investigated the relationship between particle size and the color of the solution. Today, metal nanoparticles, including semiconductor nanoparticles, play extremely important roles in the progression of nanotechnology [3–5], and several excellent reviews of their applications as chemical/biological sensors and as electronics components, together with details of their preparation, have been published during the past decade [6–12].

One of the most outstanding properties of AuNPs is their extreme visibility; for example, 40 nm AuNPs exhibit an extinction coefficient of $2 \times 10^9 M^{-1} cm^{-1}$, which is far greater than a typical fluorescent molecule such as fluorescein ($9.2 \times 10^4 M^{-1} cm^{-1}$ in ethanol at 483-nm), This high extinction coefficient results in an enhanced detection sensitivity of approximately 2.2×10^4-fold [13]. In

---

Nanomaterials for the Life Sciences Vol.5: Nanostructured Thin Films and Surfaces.
Edited by Challa S. S. R. Kumar
Copyright © 2010 WILEY-VCH Verlag GmbH & Co. KGaA, Weinheim
ISBN: 978-3-527-32155-1

addition, AuNPs exhibit properties of surface modification through their self-assembly with thiols and/or amino groups, which in turn adds surface functionality; indeed, such particles have been widely used as functional probes.

Another striking feature of AuNPs is their ability to change color as they aggregate [2]. AuNPs often carry a negative charge that leads to their particle dispersion being stable in solution. The color changes that result from aggregation can be used to great effect in the recognition of interactions between receptor-modified nanoparticles and analytes, either by the removal of particle charge or by the formation of an interconnected structure, AuNP–analyte–AuNP. Such interactions produce a measurable color change due to long-range plasmon coupling. Since the mid-1990s, a host of research groups have exploited these valuable characteristics of AuNPs to develop a wide range of applications, many of which have focused on bioanalysis.

The enhanced sensitivity of AuNPs as markers has been utilized not only for microscopic observations but also for a variety of spectroscopic techniques; the latter include fluorescence spectroscopy and surface-enhanced Raman scattering (SERS), both of which are sufficiently sensitive to report a single molecular event. Very intense SERS signals have been observed from molecules that have been adsorbed onto the aggregated particle assembly [14]. The increased mass that results from the binding of an analyte with an AuNP-immobilized electrode can be efficiently monitored using a quartz crystal microbalance (QCM).

Recently, a new sensing technique has been developed by the present authors, using an AuNP array that consists of AuNPs interconnected with a dithiol linker, and is deposited on a glass substrate. The inter-particle separation created by the dithiol is ~1 nm, and total resistances of 100–100 MΩ have been obtained with this array. These arrays can also be formed using a bulky molecule, such as AuNPs capped with dendrimers, where the nanospace created between the particles provides a receptacle of molecular dimensions that can be optimally utilized to immobilize a single receptor molecule. In this way, molecular conductivity can be measured directly, and changes in array resistance can be evaluated when an analyte binds to the inserted receptor. In fact, the insertion of an appropriate receptor molecule (e.g., a single-stranded DNA oligomer) between the particles will result in a substantial decrease in resistance when a base-matched oligomer is added, due mainly to the higher conductivity of the hybridized oligomers. Such an electrical detection constitutes a novel assay technique, and affords a cost-effective alternative to the fluorescence technique.

The first part of this chapter includes a discussion of the basic characteristics of metal nanoparticles (with special emphasis on gold), in addition to a brief description of recent developments in analytical applications. Following this, the details are provided of nonlabeling sensing systems that use AuNP arrays deposited on an insulator material. Whilst these systems are used to quantify organic gases, antigens, and DNA, particular emphasis is placed here on the DNA assay, and on efforts to enhance the sensitivity of the array sensors.

5.2
Preparation of Various AuNPs

A wide variety of spherical metal nanoparticles, including functionalized nanoparticles, have become commercially available with the growth of applications in the biological field. In comparison with nanoparticles of other metals, AuNPs can be synthesized very easily. Moreover, in the case of AuNPs control can be exercised not only on the particle shape, to yield spherical, planar (disk or trigonal), hexagonal, rod, and core–shell nanoparticles, but also on the particle size, with a variety of diameters having been reported [15–18]. To date, the control of particle size has been achieved over the range of ~100 nm. The typical preparation methods employed are described later in the chapter [19].

Among the many "conventional" methods used for AuNP synthesis, citrate reduction is perhaps the best-known [7]. The most popular of the reports appears to be that of Turkevitch which, in 1951, described the synthesis of AuNPs that were approximately 20 nm in diameter. The synthesis of AuNPs in this way is simple, and leads to an easy complex formation with DNA and antibodies. The Brust method (a two-phase synthesis and stabilization with thiol), which dates back to 1994, also ranks high among synthesis techniques as it permits not only the control of particle diameter and grain-size distribution, but also a functionalization of the particle surface with thiol molecules. The Brust method also has had a considerable impact on the thermal and air stabilities of synthesized particles with mean particle diameters ranging from 1.5 to 5.2 nm [20, 21].

Synthetic techniques using physical energy, such as gamma beams, heat, and ultrasonic waves, have also been reported [22, 23]. An interesting report was also made on the production of AuNPs via biological activity, when Paradeep and coworkers reported the formation of submicrometer-diameter Au, Ag, and Au/Ag alloy nanoparticles in bacteria, by mixing $HAuCl_4$ and $AgNO_3$ with *Lactobacillus* strains [24]. Indeed, a preparation method that uses living organisms might represent an attractive approach for an "environmentally conscious" synthesis.

Common reducing agents, such as citric acid and ascorbic acid, impart a negative surface charge to the AuNPs. However, in many applications of AuNPs, and in many fields, it is necessary to prepare positively charged AuNPs, especially for interactions with biological molecules, such as DNA. Weare *et al.* have developed a simple method to synthesize phosphine-stabilized AuNPs with a core size of approximately 1.5 nm [25]. Their procedure was as follows: Hydrogen tetrachloroaurate trihydrate (2.54 mmol) and tetraoctylammonium bromide (2.93 mmol) were dissolved in a water/toluene mixture (50:65, v/v) with nitrogen bubbling. As the golden color of the solution transferred into the organic phase, triphenylphosphine (8.85 mmol) was added, and the solution stirred for at least 10 min until the color of the organic phase changed. Aqueous $NaBH_4$ (37.3 mmol, dissolved in 10 ml water) was added rapidly, and the solution stirred for 3 h under nitrogen.

As another general scheme, these authors developed a one-step method for preparing positively charged AuNPs using aniline as a reducing agent [26]. Briefly, aqueous aniline solution was added to a 0.03% chloroauric acid aqueous solution (200 ml) and stirred at 65 °C for 30 min. The resultant solution was centrifuged at 8500 r.p.m. (7300×g) at 5 °C. The first supernatant was removed and the precipitate redispersed in 30 ml of ultrapure water; the dispersal process was repeated three times.

Typical synthesis procedures for AuNPs with various grain sizes were as follows:

2 nm AuNPs: 375 µl of 4.0% chloroauric acid and 500 µl of 0.2 M K_2CO_3 were added to 100 ml of deionized water and cooled to 4 °C. Five 1 ml aliquots of sodium borohydride solution were added to the chloroauric acid/carbonate suspension and stirred for 5 min on ice.

5 nm AuNPs: 6.25 ml of 1.0% chloroauric acid and 5.8 ml of 0.1 M K_2CO_3 were added to 500 ml of deionized water. A 4.16 ml aliquot of diluted phosphorus solution (one-fourth concentration of saturated white phosphorus in diethylether) was then added to the chloroauric acid/carbonate suspension, followed by stirring for 15 min at ambient conditions. The solution was brought to a boil and refluxed until the color changed.

12 nm AuNPs: A 5.2-ml aliquot of 2.0% L(+)-ascorbic acid (sodium salt; used as a reducing agent) and 6 ml of potassium carbonate were added to 200 ml of 0.03% aqueous chloroauric acid, and the mixture stirred at 278 K. At this point, the color of the solution was purple-red. The mixture was stirred at 353 K for 20 min until the color of the suspension changed from purple-red to red.

30 nm AuNPs: A 4.5 ml aliquot of 2.0% sodium citrate as a reducer was added to 200 ml of 0.03% aqueous chloroauric acid, and the mixture stirred at 353 K for 20 min.

50 nm AuNPs: 10 ml of 3.0% citric acid (used as a reducing agent) was added to 200 ml of 0.03% aqueous chloroauric acid, and the mixture stirred at 353 K for 20 min, producing an Au dispersion of 0.14 g l^{-1}.

5.3
Functionalization of AuNPs and their Applications through Aggregation

The ease by which AuNPs can be functionalized by surface modification is one reason why they have attracted so much attention in the field of analytical chemistry. Thiol molecules are used most frequently for the immobilization of functional materials on the surface of AuNPs. The free energy of binding between the thiol group and Au is approximately -20 kJ mol^{-1} [27]. Although the value for platinum (Pt) is very similar, it has been reported that when thiols are chemisorbed onto Pt they are prone to degradation via oxidation with adsorbed and dissociated oxygen [28, 29]. At the same time, the fact that AuNPs are less active than Pt nanoparticles during catalysis makes them more advantageous for thiol modification. For these reasons, gold is the material of choice in analytical chemistry and related

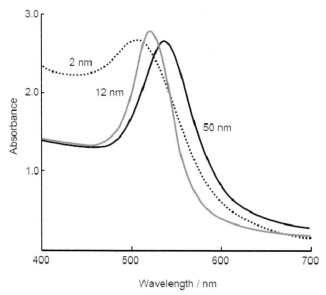

Figure 5.1 The UV-visible spectra of gold nanoparticles 2, 12, and 50 nm in diameter, dispersed in water. A red shift of the surface plasmon band is observed with increasing particle size.

fields. As mentioned above, metal nanoparticles – and AuNPs in particular – continue to attract attention as the fundamental building blocks for the formation of one- to three-dimensional, nanometer-sized architectures, sensors, and electric devices engineered for their specific shape and functionality.

AuNPs have a broad absorption band in the visible region of the electromagnetic spectrum. The first investigator to formulate the nature of the optical band as a surface plasmon effect was Mie – hence the so-called "Mie theory" [30]. The characteristics of the band arise from the collective oscillation of free-conduction electrons induced by an interacting electromagnetic field, and their resonances are noted as surface plasmon. The specific resonance frequency depends on a number of parameters, such as nanoparticle composition, morphology, concentration, solvent refractive index, surface charge, and temperature [31–38]. As for the relationship between particle diameter and plasmon absorbance, particle mean diameters of 9, 22, 48, and 99 nm give rise to surface plasmon absorbance maxima of 517, 521, 533, and 575 nm, respectively. An increase in the particle diameter results in the red shift of the plasmon band, as shown in Figure 5.1.

Although AuNPs are used as biomarkers in various types of bioassay, their surface plasmon properties have only recently been utilized in real applications [39–46]. Figure 5.2 represents some techniques based on the particle aggregation induced by the addition of target analyte, which is visually recognized as a plasmon shift from red to blue. Among these, Mirkin and coworkers first reported a technique based on the aggregation of DNA-modified nanoparticle probes for

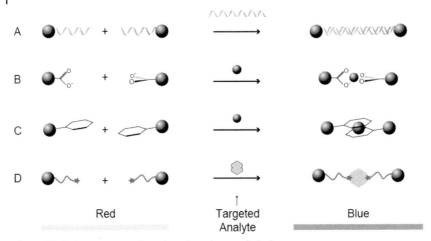

Figure 5.2 Various target analyses based on the crosslinked aggregation of gold nanoparticles. The target analytes are: (A) DNA; (B) heavy metal ions; (C) potassium ions; and (D) protein. Reprinted with permission from the Japanese Society for Analytical Chemistry.

application to DNA detection (Figure 5.3) [47]. Due to their high surface energy, AuNPs are extremely sensitive to environmental conditions. Mirkin et al. successfully applied this property to realize a simple application for the detection of nucleic acid. This method is basically a "sandwich assay" in which a target DNA molecule crosslinks two AuNPs by hybridization. Dispersed free AuNPs have a red color as a consequence of their plasmon absorption; however, upon aggregation, interparticle plasmon coupling gives rise to a lower energy state, which makes the optical observation of the hybridization event possible. Consequently, the shorter interparticle distance caused by complexation between DNA strands causes a change in the color of the AuNP dispersion. Since the early studies of Mirkin and colleagues, colorimetric detection using AuNPs has attracted considerable interest as a straightforward detection technique and an inexpensive alternative to the conventional fluorescence-type detection, owing to the distinct color change (Figure 5.2).

As a variant, Maeda and colleagues recently reported that the color change occurred even between a single DNA-modified probe nanoparticle and target DNA [48, 49]. These authors found that the color change arose very rapidly in their framework, compared to the crosslink method. A mismatch in the forefront of the probe anchored on the AuNPs was sensitive to hybridization in their system, and the existence of such a mismatch barely gave rise to precipitation, even for a high salt concentration. However, in the case of a mismatch in the probe sequence, a significant difference in the color was not observed due to particle aggregation.

It was suggested that the modified single stranded DNA (ssDNA) was adsorbed onto the AuNP surface to form a lying-down structure, leading to a nonspecific

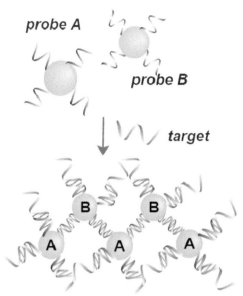

Figure 5.3 Aggregation-based DNA detection using two different gold nanoparticle probes modified by respective oligonucleotides. The probes are aggregated by hybridization with the added target complementary DNA, which induces the color change of the dispersed colloidal solution resulting in a red shift of the UV spectrum [47].

adsorption of the probe DNA. This adsorption was attributed to the interaction between the particle surface and one of the four bases constituting the probe DNA (adenine, cytosine, guanine, and thymine), thus giving rise to particle aggregation. Nonspecific adsorption occurs to different extents depending on the bases in the DNA sequence; a sequence of only thymine, for example, has less incidence [50]. Nonspecific adsorption on the gold substrate can be circumvented by inserting an appropriate alkanethiol at the 3'- or 5'-ends of a ssDNA sequence in order to reduce the density of DNA on the AuNP surface [51].

Zhao and colleagues examined the feasibility of target DNA analysis by changing the interparticle distance of AuNPs on paper substrates [52]. They conducted two assays for an endonuclease (DNase I) and aptamer-based adenosine detection. In this procedure, AuNPs crosslinked by DNA were first spotted onto the paper (purple), after which DNase I or adenosine was added to redisperse the AuNPs and change the color of the spot to red, within 1 min. This simple assay would be expected to provide a marked improvement in the practical analysis of pathogen tests, environmental analysis, and disease diagnosis.

Immunoassay techniques have attracted considerable attention in environmental monitoring, food safety, and clinical diagnosis [53]. Liu and colleagues have reported a one-step immunoassay technique for the detection of prostate cancer and free prostate-specific antigen (f-PSA) [54]. This method utilized an optical

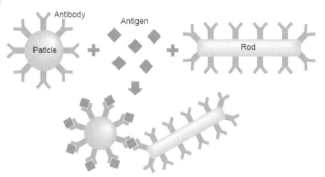

Figure 5.4 One-step immunoassay technique, reported by Liu et al., for the detection of a prostate cancer biomarker and free prostate-specific antigen. Detector antibody-modified AuNPs and capture antibody-immobilized Au nanorods are used as an optical probe. The nanoparticle and nanorod conjugate upon the addition of the target antigen through an antibody–antigen–antibody linker, leading to an increase in the particles' diameter, which is measured using dynamic light scattering. Adapted from Ref. [54] with permission from American Chemical Society.

probe that comprised a detector AuNP (37 nm) modified with the antibody and a captured Au nanorod (40 × 10 nm) immobilized with another antibody (Figure 5.4). The addition of antigen f-PSA enabled the conjugation of nanoparticle and nanorod through the antibody–antigen–antibody reaction, and the linked assembly was successfully detected using dynamic light scattering. The f-PSA molecule was detected over a concentration range from 0.1 to 10 ng ml^{-1} without the washing step, which is highly advantageous from the viewpoint of practical applications. Using Au nanoshells composed of a 96 nm silica core covered by a 22 nm gold shell, Hirsch et al. used immunoassays to detect a variety of clinically relevant blood-borne analytes that ranged in concentration from 0.8 to 88 ng ml^{-1}, within 10–30 min [55].

5.4
AuNP Assemblies and Arrays

5.4.1
AuNP Assemblies Structured on Substrates

Many research groups have attempted to provide building blocks for engineering substances of angstrom-scale and nanoscale dimensions by using metal nanoparticles. These can be fashioned on various materials with a wide functional diversity. As a result of their quantum-scale dimensions, these materials can be very different from the corresponding bulk materials with respect to their electronic, optical, and catalytic properties [56–65]. For example, AuNPs give rise to unique spectroscopic properties, such as surface plasmon resonance (SPR), and electrocatalytic properties due to their high surface area, high edge concentration, and

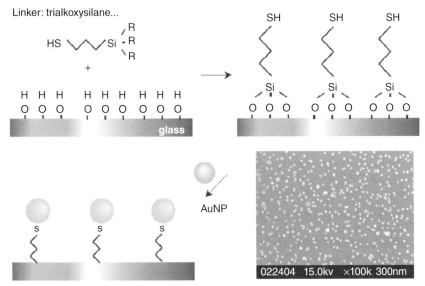

Figure 5.5 A schematic representation of the procedures for the fabrication of the gold nanoparticle assembly on a glass substrate in combination with a trialkoxysilane as a linker molecule. Lower right: Scanning electron microscopy image of a prepared AuNP monolayer film [70].

different electronic states. Assembling AuNPs onto nanoscale objects has been extensively studied due to their expected novel properties; this has been achieved through aerosol-gas depositing, inkjet printing, and the self-assembly of molecules [66]. Many reports exist on the fabrication and functionalization of 2-D or 3-D cumulative films using AuNPs, and their application to chemical sensors [67–69].

Since the mid-1990s, nanoparticles have been assembled on many substrates such as quartz, soda glass, and indium tin oxide (ITO). Willner *et al.* utilized trialkoxysilane functionalized with thiol or amino groups to modify a glass surface with AuNPs (Figure 5.5) [70]. The Au colloidal films thus prepared exhibited sufficient stability to withstand a series of electrochemical measurements, and revealed SPR bands resulting from the AuNP particles on the substrate [71–73].

The assemblies built on the conductive substrate functioned well in electrochemical experiments due to the high conductivity between the substrate and the metal nanoparticles (as discussed above). However, when the assembly was fabricated on an insulated substrate, the small particle density on the substrate surface precluded the development of electronic applications that required lateral, particle-to-particle conductivity. In order to circumvent this problem, the authors have successfully assembled AuNPs on a plastic plate and on microbeads, constructed from polystyrene and polyethylene, and using alkylthiol as a binder [74–78].

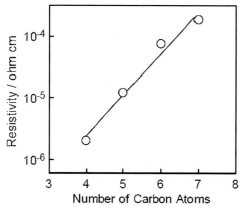

Figure 5.6 Effect of alkyl chain length in thiol binder molecules on a fabricated AuNP assembled film. The resistivity of the film changes exponentially with respect to the number of carbon atoms in the binder molecule [77].

The AuNPs assembled in this manner produced a single-layer film, the conductivity of which ($4.8 \times 10^5 \, \Omega^{-1} \, cm^{-1}$) was comparable to that of metallic bulk gold ($7.7 \times 10^5 \, \Omega^{-1} \, cm^{-1}$), when it was fabricated with a short alkyl chain binder (butanethiol); in contrast to the authors' technique, more than 15 layers of AuNPs have been required to construct a conductive AuNP array [79]. Moreover, by changing the length of the alkyl chain in the binder molecule, the resistance of the film could easily be varied from low values (typical of conductors) to high values (typical of insulators) (Figure 5.6). Electrocatalysis on the particle-deposited plastic surface was successfully monitored for H_2O_2, molecular oxygen, and organic amines [77]. Furthermore, the film prepared with butanethiol substantially attenuated the oxidation of ascorbic acid, indicating that this technique could be applied effectively for the electroanalysis of biochemical fluids.

5.4.2
AuNP Assembly on Biotemplates

For the precise positioning of AuNP nanostructures, the interaction between the particles and substrate is a key factor, and the surface charge of AuNPs, therefore, plays an important role. Most of the surface charges of nanoparticles are fixed on the particle surface through the adsorption of a reducing agent in the preparation. Organic acid reductants (e.g., citric acid and ascorbic acid) yield negatively charged AuNPs; the anionic carboxylate added in the synthetic process is adsorbed onto the surface of AuNPs as a protective layer, thereby imparting a negative charge. However, applications often need not only negatively charged but also positively charged AuNPs. For example, positively charged AuNPs are required for interaction with DNA, a polyanionic biomolecule.

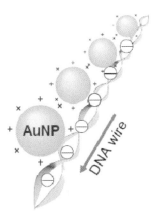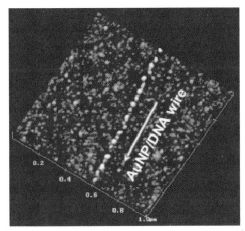

Figure 5.7 Schematic illustration of a gold nanoparticle highly ordered on a substrate by using electrostatic interaction of straightened DNA and aniline-coated, positively charged gold nanoparticles. The AFM image shows that gold nanoparticles are highly arranged in one dimension on the substrate. Illustration courtesy of Dr Nakao [86].

Typically, positively charged AuNPs are prepared by the ligand exchange technique, in which the adsorbed anion is exchanged with cationic thiol through the following controlled two-step protocol:

- Step 1: AuNPs are synthesized in an aqueous/organic solution interface and then adsorbed with the thiol molecule in organic media.
- Step 2: The protectant layer is then exchanged or modified with a cationic thiol [80, 81].

The layer thus treated, however, provides only a low coverage of the cationic thiol on the surface of the nanoparticles, so that the nanoparticle must be modified with alkylthiols before the first step. In recent years, several studies have reported a simpler preparation technique for positively charged AuNPs that employs aniline as the reducing agent; the technique is a one-step process that does not require extra control, organic solvents, and ligand exchange [82–84]. It was proposed that an aniline oligomer would act as a protectant of AuNPs from the observation of UV-visible spectra and conductivity measurements of an AuNPs film cast on the microelectrode [26].

It has been suggested that, as positively charged nanoparticles are easily carried into a living cell, this would provide a means for gene delivery [83]. Nakao *et al.* have succeeded in fixing DNA strands straight onto a glass surface without any sophisticated equipment [85]. By using DNA as a template, they created a DNA nanowire from aniline-protected and positively charged AuNPs, as described above (Figure 5.7) [86]. Warner and coworkers assembled AuNPs into lines, ribbons, and junctions (such as Y-shaped 2-D structures) templated with DNA, using positively charged AuNPs synthesized from a triphenylphosphine-passivated precursor-nanoparticle by a biphasic ligand exchange reaction with thiocholine (N,N,N-trimethylaminoethanethiol iodide) [87]. An interesting report on 2-D AuNP arrays fabricated via DNA hybridization may serve as another example, where the programmable DNA sequences provide multiple 2-D nanocomponents consisting of AuNPs with different particle sizes. This technique differed from the two reports mentioned above, in that it utilized the hybridization reaction rather than the charge interaction between AuNPs and DNA [88]. First, a scaffold was assembled that consisted of a set of 22 DNA strands. The hybridization of DNA that was immobilized on the 5 nm and 10 nm AuNPs, with the hybridization sites arranged on the DNA scaffold, provided an ordered array of AuNPs with lines of large and small nanoparticles.

5.4.3
AuNP Arrays for Gas Sensing

As noted above, over the past few decades many research groups have utilized AuNPs to develop novel biochemical and chemical sensors. For example, a new type of electrical sensor has been fabricated from an electroconducting particle array by depositing ligand-capped AuNP onto appropriate insulating substrates. Thiol was selected as the ligand and acted as a tunneling barrier between the adjacent nanoparticles, such that the adsorption (insertion) of analyte into the ligand caused the resistance to change (Figure 5.8). The array, which consisted of a networked particle film, could be prepared from a variety of chemical linkers, such as organic dithiols [89–94] and polyfunctional dendrimers [95, 96]. By using these ligand layers, several groups have fabricated gas-sensing devices, which were similar to chemiresistor sensors [89–99]. A metal nanoparticle array sensor based on a monolayer was first described by Snow et al. in 1998 [100]. Here, the thin transducer film developed was composed of 2 nm AuNPs encapsulated by octanethiol monolayers, and was deposited on an interdigital microelectrode [100, 101]. When the group applied the film to organic vapor sensing, it was discovered that its responses were large (resistance changes up to twofold or more), fast (90% response in less than 1 s), reversible, and selective. Subsequent to these studies, a number of nanoparticle arrays have been shown to serve as viable sensing and transducer layers for chemical sensing.

Zamborini and coworkers fabricated a film of monolayer-protected AuNPs with a monolayer prepared from alkanethiolate and ω-carboxylate alkanethiolate; the group observed the existence of conductivity when the AuNPs were linked together in a networked array with carboxylate-Cu^{2+}-carboxylate [102]. The chain length of the alkanethiolate ligands exponentially affected the conductivity, and the result was consistent with observations of both electron tunneling through the alkanethiolate molecule and with nonbonded contact between the particles through the ligands. The exposure of the array to organic vapors, such as ethanol, CH_2Cl_2, and nitrogen, decreased the conductivity and increased the mass of the device, as observed by quartz crystal microgravimetry. Zamborini et al. also fabricated the

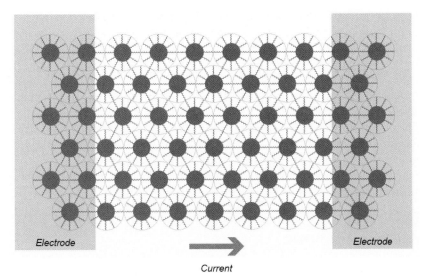

Figure 5.8 Schematic image of a 2-D nanoparticle film, consisting of repeated gold nanoparticle–insulator–gold nanoparticle sequences on a microelectrode. Reprinted with permission from the Japanese Society for Analytical Chemistry.

arrays with other metal nanoparticles, such as Pd protected with alkylamine or tetraoctylammonium and PdAg nanoparticles. The prepared films were highly reactive to H_2, even though an O_3 atmosphere and thermal treatment were not required; this result differed from that in a previous report on hexanethiol-coated Pd monolayer-protected clusters [103].

Raguse et al. developed a chemiresistor sensor to detect small organic molecules in aqueous solution; the sensor comprised an array of AuNPs capped with a 1-hexanethiol monolayer that was deposited onto a pair of microelectrodes with an inkjet printing device. The sensor allowed the detection of toluene, dichloromethane, and ethanol with detection limits of 0.1, 10, and 3000 p.p.m., respectively [104].

With respect to these sensing devices, it has been suggested that the change in permittivity and/or swelling of a vapor would induce the sensor response according to the expression for the conductivity of the film, σ [105, 106]:

$$\sigma \propto \exp(-\beta\delta)\exp\left(\frac{-E_a}{k_B T}\right) \tag{5.1}$$

where β is the tunneling decay constant, δ is the edge-to-edge separation of metal cores, E_a is the activation energy, k_B is the Boltzmann constant, and T is the absolute temperature. The swelling of an organic vapor increases resistance due to the increased tunnel distance (δ). The permittivity of the organic matrix in which the particles are embedded decreases the resistance because of the decrease in E_a and the reduction in the heights of the potential well barriers between the particle cores, thus decreasing the tunneling decay constant β. However, two conflicting reports have appeared in which directly opposite responses were reported using arrays prepared in the same way [107, 108]. To clarify the working mechanism, Joseph and colleagues have recently studied the sensor response for dodecylamine-stabilized AuNP assemblies networked through a dodecanedethiol linker [109]. They observed different sensor responses for arrays fabricated with different amounts of AuNP, and compared the results for nanoparticle-dominated, island-dominated, and 3-D close-packed arrays. Ultimately, they suggested that the preparation of a well-controlled and uniformly distributed array was the key to obtaining reproducible results.

5.4.4
AuNP Arrays for Biosensing

AuNP assemblies interlinked with organic and/or inorganic media have been fabricated for further application, not only to vapor sensing but also to biological sensing. An electric DNA-detection device has been reported by Mirkin et al. (see Figure 5.9 [110]). In this approach, small microelectrodes with a 20 μm gap were used, and a capture strand was immobilized on the glass substrate between the gaps. Based on the three-oligonucleotide sandwich approach, the hybridized target strand was captured in the electrode gap (as shown in the middle panel of Figure 5.9). The nanoparticle can function as the nucleus for metallization, and this leads to a deposition of metallic silver onto the AuNPs when the device is submerged in the silver developing solution. The deposition of silver leads to the formation of an electrical connection between the microelectrodes, and target capture is detected by a sharp decrease in the resistance of the circuit. Although this technique has provided a simple and cost-effective solution for the detection of the DNA sequences, a post-processing is required to make the deposit conductive and ready for measurement.

In an effort to develop a straightforward and cost-effective detection technique in a label-free format, many research groups have begun to actively explore the feasibility of using molecular conductivity as a primary mechanism. The conductivity of a single molecule is attracting considerable attention in many disciplines, from the viewpoint of developing nanosized electronic devices for computing, sensing, and so on. In many studies, DNA is being utilized to develop the device components, such as nanowires, nanodimensional memory, and nanosized sophisticated arithmetic units; consequently, the charge transport through single DNA strands has been studied using photochemical [111–114], biochemical [115–118], electrochemical [119–121], and direct electrical measurements [122–127].

Many studies have already investigated the use of DNA in the development of nanosized sensing devices for future nanosized applications. The fabrication of sensing devices using this property requires a molecule of choice to be wired onto a pair of probing electrodes [128]. However, there is no conformity between

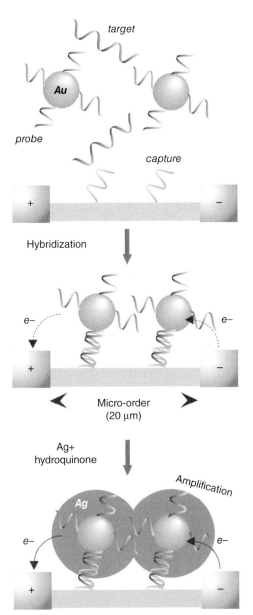

Figure 5.9 Electrical DNA detection using gold particles based on a "sandwich hybridization" of capture, probe, and target DNA with metallization [110].

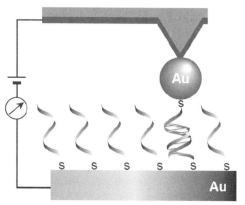

Figure 5.10 Direct measurement of current through a DNA molecule using the atomic force microscopy technique [130].

different studies with regards to the conductivities of single molecules measured in this manner, moreover, the reported values range from those of conductors to those of insulators in terms of magnitude. This large divergence is believed to be due to the difficulty in forming good molecule–electrode contacts, and also to variations in the experimental conditions [129]. Recently, experimental data for single-molecule conductivity have become available via atomic force microscopy (AFM) and scanning tunneling microscopy (STM) techniques, in which a molecule is covalently attached between a probe tip and a gold substrate (Figure 5.10) [130–133]. According to these studies, alkane and DNA molecules would not be very good conductors (10 MΩ–100 GΩ), but it is true that the molecules would exhibit conductivity, at least to some degree.

Many technical difficulties may be encountered when constructing practical electronic molecular nanodevices based on AFM/STM techniques and other nanogap techniques (e.g., the break junction method), although the techniques do not require intractable labeling molecules. In this connection, fabricating devices with a AuNP array comprising nanometer-sized gaps created through the self-assembly process appear to be a more realistic choice for the electrical detection of molecules in a high-density format, as the device size can be miniaturized economically when a mass-produced, high-density die becomes readily available.

The impedance of the AuNP array can be understood by Ohm's law. The array can be defined by the assembly of series and parallel connections of a unit impedance cell, the impedance of which is defined as Z_0. The total impedance of the array, Z, can simply be expressed as:

$$Z = \frac{N}{M} Z_0 \quad (5.2)$$

where N and M are the number of cells in the array along the x and y directions, respectively (Figure 5.11). When a voltage is applied to the array, the voltage of the

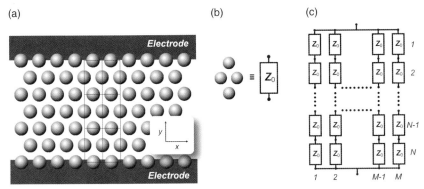

Figure 5.11 Equivalent circuit diagrams for an AuNP array. (a) An array structure; (b) A unit cell defined by an impedance of Z_0; (c) A circuit diagram equivalent to the array.

cell at the same y location should be the same throughout the array, and hence, the lateral current can be ignored. Accordingly, the array impedance is equivalent to that of an assembly of serially connected cells, each independent of the next series unit, located at $(x - 1)$ and $(x + 1)$. Each series unit is composed of serially connected N-unit cells and, at the same time, the series units are connected in parallel, where the number of the unit involved in the array is defined in the figure. Thus, the total impedance of the array is expressed in terms of Equation 5.2, and is equal to the impedance of the nanosized unit cell when $M = N$. Although this simple relationship holds only for a well-defined particle array system, it does provide some idea of how a nanodimensional event is reflected in a macro-scale system.

The present authors' group has recently succeeded in developing an electrical DNA sensor system based on the molecular conductivity of DNA using an array fabricated with a 50 nm AuNP and 1,10-decanedithiol, the latter of which was used to create a nanometer-sized-gap between the nanoparticles. The array was composed of repeated AuNP–alkyl chain–AuNP sequences on an insulator substrate (similar to that discussed in Section 5.4.3; see also Figure 5.8). The assembly process used here has received much attention as one of the most promising techniques to construct such a nanometer-sized gap in nanoparticle networks [26, 76–78]. The array was deposited on an interdigital microelectrode by submerging the electrode alternately in decanedithiol solution and an aqueous AuNP dispersion. The interdigital electrode consists of two rows each of 65 fingers of Pt electrodes spaced 5 μm apart, and is fabricated on a quartz glass plate (Figure 5.12). These repeated deposition cycles produced an array film with an equally spaced nanometer-sized gap created by the binder molecule. The gap interval between each particle can be adjusted by the length of alkyl chain of dithiol (ca. 1.3 nm), and the resistance of the array (30 Ω·cm) is comparable to the value of 20 Ω·cm reported by Han et al. for nonanedithiol (ca. 1.1 nm) [89, 134].

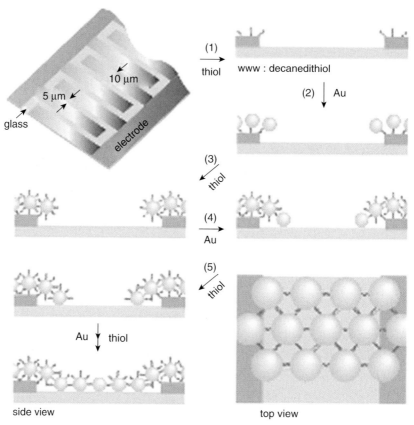

Figure 5.12 Procedures for the fabrication of a gold nanoparticle array. The sequential immersion of an interdigital electrode into thiol and gold nanoparticle solutions leads to an AuNP array in which particles are interlinked with an alkyl chain of the binder. Adapted from Ref. [138] with permission from the American Chemical Society.

A ssDNA probe (12-mer) with a sulfur head group at the 5′-phosphate end was fixed between the two nanoparticles to bind the sample DNA by hybridization. Following the probe DNA modification onto AuNPs, target ssDNA dissolved in TE buffer (10 mM Tris-HCl, 1.0 mM EDTA and 1.0 M NaCl; pH 7.4) was added to the probe-modified AuNP array. Upon the sample addition, a resistance decrease was immediately observed with a signal-to-noise ratio (SNR) >40, followed by a steady-state resistance within 2 min (Figure 5.13a). After sample addition, hybridization occurred with the probe DNA, modified on the AuNP film, between adjacent AuNPs (as shown in Figure 5.13b). The formation of a DNA double strand between nanoparticles accelerated the electron transfer, which may have induced a decrease in film resistance. The magnitude of the response was dependent on the number of mismatched base pairs (bp) in the DNA strands, and the largest response among the samples was produced by the complementary strand. An

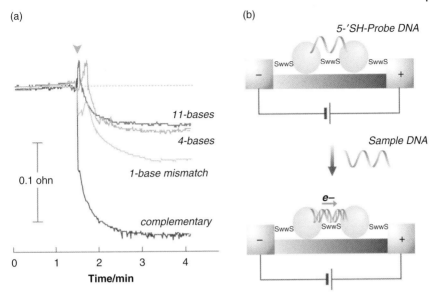

Figure 5.13 (a) Time course of resistance changes in the AuNP film caused by hybridization of the probe, modified on the film, and target DNA, indicating the possibility of single nucleotide polymorphism detection; (b) Schematic images of the sensor electrodes before and after hybridization. Adapted from Ref. [138] with permission from the American Chemical Society.

increase in the number of mismatches led to a decrease in the response, and the 11-bp mismatched DNA showed the smallest response. As seen in Figure 5.14, the resistance changed in a nonlinear fashion with respect to the number of mismatches, which implied that the presence of a 1 bp mismatch, which is an important diagnostic criterion for detecting single nucleotide polymorphisms (SNPs), was amplified [135–137]. Such differences in resistance change may contribute to the conductivity of double-stranded DNA.

Hihath et al. have measured the conductivity of fully matched and mismatched strands located between a Au substrate and a STM tip [133]. The conductivity of the 1 bp mismatched strand was one order of magnitude lower than that of the fully matched strand. Furthermore, a 2 bp mismatch decreased the conductivity of a DNA sequence by two orders of magnitude. These results agreed with the above-described observation, and confirmed the role of DNA conductivity in resistance changes with respect to the number of mismatched base pairs. In order to verify that the hybridization caused the resistance changes, a study was also made of the effect of DNase I (from bovine pancreas), an enzyme that breaks DNA strands, in random fashion, at their phosphodiester bonds [134].

In order to increase the sensor sensitivity, a bridge substructure was introduced into the AuNP array which consisted of 46 nm and 12 nm AuNPs (Figure 5.14). The substructure was formed between the 46 nm particles to anchor 12 nm AuNP probes (A and B), on which 12 nm oligomers were immobilized. The feasibility of

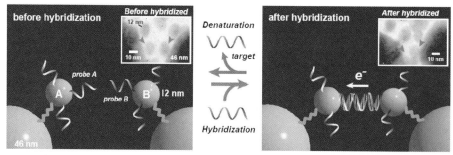

Figure 5.14 Conceptual illustrations of an open-bridge structured gold nanoparticle array for label-free DNA detection. The array comprises 46 nm particles and the DNA-capped 12 nm AuNP probes, A and B, immobilized on the 46 nm particles with dithiol. Probe DNA and complementary target DNA hybridization results in a complete bridge structure between 12 nm nanoparticles. The insets show transmission electron microscopy images of the AuNP array, revealing 46 nm particles before and after hybridization. Adapted from Ref. [138] with permission from the American Chemical Society.

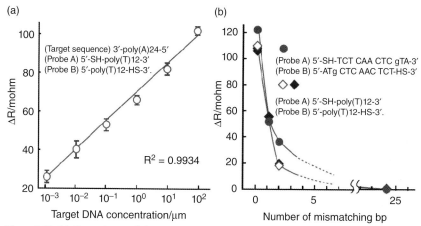

Figure 5.15 (a) Dependence of the sensor response on the target DNA concentration; (b) Sensor response versus the number of mismatched base pairs in the target DNA. The sensor array enables detection of random sequences and SNPs. Adapted from Ref. [138] with permission from the American Chemical Society.

this array structure as a new electrical sensing platform was examined. Upon the addition of a 24-mer target oligomer, hybridization of the target with two 12-mer probes occurred to close the bridge, creating an additional circuit [138]. This sensor functioned over a wide concentration range, with a detection limit of 5.0 fmol, accomplishing a sensitivity which was 1000-fold that of the single particle system (Figure 5.15). Bridged hybridization occurred between the adjacent particles, and the resultant conductivity change may account for the direct electron transfer between the probe particles as well as the molecular conductivity of DNA [124–133]. These coupled effects would result in a dramatic improvement in the detection limit, by an order of three, when the sensitivity was compared for the single-AuNP array system [134–137]. Based on these findings, the technique discussed here shows potential for use in a wide variety of applications, not only for DNA analysis but also for the quantification of other molecules, using their conductivities as a new class of sensing platform.

5.5
Conclusions

Metal nanoparticles have been used as the building blocks of structures such as nanowires, 2-D films, and stereoscopic supercrystals. As the chemical, electrical, and optical properties of the above structures are determined by their constitutional units such as size, shape, and regularity, the control of these structures is important for the elucidation of structure-dependent properties and the application of regularly ordered nanoparticles to nanodevices. From the viewpoint of straightforward preparation and surface modification techniques, AuNPs have been preferred for use in nanoarchitecture and biological areas, as well as in their integrated field. In this chapter, we have summarized the evolving research in a field that combines assembled AuNPs and molecular biotechnology. Along these lines, sensing technologies that utilize AuNPs were reviewed, with emphasis placed on recent developments in analytical applications for particle assemblies. New techniques using AuNP assemblies on glass and other substrates, together with techniques employing biomolecular templates, have been discussed in detail. Such regularly arrayed AuNPs have been applied to various organic-vapor and biomolecule sensors, as they provided the nanospace for the analyte receptor with the thiol- or amino group-derivatized linker molecule. These developments may enable the realization of high-density and high-throughput sensing platforms for bioassociated materials such as DNA, antigens, and enzymatic reactions.

References

1 Leonhardt, U. (2007) Optical metamaterials: invisibility cup. *Nature Photonics*, **1**, 207–8.
2 Faraday, M. (1857) The Bakerian lecture—experimental relations of gold (and other metals) to light. *Philosophical Transactions*, **147**, 145–81.
3 Benniston, A.C., Harriman, A. and Lynch, V.M. (1995) Photoactive [2] Rotaxanes structure and photophysical properties of anthracene- and ferrocene-stoppered [2]Rotaxanes. *Journal of the American Chemical Society*, **117**, 5275–91.
4 DeSilva, A.P., Gunazatne, H.Q.N. and McCoy, C.P. (1993) A molecular photoionic AND gate based on fluorescent signaling. *Nature*, **364**, 42–4.
5 Gopel, W. (1998) Bioelectronics and nanotechnologies. *Biosensors and Bioelectronics*, **13**, 723–8.
6 Brown, D.H. and Smith, W.E. (1980) The chemistry of the gold drugs used in

the treatment of rheumatoid arthritis. *Chemical Society Reviews*, **9**, 217–40.

7 Turkevitch, J., Stevenson, P.C. and Hillier, J. (1951) Nucleation and growth process in the synthesis of colloidal gold. *Discussions of the Faraday Society*, **11**, 55–75.

8 Frens, G. (1973) Controlled nucleation for the regulation of the particle size in monodisperse gold suspensions. *Nature Physical Sciences*, **241**, 20–2.

9 Hayat, M.A. (1989) *Colloidal Gold, Principles, Methods and Applications*, Academic Press, New York, pp. 1–8.

10 Schmid, G. and Chi, L.F. (1998) Metal clusters and colloids. *Advanced Materials*, **10**, 515–26.

11 Bradley, J.S. (1994) *Clusters and Colloids* (ed. G. Schmid), Wiley-VCH Verlag GmbH, Weinheim, Chapter 6, pp. 459–544.

12 Schmid, G. (1992) Large clusters and colloids. Metals in the embryonic state. *Chemical Reviews*, **92**, 1709–27.

13 Hainfeld, J.F., Powell, R.D. and Hacker, G.W. (2004) *Nanobiotechnology* (eds C.M. Niemeyer and C.A. Mirkin), Wiley-VCH Verlag GmbH, Weinheim, pp. 353–4.

14 Kneipp, K., Kneipp, H., Kartha, V.B., Manoharan, R., Deinum, G., Itzkan, I., Dasari, R.R. and Feld, M.S. (1998) Detection and identification of a single DNA base molecule using surface-enhanced Raman scattering (SERS). *Physical Review E*, **57**, R6281–4.

15 Grzelczak, M., Perez-Juste, J., Mulvaney, P. and Liz-Marzan, L.M. (2008) Shape control in gold nanoparticle synthesis. *Chemical Society Reviews*, **37**, 1783–91.

16 Perez-Juste, J., Pastoriza-Santos, I., Liz-Marzan, L.M. and Mulvaney, P. (2005) Gold nanorods: synthesis, characterization and applications. *Coordination Chemistry Reviews*, **249**, 1870–901.

17 Sun, Y. and Xia, Y. (2002) Shape-controlled synthesis of gold and silver nanoparticles. *Science*, **298**, 2176–9.

18 Yamamoto, Y. and Hori, H. (2006) Direct observation of the ferromagnetic spin polarization in gold nanoparticles. *Reviews on Advances in Materials Science*, **12**, 23–32.

19 Hermanson, G.T. (1996) *Bioconjugate Techniques*, Academic Press, San Diego, pp. 593–9.

20 Brust, M., Walker, M., Bethell, D., Schiffrin, D.J. and Whyman, R.J. (1994) Synthesis of thiol-derivatised gold nanoparticles in a two-phase Liquid–Liquid system. *Journal of the Chemical Society, Chemical Communications*, 801–2.

21 Brust, M., Fink, J., Bethell, D., Schiffrin, D.J. and Kiely, C.J. (1995) Synthesis and reactions of functionalised gold nanoparticles. *Journal of the Chemical Society, Chemical Communications*, 1655–6.

22 Seshadri, R. (2004) *The Chemistry of Nanomaterials*, Vol. 1 (eds C.N. Rao, A. Müller and A.K. Cheetham), Wiley-VCH Verlag GmbH, Weinheim, pp. 94–112.

23 Sastry, M., Ahmad, A., Khan, M.I. and Kumar, R. (2004) *Nanobiotechnology*, Vol. 1, (eds C.M. Niemeyer and C.A. Mirkin), Wiley-VCH Verlag GmbH, Weinheim, pp. 126–35.

24 Nair, B. and Pradeep, T. (2002) Coalescence of nanoclusters and formation of submicron crystallites assisted by *Lactobacillus* strains. *Crystal Growth Design*, **2**, 293–8.

25 Weare, W.W., Reed, S.M., Warner, M.G. and Hutchison, J.E. (2000) Improved synthesis of small ($d_{CORE} \approx 1.5$ nm) phosphine-stabilized gold nanoparticles. *Journal of the American Chemical Society*, **122**, 12890–1.

26 Shiigi, H., Yamamoto, Y., Yoshi, N., Nakao, H. and Nagaoka, T. (2006) One-step preparation of positively-charged gold nanoraspberry. *Chemical Communications*, 4288–90.

27 Schlenoff, J.B., Li, M. and Ly, H. (1995) Stability and self-exchange in alkanethiol monolayers. *Journal of the American Chemical Society*, **117**, 12528–36.

28 Shimazu, K., Sato, Y., Yagi, I. and Uosaki, K. (1994) Packing state and stability of self-assembled monolayers of 11-ferrocenyl-1-undecanethiol on platinum electrodes. *Bulletin of the Chemical Society of Japan*, **67**, 863–5.

29 Petrovykh, D.Y., Kimura-Suda, H., Opdahl, A., Richter, L.J., Tarlov, M.J.

and Whitman, L.J. (2006) Alkanethiols on platinum: multicomponent self-assembled monolayers. *Langmuir*, **22**, 2578–87.
30. Mie, G. (1908) Beitrage zur Optik truber Medien, speziell kolloidaler Metallosungen. *Annalen der Physik*, **25**, 377–445.
31. Link, S., Wang, Z.L. and El-Sayed, M.A. (1999) Alloy formation of gold-silver nanoparticles and the dependence of the plasmon absorption on their composition. *The Journal of Physical Chemistry B*, **103**, 3529–33.
32. Rodríguez-González, B., Sánchez-Iglesias, A., Giersig, M. and Liz-Marzán, L.M. (2004) AuAg bimetallic nanoparticles: formation, silica-coating and selective etching. *Faraday Discussions*, **125**, 133–44.
33. Liz-Marzan, L.M. (2004) Nanometals formation and color. *Materials Today*, **7**, 26–31.
34. Hao, E., Schatz, G.C. and Hupp, J.T., (2004) synthesis and optical properties of anisotropic metal nanoparticles, *Journal of Fluorescence*, **14**, 331–41.
35. Liz-Marzan, L.M. and Mulvaney, P. (2003) The assembly of coated nanocrystals. *Journal of Physical Chemistry B*, **107**, 7312–26.
36. Underwood, S. and Mulvaney, P. (1994) Effect of the solution refractive index on the color of gold colloids. *Langmuir*, **10**, 3427–30.
37. Ung, T., Dunstan, D., Giersig, M. and Mulvaney, P. (1997) Spectroelectrochemistry of colloidal silver. *Langmuir*, **13**, 1773–82.
38. Liz-Marzan, L.M. and Mulvaney, P. (1998) Au@SiO$_2$ colloids: effect of temperature on the surface plasmon absorption. *New Journal of Chemistry*, **22**, 1285–8.
39. Xia, Y. and Whitesides, G.M. (1998) Soft lithography. *Angewandte Chemie, International Edition in English*, **37**, 550–75.
40. Chou, S.Y., Krauss, P.R. and Renstrom, P.J. (1995) Imprint of sub-25 nm vias and trenches in polymers. *Applied Physics Letters*, **67**, 3114–16.
41. Zankovych, S., Hoffmann, T., Seekamp, J., Bruch, J.-U. and Sotomayor-Torres, C.M. (2001) Nanoimprint lithography: challenges and prospects. *Nanotechnology*, **12**, 91–5.
42. Guo, L.J. (2004) Recent progress in nanoimprint technology and its applications. *Journal of Applied Physics*, **37**, R123–41.
43. Haley, C. and Weaver, J.H. (2002) Buffer-layer-assisted nanostructure growth via two-dimensional cluster–cluster aggregation. *Surface Science*, **518**, 243–50.
44. Kerner, G. and Asscher, M.(2004) buffer layer assisted laser patterning of metals on surfaces. *Nano Letters*, **4**, 1433–7.
45. Schildenberger, M., Bonetti, Y., Gobrecht, J. and Prins, R. (2000) Nano-pits: supports for heterogeneous model catalysts prepared by interference lithography. *Topics in Catalysis*, **13**, 109–20.
46. Matsui, S. and Ochiai, Y. (1996) Focused ion beam applications to solid state devices. *Nanotechnology*, **7**, 247–58.
47. Mirkin, C.A., Letsinger, R.L., Mucic, R.C. and Storhoff, J.J. (1996) A DNA-based method for rationally assembling nanoparticles into macroscopic materials. *Nature*, **382**, 607–9.
48. Sato, K., Hosokawa, K. and Maeda, M. (2005) Non-cross-linking gold nanoparticle aggregation as a detection method for single-base substitutions. *Nucleic Acids Research*, **33**, e4.
49. Sato, K., Hosokawa, K. and Maeda, M. (2003) Rapid aggregation of gold nanoparticles induced by non-cross-linking DNA hybridization. *Journal of the American Chemical Society*, **125**, 8102–3.
50. Storhoff, J.J., Elghanian, R., Mirkin, C.A. and Letsinger, R.L. (2002) Sequence-dependent stability of DNA-modified gold nanoparticles. *Langmuir*, **18**, 6666–70.
51. Herne, T.M. and Tarlov, M.J. (1997) Characterization of DNA probes immobilized on gold surfaces. *Journal of the American Chemical Society*, **119**, 8916–20.

52 Zhao, W., Monsur Ali, M., Aguirre, S.D., Brook, M.A. and Li, Y. (2008) Paper-based bioassays using gold nanoparticle colorimetric probes. *Analytical Chemistry*, **80**, 8431–7.

53 Tsagkogeorgas, F., Ochsenkuhn-Petropoulou, M., Niessner, R. and Knopp, D. (2006) Encapsulation of biomolecules for bioanalytical purposes: preparation of diclofenac antibody-doped nanometer-sized silica particles by reverse micelle and sol–gel processing. *Analytica Chimica Acta*, **573–574**, 133–7.

54 Liu, X., Dai, Q., Austin, L., Coutts, J., Knowles, G., Zou, J., Chen, H., Huo, Q. and One-Step, A. (2008) Homogeneous immunoassay for cancer biomarker detection using gold nanoparticle probes coupled with dynamic light scattering. *Journal of the American Chemical Society*, **130**, 2780–2.

55 Hirsch, L.R., Jackson, L.B., Lee, A., Halas, N.J. and West, J. (2003) A whole blood immunoassay using gold nanoshells. *Analytical Chemistry*, **75**, 2377–81

56 Khairutdinov, R.F. (1997) Physical chemistry of nanocrystalline semiconductors. *Colloid Journal*, **59**, 535–48.

57 Tiwari, S., Rana, F., Hanafi, H., Hartstein, A., Grabbe, E.F. and Chan, K. (1996) A silicon nanocrystals based memory. *Applied Physics Letters*, **68**, 1377–9.

58 Glotzer, S.C. and Solomon, M.J. (2007) Anisotropy of building blocks and their assembly into complex structures. *Nature Materials*, **6**, 557–62.

59 Mulvaney, P. (1996) Surface plasmon spectroscopy of nanosized metal particles. *Langmuir*, **12**, 788–800.

60 Alvarez, M.M., Khoury, J.T., Schaaff, T.G., Shafigullin, M.N., Vezmar, I. and Whetten, R.L. (1997) Optical absorption spectra of nanocrystal gold molecules. *Journal of Physical Chemistry B*, **101**, 3706–12.

61 Alivisatos, A.P. (1996) Perspectives on the physical chemistry of semiconductor nanocrystals. *Journal of Physical Chemistry*, **100**, 13226–39.

62 Brus, L.E. (1991) Quantum crystallites and nonlinear optics. *Applied Physics A*, **53**, 465–74.

63 Lewis, L.N. (1993) Chemical catalysis by colloids and clusters. *Chemical Reviews*, **93**, 2693–730.

64 Ishida, T., Kinoshita, N., Okatsu, H., Akita, T., Takei, T. and Haruta, M. (2008) Influence of the support and the size of gold clusters on catalytic activity for glucose oxidation. *Angewandte Chemie, International Edition*, **47**, 9265–8.

65 Alivisatos, A.P. (1996) Semiconductor clusters, nanocrystals, and quantum dots. *Science*, **271**, 933–7.

66 Mott, M., Song, J.H. and Evans, J.R.G. (1999) Microengineering of ceramics by direct Ink-Jet printing. *Journal of the American Ceramic Society*, **82**, 1653–8.

67 Blonder, R., Sheeney, L. and Willner, I. (1998) Three-dimensional redox-active layered composites of Au–Au, Ag–Ag and Au–Ag colloids. *Chemical Communications*, 1393–4.

68 Shipway, A.N. and Willner, I. (2001) Nanoparticles as structural and functional units in surface-confined architectures. *Chemical Communications*, 2035–45.

69 Cassagneau, T.P. (2004) *Colloids and Colloid Assemblies* (ed. F. Caruso), Wiley-VCH Verlag GmbH, Weinheim, pp. 398–436.

70 Doron, A., Katz, E. and Willner, I. (1995) Organization of Au colloids as monolayer films onto ITO glass surfaces: application of the metal colloid films as base interfaces to construct redox-active monolayers. *Langmuir*, **11**, 1313–17.

71 Lahav, M., Gabai, R., Shipway, A.N. and Willner, I. (1999) Au-colloid–"molecular square" superstructures: novel electrochemical sensing interfaces. *Chemical Communications*, 1937–8.

72 Lahav, M., Heleg-Shabtai, V., Wasserman, J., Katz, E., Willner, I., Durr, H., Hu, Y.-Z. and Bossmann, S.H. (2000) Photoelectrochemistry with integrated photosensitizer-electron acceptor and Au-nanoparticle arrays. *Journal of the American Chemical Society*, **122**, 11480–7.

73. Freeman, R.G., Grabar, K.C., Allison, K.J., Bright, R.M., Davis, J.A., Guthrie, A.P., Hommer, M.B., Lackson, M.A., Smith, P.C., Walter, D.G. and Natan, M.J. (1995) Self-assembled metal colloid monolayers: an approach to SERS substrates. *Science*, **267**, 1629–32.

74. Shiigi, H., Yoshi, N., Yamamoto, Y., Iwamoto, M. and Nagaoka, T. (2008) Completely green one-step fabrication of gold patterned-flexible film. *Microelectronic Engineering*, **85**, 1214–17.

75. Yamamoto, Y., Takeda, S., Shiigi, H. and Nagaoka, T. (2007) An electroless plating method for conducting microbeads using gold nanoparticles. *Journal of the Electrochemical Society*, **154**, D462–6.

76. Yamamoto, Y., Yoshi, N., Shiigi, H. and Nagaoka, T. (2006) Electrical properties of a nanoparticle-networked film. *Solid State Ionics*, **177**, 2325–8.

77. Yamamoto, Y., Shiigi, H. and Nagaoka, T. (2005) Characterization of Au nanoparticle film electrodes prepared on polystyrene. *Electroanalysis*, **17**, 2224–30.

78. Shiigi, H., Yamamoto, Y., Yakabe, H., Tokonami, S. and Nagaoka, T. (2003) Electrical property and water repellency of a networked monolayer film prepared from Au nanoparticles. *Chemical Communications*, 1038–9.

79. Liu, Y., Wang, Y. and Claus, R.O. (1998) Layer-by-layer ionic self-assembly of Au colloids into multilayer thin-films with bulk metal conductivity. *Chemical Physics Letters*, **298**, 315–21.

80. Hussain, I., Wang, Z., Cooper, A.I. and Brust, M. (2006) Formation of spherical nanostructures by the controlled aggregation of gold colloids. *Langmuir*, **22**, 2938–41.

81. McIntosh, C.M., Esposito, E.A. III, Boal, A.K., Simard, J.M., Martin, C.T. and Rotello, V.M. (2001) Inhibition of DNA transcription using cationic mixed monolayer protected gold clusters. *Journal of the American Chemical Society*, **123**, 7626–9.

82. McIntosh, C.M., Esposito, E.A., Boal, A.K., Simard, J.M., Martin, C.T. and Rotello, V.M. (2001) Inhibition of DNA transcription using cationic mixed monolayer protected gold clusters. *Journal of the American Chemical Society*, **123**, 7626–35.

83. Yonezawa, T., Onoue, S. and Kimizuka, N. (2002) Metal coating of DNA molecules by cationic, metastable gold nanoparticles. *Chemistry Letters*, **31**, 1172–3.

84. Gandubert, V.J. and Lennox, R.B. (2005) Assessment of 4-(dimethylamino) pyridine as a capping agent for gold nanoparticles. *Langmuir*, **21**, 6532–9.

85. Nakao, H., Gad, M., Sugiyama, S., Otobe, K. and Ohtani, T. (2003) Transfer-printing of highly aligned DNA nanowires. *Journal of the American Chemical Society*, **125**, 7162–3.

86. Nakao, H., Shiigi, H., Yamamoto, Y., Tokonami, S., Nagaoka, T., Sugiyama, S. and Ohtani, T. (2003) Highly ordered assemblies of Au nanoparticles organized on DNA. *Nano Letters*, **3**, 1391–4.

87. Warner, M.G. and Hutchison, J.E. (2003) Linear assemblies of nanoparticles electrostatically organized on DNA scaffolds. *Nature Materials*, **2**, 272–7.

88. Pinto, Y.Y., Le, J.D., Seeman, N.C., Musier-Forsyth, K., Taton, T.A. and Kiehl, R.A. (2005) Sequence-encoded self-assembly of multiple-nanocomponent arrays by 2D DNA scaffolding. *Nano Letters*, **5**, 2399–402.

89. Han, L., Daniel, D.R., Maye, M.M. and Zhong, C.J. (2001) Core-shell nanostructured nanoparticle films as chemically sensitive interfaces. *Analytical Chemistry*, **73**, 4441–9.

90. Joseph, Y., Krasteva, N., Besnard, I., Guse, B., Rosenberger, M., Wild, U., Knop-Gericke, A., Schloegl, R., Krustev, R., Yasuda, A. and Vossmeyer, T. (2004) Gold-nanoparticle/organic linker films: self-assembly, electronic and structural characterisation, composition and vapour sensitivity. *Faraday Discussions*, **125**, 77–97.

91. Joseph, Y., Besnard, I., Rosenberger, M., Guse, B., Nothofer, H.G., Wessels, J., Wild, U., Knop-Gericke, A., Su, D., Yasuda, A. and Vossmeyer, T. (2003) Self-assembled gold nanoparticle/alkanedithiol films: preparation, electron microscopy, XPS-analysis, charge

92 Joseph, Y., Guse, B., Yasuda, A. and Vossmeyer, T. (2004) Chemiresistor coatings from Pt- and Au-nanoparticle/ nonanedithiol films: sensitivity to gases and solvent vapors. *Sensors and Actuators, B*, **98**, 188–95.

93 Vossmeyer, T., Joseph, Y., Besnard, I., Harnack, O., Krasteva, N., Guse, B., Nothofer, H.-G. and Yasuda, A. (2004) Gold-nanoparticle/dithiol films as chemical sensors and first steps toward their integration on chip. *Proceedings of SPIE*, **5513**, 202–12.

94 Ibañez, F.J., Gowrishetty, U., Crain, M.M., Walsh, K.M. and Zamborini, F.P. (2006) Chemiresistive vapor sensing with microscale films of gold monolayer protected clusters. *Analytical Chemistry*, **78**, 753–61.

95 Krasteva, N., Besnard, I., Guse, B., Bauer, R.E., Müllen, K., Yasuda, A. and Vossmeyer, T. (2002) Self-assembled gold nanoparticle/dendrimer composite films for vapor sensing applications. *Nano Letters*, **2**, 551–5.

96 Krasteva, N., Guse, B., Besnard, I., Yasuda, A. and Vossmeyer, T. (2003) Gold nanoparticle/PPI-dendrimer based chemiresistors: vapor-sensing properties as a function of the dendrimer size. *Sensors and Actuators, B*, **92**, 137–43.

97 Grate, J.W., Nelson, D.A. and Skaggs, R. (2003) Sorptive behavior of monolayer-protected gold nanoparticle films: implications for chemical vapor sensing. *Analytical Chemistry*, **75**, 1868–79.

98 Grate, J.W., Klusty, M., Barger, W.R. and Snow, A.W. (1990) Role of selective sorption in chemiresistor sensors for organophosphorus detection. *Analytical Chemistry*, **62**, 1927–34.

99 Leopold, M.C., Donkers, R.L., Georganopoulou, D., Fisher, M., Zamborini, F.P. and Murray, R.W. (2004) Growth, conductivity, and vapor response properties of metal ion-carboxylate linked nanoparticle films. *Faraday Discussions*, **125**, 63–76.

100 Wohltjen, H. and Snow, A.W. (1998) Colloidal metal-insulator-metal ensemble chemiresistor sensor. *Analytical Chemistry*, **70**, 2856–9.

101 Foos, E.E., Snow, A.W., Twigg, M.E. and Ancona, M.G. (2002) Thiol-terminated di-, tri-, and tetraethylene oxide functionalized gold nanoparticles: a water-soluble, charge-neutral cluster. *Chemistry of Materials*, **14**, 2401–8.

102 Zamborini, F.P., Leopold, M.C., Hicks, J.F., Kulesza, P.J., Malik, M.A. and Murray, R.W.(2002) Electron hopping conductivity and vapor sensing properties of flexible network polymer films of metal nanoparticles. *Journal of the American Chemical Society*, **124**, 8958–64.

103 Ibañez, F.J. and Zamborini, F.P. (2008) Reactivity of hydrogen with solid-state films of alkylamine- and tetraoctylammonium bromide-stabilized Pd, PdAg, and PdAu nanoparticles for sensing and catalysis applications. *Journal of the American Chemical Society*, **130**, 622–33.

104 Raguse, B., Chow, E., Barton, C.S. and Wieczorek, L. (2007) gold nanoparticle chemiresistor sensors: direct sensing of organics in aqueous electrolyte solution. *Analytical Chemistry*, **79**, 7333–9.

105 Andres, R.P., Bielefeld, J.D., Janes, D.B., Kolagunta, V.R., Kubiak, C.P., Mahoney, W.J. and Osifchin, R.G. (1996) Self-assembly of a two-dimensional superlattice of molecularly linked metal clusters. *Science*, **273**, 1690–3.

106 Wuelfing, W.P., Green, S.J., Pietron, J.J., Cliffel, D.E. and Murray, R.W. (2000) Electronic conductivity of solid-state, mixed-valent, monolayer-protected Au clusters. *Journal of the American Chemical Society*, **122**, 11465–72.

107 Evans, S.D., Johnson, S.R., Cheng, Y.L. and Shen, T. (2000) Vapour sensing using hybrid organic–inorganic nanostructured materials. *Journal of Materials Chemistry*, **10**, 183–8.

108 Zhang, H.L., Evans, S.D., Henderson, J.I., Miles, R.E. and Shen, T. (2002) Vapour sensing using surface functionalized gold nanoparticles. *Nanotechnology*, **13**, 439–44.

109 Joseph, Y., Guse, B., Vossmeyer, T. and Yasuda, A. (2008) Gold nanoparticle/

organic networks as chemiresistor coatings: the effect of film morphology on vapor sensitivity. *Journal of Physical Chemistry, C*, **112**, 12507–14.
110 Park, S.J., Taton, T.A. and Mirkin, C.A. (2002) array-based electrical detection of DNA with nanoparticle probes. *Science*, **295**, 1503–6.
111 Murphy, C.J., Arkin, M.R., Jenkins, Y., Ghatlia, N.D., Bossmann, S.H., Turro, N.J. and Barton, J.K. (1993) Long-range photoinduced electron transfer through a DNA helix. *Science*, **262**, 1025–9.
112 Lewis, F., Wu, T., Zhang, Y., Letsinger, R., Greenfield, S. and Wasielewski, M. (1997) Distance-dependent electron transfer in DNA hairpins. *Science*, **277**, 673–6.
113 Kelley, S.O., Holmlin, R.E., Stemp, E.D.A. and Barton, J.K. (1997) Photoinduced electron transfer in ethidium-modified DNA duplexes: dependence on distance and base stacking. *Journal of the American Chemical Society*, **119**, 9861–70.
114 Meade, T.J. and Kayyem, J.F. (1995) Electron transfer through DNA: site-specific modification of duplex DNA with ruthenium donors and acceptors. *Angewandte Chemie, International Edition in English*, **34**, 352–4.
115 Hall, D.B., Holmlin, R.E. and Barton, J.K. (1996) Oxidative DNA damage through long-range electron transfer. *Nature*, **382**, 731–5.
116 Ly, D., Sanii, L. and Schuster, G.B. (1999) Mechanism of charge transport in DNA: internally-linked anthraquinone conjugates support phonon-assisted polaron hopping. *Journal of the American Chemical Society*, **121**, 9400–10.
117 Meggers, E., Kusch, D., Spichty, M., Wille, U. and Giese, B. (1998) Electron transfer through DNA in the course of radical-induced strand cleavage. *Angewandte Chemie, International Edition in English*, **37**, 460–2.
118 Saito, I., Nakamura, T., Nakatani, K., Yoshioka, Y., Yamaguchi, K. and Sugiyama, H. (1998) Mapping of the hot spots for DNA damage by one-electron oxidation: efficacy of GG doublets and GGG triplets as a trap in long-range hole migration. *Journal of the American Chemical Society*, **120**, 12686–7.
119 Kelley, S.O., Barton, J.K., Jackson, N.M. and Hill, M.G. (1997) Electrochemistry of methylene blue bound to a DNA-modified electrode. *Bioconjugate Chemistry*, **8**, 31–7.
120 Kelley, S.O., Jackson, N.M., Hill, M.G. and Barton, J.K. (1999) Long-range electron transfer through DNA films. *Angewandte Chemie, International Edition*, **38**, 941–5.
121 Hartwich, G., Caruana, D.J., De lumley-Woodyear, T., Wu, Y., Campbell, C.N. and Heller, A. (1999) Electrochemical study of electron transport through thin DNA films. *Journal of the American Chemical Society*, **121**, 10803–12.
122 Okahata, Y., Kobayashi, T., Tanaka, K. and Shimomura, M. (1998) Anisotropic electric conductivity in an aligned DNA cast film. *Journal of the American Chemical Society*, **120**, 6165–6.
123 Braun, E., Eichen, Y., Sivan, U. and Ben-Yoseph, G. (1998) DNA-templated assembly and electrode attachment of a conducting silver wire. *Nature*, **391**, 775–8.
124 Porath, D., Bezryadin, A., De Vries, S. and Dekker, C. (2000) Direct measurement of electrical transport through DNA molecules. *Nature*, **403**, 635–8.
125 Fink, H.-W. and Schonenberger, C. (1999) Electrical conduction through DNA molecules. *Nature*, **398**, 407–10.
126 Kasumov, A.Y., Kociak, M., Gueron, S., Reulet, B., Volkov, V.T., Klinov, D.V. and Bouchiat, H. (2001) Proximity-induced superconductivity in DNA. *Science*, **291**, 280–2.
127 Zhang, Y., Austin, R.H., Kraeft, J., Cox, E.C. and Ong, N.P. (2002) Insulating behavior of λ-DNA on the micron scale. *Physical Review Letters*, **89**, 198102-1–4.
128 Dekker, C. and Ratner, M.A. (2001) Electronic properties of DNA. *Physics World*, **14**, 29–33.
129 Hipps, K.W. (2001) Molecular electronics: it's all about contacts. *Science*, **294**, 536–7.

130 Cohen, H., Nogues, C., Naaman, R. and Porath, D. (2005) Direct measurement of electrical transport through single DNA molecules of complex sequence. *Proceedings of the National Academy of Sciences of the United States of America*, **102**, 11589–93.

131 Wierzbinski, E., Arndt, J., Hammond, W. and Slowinski, K. (2006) In situ electrochemical distance tunneling spectroscopy of ds-DNA molecules. *Langmuir*, **22**, 2426–9.

132 Xu, B., Zhang, P., Li, X. and Tao, N. (2004) Direct conductance measurement of single DNA molecules in aqueous solution. *Nano Letters*, **4**, 1105–8.

133 Hihath, J., Xu, B., Zhang, P. and Tao, N. (2005) Study of single-nucleotide polymorphisms by means of electrical conductance measurements, *Proceedings of the National Academy of Sciences of the United States of America*, **102**, 16979–83.

134 Tokonami, S., Iwamoto, M., Hashiba, K., Shiigi, H. and Nagaoka, T. (2006) Fabrication of a highly sensitive sensor electrode using a nano-gapped gold particle film. *Solid State Ionics*, **177**, 2317–20.

135 Tokonami, S., Shiigi, H. and Nagaoka, T. (2008) Label-free DNA detection of nano-gapped gold nanoparticle array electrode. *Journal of the Electrochemical Society*, **155**, J105–9.

136 Tokonami, S., Shiigi, H. and Nagaoka, T. (2008) Preparation of nanogapped gold nanoparticle array for DNA detection. *Electroanalysis*, **20**, 355–60.

137 Shiigi, H., Tokonami, S., Yakabe, H. and Nagaoka, T. (2005) Label-free electronic detection of DNA-hybridization on nanogapped gold particle film. *Journal of the American Chemical Society*, **127**, 3280–1.

138 Tokonami, S., Shiigi, H. and Nagaoka, T. (2008) Open bridge-structured gold nanoparticle array for label-free DNA detection. *Analytical Chemistry*, **80**, 8071–5.

6
Thin Films on Titania, and Their Applications in the Life Sciences

Izabella Brand and Martina Nullmeier

6.1
Introduction

When Brånemark introduced titanium as an implant material almost 40 years ago [1], it very quickly became the most favored and one of the most widely used medical implant and prosthetic materials [2–5]. The major success of titanium implants is due to their very high acceptance by host organisms [2, 5]. Titanium implants fulfill the requirements of biocompatibility; namely, they are not corrosive in the environment of an organism, nor do they cause a number of inflammatory reactions [1–3]. Much histological information is available relating to deposits which build up *in vivo* on the surface of titanium implants following implantation [2, 3, 6–8]. The interface between the titanium implant and the bone is filled by a *cement layer*, which is composed of biomaterial from the extracellular matrix (ECM) such as polysaccharides, proteins, inorganic mineral material containing calcium and phosphate ions [2, 3, 6, 7, 9]. This very stable *cement coating* is integrated between the titanium implant and bone. Although cells rarely occur within this layer, the surface of the layer provides ideal conditions for cell and tissue growth. In contrast to a titanium implant, the surface of a steel implant will attract inflammatory cells that are in direct contact with the implant material and, as a consequence, the adhesion of the steel implant to the bone surface will be poor [2, 9]. Interestingly, for titanium implants the composition and thickness of the layer of adsorbing biomolecules differ from those on the surfaces of other implants, and it is for these reasons that titanium is such a unique and valuable material in medicine [5, 10, 11]. Whilst the excellent performance of titanium implants within organisms is not understood at the molecular level, the situation was summarized by Parsegian, who suggested that: "It might be wild luck that makes titanium work so well" [12]. Despite the many studies aimed at determining the molecular level of interactions, including the structure of adsorbing water [13, 14], proteins [15–18], lipids [19–21], cell membranes [22, 23], inorganic layers containing calcium ions [2, 24, 25], many details remain unknown, with the following structural data still to be derived:

Nanomaterials for the Life Sciences Vol.5: Nanostructured Thin Films and Surfaces.
Edited by Challa S. S. R. Kumar
Copyright © 2010 WILEY-VCH Verlag GmbH & Co. KGaA, Weinheim
ISBN: 978-3-527-32155-1

- The exact composition, structure and hydration of various proteins in the adsorbed layer.
- The exact composition of the *cement* coating on the titania surface formed from the ECM, and its structure.
- The structure and hydration of lipid molecules in the lipid membrane that is in contact with the titania.

In practical terms, many surface-characterizing techniques have been used to study the structure of the biomolecules adsorbed onto the surface of the titanium implant [10]. Unfortunately, however, the requirements of these techniques may introduce certain limitations to the system being analyzed, including: (i) a uniform orientation of the molecules in the layer; (ii) a limited thickness of the adsorbed layer; (iii) the introduction of a third component to the system (e.g., a fluorescent probe); and (iv) that the signals measured from various molecules can be overlapped. In order to overcome these problems, and to identify the problems from medical and technological points of view, several important structural details and models that mimic the assembly on a titanium implant have been examined [10, 26]. The aim of this chapter is to present the available information relating to the structure of biological materials that are adsorbed directly onto the titania surface, thus mimicking the situation within an organism. First, the composition of the biological layer created *in vivo* on the surface of the titania implant is briefly described. As cell membranes and proteins may come into direct contact with the implant surface, the discussion here will be limited to lipid bilayers and to two proteins of the ECM, namely collagen and fibronectin. Interactions with cells occur via a cell membrane that is composed mainly of lipids, with structures being assembled on the implant surface as and when required. Since, *in vivo*, most proteins will be in direct contact with a titania surface, the recognition of structural details is important in areas of both medicine and technology.

6.2
Titanium in Contact with a Biomaterial

The titanium surface, when in contact with air, is very rapidly covered by a thin native oxide layer [5, 27]; in fact, in less than 1 ms a layer of titanium oxide which is approximately 1 nm thick will be built up on the metal surface [27]. On a titanium implant, however, the native oxide layer will range from 3 to 10 nm in thickness [10, 27], and incorporate the stable titanium oxides of TiO_2, TiO and Ti_2O_3 [27, 28], with TiO_2 being the most common [28]. The chemical properties – and thus the interfacial chemistry – of the titanium implant are determined by the thin oxide layer, and not by the metal itself [29, 30]. Those biomolecules which are adsorbed onto the surface of the titanium implant will make direct contact with the oxide and, under physiological conditions, the TiO_2 surface will have a negative net charge [25, 31].

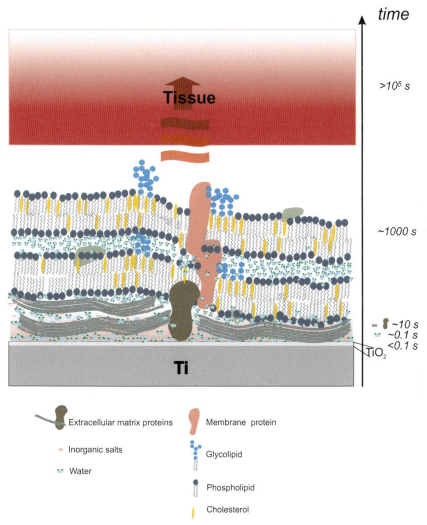

Figure 6.1 Scheme of the adsorption of various biomolecules (listed below the main diagram) on the surface of the titania implant. The size of individual molecules is not to scale. The approximate time scale is shown.

Within an organism, the adsorption of biomolecules on an implant surface will occur via a strictly ordered procedure. First, water molecules and small inorganic ions (such as OH^-, Ca^{2+}, Na^+, Cl^- or PO_4^{3-}) will contact the titania surface (see Figure 6.1). Then, in less than 1 s, water molecules and hydroxyl ions will be adsorbed onto the titania surface to form an interfacial water layer that is 0.5–1.5 nm thick [32, 33].

Body fluids such as blood and serum contain a variety of inorganic ions which, together with water, will be adsorbed onto the titania surface. A high negative charge which is accumulated on the TiO_2 surface will attract positively charged species, especially polyvalent cations [25, 34]. The Ca^{2+} cations from physiological fluids will be attracted by oxygen atoms at the titania interface by electrostatic interactions [2, 25]. Within an organism, those molecules that bind calcium contain functional groups of carboxyl, phosphate, or sulfate [25, 27], while the proteoglycans and polysaccharides from the ECM contain carboxyl and/or sulfate moieties [25]. Hydroxyapatite, the main inorganic component of bone, contains phosphate groups [18]. Hence, for successful osteointegration to occur, all of these molecules will need to be adsorbed onto the implant surface [2, 7].

Before cells approach the surface of the titania implant, proteins are adsorbed at the interface [5, 35]. Then, only a few seconds after implantation, the proteins (in parallel with other components of the ECM) will begin to adsorb to the titania surface. As the titanium implant normally serves as a bone substitute, collagen (due to its abundance in hard tissues) would be expected to adsorb in large quantities at the interface. Indeed, collagen has been shown to form a three-dimensional (3-D) lattice surrounding the implant [2]. From a physiological point of view, the adsorption of proteins other than collagen is critical for cell adhesion [36]. At approximately 1 h after implantation, the first cells are ready to assemble on the protein layer that has been built on the titania implant surface [5]. If this process of cell adhesion is successful, the cells will start to grow such that, within one month a new tissue will be created on the implant surface. Figure 6.1 illustrates, schematically, the structurally very complex process that occurs *in vivo* at the implant surface. Since an analysis at the molecular and atomic levels of the biomaterials involved in the interaction (and also of their conformation and structures) would be extremely difficult to conduct, simpler systems (e.g., using a lipid bilayer or protein coat) can be used as a compromise. The morphology and structure of the lipid bilayers and of collagen and fibronectin, as adsorbed onto the surface of the implant, are discussed in Sections 6.3 and 6.4, respectively.

6.3
Lipid Bilayers at the Titania Surface

The preparation and detailed morphological and structural characteristics of phospholipid bilayers on the titania surface are of major importance to understand the responses of an implant material in a biological host. Due to similarities in its organization, physical state and structure, the phospholipid bilayer mimics very well a cell membrane [37]. In addition, a variety of biologically important molecules, including other lipids, polysaccharides, proteins, and polypeptides can easily be incorporated into the phospholipid bilayer [38]. Phosphatidylcholine (PC), one of the most extensively studied lipids used to model natural membranes, is known to preserve its structure in crystal, gel and liquid phases, while the polar head group is aligned with the bilayer plane. Long-chain diacyl PCs form predominantly lamellar phases over a wide range of temperature and concentration [37, 39]. Such properties have led to the PCs becoming universal and fundamental models of lipid bilayers.

6.3.1
Formation of Lipid Bilayers on the Titania Surface

A lipid bilayer can be successfully transferred onto the titanium surface by the combined Langmuir–Blodgett and Langmuir–Schaefer (LB-LS) method [21, 40, 41]. The spreading of lipid vesicles represents an alternative method to form an organized lipid bilayer [42, 43]. However, the experimental results showed that, on the titania surface, lipid vesicles did not always spread to form a planar bilayer. Details relating to the formation of lipid bilayers on TiO_2 surfaces from vesicles are outlined below.

6.3.1.1 Spreading of Vesicles on a TiO_2 Surface: Comparison to a SiO_2 Surface

Although, on the surface of a hydrophilic material such as SiO_2 or mica, unilamellar phospholipid vesicles spread to form a compact, well-organized bilayer [42, 43], on the titania surface the preparation of a lipid bilayer from vesicles is both problematic and challenging. It has been shown that, on the TiO_2 surface, the lipid vesicles will indeed be adsorbed, but will remain in the form of a supported vesicles layer (SVL) rather than spread to form a bilayer [44–46]. When the adsorption and spreading of egg-yolk PC unilamellar vesicles of various sizes on silica and titania surfaces was investigated using a quartz crystal microbalance (QCMB) [44–46], the lipid concentration was shown to range from 10 to 15 $mg\,ml^{-1}$, and the size of the filtered vesicles from 25 to 200 nm diameter [44, 45]. Independently of vesicle size, the measured frequency of the microbalance decreased to a minimum after only 4–8 min adsorption onto the silica surface. However, both the minimum frequency and time at which the minimum was reached were greater for larger vesicles. With longer adsorption times, the frequency increased to a constant value of approximately −25 Hz [44], and the larger the vesicle size, the lower was the frequency minimum. The minimum value was also attained over a longer time. By contrast, on the TiO_2 surface, the frequency values decreased monotonically to reach a plateau after 40–60 min of adsorption [44]. Here, the larger the vesicles size, the lower were the frequencies measured. The change in the frequency during egg-yolk PC vesicle adsorption was seen to be greater on a titania than on a silica surface. For example, when egg-yolk PC vesicles of 200 nm diameter were applied to silica a frequency minimum of approximately −80 Hz was reached after 6 min adsorption. In contrast, for TiO_2 the frequency plateaued after about 50 min, and decreased to about −350 Hz [44]. It was suggested that, initially, the silica surface had adsorbed only unruptured vesicles. The critical coverage was achieved at a time which corresponded to the measured minimum of the QCMB frequency. At longer times, the rupture and spreading of vesicles and formation of an egg-yolk PC bilayer was observed. An increase in the frequency, corresponding to a decrease in the mass, accompanied the process of lipid bilayer formation. The mass reduction was considered due to the release of water trapped inside the vesicles. On the titania surface, only a monotonic increase in mass was observed. The lipid vesicles were adsorbed onto the TiO_2 surface until a saturated coverage was achieved, and they remained in the form of a SVL [45].

Additional, more intensive, studies showed that the phospholipid vesicles may spread and form an organized bilayer on the titania surface under the following conditions:

- the lipid concentration in the vesicles of the suspension was changed;
- other lipid/lipids was/were introduced into the PC vesicles;
- certain specific peptides are used;
- the ionic composition and strength are changed; and
- the pH is changed.

6.3.1.1.1 Concentration-Related Factors

The PC vesicles will fuse to form a well-ordered bilayer on the titania surface over a certain concentration range of lipid vesicles in solution. The spreading of PC vesicles on an atomically flat TiO_2 (100) surface, composed of monoatomic and biatomic steps, led to the formation of lipid bilayers [47]. In this study, the vesicles were composed of two PCs: 1,2-dipalmitoleoyl-sn-glycero-3-phosphocholine (DPoPC); and 1,2-dipalmitoyl-sn-glycero-3-phosphocholine (DPPC). At a temperature of 5 °C, at which the process of vesicle adsorption and spreading was monitored, the lipids used exist in two phases, namely DPoPC in liquid phase, and DPPC in gel phase [47]. When the concentration of the vesicles in the solution was less than 0.025 mg ml^{-1}, only the SVL was formed, in agreement with previous reports [44–46]. However, at higher vesicle concentrations, ranging from 0.025 to 0.05 mg ml^{-1}, a lipid bilayer was formed on the titania surface [47]. Images produced using atomic force microscopy (AFM) showed that the lipid molecules had spread to follow the surface structure of the TiO_2 (100), while the formation of lipid domains in the gel phase was also observed. Such gel-phase domains were ~200 nm wide (comparable to the terrace width), and were able to grow such that they crossed monomolecular or bimolecular steps on the TiO_2 (100) surface [47]. The DPPC gel state domains were higher by 1.3 nm relative to the surrounding DPoPC matrix existing in the liquid state. Although the formation of well-ordered lipid bilayers on the titania surface was proven [47], if the vesicle suspension concentration exceeded 0.05 mg ml^{-1}, then second bilayers and more of adsorbed vesicles were observed [47].

6.3.1.1.2 Composition-Related Factors

The preparation of vesicles from a mixture of different lipids also results in the formation of a well-organized lipid bilayer on the titania surface [40, 48, 49]. Epifluorescence images of lipid assemblies prepared from 1-palmitoyl-2-oleoyl-sn-glycero-3-phosphocholine (POPC) vesicles, from vesicles containing POPC and 30 mol% cholesterol, and from a POPC bilayer prepared using the LB-LS method on the titania surface, were compared [40]. Pure POPC vesicles did not spread to form a uniform bilayer, but bright spots were observed on epifluorescence images. In contrast, the LB-LS POPC bilayers formed a uniformly oriented lipid bilayer on the titania surface [40]. However, the addition of 30 mol% cholesterol to POPC vesicles resulted in the formation of a uniform lipid bilayer on the titanium oxide surface [40]. Cholesterol facilitates formation of the lipid bilayer on the titania surface, either by enhancing vesicle fusion or by inhibiting their adhesion. Phosphatidylserine (PS) was also reported to facilitate vesicle spreading on the titania surface [48, 50]. Mixed vesicles containing 0–30% by weight of 1,2-dioleoyl-sn-glycero-3-[phospho-L-serine] (sodium salt) (DOPS) in a 1,2-dioleoyl-sn-glycero-3-phosphocholine (DOPC) matrix were used to study the adsorption process on the titania surface. At DOPS concentrations below 10% (by weight), the intact vesicles remained adsorbed on the TiO_2 surface; however, if the DOPS concentration was raised above 20% and Ca^{2+} ions were present in the electrolyte solution, a supported lipid bilayer was formed [50]. Moreover, lipid molecules in the bilayer were formed from DOPS and the DOPC vesicles were distributed unsymmetrically [50]. The negatively charged DOPC molecules were concentrated in the leaflet that was turned toward the titania surface. This finding was confirmed by studying the adsorption of undiluted human serum protein (concentration ~52 ng ml^{-1}) on DOPC and mixed DOPC:DOPS bilayers [50], when far less protein was found in 20% DOPS/80% DOPC bilayers on titania than on silica. These findings led to the conclusion that much less PS was present in the leaflet facing the solution than in the leaflet which was turned towards the titania surface. It was postulated that the unsymmetric bilayer was formed during the bilayer formation process [50], since lipid redistribution is favored by the presence of defects which are abundant during supported phospholipid bilayer (SPB) formation [51]. Interestingly, the unsymmetric distribution of PS molecules in the lipid bilayer mimics the membranes of eukaryotic cells [52, 53].

It should be noted at this point that the formation of SPBs on the titania surface was observed only in the presence of calcium ions in solution [46, 48, 50]. The importance of ionic strength and composition in relation to the formation of planar lipid bilayers from vesicles is detailed in the next section.

6.3.1.1.3 Ionic Composition- and Ionic Strength-Related Factors

As noted above, the presence of divalent cations in an electrolyte solution facilitates the spreading of lipid vesicles and the formation of planar lipid bilayers on a titania surface [46, 48, 50]. In the presence of Ca^{2+} cations in the solution, vesicles composed of POPC and phosphatidylglycerol spread to form a bilayer. However, in 44% of the experiments the adsorbed mass detected using the QCMB was ~1.8-fold greater than was expected to form a SPB. Thus, it was suggested that the POPC bilayer might contain adsorbed, unruptured vesicles [46]. Indeed, Rossetti et al. [48, 50] later confirmed that lipid vesicles would spread over the titania surface to form a planar lipid bilayer when Ca^{2+} ions were present in solution. Although, in the absence of Ca^{2+}, vesicles containing DOPC and DOPS did not form a planar bilayer on the TiO_2 surface, the addition of 2 mM $CaCl_2$ to a buffer containing 10 mM N-(2-hydroxyethyl)piperazine-N'-2-ethanesulfonic acid (HEPES) and 100 mM NaCl resulted in the formation of a planar lipid bilayer [48, 50]. Phosphatidylserine is a negatively charged lipid that is known to interact with Ca^{2+} ions [54–56]. When vesicles contain more than 20% (by weight) of DOPS, fluorescence microscopy images demonstrated the formation of a well-ordered lipid bilayer [48]. Ca^{2+} ions mediate the adsorption of vesicles containing PS and, through an interaction with the lipid and the negatively charged TiO_2 surface, facilitate the formation of a planar bilayer [48, 50]. It was postulated that "... *the role of Ca^{2+} ions is to reduce or abolish the energy barrier which arises due to electrostatic repulsion between the negatively charged titania and vesicles surfaces*" [48]. A sufficient number of lipid vesicles can be adsorbed onto the titania surface due to Ca^{2+} mediation in ion–vesicle and ion–metal oxide interactions. The presence of densely packed vesicles layer is required to form a bilayer via either a fusion or fusion-decomposition pathway [57–59]. Vesicles containing only PC do not form a bilayer, even in the presence of Ca^{2+} ions in solution.

Vesicles containing the cationic lipid 1,2-dioleoyl-3-trimethylammonium propane (DOTAP) and DOPC were found to fuse on the TiO_2 surface and to form bilayers, even in the absence of divalent cations [49]. DOTAP vesicles (100 nm diameter) were adsorbed onto the titania surface in NaCl solutions of various concentrations (0–300 mM), although three different lipid assemblies were found on the TiO_2 surface depending on the NaCl concentration. No spreading of the DOTAP vesicles was observed in the absence of NaCl, [49], which was consistent with studies using neutral PCs [40, 46, 47] and PS [48, 50]. At NaCl concentrations approaching 100 mM, a patchy bilayer was formed [49], while further increases in NaCl concentration (to 150–200 mM) led to the formation of a tubule network. As the titania surface is negatively charged, an interaction with positively charged lipid molecules can facilitate the adsorption of vesicles and also their fusion, with the resultant formation of a lipid bilayer.

6.3.1.1.4 Use of Some Specific Peptides Recently, the addition of an amphipathic α-helix (AH) peptide (concentration 0.05 mg ml^{-1}) to the electrolyte solution was found to facilitate the spreading of lipid vesicles and the formation of a planar POPC bilayer on the titania surface [60]. In the absence of the AH peptide, however, the POPC vesicles were adsorbed onto the titania surface. The AFM image showed the titania surface to be covered by grains, the size and height of which corresponded to those of the used unilamellar vesicles. The vesicles layer (30 nm diameter) was transferred into a planar compact lipid bilayer after a 50–70 min exposure to the AH solution, such that a complete POPC bilayer was formed, and the roughness of the POPC bilayer corresponded to the roughness of the titania surface. However, when larger vesicles (100 nm diameter) were used in the presence of the AH peptide, an incomplete lipid bilayer was formed on the titania surface, with some unruptured lipid vesicles being observed on the top of the lipid bilayer [60]. The AH peptide created, through electrostatic interaction, an instability on the surface of the lipid vesicle, and this led to an expansion of the vesicle and finally to its spreading on the titania surface. The interesting point here was that a biological material which could interact with a lipid molecules may lead to the fusion of adsorbed vesicles and the formation of a well-organized lipid bilayer.

6.3.1.1.5 pH Factor

An interesting study was carried out to investigate the adsorption of small unilamellar vesicles of DPPC (diameter 50 nm) onto a titania surface from solution over a range of pH values [31]. The adsorption phenomena and process of bilayer formation was compared for three substrates, namely α-SiO$_2$ (quartz), α-TiO$_2$ (rutile), and α-Al$_2$O$_3$ (corundum). The experiments were conducted at pH 5.0, 7.2, and 9.0 in the presence and absence of Ca^{2+} ions in the vesicle solution. Each oxide has a characteristic pH at which the net surface charge is equal to zero; this is the so-called "potential of zero charge" (pzc). Quartz, rutile and corundum have pzc-values of 3, 5.8, and 9.4, respectively [61, 62]. When the pH value is below the pzc, the surface of an oxide will bear a positive charge, but when the pH is above the pzc it will bear a negative charge. The adsorption isotherms of DPPC vesicles for three investigated substrates at various pH values are shown in Figure 6.2 [31].

On all surfaces, in solutions at an initial pH of 5.0, 7.2, and 9.0, and at low concentrations of DPPC vesicles (0.3–0.5 mM), vesicle adsorption occurred in similar manner. However, at higher lipid concentrations greater differences were observed in the adsorption of negatively and positively charged oxides. For α-SiO$_2$ at all pH-values, and for α-TiO$_2$ at pH ≥ 7.2, a plateau corresponding to 10–13 μmol m^{-2} of adsorbed lipid was observed (Figure 6.2). On α-Al$_2$O$_3$ at all pH-values, and on α-TiO$_2$ at pH 5.0, an increase in the concentration of adsorbed lipid (to ~20–25 μmol m^{-2}) was observed. On the corundum surface at pH 9.0, however, less lipid was adsorbed than at pH 5.0 and 7.2 (Figure 6.2c). The adsorption process was studied at 55 °C, at which temperature DPPC exists in the liquid state [31]. In the liquid state, the surface area of a lipid molecule is determined by the area of the PC head group, and is equal to 0.64 nm^2 [39]. Thus, a simple calculation showed that formation of the DPPC bilayer required 5.2 μmol m^{-2} of the lipid. The experimentally determined quantities were twofold times higher for oxides having a negative net charge, which indicates that two lipid bilayers are formed on α-SiO$_2$ at all pH-values, and on α-TiO$_2$ at pH ≥ 7.2. The results of an earlier QCMB experiment showed that the quantity of adsorbed POPC vesicles on a titania surface in solution at pH 7.3 was ~1.8-fold higher than for a bilayer [46]; thus it was postulated that, on a negatively charged TiO$_2$ surface, a planar bilayer with unruptured vesicles would be formed. Recently, Tero et al. [47] observed the formation of a disturbed second bilayer of DPPC and POPC at lipid suspension concentrations >0.05 mg ml^{-1}. These results corresponded very well with the findings of Oleson and Sahai [31] who showed that, on surfaces bearing a positive charge, the concentration of DPPC molecules would reach ~20–25 μmol m^{-2}. This indicated that up to four lipid bilayers may be formed on α-Al$_2$O$_3$ at all pH-values, and on α-TiO$_2$ at pH 5.0.

An increase in pH resulted in a lesser amount of lipid being adsorbed on all of the studied substrates (see Figure 6.2). An increased pH indicates a more negative surface charge on each oxide. The PC headgroup is a zwitterion bearing a negative charge on the phosphate ester moiety; indeed, the negatively charged phosphate ester group is electrostatically attracted by the positively charged sites on α-TiO$_2$ and on α-Al$_2$O$_3$. The phosphate group will be repelled from α-SiO$_2$ at all pH-values,

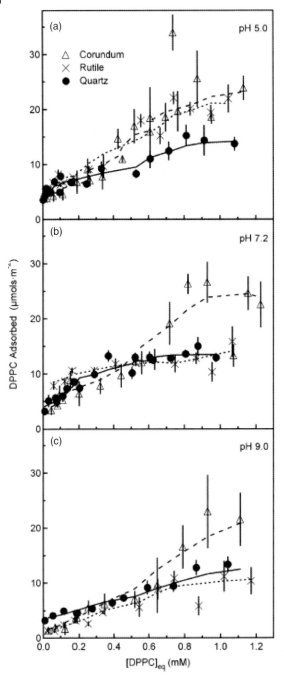

Figure 6.2 Equilibrium adsorption of DPPC from bulk solution at pH (a) 5.0, (b) 7.2 and (c) 9.0 onto quartz, rutile, and corundum surfaces at 55 °C in 50 mM HEPES buffer. Error bars represent one standard deviation, based on spectrophotometric analysis of triplicate solution samples. The lines are shown to guide the eye. Reproduced with permission from Ref. [31]; © American Chemical Society.

and from α-TiO$_2$ at pH ≥ 7.2. However, as adsorption occurred at pH ≥ pzc for the oxides, this indicated that the adsorption was controlled by attractive van der Waals forces, and modified by electrostatic interactions of the oxide surface with charged phosphate moieties in the lipid headgroup [31].

As noted above, the adsorption and spreading of lipid vesicles on titania surfaces is a complicated process that is influenced by many factors. The formation of a planar lipid bilayer is dependent on the lipid (or lipids) composition used, the ionic composition of the vesicles suspension, and also on other biological molecules present in the solution. The spreading of lipid vesicles on a titania surface represents a sum of various attractive and repulsive interactions that take place at the interface. Interactions leading to the formation of planar lipid bilayers on titania surfaces are discussed in the following section.

6.3.2
Interactions: Lipid Molecule–Titania Surface

The formation of a planar lipid bilayer from adsorbed vesicles is the result of interactions between the surface and the vesicle, between the electrolyte and the surface, and between the electrolyte and the vesicle. The size of lipid vesicles ranges between 20 and 200 nm, which corresponds to the size of colloidal particles; consequently, the interactions of adsorbing vesicles with the titania surface may be described using the Derjaguin–Landau–Verwey–Overbeek (DLVO) theory [63]. According to this theory, interactions between colloidal particles are controlled by both van der Waals and electrostatic forces. In addition, short-range non-DLVO interactions between two surfaces in a solvent, such as hydration, double layer and steric forces, should also be taken into account [47]. In general, the interaction energy between lipids and the TiO$_2$ surface is the sum of the van der Waals forces, double layer and hydration interactions [63]. Moreover, the total interaction energy can be modified by electrostatic interactions. Each of these interaction energies can be expressed as a function of the separation distance. Recently, interaction energies between PC and the titania surface were discussed in detail by Tero et al. [47] and by Oleson and Sahai [31]. It is well recognized that a thin water layer (about 0.5–2.0 nm thick) is present between the titania surface and an adsorbed lipid bilayer. When calculating the interaction energy, it was necessary to introduce three media, namely titania–water layer–lipid bilayer [47]. In order to provide the total interaction energy, the van der Waal, double layer and hydration forces were each taken into account

The van der Waals interaction energy is a function of the Debey length, the static dielectric constant of each medium, the adsorption frequency of each medium in the UV range, and the temperature [47]. The adsorption frequency is a function of the refractive index of the medium, while the Tabor–Winterton (TW) approximation is a simplification which assumes that the adsorption frequencies of all media are identical [64–66]. However, it was reported that the TW approximation for the TiO$_2$ was overestimated due to a significantly lower absorption frequency than that of the SiO$_2$, water, or hydrocarbons [47]. An alternative to the TW

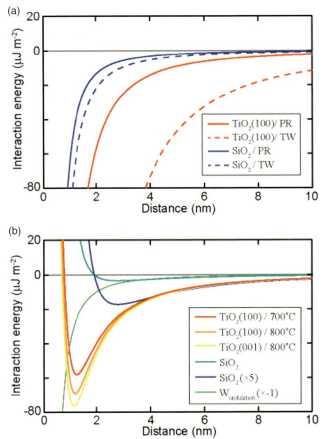

Figure 6.3 (a) The van der Waals interaction energy of solid–water–lipid systems as a function of separation distance calculated for TiO_2 (100) and SiO_2 using the PR (solid lines) and TW (dashed lines) approximations; (b) Total interaction energy of solid–water–lipid systems calculated for TiO_2 (100), TiO_2 (001) and SiO_2 as a function of separation distance. The plot for the silica is magnified fivefold to facilitate visualization. The undulation energy is plotted for comparison. Reproduced with permission from Ref. [47]; © American Chemical Society.

approximation is the Prieve–Russel (PR) approximation, which utilizes experimentally obtained absorption frequencies for each media [47]. The van der Waals interaction energies (calculated using both approaches) for the TiO_2–water layer–lipid bilayer and the SiO_2–water layer–lipid bilayer interfaces are shown in Figure 6.3a.

On the silica surface, the TW and PR approaches produce similar results, whereas on the titania surface they differ significantly (Figure 6.3a). However, a general trend in the shape of two plots is preserved, which shows that the van der Waals interaction is stronger on the titania surface than on the silica surface. On the titania surface, a lipid molecule at 1–2 nm distance (the thickness of the water layer) gains more energy than on the silica surface. However, the van der Waals interaction is still attractive and operates to a distance of ~6 nm from the titania surface.

The double layer energy for a neutral lipid such as PC is equal to zero, and has no influence on the total interaction energy.

The third term, namely the hydration energy between the oxide and the lipid headgroup is repulsive, as it occurs between two hydrophilic particles in aqueous solution. At the TiO_2–lipid bilayer interface the hydration energy is attributed to the structured water layer bound to the solid substrate. The hydration repulsive energy between two lipid bilayers is dominated by thermal fluctuations [47].

The sum of the three forces described above provides the total interaction energy for both the TiO_2–water–PC bilayer and the SiO_2–water–PC bilayer systems, which depend on the distance from the interface (as shown in Figure 6.3b). The total interaction energy is presented for two different TiO_2 surfaces: TiO_2 (100) and TiO_2 (001) annealed at 700 °C and at 800 °C, respectively [47]. The minimum energy is equal to $-58.2\,\mu J\,cm^{-2}$ on the TiO_2 (100) surface, and to $-76.7\,\mu J\,cm^{-2}$ on the TiO_2 (001) surface. A sharp energy minimum indicates that the distance between the bilayer and the titania surface is precisely determined. This distance, of 1.2 nm, corresponds to the water layer assembled between the substrate and the lipid bilayer. The blue line in Figure 6.3b corresponds to the total interaction energy for the SiO_2–water–PC bilayer system. In order to facilitate reading, this plot was magnified fivefold. Indeed, on the SiO_2 surface the total energy of interaction has a very weak minimum at $-3.41\,\mu J\,cm^{-2}$ at a distance of 2.8 nm. According to the authors, the calculated distance for energy minimum was overestimated [47]. The thickness of the water layer between the silica surface and the lipid bilayer, as obtained experimentally, ranged from 1.3 to 1.7 nm [67]. In conclusion, the total interaction energy of the SiO_2–water–PC bilayer system was much weaker than that of the TiO_2–water–PC bilayer system. On the silica surface, the energy gain from van der Waals interactions was compensated by the energy loss from double layer and hydration energies at a distance of ~4 nm. On the TiO_2 surface, however, the attractive van der Waals interaction energy was significantly larger than the repulsive hydration energy. At a distance from the interface where the second bilayer may be formed (~7 nm), the total interaction energy had a value of $-4.7\,\mu J\,cm^{-2}$ (Figure 6.3b). It should be noted that this value was lower than the interaction energy calculated for adsorption of the first bilayer on the silica surface ($-3.4\,\mu J\,cm^{-2}$) [47]. Apparently, the strong van der Waals attraction on the titania surface promotes the formation of a second lipid bilayer (or at least its fragments). Indeed, the results of earlier QCMB studies pointed to an adsorption of lipid mass onto the titania surface which corresponded to the formation of about 1.8 bilayers [46]. In the study conducted by Tero *et al.* [47], the AFM images demonstrated the formation of a second bilayer on the titania surface [47], a result which agreed very well with more recent investigations into the adsorption of DPPC vesicles on rutile, silica, and corundum surfaces [31]. In these studies, two contributions to the total energy – the van der Waals and electrostatic interactions – were taken into account [31]. Whilst the van der Waals interaction was much weaker on silica and corundum surfaces than on titania surfaces, the electrostatic interaction was alternated or even diminished by altering the pH of the solution. On negatively charged surfaces, the DPPC concentration corresponded to the formation of two bilayers. As shown in Figure 6.3b, the van der Waals forces did indeed operate at a distance of two bilayers, allowing for formation of the second DPPC bilayer on the titania surface [31, 47]. The adsorption is halted at this stage. On positively charged surfaces, however, the determined amount of DPPC corresponded to the formation of four lipid bilayers. Apparently, at distances in excess of two bilayers the van der Waals interaction is modified by the electrostatic interaction [31]. Repulsion on negatively charged surfaces hinders further adsorption, whereas attraction on positively charged surfaces allows further adsorption. As PC is a zwitterion with negative charge accumulated on the phosphate ester group, and positive charge on the choline moiety, Oleson and Sahai suggested that the negatively charged phosphate ester group might be involved in an electrostatic attraction with positively charged TiO_2 and Al_2O_3 surfaces. Based on this interaction, two additional bilayers could be formed on the positively charged oxide surface [31].

The above-discussed studies are complementary, and point to a specific interaction between the titania surface and an adsorbing lipid molecule [31, 47]. On the titania surface, the strong attractive van der Waals interaction serves as the driving force for adsorption and the spreading of lipid vesicles, and this results in the formation of a planar bilayer. At large distances from the interface, however, the van der Waals attraction compensates the repulsive hydration and double layer forces, facilitating the adsorption of two lipid bilayers. At even larger distances (>7 nm), the van der Waals forces can be further modified by electrostatic interaction, thus facilitating the adsorption of larger amounts of biological material. On the one hand, these very interesting and important findings show that, under strictly defined conditions, the adsorption of well-organized lipid bilayers on the titania surface may be achieved. This property of titania may result in a high acceptance of this material for implantation *in vivo*. On the other hand, the presence of PS in the leaflet facing the titania (which corresponds to the outer leaflet of the natural cell membrane) may activate macrophages, initiate blood coagulation at the site of injury or at the implant surface, and also activate the inflammatory response. All of these reactions are characteristic of implant rejection [50].

The results discussed above give rise to the question of whether the strong attraction on a titania surface may influence the hydration, lateral distribution, and structure of lipid molecules in the bilayer assembly. Current data relating to the structure and conformation of lipid molecules in the bilayer on titania surfaces, both as predicted by calculations and as found experimentally, are discussed below.

6.3.3
Structure and Conformation of Lipid Molecules in the Bilayer on the Titania Surface

As noted above, the interaction between a lipid molecule and the titania surface is a complicated issue that involved van der Waals, hydration, double layer, and electrostatic interactions [31, 47]. On the one hand, this strong attraction may support the excellent biocompatibility of titania as an implant material. On the other hand, strong forces operating at large distances (~7 nm) from the interface may significantly influence the conformation and structure of a lipid assembly on the titania surface. Recently, a number of investigations, both theoretical [19] and experimental [20, 21], have been conducted to identify the molecular-scale picture of a lipid bilayer as it is adsorbed on the titania surface. A series of molecular dynamics studies provided the atomic-scale picture of three lipid molecules adsorbed onto the titania surface, namely DOPC, DOPS, and 1,2-dimyristoyl-3-trimethylammonium-propane (DMTAP). The DOPS and DMTAP lipids are both charged under physiological conditions, with Na^+ and Cl^- counterions (respectively) being used to neutralize the charge [19]. Interaction of the lipid headgroup with the substrate was investigated for neutral nonhydroxylated, fully hydroxylated, and partially hydroxylated, negatively charged titania surfaces.

The structure of the PC headgroup, as obtained by theoretical means, was at variance with experimental data obtained using infrared reflection absorption spectroscopy (IRRAS) [20, 21]. Moreover, the lack of experimental data on the structure of lipid bilayers composed of PS and TAP on the titania surface raised the need for experimental confrontation.

6.3.3.1 Structure of Phosphatidylcholine on the Titania Surface

6.3.3.1.1 Structural Data Available from Theoretical Considerations

In the headgroup region of PC (the most widespread lipid in a cell membrane), four oxygens can interact directly with Ti atoms on the surface, while eight possible acceptors for hydrogen bonds are present [19]. The atom–surface distance and probability distributions of oxygen and nitrogen atoms as a function of the simulation time for two DOPC molecules adsorbing on the nonhydroxylated, partially hydroxylated and hydroxylated titania surfaces, are shown in Figure 6.4.

On the nonhydroxylated and partially hydroxylated TiO_2 surfaces, two oxygen atoms – one belonging to the phosphate nonesterified moiety (O9) and one carbonyl oxygen atom (O16) in the sn2 chain – are seen to interact directly with Ti atoms on the surface. The structure of the PC headgroup on both surfaces is similar. The phosphate oxygens were found to stay in close contact with the Ti atom of the interface; the average distance O(9)–Ti was equal to 1.94 Å (Figure 6.4a,b) [19]. This conformation of the headgroup favored an interaction of the carbonyl oxygen (O16) of the sn2 chain. The distance between O(16)–Ti atoms is equal to 2.01 Å [19]. The oxygen atom in the ester group (O37) in the sn1 chain has greater flexibility, and is exposed to water molecules; this oxygen atom may be involved in hydrogen bonding. On the nonhydroxylated TiO_2 surface, the PC headgroup is stiff and strongly bonded to the surface. On the partially hydroxylated titania surface, the structure of the PC remains the same as on the nonhydroxylated surface, but the lipid molecule gains more flexibility, due to the loss of some Ti–O binding sites; thus, a weaker interaction with the surface is observed [19]. On the nonhydroxylated titania surface, the angle between the P–N vector and the

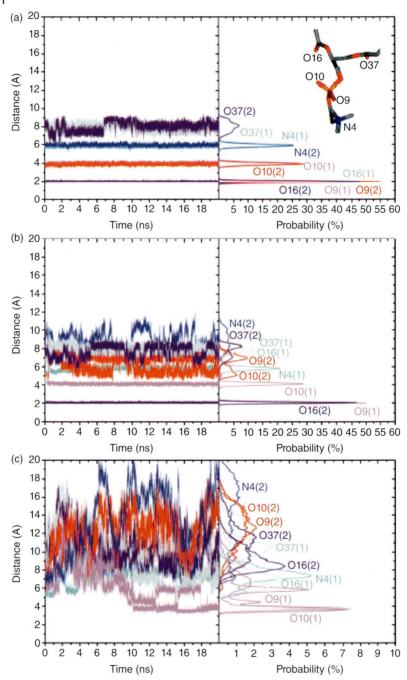

Figure 6.4 Plots of atom–surface distances and probability distributions as a function of the simulation time for DOPC molecules adsorbed on: (a) nonhydroxylated; (b) partially hydroxylated; and (c) hydroxylated titania surfaces. The number in parenthesis after each atom refers to two lipid molecules. The inset to panel (a) shows the structure of the DOPC headgroup, with marked and numbered O and N atoms. Reproduced with permission from Ref. [19]; © American Chemical Society.

surface normal is close to 136°, pointing away from the surface. The P=O bonds create an angle of 44° with the surface normal [19]. On the partially hydroxylated TiO_2 surface, the P–N vector creates an angle of ~80° and 94° with the surface normal. A turn of the P–N vector in the direction of the surface normal indicates a weaker adsorption and an increased mobility of the PC group. Indeed, the simulations confirmed a weakening of the Ti–O binding in the phosphate and carbonyl groups [19]. An interaction with nitrogen atom of PC was detected. On the hydroxylated titania surface, a much weaker interaction with PC was reported [19]. The phosphate oxygens were found to create hydrogen bonds to OH groups adsorbed on the titania surface. As seen in Figure 6.4c, the distance between O in the phosphate group and the TiO_2 surface is large (~4 Å), and the probability of O atoms approaching the titania surface is significantly decreased to below 10%, compared to ~50% on the nonhydroxylated and partially hydroxylated titania surfaces [19].

6.3.3.1.2 Structure Obtained from Experiments: IR Spectroscopy
The structure of the DPPC LB monolayer and multilayers, and also of the film formed from vesicles, was investigated using attenuated total reflection infrared spectroscopy (ATR IRS) [20]. The structure of the 1,2-dimyristoyl-sn-glycero-3-phosphocholine (DMPC) LB-LS bilayers transferred onto the titania in the liquid expanded (LE) and liquid condensed (LC) states were studied using polarization modulation-infrared reflection absorption spectroscopy (PM-IRRAS) [21]. First, the structure and hydration of the polar PC headgroup are reported; subsequently, the conformation and structure of the hydrophobic hydrocarbon chains are presented.

6.3.3.1.3 Choline Group
The choline moiety is the most outer fragment of the PC headgroup, which bears a positive net charge on the nitrogen atom. The choline group gives rise to absorptions arising from two asymmetric methyl bending modes $\delta_{as}(N^+(CH_3)_3)$ located at ~1491 cm^{-1} and ~1479 cm^{-1}, and to a weak symmetric methyl bending mode $\delta_s(N^+(CH_3)_3)$ at ~1400 cm^{-1} [20, 21]. In the DMPC bilayer in the LE state, the PM IRRA spectrum shows clearly two overlapping $\delta_{as}(N^+(CH_3)_3)$ bands. However, the $\delta_{as}(N^+(CH_3)_3)$ mode in the LC DMPC bilayer has an ill-defined band with the overall maximum found at 1490 cm^{-1}. It was reported that an electrostatic interaction of the ammonium cation in the cetyltrimethylammonium bromide (CTAB) molecule with sodium dodecyl sulfate (SDS) anion gives rise to a broad, badly defined $\delta_{as}(N^+(CH_3)_3)$ mode observed at 1482 cm^{-1}, instead of two well-defined bands centered at 1490 cm^{-1} and 1479 cm^{-1} [68]. A lateral interaction between positively and negatively charged molecules is responsible for the described changes in the shape and position of the $\delta_{as}(N^+(CH_3)_3)$ mode. To conclude, in the zwitterionic PC the ill-defined $\delta_{as}(N^+(CH_3)_3)$ mode is due to lateral interactions between the positively charged $N^+(CH_3)_3$ choline and the negatively charged phosphate PO_2^- moieties in neighboring lipid molecules [20, 21]. The fact that the $\delta_{as}(N^+(CH_3)_3)$ mode is less-defined in the DMPC bilayer in the LC state supports stronger interactions between choline and phosphate groups than in the loosely packed bilayer that exists in the LE state.

The advantage of the PM-IRRAS technique is the possibility of performing a quantitative analysis of selected parts of an organic molecule adsorbed onto the solid surface [69, 70]. From the integral intensity of a given IR absorption band, the angle between the transition dipole moment of a given vibration and the electric field vector (being perpendicular to the substrate) can be calculated. This angle allows conclusions to be drawn regarding the orientation of a particular group in the lipid bilayer [69–71].

The $\delta_{as}(N^+(CH_3)_3)$ mode at 1492 cm^{-1} has the transition dipole moment located along the C_{3v} axis [21]. The angle calculated for this band corresponds directly to the inclination of the C–N bond with respect to the surface normal. In the DMPC bilayer prepared in the LE state, the C–N bond makes an angle of 70 ± 2°. In the DMPC bilayer prepared in the LC state, the C–N bond makes a smaller angle of 54.5 ± 1.5° with respect to the surface normal [21]. The tilt angle of the C–N bond in the choline moiety corresponds to the average tilt in the bilayer assembly, in which statistically half of the molecules are in direct contact with the titania surface. The analysis does not provide directly the tilt of the C–N bond in the leaflet turned towards the substrate. However, in the LE state of the bilayer a large tilt (70° with respect to the surface normal) suggests that this bond is more inclined towards the bilayer plane than in the LC state. Interestingly, only in the LC bilayer is a $\delta_s(N^+(CH_3)_3)$ mode at 1399 cm^{-1} observed [21]. Earlier, Jiang et al. [20] reported that, on the titania surface in a DPPC monolayer, the symmetric $\delta_s(N^+(CH_3)_3)$ band is more intense than in three or five DPPC layers. The more DPPC LB layers were transferred onto the titania surface, the weaker the $\delta_s(N^+(CH_3)_3)$ mode at 1396 cm^{-1} was observed [20]. Similarly, the IR spectra of a positively charged CTAB adsorbing onto the titania surface showed a higher intensity of the $\delta_s(N^+(CH_3)_3)$ mode, with an increase of the surface concentration of the salt adsorbing directly on the surface [72]. In the DMPC bilayer transferred in the LC state, the average area per lipid molecule was equal to 0.48 nm^2 and provided more sites for choline–titania interactions than in the bilayer transferred in the LE state, where the surface area per lipid molecule was equal to 0.72 nm^2 [21]. In addition, an electrostatic interaction with the surface supported an oriented arrangement of the choline group in the LC bilayer.

In DPPC monolayer and multilayers, the ratio (A_{1396}/A_{2850}) of the absorption band at 1396 cm^{-1} ($\delta_s(N^+(CH_3)_3)$) (providing the number of lipid molecules directly interacting with the titania surface) to the absorption at 2850 cm^{-1} ($v_s(CH_2)$) (corresponding to the methylene stretching mode in the acyl chains, thus being not sensitive to interaction with the substrate) was calculated (Figure 6.5) [20].

The intensity ratio A_{1396}/A_{2850} decreases monotonically with the increase of the number of the LB bilayers transferred onto the titania surface. This ratio shows the percentage of lipid molecules which directly interact with the charge sites on the titania surface. The A_{1396}/A_{2850} ratios for the DPPC LB films and the film obtained from the DPPC vesicles are compared in Figure 6.5. Interestingly, the A_{1396}/A_{2850} ratio in the film formed from vesicles was close to 0.043, and this value was close to that of 0.028 obtained for five LB layers. Five LB layers corresponds to two-and-a-half of the lipid bilayers. As reported by others [20], the ATR spectra of five LB layers of DPPC, and of the film obtained from vesicles, are very similar. These results, together with those discussed in Section 6.3.2, confirm indeed,

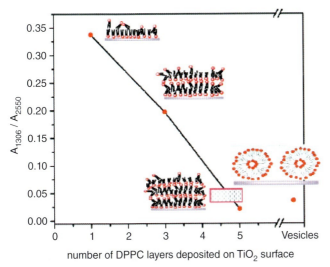

Figure 6.5 Relative intensity of the $\delta_s(N^+(CH_3)_3)$ band (A_{1396}/A_{2850}) as a function of the number of DPPC layers and DPPC vesicles deposited on the titania surface. The pictures are provided as a guide to show the percentage of DPPC molecules directly interacting through the positively charged choline moiety as a function of the number of transferred layers. Reproduced with permission from Ref. [20]; © American Chemical Society.

that PC vesicles spread on the titania surface form more than a single lipid bilayer.

6.3.3.1.4 Phosphate Group

The 1400–1000 cm^{-1} spectral region corresponds to phosphate-stretching modes and to methylene-wagging modes. The polarization modulation-infrared reflection absorption (PM-IRRA) spectra of DMPC bilayers existing in LE and LC states, and transferred onto the titania surface, are shown in Figure 6.6.

The number, positions, and intensities of absorption bands observed in this region differ significantly in those two DMPC bilayers. In general, the phosphate stretching modes of DMPC bilayers transferred in the LE state are very weak and red-shifted compared to the bilayer transferred in the LC state. In the DPPC monolayer, the $\nu_{as}(PO_2^-)$ mode is observed at 1223 cm^{-1}, and is so broad that the band almost disappears from the ATR spectra [20]. To conclude, in the LC DMPC bilayer as well as in the DPPC monolayer, the phosphate group is well hydrated and the 1223–1221 cm^{-1} position of the $\nu_{as}(PO_2^-)$ mode and its low intensity combined with a large width, point to hydrogen bonding occurring between the water molecules and the phosphate moiety [20, 21]. In DPPC multilayers, absorption arising from the $\nu_{as}(PO_2^-)$ mode is stronger [20]. In the DMPC bilayer transferred onto the titania surface in the LC state, the $\nu_{as}(PO_2^-)$ mode is overlapped with CH$_2$ wagging modes progression. The deconvolution shows that the $\nu_{as}(PO_2^-)$ band is located at 1258 cm^{-1} [20]. An increase of packing of DMPC molecules, and an increased

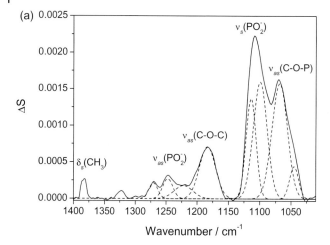

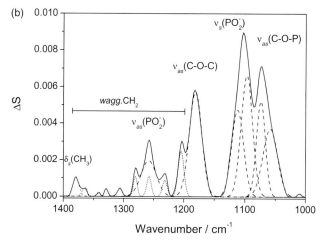

Figure 6.6 The PM-IRRA spectra of phosphate-stretching modes of the DMPC bilayer transferred onto titania surface in the (a) LE state and (b) LC state. Reproduced with permission from Ref. [21]; © American Chemical Society.

number of LB layers transferred onto the titania surface, are due to a blue shift of the $v_{as}(PO_2^-)$ mode. This indicates a lower hydration of the phosphate group.

As shown in Figure 6.6, the major difference in the PM IRRA spectra of two DMPC bilayers arises from the different intensities of absorption of the phosphate stretching modes. In the LE bilayer, the phosphate stretching modes are weak. The reduction of the $v(PO_2^-)$ band intensity in the LE state is partially due to the increased hydration of the phosphate group [73, 74]. However, a four- to fivefold change in the integral intensities cannot originate from changes in hydration degree only, but rather is related to a different orientation of the phosphate moiety in the LE and LC states of the lipid bilayer. Indeed, the esterified O–P–O part of the PC headgroup has a different tilt in both bilayers. In the LE state of the DMPC bilayer, the in-plane vector of the O–P–O group makes an angle of only 15° with respect to the surface normal. However, in the densely packed DMPC bilayer (LC state), the in-plane vector of the O–P–O ester group is tilted by 28° with respect to the surface normal. These results are counterintuitive; on the one hand there is an almost vertical orientation of the in-plane vector of the O–P–O ester group, together with large intermolecular distances in the LE state provides space for water of hydration. On the other hand, in a loosely packed bilayer, a larger tilt would be expected than in a densely packed bilayer.

6.3.3.1.5 Ester Carbonyl Group

In PC, the C=O group in the intermediate ester moiety strongly absorbs IR light in the 1760–1700 cm^{-1} region. The ν(C=O) stretching mode in the DMPC bilayers and the DPPC LB films is unsymmetric [20, 21]. This unsymmetric ν(C=O) band is composed of two absorptions: one centered around 1743–1742 cm^{-1}, and one around 1733–1726 cm^{-1}. In PC, two carbonyl C=O groups in the ester moiety (sn1 and sn2) are not equivalent for the IR spectroscopy, and this leads to a splitting of the ν(C=O) stretching mode into two bands that correspond to the in-phase and out-of-phase motions of oxygen atoms. However, in DMPC and other PCs, this split is small and varies from 1 cm^{-1} to 5 cm^{-1}, depending on the structure of the PC headgroup [75, 76]. The large split observed for the DMPC and DPPC bilayers cannot originate from in-phase and out-of-phase stretching of the C=O bond. It has already been proved that the position of the ν(C=O) mode is heavily dependent on the hydration of this group [76]. Indeed, for the above mentioned PC films, the high-frequency band corresponds to a dry ester group, and the low-frequency band to a well-hydrated ester group [21, 75, 76]. To conclude, independent of the packing density and the number of transferred layers, the ester group in PC has two different degrees of hydration. The C=O bond makes an angle of 74.0 ± 0.5° in the LE DMPC bilayer, and 62.6 ± 0.5° in the LC bilayer. The C=O bond of the ester group tends to lie in the bilayer plane.

6.3.3.1.6 Hydrocarbon Chains

The physical state of the hydrophobic hydrocarbon chains in the lipid molecule is provided by the position and full-width at half maximum (fwhm) of the methylene stretching modes [76]. The positions and fwhm values of the asymmetric $\nu_{as}(CH_2)$ and symmetric $\nu_s(CH_2)$ methylene stretching modes in DPPC and DMPC assemblies on the titania surface are listed in Table 6.1.

High frequencies of the methylene stretching modes show that, in the LE DMPC bilayer, the hydrocarbon chains are melted and contain many *gauche* conformations. In the LC DMPC bilayer and in all assemblies of DPPC (this lipid has two methylene units more in each chain compared to DMPC), the hydrocarbon chains exist in the all-*trans* conformation. In the DMPC LC bilayer, however, the $\nu_{as}(CH_2)$ mode has a higher frequency than 2920 cm^{-1} and the $\nu_s(CH_2)$ than 2850 cm^{-1}, indicating some gauche defects in the acyl chain region. Indeed, the CH_2 wagging modes confirm the existence of the hydrocarbon chain in a ripple

Table 6.1 Positions and full width at half-maximum (fwhm) of the asymmetric $v_{as}(CH_2)$ and symmetric $v_s(CH_2)$ methylene stretching modes in 1,2-dipalmitoyl-sn-glycero-3-phosphocholine (DPPC) and 1,2-dimyristoyl-sn-glycero-3-phosphocholine (DMPC) films on the titania surface.

Lipid sample and its arrangement	$v_{as}(CH_2)/cm^{-1}$ frequency	fwhm	$v_s(CH_2)/cm^{-1}$ frequency	fwhm	Technique	Reference
DPPC						
1 LB layer	2919	20	2851	14	ATR-FTIRS	[20]
3 LB layers	2919	19	2850	13	ATR-FTIRS	[20]
5 LB layers	2918	18	2849	12	ATR-FTIRS	[20]
Vesicles	2916	15	2848	11	ATR-FTIRS	[20]
DMPC						
LE bilayer	2926.3	22	2855.7	16	PM-IRRAS	[21]
LC bilayer	2921.1	19	2852.8	15	PM-IRRAS	[21]

ATR-FTIRS: Attenuated total reflection-Fourier transform infrared spectroscopy; PM-IRRAS: Polarization modulation-infrared reflection absorption spectroscopy.

phase. In the ripple phase the hydrocarbon chain has predominantly all-*trans* conformation with some *gauche-trans-gauche* (gtg) kinks and end-gauche conformations [21, 77–79]. The all-stretched hydrocarbon chain in the LC DMPC bilayer makes an angle of 26° with respect to the surface normal [21].

6.3.3.1.7 Conclusions The results of simulations and experimental investigations into the structure, conformation and hydration of PCs adsorbing onto partially hydroxylated titania surfaces are compatible. The partially hydroxylated titania surface correspond most to the surface of the titania existing under experimental conditions, and *in vivo* after implantation. The calculations point to an electrostatic interaction between the nitrogen of the choline group and the titanium atoms [19]. Indeed, IR spectroscopy based on the analysis of the $\delta_s(N^+(CH_3)_3)$ mode in loosely and densely packed DMPC and DPPC films confirmed this interaction. The interaction between N–Ti atoms influences the tilt of the C–N bond in the choline group, which is most pronounced in the loosely packed DMPC bilayer transferred in the LE state. Moreover, in the LC DMPC bilayer, IR spectroscopy pointed to increased interactions between positively charged choline and negatively charged phosphate groups of adjacent molecules. Interestingly, the line of the esterified O–P–O group has a small angle of 15° and 25° with respect to the surface normal in LE and LC bilayers, respectively. These values are by 15–20° smaller than the tilt of the O–P–O vector in DMPC bilayers assembled on Au and SiO_2 surfaces [80, 81]. In the loosely packed DMPC bilayer, an almost vertical orientation of the esterified phosphate group provides space for hydration and hydrogen bonding. Indeed, hydrogen bonding of the phosphate group to water has been detected using both PM-IRRAS [21] and ATR IRS [20], and also predicted by simulations [19]. The small tilt of the O–P–O esterified line, and large tilt of the C–N bond (average tilt of 70° with respect to the surface normal), suggest an open conformation of the PC headgroup. X-ray diffraction (XRD) studies have proven the existence of two crystal structures of DMPC, namely A and B [39, 82]. The B structure has a more extended conformation due to a tighter binding of water (relevant here) and its interaction with charged groups of the adjacent bilayers (not relevant for a single bilayer studies) [82]. In this structure, the ammonium group does not fold back under the phosphate group. In the crystal form, the polar headgroup tends to orient the P–N dipole parallel to the bilayer plane and does not separate opposite charges. In the PC headgroup adsorbed onto a partially hydroxylated titania surface, the P–N vector is oriented perpendicularly to the surface normal. To conclude, the orientation of P–N found by XRD studies, and predicted for the adsorbed lipid molecule on the titania surface, is the same. From an IR spectroscopic experiment, the tilt of the P–N vector cannot be provided. An agreement of the IR spectroscopic studies [20, 21] with XRD data [82] on the one hand, and of the theoretical predictions [19] with XRD [82] on the other hand, indicate that in a loosely packed PC bilayer, the B structure dominates and that the P–N vector lies in the bilayer plane.

In a densely packed DMPC bilayer, the tilt of the C–N bond decreases to ~55° and the in-plane vector of the O–P–O line increases to 28° with respect to the surface normal [21]. These rearrangements result in a reduction of the space available for a single lipid molecule, and thus the hydration of the polar headgroup decreases. In addition, the C=O bond makes an angle of 62° and the acyl chains are tilted by 26° with respect to the surface normal. In the A structure of the DMPC molecule, the C=O bond makes a normal angle to the tilt of the hydrocarbon chain; moreover, the C–N bond of the choline groups makes an angle of ~55° with the surface normal [39, 82]. In a densely packed PC film, the A structure of the lipid is found. In the crystal, the P–N vector is inclined by 17° towards the bilayer plane. Simulations show that the P–N vector is perpendicular to the surface normal, and thus indeed lies in the bilayer plane [19]. The IR spectroscopy experiments showed an increase in the coulombic interactions between the choline and the phosphate group. An increased interaction is justified by a weaker hydration of the phosphate moiety, and by the parallel to the bilayer plane orientation of the P–N vector.

Interestingly, IR spectroscopy also showed that some of the carbonyl ester groups in the DMPC and DPPC films are dry, while others are well hydrated. Indeed, the calculations showed that the sn2 C=O(16) oxygen is involved in the direct interaction with Ti surface atom, while the sn1 C=O(37) oxygen is flexible and forms a hydrogen bond to water.

The degrees of hydration and hydration sites predicted by simulations and observed experimentally, are in agreement. Depending on the packing of the lipid molecules on the titania surface, reorientations and conformational changes in the lipid films were observed. The polar headgroup rearranges depending on the strength of the interactions between N–Ti, PO_2^-–Ti, PO_2^-–H_2O, C=O–H_2O and N–PO_2^-. To conclude, it is worth underlining here that these experimental results agree very well with theoretical predictions. An analysis of the orientation of various bonds in the polar headgroup, and of the hydrocarbon chains of PC, provides a detailed molecular-scale picture of the lipid bilayer adsorbed onto the titania surface.

6.4
Characteristics of Extracellular Matrix Proteins on the Titania Surface

On an implant surface, the assembly of cells and tissues is followed by the adsorption of proteins. Those proteins which are exposed to an implant material belong to the ECM, which is composed of gels of polysaccharides and fibrous proteins [83]. Among the proteins of the ECM, the most abundant is collagen; other ECM proteins include: fibronectin (FN), elastin, and laminin [84]. In an organism, the ECM has the following functions: (i) to support the adhesion of cells; (ii) to segregate tissues from one another; and (iii) to regulate intercellular communication [84]. The proteins of the ECM are essential for the growth of cells and tissues, for the healing of wounds, and for fibrosis (the formation and development of excess fibrous connective tissue).

A fibrillous collagen is the main protein found in the bones, skin, cartilage, and the walls of blood vessels; in fact, in the human bone matrix, collagen constitutes almost 90% of the protein present. Fibrillar collagen (types I, II, and III), due to its abundance and importance in nature, is the most studied structure of collagen. Within the collagen molecule, tropocollagen forms the subunit of a larger collagen aggregate (fibrils); tropocollagen is approximately 300 nm long and ~1.5 nm wide [85]. Collagen is composed of three polypeptide strands, each possessing the conformation of a left-handed helix; these are twisted together into a right-handed coil, a triple helix. Collagen has a very specific structure and composition that is not found in other proteins; glycine (Gly) constitutes one-third of all the amino acids contained in collagen, while two other amino acids – proline (Pro) and hydroxyproline (Hyp) – constitute around one-sixth. In the collagen fibril, the sequence of the three amino acids, $(Gly-Pro-Pro)_x$ and $(Gly-Pro-Hyp)_y$, is repeated. The relatively high content of Pro and Hyp rings, which have geometrically constrained carbonyl and secondary amino groups, combined with a large content of Gly, creates a spontaneous left-hand turn of a peptide strand. The assembly of the triple helix places the Gly moiety in the interior of the helix, while the Pro and Hyp rings are oriented outside. Neutral Gly, negatively charged glutamic acid (Glu) and the positively charged lysine (Lys) are present in the side chains of collagen [84].

In parallel with collagen, the presence of the second important ECM protein, FN is essential for cell healing and development processes, as well as in the binding of proteins from the cell membrane proteins [36, 86, 87]. The presence of FN is of major importance in blood coagulation and bacterial adherence reactions [88]. Thus, FN may be responsible for inflammation and lead to the rejection of a newly implanted material.

Fibronectin is a large, adhesive, dimer glycoprotein; each monomer has a molecular weight of 220–260 kDa [87, 89]. The FN protein is composed of three repeating homologies: FN I, FN II, and FN III. Close to the C-terminus, the FN I homologies contain Cys residues that form two disulfide bonds which join two monomers of the protein [87, 90]. The FN protein contains the highest number of FN III homologies, which have a common secondary structure, namely, a β-sandwich formed by two antiparallel β sheets [87, 90]. The FN III$_{10}$ homologue contains a loop composed of three amino acids, arginine-glycine-aspartic acid (Arg-Gly-Asp; this is also known as RGD). The RGD loop is responsible for the binding of integrin proteins present in a cell membrane. The RGD loop has a well-preserved structure and extends by ~10 Å above the core of the molecule [36, 90, 91].

The secondary structure of FN is composed of β sheet structures. The amino acid sequence, and particularly the presence of charged Arg and Asp amino acids, make FN one of the most flexible proteins studied to date [87, 89, 92]. Under physiological conditions, FN can exist in a folded and flexible conformation, but an increase in ionic strength (e.g., at NaCl >0.3 M) [93], in pH (to 11) [89], or the binding of collagen [89] will cause the protein to unfold. The structural changes in the FN molecule that occur in solutions of various pH and ionic strengths suggest that electrostatic interactions are very important for the ability of FN to fold. The charge which accumulates on the surface of an implant may also influence the structure of the adsorbing FN and, in the adsorbed state, also modify the properties of the protein as compared to the native state.

The molecular-scale recognition of the changes that take place in collagen and fibronectin, when adsorbed onto the titania surface as a perfect implant material, remain largely unknown. Current knowledge regarding the adsorption, morphology, and structural changes that accompany collagen and fibronectin when they are adsorbed onto titania surfaces is presented in Sections 6.4.1 and 6.4.2, respectively.

6.4.1
Collagen Adsorption on Titania Surfaces

As an implant material, titanium is used extensively in bone replacement procedures, notably for dental purposes, and in hip and knee implants. It follows, therefore, that the protein which most frequently comes into contact with titania surfaces is collagen. Collagen layers, when examined *ex situ*, are unstable on a titanium surface, which itself is covered by a 3–10 nm thick layer of native oxide [5]. The adsorption of collagen onto an oxidized titanium surface is slow, and the protein films formed are generally unstable [94–96]. Investigations using an AFM-separation approach have shown that collagen, when in direct contact with a titania surface, has a reduced elasticity and poor mechanical properties [95]. In order to facilitate the adsorption of collagen that is so crucial to osteointegration, the titania surface must either be preadsorbed with other biomolecules, or the metal surface must be activated [11]. Although the predeposition of biomolecules on the titania surface may facilitate further collagen adsorption, it will not provide any direct contact between the collagen protein and the TiO$_2$ surface. Moreover, the morphology and structure of the adsorbed protein may differ from that of a protein adsorbed directly from the ECM. In order to facilitate and strengthen collagen adsorption, the titania surfaces can be activated in several ways, including:

a) a thermal or anodic oxidation process can be applied to generate a thick surface titania layer [18, 97, 98];
b) the oxidation can be conducted under corrosive conditions (with the use of H_2O_2 and HCl) [96];
c) a calcium coating can be created, to serve as an anchor layer for the collagen [18, 99].

6.4.1.1 Morphology of Collagen Adsorbed on an Oxidized Titanium Surface

A titania surface which is covered by a 10 to 180 nm thick titanium oxide layer has been shown to provide a good matrix for collagen adsorption [18, 97, 98]. The electrochemical oxidation of the titanium facilitated the adsorption of type I collagen [18, 97, 98]. In the presence of an electric field, collagen molecules containing positive charges were attracted by the cathode, thus facilitating protein deposition on the TiO_2 surface. The adsorbed collagen fibrils, with a diameter of 400–500 nm, were much larger than their native counterparts [18]. The adsorbed collagen fibrils did not form a dense monolayer on the titania surface, but a similar structure of type I collagen adsorbed on an oxidized titanium surface was reported by Born et al. [97]. On the TiO_2 surface, the collagen fibrils were seen to overlap each other and to be randomly oriented. On an electrochemically oxidized titanium surface, the collagen molecules were partially integrated into the TiO_2 layer [97].

Another electrochemical study showed that collagen adsorbed onto a 50 µm thick titanium oxide surface layer covered the surface uniformly, with the open circuit potential remaining constant for almost three months [98]. The open circuit potential in a buffer composed of 8.74 g l^{-1} NaCl, 0.35 g l^{-1} NaHCO$_3$, 0.06 g l^{-1} Na$_2$HPO$_4$·12H$_2$O and 0.06 g l^{-1} NaH$_2$PO$_4$ (pH 6.5) was close to −25 mV (versus Ag/AgCl reference electrode) [98]. It appeared that the collagen had formed a dense layer, thus preventing corrosion of the titania surface. It must be noted, however, that these results should not be compared due to the different collagen preparations, and the different adsorption times and collagen concentrations used [18, 98]. The measurements of open circuit potential were also conducted in an electrolyte solution containing Ca^{2+} and PO_4^{3-} ions [98], both of which may crystallize on the titania surface and facilitate collagen adsorption [18, 99]. In summary, whilst an initial adsorption of collagen results in a surface coverage of less than a monolayer, the formation of a densely packed protein film over a longer time cannot be excluded.

To conclude, our existing knowledge of the morphology of collagen adsorbed onto oxidized titanium surface remains very limited. Likewise, the structure of the protein, and the eventual changes that accompany the adsorption process, are essentially unknown.

6.4.1.2 Adsorption of Collagen on a Hydroxylated Titania Surface

It has been reported that collagen fibrils may be adsorbed successfully on a hydroxylated titania surface [96] such that, after a 30 min adsorption period a frequency decrease of 630 Hz was measured using QCMB. Bearing in mind the uniform distribution of adsorbed collagen on a titania surface, this frequency change corresponded to a 12 nm thick protein layer [96]. Quantitatively, the assembly of collagen on titania surfaces has been monitored using ATR IRS and X-ray photoelectron spectroscopy (XPS). The ATR IR spectra recorded after collagen adsorption showed a broad absorption band in the range of 3400–3000 cm^{-1}, originating from the OH and NH stretching modes, and sharp absorption bands at about 1650 cm^{-1} and 1550 cm^{-1}, arising from the amide I and II modes, respectively [96]. An XPS spectrum of collagen adsorbed onto a titania surface displayed O1s, C1s, N1s and Ti2p signals [96]; however, the presence of O1s and Ti2p indicated that the collagen layer was not distributed uniformly on the titania surface.

The stability of collagen adsorbed onto titania surfaces was studied over wide temperature (10–60 °C) and pH (1–13) ranges [96]. Interestingly, up to 50 °C no change in the mass of adsorbed collagen was detected with the QCMB. Collagen also remained adsorbed on the titania surface over a wide pH range, from 1.3 to 13.0. Taken together, these results point to a high stability of collagen when adsorbed onto the titania surface, and suggested a strong interaction with the substrate. Over the pH range studied, the surface charge of the titania changed from positive to negative [31, 96, 100]. The fact, that collagen is strongly adsorbed onto the positively and negatively charged surfaces, suggests that the net charge on the titania does not influence the adsorption process [96]. Unfortunately, in the study reported by Acharya and Kunitake [96], no comparison was provided of the ATR IR spectra of the collagen layer at various temperatures and pH-values. Consequently, no conclusions can be drawn relating to changes in the secondary structure and hydration during protein desorption and/or denaturation.

6.4.1.3 Morphology and Structure of Collagen Adsorbed on a Calcified Titania Surface

As noted above, titanium implants serve in the majority of cases as bone substitution materials. *In vivo*, hydroxyapatite [HA; $Ca_{10}(PO_4)_6x(OH)_2$], which is the main inorganic component of many human hard tissues [99], forms crystals that grow in contact with collagen fibrils [18, 99]. As the coexistence of these two compounds is crucial in the process of bone mineralization, it was proposed that a coating of HA might further increase the biocompatibility of titanium implants, and this in turn would result in an increased (and stronger) adsorption of collagen onto the TiO_2 surface [18, 94, 99, 101]. The process of bone mineralization requires the presence of two components, namely inorganic HA and organic collagen. Methods that have been reported for the surface coating of titania include spin-coating [99] and electrochemical deposition [18, 94, 97], and the morphology of the created collagen/HA coating was found to depend on the method used for the surface modification. Spin-coating of the titanium plate, using sols that contained 10%, 20% and 30 wt% HA showed an increase in the surface roughness and also an increase of the HA content in the film [99]. The layer containing collagen/30% HA was found to be 7.5 μm thick, while scanning electron microscopy images showed the collagen/HA coatings to be tightly connected to the titania surface, with both components being homogeneously distributed [99].

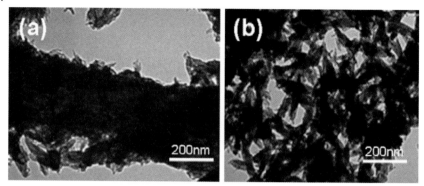

Figure 6.7 Transmission electron microscopy images of collagen fibrils/HA nanocrystals coating obtained by electrodeposition on the titania surface for 30 min. (a) The collagen fibril; (b) The HA nanocrystals, shown between the collagen fibrils. Film was deposited from 42 mM Ca^{2+} and 25 mM PO_4^{-3} solution containing 0.012 wt% collagen, pH 7.4 Reproduced with permission from Ref. [18]; © Elsevier.

The electrochemical deposition of apatite [from $Ca(NO_3)_2$, $NH_4H_2PO_4$] and type I collagen facilitated collagen adsorption on the titania surface [18, 94, 97]. Transmission electron microscopy (TEM) images of collagen fibrils/HA nanocrystals assembled on the titania surface after a 30 min period of electrodeposition is shown in Figure 6.7 [18]. Figure 6.7a shows a collagen fibril surrounded by small HA crystals, while Figure 6.7b shows that, on the titania surface, certain areas are present that are covered only by HA nanocrystals that, typically, were 160 ± 20 nm long and 60 ± 10 nm wide [18]. Interestingly, the HA crystals mimicked perfectly the morphology and dimensions of bone inorganic crystals [18]. To conclude, both components of the coating were adsorbed onto the titania surface, and the collagen fibrils were surrounded by HA crystals. However, as seen in Figure 6.7, phases that either contained or were rich in only one component were present on the titania surface. The IR microscopy spectra of collagen/HA coatings obtained after 30 s, and 5 and 30 min of electrodeposition were recorded and analyzed [18]. Whilst the formation of nonhomogeneous collagen-rich or HA-rich domains was confirmed at short times of adsorption, an increased collagen/HA deposition time resulted in the formation of more homogeneous coatings. IR microscopy spectra collected after 30 s of electrodeposition showed bands that were dominated by the absorption of functional groups originating from collagen or from HA crystals. The IR spectra typical of aggregates rich in collagen showed absorption bands around 3500–3000 cm^{-1}, 1650 cm^{-1} and 1550 cm^{-1} that originated from the amide A, amide I, and amide II modes, respectively [18]. In these spectra no (or minimal) absorption originating from the phosphate stretching modes in the HA crystal around 1100–1000 cm^{-1} was detected [18]. An increase in the adsorption time resulted in a more homogeneous distribution of two phases in the coating. IR spectra recorded at various locations on the titania surface had absorption bands that were characteristic of collagen and hydroxyapatite, while the intensities of each band were comparable.

The first attempts to interpret the interactions and structure of the collagen/HA nanocrystal coating on the titania surface were based on analyses using XPS and IR spectroscopy [18, 94]. In the XPS spectra, signals originating from Ca2p, C1s, N1s were observed [94], with the major contribution to the C1s signal at 285.0 eV being due to the aliphatic carbon atoms of collagen. Weak signals at 286.5, 288.0, and 290.5 eV corresponded to C–N, C=O, and O–C=O bonds in the protein, respectively [94]. The calcium signal Ca2p was split into a doublet ($Ca2p_{1/2}$ and $C2p_{2/3}$) typical for Ca^{2+} ions in inorganic phosphate compounds [94, 102]. The exact positions of the Ca2p signals depended on the compound containing Ca^{2+} cations and being deposited with collagen (HA and β-tricalcium phosphate) [94]. The shift of the binding energy pointed to an interaction between the collagen fibrils and the Ca^{2+} ions in the coating. It was concluded, that the carboxylic group COO^- in the side chains interacted with Ca^{2+} cations [94], and this was confirmed using IR spectroscopy [18, 103]. The symmetric COO^- stretching mode ($v_s(COO^-)$) in collagen adsorbed onto the titania surface was found at 1347 cm^{-1} [18]. Already, in the initial stages of adsorption of the collagen/HA coating, the $v_s(COO^-)$ mode was red-shifted to 1337 cm^{-1} [18]; a red-shift of the $v_s(COO^-)$ mode was also reported for calcified collagen [103]. A 10 cm^{-1} red shift of the $v_s(COO^-)$ mode indicated a strengthening of the binding of the carboxylic group, which mirrored an interaction of the carboxylic acid group in the side chain of the collagen fibril with Ca^{2+} ions in the HA crystal.

To summarize, the morphology of the collagen/HA coatings is known and, importantly, corresponds with the morphology of the mineralized bone material [18]. Long collagen fibrils are distributed randomly on the titania surface and are surrounded by HA nanocrystals. At the molecular scale, an interaction between the carboxylic group of the amino acid side chain of collagen and Ca^{2+} ions was detected.

6.4.1.4 Conclusions

Despite the fact that the IR spectra of collagen and collagen/HA coatings were recorded, no attempt was made to discuss the structure of the protein and its eventual changes when adsorbed on the titania surface. The IR spectra of collagen, collagen/HA, HA and turkey calcified tendon coatings are compared in Figure 6.8.

The amide I and amide II modes of collagen adsorbed onto the titania surface have maxima of absorption at 1667 cm^{-1} and 1545 cm^{-1}, respectively. Neither the scale of the IR spectra in Figure 6.8 nor the spectral resolution allow for any detailed analysis of the amide I region. However, in Figure 6.8 it is clearly visible that the amide I band of pure collagen is composed at least of two absorptions. In aqueous solution, the amide I band of type I collagen has an overall maximum of absorption at 1656 cm^{-1} and is composed of three overlapping bands centered at 1637, 1656, and 1678 cm^{-1} [104]. Three absorptions that compose amide I band originate from a unique amino acid sequence and structure of collagen [104, 105]. Moreover, as shown in Figure 6.8, a red-shift of the maximum of absorption of

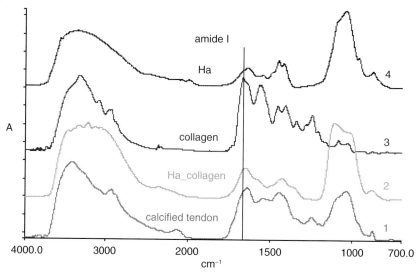

Figure 6.8 The IRRA spectra of the HA, collagen, collagen/HA and turkey calcified tendon coatings, prepared on the titania surface. Spectral resolution 4 cm^{-1}, average of 32 scans. Reproduced with permission from Ref. [18]; © Elsevier.

the amide I band of the collagen/HA and calcified tendon coatings compared to pure collagen film was observed. This change may be induced by: (i) the denaturation process of collagen; (ii) the reduction in the degree of hydration of the amide group upon interaction with Ca^{2+}; and/or (iii) a change in the secondary structure of the protein, namely in an increase of the unordered collagen structure [106]. The problems listed above are open and require further investigation. The first steps to provide a molecular-scale image of the collagen molecule adsorbed on the titania surface were taken based on theoretical calculations. Information relating to the hydration, molecular orientation and structural changes that accompany the adsorption of collagen onto the titania surface are detailed in the following subsection.

6.4.1.5 Structure of Collagen on the Titania Surface: Theoretical Predictions

Collagen, compared to other proteins, has a specific amino acid sequence and, in consequence, a unique structure. A small ratio of the molecular width to length may determine the strength of interactions with the titania surface. Moreover, adsorbed collagen fibrils may undergo structural changes. Due to the specific structure of the collagen molecule, molecular dynamics studies of protein adsorption onto the titania surface cannot be compared with theoretical studies of the adsorption of other proteins.

The process of adsorption – the interaction at the atomic level of fragments of the triple helix of type I collagen with the titania surface – has been the subject of molecular dynamics studies [13, 107], the majority of which employed a triple helix of type I collagen that was composed of two identical α1 helices and one α2 helix. In the study conducted by Monti [13], each helix was composed of 21 amino acid residues, while the side chains contained non-charged amino acids such as alanine (Ala), Gly, or serine (Ser). In a second study, Köppen et al. [107] used a triple helix of a type I collagen-like peptide composed of 30 amino acids in the two α1 helices, and of 29 amino acid residues in the α2 helix. In contrast to previous investigations, the side chains of the collagen-like peptide contained charged amino acids, including Glu, Asp and Lys [107]. In both studies, the adsorption process of collagen was modeled onto the hydroxylated titania surface in aqueous solution [13, 107].

Despite different amino acids having been placed in the side chains of collagen, as introduced by Monti [13] and Köppen et al. [107], similarities in the adsorption process at the molecular level were observed. The theoretical studies showed that, upon adsorption of the collagen-like peptide, a number of water molecules were removed from the surface, due to the stronger interaction between the TiO_2 and peptide functional groups [13, 107]. A direct interaction between the amino acids of the collagen helix and the titania surface was observed only for collagen without charged amino acids in the side chain. These modeling results showed that side chains formed the highest number of connections to the surface atoms of the titania. The noncharged amino acids found in the side chains of the collagen-like peptide formed hydrogen bonds via their hydroxyl and carbonyl groups to the oxygen atom of the TiO_2. The O atom of interacting hydroxyl group is preferentially oriented toward the Ti surface atom. The number of O–Ti coordinations and the number of hydrogen bonds was shown to depend on the initial orientation of the collagen-like peptide, in which electrostatic interactions between the charged amino acids (e.g., Asp, Glu, Lys) and the titania surface dominated. The negatively charged carboxylic groups of Asp and Glu at the C-terminus of the peptide chain formed very stable hydrogen bonds with the protonated bridging atom of the titania surface. Lys, the positively charged amino acid of the side chain at the N-terminus, created a weak hydrogen bond to the hydroxyl group on the titania surface. These calculations showed that the ammonium group NH_3^+ interacted with two adjacent oxygen atoms on the surface, while one hydrogen of the NH_3^+ group formed a stable hydrogen bond with the OH on the surface, and two other hydrogen atoms formed unstable hydrogen bonds with one hydroxyl group on the titania surface [107]. These hydrogen bonds were constantly opened and broken. The molecular dynamics studies showed that the ammonium group had rotational freedom, which pointed to its weak adsorption on the titania surface. Indeed, a threefold (C_{3v}) symmetry of the ammonium group did not find optimal orientation in the twofold symmetry of the rutile surface [108].

It should be emphasized at this point that the two simulations presented above were performed on atomically smooth [13, 107] and rough [107] titania surfaces. Interestingly, at the molecular scale, the differences in collagen adsorption were negligible, due to the formation of a relatively small number of contact points between the collagen-like protein and the titania. To summarize, the interaction between collagen and the titania surface is limited to rather weak electrostatic interactions of the carbonyl group with the Ti atom and the OH group on the surface, and the formation of hydrogen bonds between the carbonyl and hydroxyl groups of amino acids and O and OH surface groups of the titania. Interestingly, it was predicted, on a theoretical basis, that at a certain pH the chemisorption of the carboxylic group (COO$^-$) of the amino acids in the side chains to the titania surface would be possible [107, 108]. At sufficient pH, the hydroxyl group bridging two Ti atoms could be removed from the surface; the deprotonated carboxylic moiety could then be introduced into this gap [108].

6.4.2
Fibronectin Adsorption on the Titania Surface

Fibronectin is an important protein from the ECM, the adsorption of which onto an implant material is crucial for the successful binding of cells and, in consequence, for the acceptance of the implant by the recipient organism. The morphology, interactions with the titania surface and structure of FN are discussed below. The process of FN adsorption, from solutions containing between 0.1 ng ml^{-1} and 200 µg ml^{-1} of the protein on titania surfaces of various roughness, was intensively studied [91, 109–112]. To demonstrate the differences in the morphology of various titania surfaces, it should be noted that commercially pure titanium treated with H_2O_2 (TiO$_2$ cp) had a 14-fold greater surface roughness than did titanium oxide sputtered on Si (TiO$_2$ sp) [109]. The TiO$_2$ sp surface was, in effect, atomically flat; the root-mean square (RMS) parameter measured using AFM in air was close to 0.4 nm, and in water close to 1.6 nm. The RMS parameter measured for the TiO$_2$ cp sample was close to 6 nm in air, and to 16 nm in aqueous solution [109]. Independently of the surface roughness of the titania and the concentration of FN, the protein was found to adsorb easily onto TiO$_2$ surfaces. The adsorption of FN onto titania surfaces of various roughness, and degrees of hydration, were confirmed quantitatively using XPS [109, 110]. On TiO$_2$ cp and TiO$_2$ sp surfaces, the following XPS signals were detected: C1s, N1s, S2p, O1s, and Ti2p [109–111]. In the XPS spectra, the adsorption of FN was reflected by a broadening of the C1s signal. The main contribution to the C1s peak at 285.0 eV corresponded to the aliphatic C atoms (CH$_3$ and CH$_2$ groups) in the protein. The peak at 286.6 eV originated from amine groups (C–NH), at 288.2 eV from the peptide bond (C=O–NH) and at 289.6 eV from acidic carboxylic groups (COO$^-$) [111, 113]. The N1s signal at 400.1 eV was due to the peptide bond (C=O–NH). Two weak N1s signals at 401.8 eV and 389.9–399.9 eV were due to positively charged ammonium groups (NH$_3^+$) and to amine NH$_2$ groups, respectively [111]. The S2p signal at 164.1 eV originated from the S–S disulfide bond in the FN molecule. The O2p signal was composed of two peaks; one peak originated from the O atoms in carboxyl groups of the peptide and O atoms in the electrolyte solution, while a much smaller second O2p signal originated from the O bonded to Ti atoms of the substrate [111]. In concluding, XPS confirmed the adsorption of FN onto the titania surface. The presence of Ti2p and O2p signals in the XPS spectra indicated either a surface coverage of FN below a monolayer, or the presence of aggregates and regions poor in protein on the titania surface [109–111]. Independent ellipsometric [109] and surface plasmon resonance (SPR) [113] experiments showed that the FN adsorbed on the titania surface formed a 2 nm thick layer. Although, these techniques proved to be perfect for studying the structure of an assembly in the direction normal to the surface, they are not sufficiently sensitive to in-plane discontinuities.

6.4.2.1 Morphology of Fibronectin Adsorbed on the Titania Surface

Atomic force microscopy is widely used to investigate the morphology of proteins adsorbed onto various surfaces. The two-dimensional (2-D) structure of FN adsorbed onto titania surfaces of various surface roughness and hydrophilicity was studied [109, 110]. On the rough hydroxylated TiO_2 cp surface in water, a decrease in the RMS parameter, from 16 nm to 8 nm, was detected upon FN adsorption, which indicated that the protein had been adsorbed into "valleys" on the titania surface. The AFM images of FN recorded in air on the smooth TiO_2 sp surface showed an increase in surface roughness by 0.2 nm [109], whereas those recorded on both rough and smooth titania surfaces did not show a compact, densely packed FN layer. These findings agreed very well with the above-reported XPS studies of FN adsorbed onto the titania surface [109–111]. The dimensions of a single protein molecule, or its aggregates, were provided based on the cross-section analysis of AFM images. Following a 10 min period of adsorption of FN from a 50 µg ml^{-1} solution, the average dimensions of disk-shaped aggregates (10 samples) indicated a radius of 55 ± 9 nm and a height of 4.6 ± 1.6 nm. After a 30 min adsorption period, the radius of the assembled protein on the titania surface had not changed, but the height had increased to 7.1 ± 3.2 nm. The fact that only the thickness of the FN layer had increased indicated that FN had been adsorbed on the top of an already existing protein layer [109]. When FN was adsorbed onto the titania surface from a more concentrated (75 µg ml^{-1}) solution, larger aggregates were formed in which the average width was 83 ± 17 nm and the height 8.4 ± 5.3 nm [109]. An additional AFM study of FN adsorption from a diluted solution (10 ng ml^{-1} FN in 0.05 M Tris buffer and 0.15 M NaCl) showed small globular protein assemblies on the titania surface, with an average length of 16.5 ± 1.0 nm, a width of 9.6 ± 1.2 nm, and a height of 2.5 ± 0.5 nm [110]. The thickness of the protein layer measured here matched that determined by ellipsometry and SPR [109, 114]. In addition, these parameters corresponded very well to the ~13 × ~10 × 1.3 nm dimensions of unfolded FN in NaCl solution, as determined using X-ray scattering [93]. It should be noted here that the concentration of FN used by Sousa et al. [109] was 7500-fold higher than that used by MacDonald [110]. The different images of FN adsorbed onto the titania surface reported here [109, 110] may arise either from the large differences in the initial concentrations of the protein, or from structural changes of the FN molecule. When the concentration of FN was very low, the adsorption of single protein molecules was observed [110]. FN is a very flexible protein that undergoes reversible folding and unfolding, depending on the ionic strength or pH of the solution [87, 89]. This process drastically alters the dimensions of the FN molecule, with protein unfolding changing the globular FN into an elongated fibril approximately 130 nm long and 2 nm in diameter. As suggested by Sousa et al. [109], a partial unfolding of the FN protein adsorbed in the aggregate on the titania surface cannot be excluded. To support this proposal, it should be mentioned that unfolded FN was adsorbed onto a hydrophilic mica surface [115]. To conclude, when FN is adsorbed onto the titania surface it exists rather in folded state; however, the presence of fully unfolded and partially unfolded molecules must be taken into account, especially when the FN film was adsorbed from concentrated solutions.

6.4.2.2 Fibronectin–Titania Interactions

As described above, FN is easily adsorbed onto the titania surface, forming aggregates of various sizes, in which individual protein molecules have been suggested to exist in different states and orientations. These findings also suggested that interactions governing the affinity of FN to the titania surface may depend on the molecular packing, surface charge, hydrophilicity, or pH. The adsorption of FN, and the strength of interactions between the protein and the titanium covered by its native oxide layer and hydroxylated titania, have been investigated [91, 111, 112, 116]. The two titania surfaces differed in hydrophilicity; furthermore, the adsorption of FN altered the hydrophilicity of the titania surfaces. As referred to previously [111], a contact angle equal to $76 \pm 4°$ was measured for the TiO_2 cp surface treated with H_2O_2 (hydroxylated or partially hydroxylated). FN adsorption from solutions containing $20\,\mu g\,ml^{-1}$ and $200\,\mu g\,ml^{-1}$ caused decreases in the contact angle, to $35 \pm 3°$ and to only $62 \pm 2°$, respectively; on the TiO_2 sp surfaces the measured contact angles were 5–7° smaller [111]. Interestingly, the adsorption of FN from solutions containing less protein caused more drastic changes in the contact angle, producing more hydrophilic surfaces. These results suggested the existence of different orientations and/or structures of FN adsorbed from diluted solutions, where the polar groups of the protein were exposed to the electrolyte solution.

Values of measured contact angles and the surface tension of FN in aqueous solutions were used to calculate the work of adhesion of the protein on titania surfaces. This does not provide a direct measure of protein adsorption, but rather describes the change in interfacial free energy of the solid–liquid interface upon protein adsorption [117]. Values of the work of adhesion on two investigated titania surfaces were shown to decrease with an increase in FN concentration, indicating a weakening of the protein–titania interactions. In the case of loosely and densely packed monolayers, this points to differences in interactions between the adsorbing protein and the titania surface. A better understanding of protein behavior during the adsorption process was acquired by investigating the adsorption isotherm [111], using a range of FN concentrations, from 0 to $200\,\mu g\,ml^{-1}$ (in PBS buffer, pH 7.4). The adsorption isotherm created showed a gradual increase in the surface concentration of adsorbed FN with an increase in the solution concentration of the protein, but no plateau was observed [111]. The surface concentrations of FN adsorbed onto a solid surface, assuming side-on (folded state) and end-on (unfolded state) conformations, were calculated [118] as $1750\,\mu g\,m^{-2}$ and $41\,000\,\mu g\,m^{-2}$, respectively.

At low FN concentrations, the surface coverage was below the maximal coverage corresponding to the side-on orientation of FN (<1750 $\mu g\,m^{-2}$). It was concluded that, when adsorbing onto the titania surface, FN existed in a folded conformation that provided more binding sites and strengthening interactions with the surface. This proposal was supported by higher values of the work of adhesion determined at low FN concentrations. A further increase in protein concentration resulted in the formation of more dense FN layers, in which the probability of steric hindrance and electrostatic repulsion was increased. At FN concentrations >100 $\mu g\,ml^{-1}$, the amount of adsorbed protein was higher than expected for a FN monolayer oriented in the side-on position [111]. The maximum surface concentration of FN adsorbed from a 200 $\mu g\,ml^{-1}$ solution onto a TiO_2 cp surface was close to 8000 $\mu g\,m^{-2}$ [111]; this was approximately fivefold smaller than would be expected for a FN monolayer in an end-on orientation, and indicated that an exclusively end-on orientation of the protein would be unlikely. The amounts of adsorbed FN, measured using the QCMB, were close to ~11 000 $\mu g\,m^{-2}$ (solution concentration 100 $\mu g\,ml^{-1}$, Tris buffer, pH 7.4) [91] and ~5800 $\mu g\,m^{-2}$ (solution concentration 20 $\mu g\,ml^{-1}$) [116]; these values were higher by a factor of 4–5 than reported by Sousa et al.[111]. Indeed, several-fold higher concentrations determined with the QCMB were expected due to the sensitivity of this technique water that had been coadsorbed and/or was bound to the protein [91]. In the study of Hemmersam et al. [91], the calculated average surface area for a single FN molecule was approximately 62 nm^2, compared to the value of 348 nm^2 found by Tooney for FN adsorbed onto carbon tape [119].

To conclude, in a densely packed FN monolayer two different protein orientations coexist, namely side-on and end-on adsorbed molecules. This indicates that a folded FN may exist next to an unfolded and/or partially folded molecule, and suggests a degree of disorder in the monolayer. Molecular reorientations in these densely packed FN monolayers may lead to a weakening of the interactions with the titania surface, and this is observed experimentally as a decrease in the work of adhesion. Such weakening of interactions may also be demonstrated by the formation of FN adlayers on the top existing monolayer, as observed using AFM [111].

Sousa et al. observed that, at each FN concentration in solution, the work of adhesion calculated for the TiO_2 cp treated with an H_2O_2 surface was larger than for Si TiO_2 sp [111]. Higher values of the work of adhesion on hydroxylated or partially hydroxylated TiO_2 cp surfaces indicated stronger interactions with the protein and, in consequence, a stronger binding. A higher affinity of FN for the TiO_2 cp surface, with its greater number of hydroxyl groups, points to the importance of electrostatic interactions in the adsorption process. MacDonald et al. [112] showed that the zeta potential (a measure of accumulated surface charge) of the TiO_2 shifted towards more negative values when the sample was heated, or when treated with H_2O_2. The adsorption of FN onto the titania surface also caused a further negative shift of zeta potential [100]. The reported potential of zero charge for FN (range: 5.5 to 6.3) [120] indicated that, under physiological conditions, FN has a net negative charge. The accumulation of a charge on the protein facilitates electrostatic interactions with hydroxyl groups adsorbed on the titania surface, or with Ti surface atoms; this explains the higher observed affinity of FN to this surface [111].

Approximately 70% of the FN adsorbed onto the titania surface was removed by washing with a buffer [100]; this not fully reversible desorption process indicated a moderate affinity of FN for the titania surface. A quite strong adsorption of FN onto the titania surface was explained either by a high number of binding sites, or by a partial denaturation of the protein in the adsorbed state [100]. Indeed, the electrostatic interactions and large number of binding sites appears to determine the strength of FN adsorption on the titania surface. A surface coverage that was higher than in a side-up oriented monolayer pointed to the presence of folded and unfolded molecules in the adsorbed layer, and suggested that structural changes might accompany the adsorption process. Details of the FN structure on the titania surface are presented in the following section.

6.4.2.3 Structure of Fibronectin Adsorbed onto the Titania Surface

Spectroscopic methods such as Raman spectroscopy (RS) and infrared spectroscopy (IRS) are well established procedures used to investigate the secondary structures of proteins [106, 121]. The structure of the bulk FN, and of FN adsorbed onto TiO_2 nanoparticles, was studied using RS [17]. In another investigation, the structure of two peptides which may serve as models of FN homologies – Peptide A, composed of H-RGDAEAEAKAKAEAEAKAK-NH$_2$, and Peptide B, composed of H-RGDAAKAEAEAAEKAKAEK-NH$_2$ – were each adsorbed onto the TiO_2 surface and studied using IRS [16]. Within their C-termini, the two peptides each contained the tripeptide RGD that was found in FN and was responsible for the binding of cell membrane proteins. Here, RGD corresponds to the amino acid sequence Arg-Gly-Asp, while A = Ala, E = Glu, and K = Lys [16]. The amino acid sequence in both peptides can mimic the structure of FN. The presence of charged amino acids, as in FN, provides locations for the formation of hydrogen bonds, which in turn influences the secondary structure of the peptide. This simplified FN model was used to identify the individual groups of amino acids involved directly in interactions with the titania surface, and to determine their orientation in the adsorbed state [16].

In Figure 6.9, the Raman spectra of FN adsorbed onto the TiO_2 surface (upper part) and in the bulk (lower part) are compared [17]. The main differences between the two spectra are apparent in the 1800–1400 cm^{-1} region. A comparison of the bulk and adsorbed FN Raman spectra showed two main differences: (i) the appearance of a new band at 1748 cm^{-1} in the adsorbed state; and (ii) a change of the Raman intensity ratio of amide I band in relation to the CH_2 and CH_3 deformation modes around 1450–1460 cm^{-1} (Figure 6.9) [17].

As seen in the upper part of Figure 6.9, the spectrum of FN adsorbed onto the titania surface has a weak band at 1748 cm^{-1} [17]. Similarly, Peptide B adsorbed onto the titania surface has a weak absorption in an IR spectrum around 1710 cm^{-1} [16]. The appearance of a new absorption band in the 1760–1700 cm^{-1} region originates from the $\nu(C=O)$ stretching mode of a carboxylic group, which suggests

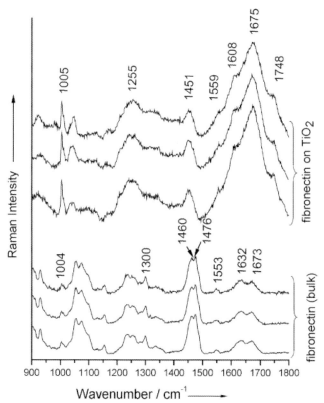

Figure 6.9 Raman spectra after background correction of fibronectin. Upper panel: adsorbed onto TiO$_2$ nanoparticles; Lower panel: bulk dry sample. Excitation wavelength 514.5 nm, spectral resolution 3 cm^{-1}. Reproduced with permission from Ref. [17]; © the Royal Society of Chemistry.

that the side chains containing the carboxylic group are directly involved in the interaction with the titania surface [16, 17]. Indeed, a carboxylic group is known to form a bidentate or monodentate binding with the TiO$_2$ surface [122].

As mentioned above, the second difference between the bulk and adsorbed FN Raman spectra was the ratio of the Raman intensities of CH$_2$ and CH$_3$ deformation modes (at 1450 cm^{-1}) to the amide I mode [17]. In the bulk FN, this ratio was equal to ~2.6, but in the adsorbed state it had decreased to ~0.15. Strehle et al. ascribed the change in intensity to the involvement of FN side chains in the adsorption process [17].

The amide I band of bulk FN has two peaks located at 1673 cm^{-1} and 1632 cm^{-1}. As seen in Figure 6.9, the amide I band of FN adsorbed onto the TiO$_2$ nanoparticles dominates the Raman spectrum; this broad band has a maximum intensity at 1675 cm^{-1}, and is overlapped with a mode at 1608 cm^{-1}. The weak mode at 1608 cm^{-1}

is due to the ring breathing modes of aromatic amino acids of FN [17]. However, the presence of more than one band in the broad and intense amide I band cannot be excluded, as was the case for the bulk FN sample [17]. Unfortunately, deconvolution of the amide I band of FN adsorbed onto TiO$_2$ nanoparticles was not reported [17]. The amide I mode of Peptide A, when adsorbed onto the titania surface, has two maxima of absorption at 1684 cm^{-1} and 1620 cm^{-1} [16]. The IR spectrum of Peptide B in this spectral region is different, with two absorptions centered at 1635 cm^{-1} and 1650 cm^{-1} being observed [16]. In both RS and IRS, the amide I bands that are located around 1670–1685 cm^{-1} and around 1615–1625 cm^{-1} are characteristic of an anti-parallel β sheet, while those around 1670–1685 cm^{-1} and 1635 cm^{-1} are characteristic of a parallel β sheet secondary structure [16, 106]. To conclude, the β sheet secondary structure is found in bulk and adsorbed FN, as well as in Peptide A. The positions of the amide I bands of Peptide B point to a disturbed, unordered β sheet structure; indeed, the secondary structure of FN is dominated by β sheet structures [87]. It is also known that adsorption favors the β sheet structure of a protein, as more than 50% of the side chains in this structure are turned in the same direction, facilitating protein approach and binding to a solid substrate [17, 123]. It appears that, upon adsorption, negligible changes in the secondary structure of FN and peptides modeling FN were observed, as might be expected for a protein with a β sheet structure.

The details concerning the orientation of individual atoms of various amino acids involved in the interaction with the TiO$_2$ surface were also studied for Peptides A and B, using near edge X-ray absorption fine structure spectroscopy (NEXAFS) [16]. Depending on the angle of the incident X-ray light, the measured intensities of the NEXAF spectra of O and C atoms of Peptide A were changed. In contrast, all NEXAF spectra of Peptide B, independent of the angle of incidence, had the same intensities. The orientation of a carboxylic group, and of the chain of the Peptide A, were then calculated from the intensities ratio originating from the O and C atoms. The carboxylic group that was directly involved in binding to the TiO$_2$ surface was inclined by 40–45° with respect to the substrate surface [16], while the tilt angle of the main axis of the peptide was close to 75° with respect to the substrate surface [16]. This result was complementary to IRS, which showed an ordered structure of Peptide A adsorbed onto the titania surface. Peptide B, as determined using by IRS and NEXFAS, formed disordered monolayers where no preferred orientation was identified [16].

6.4.2.4 Atomic-Scale Picture of Fibronectin Adsorbed on the Titania Surface: Theoretical Predictions

The amino acid sequence of most homologies of FN guarantees a high plasticity of the molecule, favoring a β sheet secondary structure. FN contains the highest number of FN III homologies, which are arranged in antiparallel β sheet bilayers [92]. As FN is a large protein (molecular weight >400 kDa), the modeling of a whole protein is a complicated task. However, small protein fragments (such as homologies, or even smaller) containing functional groups are good models for theoretical studies of protein adsorption [15, 124–126].

The adsorption of a protein unit composed of two antiparallel β sheet bilayers (an analogue of the FN III homologue) onto a titania surface in an aqueous environment was monitored in a molecular dynamics study [127]. For this, each β sheet contained 16 peptides such as Ala, Gly, Asp, Lys; the peptide contained eight β sheets in each layer. The presence of charged Asp and Lys permitted the formation of intramolecular hydrogen bonds which stabilized the secondary structure of the peptide, as was found in the FN III homologies [92]. Two orientations of the protein fragment on the TiO_2 surface were taken into consideration, namely parallel and perpendicular orientations of the β sheets with respect to the titania surface.

The parallel orientation, compared to its perpendicular counterpart, provided more interaction points between the charged side chains and the titania surface. Adsorption of the peptide led to a removal of previously adsorbed water from the titania surface. In the parallel orientation, the desorbed water moved into the peptide structure, and molecular dynamics studies showed wider variations in the β sheet structure at the sides of the adsorbing peptide, with the geometry of the central part of the β sheet structure being well preserved. The side chains of amino acids located close to the titania surface showed a limited mobility, which pointed to direct interactions with the TiO_2 and/or the adsorbed water [127]. In the parallel orientation of the peptide, one bidentate binding of the COO^- group of Asp was detected, although many instances were identified of the hydrogen bonding of NH_2 groups of Arg to O atoms of the TiO_2 and adsorbed water. When the peptide was orientated perpendicularly, fewer interactions with the titania surface were predicted and the peptide had a greater mobility. It could also rotate and reorient its more hydrophilic regions towards the TiO_2 surface, in this way strengthening the interactions with the surface. Hydrogen bonds between Arg and Asp side chains with adsorbed water molecules were detected, with the Asp side chain able to form a monodentate coordination to Ti atoms at the surface [127].

In both cases, the secondary structure of the studied protein fragment, namely the β sheet structure, was not affected by the adsorption process. Two types of interaction – direct coordination to the Ti surface atoms, and the formation of hydrogen bonds to the O atoms of TiO_2 and H_2O – were observed. The strength of the binding of the protein fragment was seen to depend on the number of direct and indirect interactions of the adsorbing molecule with the titania surface. The presence of charged COO^- groups in the side chains facilitated adsorption.

6.4.2.5 Conclusions

As demonstrated above, the adsorption of FN onto the titania surface is a well-known process; in contrast to collagen, FN is easily adsorbed onto titania surfaces of various degrees of roughness and hydrophilicity [109, 111]. The adsorption of FN onto titania surfaces, as confirmed by washing experiments, is of moderate strength [100], with the electrostatic interaction being the "driving force" of the adsorption process. The strength of FN titania interactions, the number of binding sites and the protein orientation were each found to depend on the surface coverage [111]. The adsorption of FN onto titania surfaces does not affect its β sheet

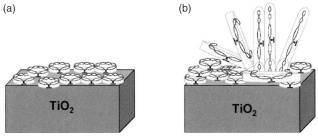

Figure 6.10 Schematic orientation of fibronectin on the titania surface in: (a) a monolayer with side-on orientation; (b) a densely packed monolayer with end-on and side-on orientations.

secondary structure; rather, the adsorbing FN protein interacts with the TiO$_2$ surface via its carbonyl groups, both in the main chain and in the side chains [16, 125, 127]. As noted experimentally [16], and confirmed by numerous theoretical studies [15, 124, 126, 127], the carboxylic groups are able to coordinate with Ti surface atoms; indeed, the carboxylic group COO$^-$ of Asp side chains make the strongest coordination with Ti atoms at the interface [127]. In the initial stages of the adsorption process, single, folded FN molecules oriented side-on are present on the titania surface (as shown schematically in Figure 6.10a).

In this orientation, the largest number of interactions with the titania surface was detected, and confirmed by the highest measured values of the work of adhesion. Increases in the time of adsorption or in FN concentrations lead to the formation of a densely packed FN monolayer, with surface coverage reaching values characteristic of a full coverage of the side-on oriented FN (Figure 6.10a). In this densely packed monolayer, both steric hindrances and electrostatic repulsions start to overcome the interactions with the titania surface. The surface concentration of adsorbing FN was also increased above values calculated for the uniform side-on orientation of the protein in the monolayer, as observed experimentally [91, 109, 111] and predicted on a theoretical basis [127] (Figure 6.10b). Some of FN molecules undergo reorientations, leading to an end-on arrangement, and this may be connected to the unfolding (or at least partial unfolding) of the FN molecules. These two processes reduce the number of binding sites between the protein and the TiO$_2$ surface, as found both theoretically and experimentally.

6.5
Conclusions

Since titanium was first used as an implant material, major progress has been made in understanding the biological processes that occur on its surface. Whilst the composition of the biomaterial which initially adsorbs onto – and in time integrates into – the titania surface has been identified, the sequence of events that accompany this osteointegration process are incompletely understood. Within an organism, a highly specific and selective adsorption of biomolecules is required before a new implant can be "adopted." Yet, the structures of the individual molecules adsorbed onto the titania surface (notably proteins) remain unknown. Despite the sophisticated techniques available, the structure of the biomaterial that is adsorbed onto the titania surface *in vivo* has not been determined at the molecular level. In an attempt to determine the molecular-level structure of the interface, simple systems such as lipid bilayers or self-assembled protein layers have been used. It has long been known that lipid bilayers closely mimic the cell membrane, and that the layers of collagen and FN approximate the ECM proteins adsorbed onto the implant surface. Recently, substantial progress was made in analyzing the structure of lipid bilayers adsorbed onto titania surfaces at the molecular and even atomic levels. Whilst the structure of the adsorbed polar headgroup of a variety of phospholipids was predicted using molecular dynamics studies, in only one case – PC – was the theoretically predicted structure supported by experimental data. In the PC bilayer, phosphate, ammonium and carbonyl groups interact directly with the titania surface, influencing the hydration and orientation of the lipid molecule's polar headgroup. The structures of other lipids important for correct cell membrane function, including charged phospholipids, cholesterol or glycolipids adsorbed onto the titania surface, remain unknown and await further investigation.

At present, the structural information relating to protein layers adsorbed onto titania surfaces is rather limited. Collagen, which is the most important ECM protein and is extensively adsorbed onto titania surfaces, continues to provide the greatest problems, as the details of collagen adsorption are poorly understood. As noted in this chapter, theoretical studies have indicated that collagen adsorption onto titania can range from a weak physisorption (with interactions occurring via hydrogen bonds) to chemisorption (where carboxylic groups are bonded covalently between two Ti surface atoms). It is of extreme importance therefore, that structural studies of collagen–titania adsorption are extended.

From medical and technological aspects, it is essential that all processes occurring at the implant surface be examined in the greatest possible detail. An in-depth recognition of molecular level behavior during osteointegration should, in time, permit the creation of modified titania surfaces that will enhance the performance of the titanium implants. For this, two approaches might be taken:

- To modify the titanium surface itself, via oxidation and/or exposure to corrosive substances; in this way the surface morphology might be altered. Alternatively, foam formation might be possible, using powder metallurgy;

- The creation of specific biochemical modifications.

It is well known that the surface properties of an implant, and notably its topology, may significantly alter the osteointegration process. Thus, biomodifications of the titania surface should aim at enhancing the required selective response from proteins or cells (i.e., their adsorption), at increasing the resistance of bacterial attachment, or at reducing inflammatory reactions. The titania surface may also be modified by the adsorption of peptides or proteins, by the formation of a monolayer with incorporated drugs or antibodies, or by coating it with hydrogels that contain therapeutic agents. However, it should be stressed that, from a technological standpoint, the challenge remains first to design a well-ordered structure at the nanolevel, and then to adapt this to function within the macroscopic world of an organism – a task which has already been perfected by Nature itself.

Acknowledgments

The authors acknowledge financial support from the Deutsche Forschungsgemeinschaft (Za 543/1).

References

1 Brånemark, P.I., Hansson, B.O., Adell, R., Breine, U., Lindström, J., Hallen, O. and Ohman, A. (1977) Osseointegrated implants in the treatment of the edentulous jaw. Experience from a 10-year period. *Scandinavian Journal of Plastic and Reconstructive Surgery, Supplement*, **16**, 1–132.

2 Albrektsson, T. and Hansson, H.A. (1986) An ultrastructural characterization of the interface between bone and sputtered titanium or stainless steel surfaces. *Biomaterials*, **7**, 201–5.

3 Hansson, H.A., Albrektsson, T. and Brånemark, P.I. (1983) Structural aspects of the interface between tissue and titanium implants. *Journal of Prosthetic Dentistry*, **50**, 108–13.

4 Brunski, J.B., Moccia, A.F. Jr., Pollack, S.R., Korostoff, E. and Trachtenberg, D.I. (1979) The influence of functional use of endosseous dental implants on the tissue-implant interface. I. Histological aspects. *Journal of Dental Research*, **58**, 1953–69.

5 Brunette, D.M., Tengvall, P., Textor, M. and Thomsen, P. (2001) *Titanium in Medicine*, Springer, Berlin, Germany.

6 Kothari, S., Hatton, P.V. and Douglas, C.W.I. (1995) Protein adsorption to titania surfaces. *Journal of Materials Science: Materials in Medicine*, **6**, 695–8.

7 Eguchi, S., Yamashita, K., Morimoto, H., Ichikawa, T., Nakajo, N. and Kitamura, S. (2004) Extracellular matrix formed by MC3T3-E1 osteoblast-like cells cultured on titanium. II. Collagen fiber formation. *Connective Tissue*, **36**, 9–15.

8 Yamashita, K., Eguchi, S., Morimoto, H., Hanawa, T., Ichikawa, T., Nakajo, N. and Kitamura, S. (2004) Extracellular matrix formed by MC3T3-E1 osteoblast-like cells cultured on titanium. I. Anchor structure. *Connective Tissue*, **36**, 1–8.

9 Höök, F., Vörös, J., Rodahl, M., Kurrat, R., Böni, P., Ramsden, J.J., Textor, M., Spencer, N.D., Tengvall, P., Gold, J. and Kasemo, B. (2002) A comparative study of protein adsorption on titanium oxide surfaces using in situ ellipsometry, optical waveguide lightmode spectroscopy, and quartz crystal microbalance/dissipation. *Colloids and Surfaces B*, **24**, 155–70.

10 Kasemo, B. (2002) Biological surface science. *Surface Science*, **500**, 656–77.

11 Schuler, M., Trentin, D., Textor, M. and Tosatti, S.G.P. (2006) Biomedical interfaces: titanium surface technology for implants and cell carriers. *Nanomedicine*, **1**, 449–63.

12 Parsegian, V.A. (1983) Molecular forces governing tight contact between cellular surfaces and substrates. *Journal of Prosthetic Dentistry*, **49**, 838–42.

13 Monti, S. and Dynamics, M. (2007) Simulations of collagen-like peptide adsorption on titanium-based material surfaces. *Journal of Physical Chemistry C*, **111**, 6086–94.

14 Madhan, B., Subramanian, V., Rao, J.R., Nair, B.U. and Ramasami, T. (2005) Stabilization of collagen using plant polyphenol: role of catechin. *International Journal of Biological Macromolecules*, **37**, 47–53.

15 Monti, S., Carravetta, V., Battocchio, C., Iucci, G. and Polzonetti, G. (2008) Peptide/TiO_2 surface interaction: a theoretical and experimental study on the structure of adsorbed ALA-GLU and ALA-LYS. *Langmuir*, **24**, 3205–14.

16 Iucci, G., Battocchio, C., Dettin, M., Gambaretto, R., Di Bello, C., Borgatti, F., Carravetta, V., Monti, S. and Polzonetti, G. (2007) Peptides adsorption on TiO_2 and Au: molecular organization investigated by NEXAFS, XPS and IR. *Surface Science*, **601**, 3843–9.

17 Strehle, M.A., Roesch, P., Petry, R., Hauck, A., Thull, R., Kiefer, W. and Popp, J.A. (2004) Raman spectroscopic study of the adsorption of fibronectin and fibrinogen on titanium dioxide nanoparticles. *Physical Chemistry Chemical Physics*, **6**, 5232–6.

18 Manara, S., Paolucci, F., Palazzo, B., Marcaccio, M., Foresti, E., Tosi, G., Sabbatini, S., Sabatino, P., Altankov, G. and Roveri, N. (2008) Electrochemically-assisted deposition of biomimetic hydroxyapatite-collagen coatings on titanium plate. *Inorganica Chimica Acta*, **361**, 1634–45.

19 Fortunelli, A. and Monti, S. (2008) Simulations of lipid adsorption on TiO_2 surfaces in solution. *Langmuir*, **24**, 10145–54.

20 Jiang, C., Gamarnik, A. and Tripp, C.P. (2005) Identification of lipid aggregate structures on TiO_2 surface using headgroup IR bands. *Journal of Physical Chemistry B*, **109**, 4539–44.

21 Zawisza, I., Nullmeier, M., Pust, S.E., Boukherroub, R., Szunerits, S. and Wittstock, G. (2008) Application of thin titanium/titanium oxide layers deposited on gold for infrared reflection absorption spectroscopy: structural studies of lipid bilayers. *Langmuir*, **42**, 7378–87.

22 Kiwi, J. and Nadtochenko, V. (2005) Evidence for the mechanism of photocatalytic degradation of the bacterial wall membrane at the TiO_2 Interface by ATR-FTIR and laser kinetic spectroscopy. *Langmuir*, **21**, 4631–41.

23 Andersson, A.-S., Glasmästar, K., Sutherland, D., Lidberg, U. and Kasemo, B. (2003) Cell adhesion on supported lipid bilayers. *Journal of Biomedical Materials Research, Part A*, **64A**, 622–9.

24 Bernardi, G. and Kawasaki, T. (1968) Chromatography of proteins on hydroxyapatite. I. Chromatography of polypeptides and proteins on hydroxyapatite columns. *Biochimica et Biophysica Acta*, **160**, 301–10.

25 Ellingsen, J.E. (1991) A study on the mechanism of protein adsorption to TiO_2. *Biomaterials*, **12**, 593–6.

26 Chang, W.-J., Ou, K.-L., Lee, S.-Y., Chen, J.-Y., Abiko, Y., Lin, C.-T. and Huang, H.-M. (2008) Type I collagen grafting on titanium surfaces using low-temperature glow discharge. *Dental Materials Journal*, **27**, 340–6.

27 Kasemo, B. (1983) Biocompatibility of titanium implants: surface science aspects. *Journal of Prosthetic Dentistry*, **49**, 832–7.

28 Cacciafesta, P., Hallam, K.R., Watkinson, A.C., Allen, G.C., Miles, M.J. and Jandt, K.D. (2001) Visualization of human plasma fibrinogen adsorbed on titanium implant surfaces with different roughness. *Surface Science*, **491**, 405–20.

29 Kasemo, B. and Lausmaa, J. (1986) Surface science aspects on inorganic biomaterials. *CRC Critical Reviews in Biochemistry*, **2**, 335–80.

30 Sittig, C., Textor, M., Spencer, N.D., Wieland, M. and Vallotton, P.H. (1999) Surface characterization of implant materials c.p. Ti, Ti-6Al-7Nb and Ti-6Al-4V with different pretreatments. *Journal of Materials Science: Materials in Medicine*, **10**, 35–46.

31 Oleson, T.A. and Sahai, N. (2008) Oxide-dependent adsorption of a model membrane phospholipid, dipalmitoylphosphatidylcholine: bulk

adsorption isotherms. *Langmuir*, **24**, 4865–73.
32. Bearinger, J.P., Orme, C.A. and Gilbert, J.L. (2001) Direct observation of hydration of TiO_2 on Ti using electrochemical AFM: freely corroding versus potentiostatically held. *Surface Science*, **491**, 370–87.
33. Boehm, H.P. (1971) Acidic and basic properties of hydroxylated metal oxide surfaces. *Discussions of the Faraday Society*, **52**, 264–75.
34. Abe, M. (1982) Oxides and hydrous oxides of multivalent metals as inorganic ion exchangers. *Inorganic Ion Exchange Materials*, 161–273.
35. Andrade, J.D. (1985) *Surface and Interfacial Aspects of Biomedical Polymers*, Plenum Press, New York.
36. Vogel, V. and Baneyx, G. (2003) The tissue engineering puzzle: a molecular perspective. *Annual Review of Biomedical Engineering*, **5**, 441–63, 2 plates.
37. Small, D.M. (1984) Lateral chain packing in lipids and membranes. *Journal of Lipid Research*, **25**, 1490–500.
38. Tiede, D.M. (1985) Incorporation of membrane proteins into interfacial films: model membranes for electrical and structural characterization. *Biochimica et Biophysica Acta*, **811**, 357–79.
39. Hauser, H., Pascher, I., Pearson, R.M. and Sundell, S. (1981) Preferred conformation and molecular packing of phosphatidylethanolamine and phosphatidylcholine. *Biochimica et Biophysica Acta*, **650**, 21–51.
40. Starr, T.E. and Thompson, N.L. (2000) Formation and characterization of planar phospholipid bilayers supported on TiO_2 an $SrTiO_3$ single crystals. *Langmuir*, **16**, 10301–8.
41. Crane, J.M. and Tamm, L.K. (2004) Role of cholesterol in the formation and nature of lipid rafts in planar and spherical model membranes. *Biophysics Journal*, **86**, 2965–79.
42. Sackmann, E. (1996) Supported membranes: scientific and practical applications. *Science*, **271**, 43–8.
43. Cremer, P.S. and Boxer, S.G. (1999) Formation and spreading of lipid bilayers on planar glass supports. *Journal of Physical Chemistry B*, **103**, 2554–9.
44. Reimhult, E., Höök, F. and Kasemo, B. (2002) Vesicle adsorption on SiO_2 and TiO_2: dependence on vesicle size. *Journal of Chemical Physics*, **117**, 7401–4.
45. Reimhult, E., Hoeoek, F. and Kasemo, B. (2003) Intact vesicle adsorption and supported biomembrane formation from vesicles in solution: influence of surface chemistry, vesicle size, temperature, and osmotic pressure. *Langmuir*, **19**, 1681–91.
46. Csucs, G. and Ramsden, J.J. (1998) Interaction of phospholipid vesicles with smooth metal-oxide surfaces. *Biochimica et Biophysica Acta, Biomembranes*, **1369**, 61–70.
47. Tero, R., Ujihara, T., Urisu, T. and Bilayer, L. (2008) Membrane with atomic step structure: supported bilayer on a step-and-terrace TiO_2 (100) surface. *Langmuir*, **24**, 11567–76.
48. Rossetti, F.F., Bally, M., Michel, R., Textor, M. and Reviakine, I. (2005) Interactions between titanium dioxide and phosphatidyl serine-containing liposomes: formation and patterning of supported phospholipid bilayers on the surface of a medically relevant material. *Langmuir*, **21**, 6443–50.
49. Yuan, J., Parker, E.R., Hirst, L.S. and Lipid, C. (2007) Absorption on titanium: a counterion-mediated bilayer-to-lipid-tubule-network transition. *Langmuir*, **23**, 7462–5.
50. Rossetti, F.F., Textor, M. and Reviakine, I. (2006) Asymmetric distribution of phosphatidyl serine in supported phospholipid bilayers on titanium dioxide. *Langmuir*, **22**, 3467–73.
51. Richter, R.P., Maury, N. and Brisson, A.R. (2005) On the effect of the solid support on the interleaflet distribution of lipids in supported lipid bilayers. *Langmuir*, **21**, 299–304.
52. Balasubramanian, K. and Schroit, A.J. (2003) Aminophospholipid asymmetry: a matter of life and death. *Annual Review of Physiology*, **65**, 701–34.
53. Yeagle, P.L. (1987) *The Membranes of Cells*, Academic Press, San Francisco.

54 Casal, H.L., Martin, A., Mantsch, H.H., Paltauf, F. and Hauser, H. (1987) Infrared studies of fully hydrated unsaturated phosphatidylserine bilayers. Effect of Li^+ and Ca^{2+}. *Biochemistry*, **26**, 7395–401.

55 Choi, S., Ware, W. Jr., Lauterbach, S.R. and Phillips, W.M. (1991) Infrared spectroscopic studies on the phosphatidylserine bilayer interacting with calcium ion: effect of cholesterol. *Biochemistry*, **30**, 8563–8.

56 Harris, W.E. (1977) Interactions between fluorescent labeled phosphatidyl serine and cations. *Chemistry and Physics of Lipids*, **19**, 243–54.

57 Keller, C.A. and Kasemo, B. (1998) Surface specific kinetics of lipid vesicle adsorption measured with a quartz crystal microbalance. *Biophysical Journal*, **75**, 1397–402.

58 Reviakine, I. and Brisson, A. (2000) Formation of supported phospholipids bilayers from unillameral vesicles investigated by Atomic Force Microscopy. *Langmuir*, **16**, 1806–15.

59 Zhdanov, V.P. and Kasemo, B. (2001) Comments on rupture of adsorbed vesicles. *Langmuir*, **17**, 3518–21.

60 Cho, N.-J., Cho, S.-J., Cheong, K.H., Glenn, J.S. and Frank, C.W. (2007) Employing an amphipathic viral peptide to create a lipid bilayer on Au and TiO_2. *Journal of the American Chemical Society*, **129**, 10050–1.

61 Bérubé, Y.G. and De Bruyn, P.L. (1968) Adsorption at the rutile-solution interface. I. Thermodynamic and experimental study. *Journal of Colloid and Interface Science*, **27**, 305–18.

62 Sverjensky, D.A. and Sahai, N. (1998) Theoretical prediction of single-site enthalpies of surface protonation for oxides and silicates in water. *Geochimica et Cosmochimica Acta*, **62**, 3703–16.

63 Israelachvili, J. (1992) *Intermolecular and Surface Forces*, Academic Press, London.

64 Bergström, L. (1997) Hamaker constants of inorganic materials. *Advances in Colloid and Interface Science*, **70**, 125–69.

65 Knowles, K.M. (2005) Dispersion forces at planar interfaces in anisotropic ceramics. *Journal of Ceramic Processing Research*, **6**, 10–16.

66 Knowles, K.M. and Turan, S. (2000) The dependence of equilibrium film thickness on grain orientation at interphase boundaries in ceramic-ceramic composites. *Ultramicroscopy*, **83**, 245–59.

67 Johnson, S.J., Bayerl, T.M., McDermott, D.C., Adam, G.W., Rennie, A.R., Thomas, R.K. and Sackmann, E. (1991) Structure of an adsorbed dimyristoylphosphatidylcholine bilayer measured with specular reflection of neutrons. *Biophysical Journal*, **59**, 289–94.

68 Li, H. and Tripp, C.P. (2004) Use of infrared bands of the surfactant headgroup to identify mixed surfactant structures adsorbed on titania. *The Journal of Physical Chemistry B*, **108**, 18318–26.

69 Zawisza, I., Lachenwitzer, A., Zamlynny, V., Horswell, S.L., Goddard, J.D. and Lipkowski, J. (2003) Electrochemical and photon polarization modulation infrared reflection absorption spectroscopy study of the electric field driven transformations of a phospholipid bilayer supported at a gold electrode surface. *Biophysical Journal*, **86**, 4055–75.

70 Lippert, R.J., Lamp, B.D. and Porter, M.D. (1998) *Modern Techniques in Applied Molecular Spectroscopy* (ed. F.M. Mirabella), John Wiley & Sons, Inc., New York, pp. 83–126.

71 Allara, D.L. and Swalen, J.D. (1982) An infrared reflection spectroscopy study of oriented cadmium arachidate monolayer films on evaporated silver. *The Journal of Physical Chemistry*, **86**, 2700–4.

72 Li, H. and Tripp, C.P. (2002) Spectroscopic identification and dynamics of adsorbed cetyltrimethylammonium bromide structures on TiO_2 surfaces. *Langmuir*, **18**, 9441–6.

73 Binder, H. (2003) The molecular architecture of lipid membranes – new insights from hydration-tuning infrared linear dichroism spectroscopy. *Applied Spectroscopy Reviews*, **38**, 15–69.

74 Holler, F. and Callis, J.B. (1989) Conformation of the hydrocarbon chains of sodium dodecyl sulphate molecules in micelles: an FTIR study. *Journal of Physical Chemistry*, **93**, 2053–8.

75 Bin, X., Zawisza, I., Goddard, J.D. and Lipkowski, J. (2005) Electrochemical and PM-IRRAS studies of potential driven transformations of phospholipid bilayers on a Au (111) electrode surface. *Langmuir*, **21**, 330–47.

76 Lewis, R.N.A.H. and McElhaney, R.N. (1996) *Infrared Spectroscopy of Biomolecules* (eds H.H. Mantsch and D. Chapman), John Wiley & Sons, Inc., New York, pp. 159–202.

77 Chia, N.C. and Mendelsohn, R. (1992) CH_2 wagging modes of unsaturated acyl chains as IR probes of conformational order in methyl alkenoates and phospholipid bilayers. *Journal of Chemical Physics*, **96**, 10543–7.

78 Senak, L., Moore, D. and Mendelsohn, R. (1992) CH_2 wagging progressions as IR probes of slightly disordered phospholipids acyl chain states. *Journal of Physical Chemistry*, **96**, 2749–54.

79 Wolfangel, P., Lehnert, R., Meyer, H.H. and Müller, K. (1999) FTIR studies of phospholipid membranes containing monoacetylenic acyl chains. *Physical Chemistry Chemical Physics*, **1**, 4833–41.

80 Zawisza, I., Wittstock, G., Boukherroub, R. and Szunerits, S. (2007) PM IRRAS investigation of thin silica films deposited on gold. Part 1. Theory and proof of concept. *Langmuir*, **23**, 9303–9.

81 Zawisza, I., Wittstock, G., Boukherroub, R. and Szunerits, S. (2008) Polarization modulation infrared reflection absorption spectroscopy investigations of thin silica films deposited on gold. 2. Structural analysis of a 1,2-dimyristoyl-sn-glycero-3-phosphocholine bilayer. *Langmuir*, **24**, 3922–9.

82 Pearson, R.H. and Pascher, I. (1979) The molecular structure of lecithin dihydrate. *Nature*, **281**, 499–501.

83 Berg, J.M., Tymoczko, J.L. and Stryer, L. (2003) *Biochemistry*, Freeman, New York.

84 Nelson, D.L. and Cox, M.M. (2005) *Principles of Biochemistry*, Freeman, New York.

85 Falini, G., Fermani, S., Foresti, E., Parma, B., Rubini, K., Sidoti, M.C. and Roveri, N. (2004) Films of self-assembled purely helical type I collagen molecules. *Journal of Material Chemistry*, **14**, 2297–302.

86 Sakai, T., Johnson, K.J., Murozono, M., Sakai, K., Magnuson, M.A., Wieloch, T., Cronberg, T., Isshikis, A., Erickson, H.P. and Fassler, R. (2001) Plasma fibronectin supports neuronal survival and reduces brain injury following transient focal cerebral ischemia but is not essential for skin-wound healing and hemostasis. *Nature Medicine*, **7**, 324–30.

87 Potts, J.R. and Campbell, I.D. (1994) Fibronectin structure and assembly. *Current Opinion in Cell Biology*, **6**, 648–55.

88 Vaudaux, P., Lerch, P., Velazco, M.I., Nydegger, U.E. and Waldvogel, F.A. (1986) Role of fibronectin in the susceptibility of biomaterial implants to bacterial infections. *Advanced Biomaterials*, **6**, 355–60.

89 Williams, E.C., Janmey, P.A., Ferry, J.D. and Mosher, D.F. (1982) Conformational states of fibronectin. Effects of pH, ionic strength, and collagen binding. *Journal of Biological Chemistry*, **257**, 14973–8.

90 Copié, V., Tomita, Y., Akiyama, S.K., Aota, S.-I., Yamada, K.M., Venable, R.M., Pastor, R.W., Krueger, S. and Torchia, D.A. (1998) Solution structure and dynamics of linked cell attachment modules of mouse fibronectin containing the RGD and synergy regions: comparison with the human fibronectin crystal structure. *Journal of Molecular Biology*, **277**, 663–82.

91 Hemmersam, A.G., Rechendorff, K., Foss, M., Sutherland, D.S. and Besenbacher, F. (2008) Fibronectin adsorption on gold, Ti-, and Ta-oxide investigated by QCM-D and RSA modelling. *Journal of Colloid and Interface Science*, **320**, 110–16.

92 Craig, D., Gao, M., Schulten, K. and Vogel, V. (2004) Tuning the mechanical stability of fibronectin type III modules through sequence variations. *Structure*, **12**, 21–30.

93 Sjöberg, B., Eriksson, M., Österlund, E., Pap, S. and Österlund, K. (1989) Solution structure of human plasma fibronectin as a function of sodium chloride concentration determined by small-angle x-ray scattering. *European Biophysics Journal*, **17**, 5–11.

94 Hosaka, M., Shibata, Y. and Miyazaki, T. (2006) Preliminary b-tricalcium phosphate coating prepared by discharging in a modified body fluid enhances collagen immobilization onto titanium. *Journal of Biomedical Materials Research, Part B*, **78B**, 237–42.

95 Morra, M., Cassinelli, C., Cascardo, G., Cahalan, P., Cahalan, L., Fini, M. and Giardino, R. (2003) Surface engineering of titanium by collagen immobilization. Surface characterization and in vitro and in vivo studies. *Biomaterials*, **24**, 4639–54.

96 Acharya, G. and Kunitake, T. (2003) A general method for fabrication of biocompatible surfaces by modification with titania layer. *Langmuir*, **19**, 2260–6.

97 Born, R., Scharnweber, D., Roessler, S., Stoelzel, M., Thieme, M., Wolf, C. and Worch, H. (1998) Surface analysis of titanium based biomaterials. *Fresenius Journal of Analytical Chemistry*, **361**, 697–700.

98 Popescu, S., Demetrescu, I., Sarantopoulos, C., Gleizes, A.N. and Iordachescu, D. (2007) The biocompatibility of titanium in a buffer solution: compared effects of a thin film of TiO_2 deposited by MOCVD and of collagen deposited from a gel. *Journal of Materials Science: Materials in Medicine*, **18**, 2075–83.

99 Teng, S.-H., Lee, E.-J., Park, C.-S., Choi, W.-Y., Shin, D.-S. and Kim, H.-E. (2008) Bioactive nanocomposite coatings of collagen/hydroxyapatite on titanium substrates. *Journal of Materials Science: Materials in Medicine*, **19**, 2453–61.

100 MacDonald, D.E., Markovic, B., Boskey, A.L. and Somasundaran, P. (1998) Physicochemical properties of human plasma fibronectin binding to well characterized titanium dioxide. *Colloids and Surfaces. B*, **11**, 131–9.

101 Schliephake, H., Scharnweber, D., Dard, M., Rossler, S., Sewing, A. and Huttmann, C. (2003) Biological performance of biomimetic calcium phosphate coating of titanium implants in the dog mandible. *Journal of Biomedical Materials Research, Part A*, **64A**, 225–34.

102 Shibata, Y., Takashima, H., Yamamoto, H. and Miyazaki, T. (2004) Functionally gradient bonelike hydroxyapatite coating on a titanium metal substrate created by a discharging method in HBSS without organic molecules. *International Journal of Oral & Maxillofacial Implants*, **19**, 177–83.

103 Kikuchi, M., Itoh, S., Ichinose, S., Shinomiya, K. and Tanaka, J. (2001) Self-organization mechanism in a bone-like hydroxyapatite/collagen nanocomposite synthesized in vitro and its biological reaction in vivo. *Biomaterials*, **22**, 1705–11.

104 Lazarev, Y.A., Grishkovskii, B.A. and Khromova, T.B. (1985) Amide I band of IR spectrum and structure of collagen and related polypeptides. *Biopolymers*, **24**, 1449–78.

105 Doyle, B.B., Bendit, E.G. and Blout, E.R. (1975) Infrared spectroscopy of collagen and collagen-like polypeptides. *Biopolymers*, **14**, 937–57.

106 Haris, P.I. and Chapman, D. (1995) The conformational analysis of peptides using Fourier transform IR spectroscopy. *Biopolymers*, **37**, 251–63.

107 Köppen, S., Ohler, B. and Langel, W. (2007) Adsorption of collagen fragments on titanium oxide surfaces: a molecular dynamics study. *Zeitschrift für Physikalische Chemie*, **221**, 3–20.

108 Langel, W. and Menken, L. (2003) Simulation of the interface between titanium oxide and amino acids in solution by first principles MD. *Surface Science*, **538**, 1–9.

109 Sousa, S.R., Bras, M.M., Moradas-Ferreira, P. and Barbosa, M.A. (2007) Dynamics of fibronectin adsorption on TiO_2 surfaces. *Langmuir*, **23**, 7046–54.

110 MacDonald, D.E., Markovic, B., Allen, M., Somasundaran, P. and Boskey, A.L. (1998) Surface analysis of human

plasma fibronectin adsorbed to commercially pure titanium materials. *Journal of Biomedical Materials Research*, **41**, 120–30.

111 Sousa, S.R., Moradas-Ferreira, P. and Barbosa, M.A. (2005) TiO_2 type influences fibronectin adsorption. *Journal of Materials Science: Materials in Medicine*, **16**, 1173–8.

112 MacDonald, D.E., Deo, N., Markovic, B., Stranick, M. and Somasundaran, P. (2002) Adsorption and dissolution behavior of human plasma fibronectin on thermally and chemically modified titanium dioxide particles. *Biomaterials*, **23**, 1269–79.

113 Lindberg, B., Maripuu, R., Siegbahn, K., Larsson, R., Goelander, C.G. and Eriksson, J.C. (1983) ESCA studies of heparinized and related surfaces. 1. Model surfaces on steel substrates. *Journal of Colloid and Interface Science*, **95**, 308–21.

114 Zenhausern, F., Adrian, M. and Descouts, P. (1993) Solution structure and direct imaging of fibronectin adsorption to solid surfaces by scanning force microscopy and cryo-electron microscopy. *Journal of Electron Microscopy*, **42**, 378–88.

115 Engel, J., Odermatt, E., Engel, A., Madri, J.A., Furthmayr, H., Rohde, H. and Timpl, R. (1981) Shapes, domain organizations, and flexibility of laminin and fibronectin, two multifunctional proteins of the extracellular matrix. *Journal of Molecular Biology*, **150**, 97–120.

116 Hayakawa, T., Yoshinari, M. and Nemoto, K. (2005) Quartz-crystal microbalance-dissipation technique for the study of initial adsorption of fibronectin onto tresyl chloride-activated titanium. *Journal of Biomedical Materials Research, Part B Applied Biomaterials*, **73B**, 271–6.

117 Ferraz, M.P., Monteiro, F.J., Serro, A.P., Saramago, B., Gibson, I.R. and Santos, J.D. (2001) Effect of chemical composition on hydrophobicity and zeta potential of plasma sprayed HA/ $CaO-P_2O_5$ glass coatings. *Biomaterials*, **22**, 3105–12.

118 De Pereda, J.M., Wiche, G. and Liddington, R.C. (1999) Crystal structure of a tandem pair of fibronectin type III domains from the cytoplasmic tail of integrin alpha6 beta4. *EMBO Journal*, **18**, 4087–95.

119 Tooney, N.M., Mosesson, M.W., Amrani, D.L., Hainfeld, J.F. and Wall, J.S. (1983) Solution and surface effects on plasma fibronectin structure. *Journal of Cell Biology*, **97**, 1686–92.

120 Vuento, M., Wrann, M. and Ruoslahti, E. (1977) Similarity of fibronectins isolated from human plasma and spent fibroblast culture medium. *FEBS Letters*, **82**, 227–31.

121 Torii, H. and Tasumi, M. (1996) *Infrared Spectroscopy of Biomolecules* (eds H.H. Mantsch and D. Chapman), John Wiley & Sons, Inc., New York, pp. 1–18.

122 Nakamoto, K. (1978) *Infrared and Raman Spectra of Inorganic and Coordination Compounds*, John Wiley & Sons, Inc., New York.

123 Creighton, T.E. (1993) *Proteins*, W.H. Freeman and Co., New York.

124 Zhang, H.-P., Lu, X., Fang, L.-M., Weng, J., Huang, N. and Leng, Y. (2008) Molecular dynamics simulation of RGD peptide adsorption on titanium oxide surfaces. *Journal of Materials Science: Materials in Medicine*, **19**, 3437–41.

125 Monti, S., Carravetta, V., Zhang, W. and Yang, J. (2007) Effects due to interadsorbate interactions on the dipeptide/TiO_2 surface binding mechanism investigated by molecular dynamics simulations. *Journal of Physical Chemistry C*, **111**, 7765–71.

126 Carravetta, V. and Monti, S. (2006) Peptide-TiO_2 surface interaction in solution by ab initio and molecular dynamics simulations. *Journal of Physical Chemistry B*, **110**, 6160–9.

127 Monti, S. (2007) RAD16II beta -sheet filaments onto titanium dioxide: dynamics and adsorption properties. *Journal of Physical Chemistry C*, **111**, 16962–73.

7
Preparation, Characterization, and Potential Biomedical Applications of Nanostructured Zirconia Coatings and Films

Xuanyong Liu, Ying Xu, and Paul K. Chu

7.1
Introduction

Zirconia (ZrO_2) has excellent mechanical strength, thermal stability, chemical inertness, lack of toxicity, and an affinity for groups containing oxygen. These and other favorable attributes make ZrO_2 ceramics, coatings and films potentially useful as biocoatings, femoral heads, orthopedic implants, and biosensors. Following the first report of the biomedical application of zirconia in 1969, its use in the ball heads of total hip replacements (THRs) was introduced in 1988 [1]. In fact, the THR femoral head remains one of zirconia's main uses, with more than 600 000 units having been implanted worldwide, mainly in the US and Europe, up until 2005 [2].

Zirconia coatings and films may also be deposited onto other materials in order to improve their surface properties, such as wear resistance, corrosion resistance, thermal barrier capability, and biocompatibility. Several techniques, including sol–gel processing, plasma spraying, anodic oxidation, magnetron sputtering, electrochemical deposition, and plasma deposition, have been used to prepare ZrO_2 coatings and films. In this chapter, we review the surface morphology, microstructure, crystallite size, phase composition and biomedical characterization of nanostructured ZrO_2 coatings and films prepared using different techniques.

7.2
Preparation and Characterization of Nano-ZrO_2 Films

7.2.1
Cathodic Arc Plasma Deposition

Among others, research groups at the City University of Hong Kong and Shanghai Institute of Ceramics, Chinese Academy of Sciences (SICCAS) have fabricated ZrO_2 thin films with nanosized surfaces on Si (100) wafers using a

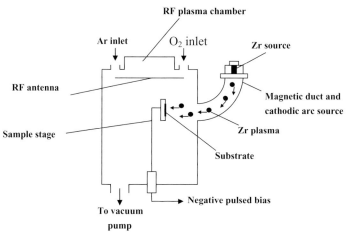

Figure 7.1 Schematic diagram of the synthesis of ZrO$_2$ films using a filtered cathodic arc system [4].

filtered cathodic arc system [3] (see Figure 7.1). The experimental apparatus includes a magnetic duct and cathodic arc plasma source, with the zirconium discharge being controlled by the main arc current between the cathode and anode. Oxygen gas is bled into the arc region, and the mixed zirconium and oxygen plasma is guided into the vacuum chamber by an electromagnetic field applied to the curved duct. The duct is biased to −20 V in order to exert a lateral electric field, while the external solenoid coils wrapped around the duct produce the axial magnetic field with a magnitude of 100 G. Before deposition, the samples that are typically positioned about 15 cm away from the exit of the plasma stream are sputter-cleaned with an argon plasma for 2 min, using a sample bias of −500 V. The base pressure in the vacuum chamber is approximately 1×10^{-5} Torr, and a radiofrequency (RF) power of 100 W is applied for a deposition time of 120 min. The as-deposited ZrO$_2$ thin films are subsequently heat-treated at either 800 °C or 1000 °C for 2 h.

The X-ray diffraction (XRD) results in Figure 7.2 show that the resultant thin film is crystallized, as indicated by the diffraction peaks at 29.83° and 34.85° which can be attributed respectively to the (101) and (002) planes of the tetragonal ZrO$_2$ phase. The small peak marked by the arrow in Figure 7.2 may stem from the (111) plane of the monoclinic phase. The surface views of the as-deposited and thermally treated ZrO$_2$ thin films observed using scanning electron microscopy (SEM) and atomic force microscopy (AFM) are shown in Figures 7.3 and 7.4. The surface of the as-deposited ZrO$_2$ thin film was very smooth, such that the surface features could not easily be distinguished using SEM (Figure 7.3a). However, some very small particles were observed on the surface of the ZrO$_2$ thin film in the AFM image (Figure 7.3b). Following heat treatment at either 800 °C or 1000 °C for 2 h, nanosized particles could be observed on the surface of the ZrO$_2$ thin film (Figure 7.4). The sizes of the particles on the ZrO$_2$ thin

7.2 Preparation and Characterization of Nano-ZrO$_2$ Films

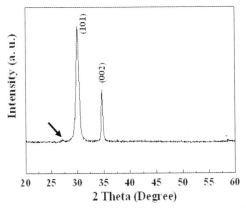

Figure 7.2 Thin film-X-ray diffraction pattern acquired from the ZrO$_2$ thin film deposited on silicon wafer [5].

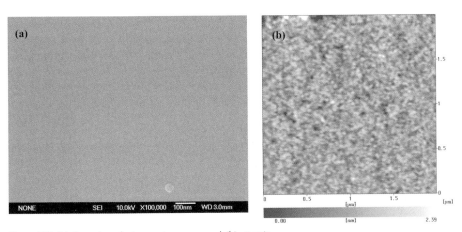

Figure 7.3 (a) Scanning electron microscopy and (b) atomic force microscopy images of the as-deposited ZrO$_2$ thin film [4].

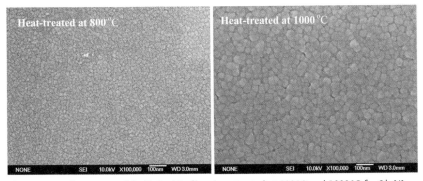

Figure 7.4 Surface views of the ZrO$_2$ thin films heat-treated at 800°C and 1000°C for 2 h [4].

film heat-treated at 800 °C or 1000 °C were approximately 20 nm and 40 nm, respectively, which indicated that the particle size could be regulated by optimizing the post-treatment processes.

7.2.2
Plasma Spraying

Plasma spraying has been used extensively to prepare ceramic coatings since its development by Union Carbide during the mid-1950s. In plasma spraying, an electrical arc is used to melt and spray the materials onto the surface of a substrate (see Figure 7.5), with the parameters of the density, temperature and velocity of the plasma beam each being important when forming the coatings. Although the temperature in the core region of the plasma beam is relatively constant at approximately 12 000 K, it increases dramatically towards the nozzle, to a point where almost any material can be melted in the plasma jet.

Nanostructured zirconia coatings have been prepared by the research group at SICCAS since 2002, using plasma spraying [7–11]. The nanostructured zirconia coating fabricated by Zeng et al. possesses two types of structure: (i) a poorly consolidated structure composed of nanosized particles; and (ii) an overlapping structure consisting of micrometer-sized particles [7]. The former structure constitutes the main component of the coating. Whilst the zirconia coating has the same phase composition as the starting powder, its thickness appears to be inconsistent, as shown in the cross-sectional view in Figure 7.6. Subsequently, others have investigated further the preparation and properties of plasma-sprayed nanostructured zirconia coatings, with the result that zirconia coatings with nanostructured surfaces and uniform thicknesses have been successfully fabricated by optimizing the spraying parameters (Figure 7.7).

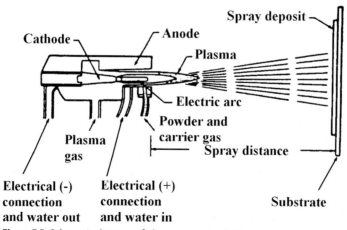

Figure 7.5 Schematic diagram of plasma spray torch [6].

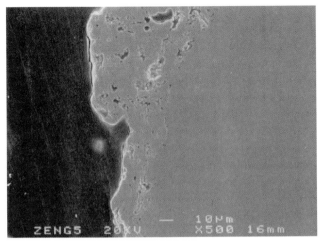

Figure 7.6 Cross-sectional views of the nanostructured zirconia coating fabricated by Zeng et al. [7].

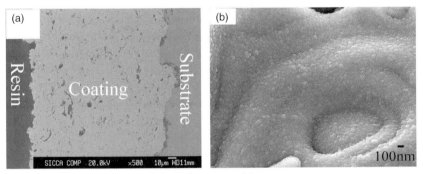

Figure 7.7 (a) Cross-sectional and (b) surface views of the nanostructural zirconia coating fabricated by Wang et al. [11].

7.2.3
Sol–Gel Methods

The sol–gel process has been used extensively to deposit thin films (<10 μm). Compared to other, conventional, thin-film processes the sol–gel system allows for a better control of the chemical composition and microstructure of the film, for the preparation of homogeneous films, for a reduction of the densification temperature, and finally the use of simpler equipment and a lower cost. The particular advantages of the sol–gel approach include an easy purification of the precursors (by distillation or crystallization), and the ability to introduce traces of other elements into the film. In addition, the precursors can be mixed at the molecular level in the solution, such that a high degree of homogeneity can be

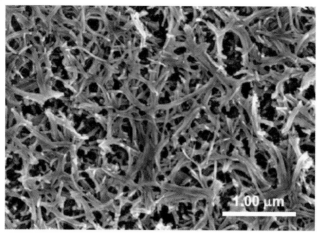

Figure 7.8 Morphology of the ZrO$_2$ film obtained at 150 g ZrCl$_4$ l^{-1} [12].

attained in the films. As a consequence, the sol–gel process permits a lower temperature to be used during the sintering stage. The resultant microstructure depends not only on the treatment of the precursors, but also on the relative rates of condensation and evaporation during film deposition.

In the sol–gel process, the thin films are normally produced using spin- or dip-coating techniques. The *spin-coating* process consists of four stages: deposition; spin-up; spin-off; and evaporation. However, for complex-shaped substrates, the most frequently used sol–gel technique is *dip-coating*, where the sample is dipped into a solution containing the precursors and then withdrawn at a constant speed, usually with the aid of a motor. The deposition of a solid film results from gravitational draining and solvent evaporation, accompanied by further condensation reactions.

Chang *et al.* [12] have proposed the use of a spin-coating sol–gel method to fabricate highly porous or smooth ultra-thin ZrO$_2$ films. The morphology and thickness of these thin films can easily be controlled by adjusting the concentrations of ZrCl$_4$ in the sol solutions. A porous film with a large specific surface area (180 m^2 g^{-1}) can be obtained using 150 g ZrCl$_4$ l^{-1}, without the addition of templates. The ZrO$_2$ film prepared under these conditions is composed of nanofibers with a typical width of approximately 100 nm. The entangled and branched nanofibers that constitute the thin film appear porous (Figure 7.8). A type II isotherm is clearly found in the N$_2$ adsorption curve, indicating the macroporous characteristics of the thin film. When the concentration of ZrCl$_4$ is reduced to 17 g l^{-1}, the porous surface turns into a smooth thin film. A linear relationship between film thickness and the concentration of ZrCl$_4$, ranging from 17 to 3 g l^{-1}, has been observed; for example, an ultrathin film with a thickness of 1.8 nm can be fabricated when a ZrCl$_4$ concentration of 3 g l^{-1} is used.

The texture of sol–gel-derived thin films is mainly controlled by the rates of hydrolysis, condensation, and phase separation. In theory, a smooth morphology

7.2 Preparation and Characterization of Nano-ZrO$_2$ Films

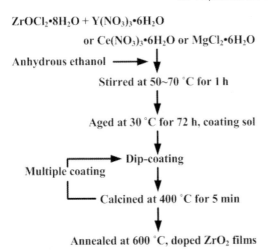

Figure 7.9 Flow diagram demonstrating the preparation of doped zirconia films [14].

should be formed when hydrolysis and condensation are conducted separately at a slow reaction rate. With decreasing ZrCl$_4$ concentrations the size of the ZrO$_2$ fibers increases, however, and the thin films have a smoother morphology due to the slower hydrolytic rate.

Lucca et al. [13] prepared multilayered ZrO$_2$ thin films by dip-coating and subsequent heat treatment at temperatures ranging from 400 to 700°C, and then investigated the near-surface mechanical responses of the films to nanoindentation. The elastic modulus and hardness are measured at depths which correspond to 7–10% of the film thickness. The temperature of the heat treatment is found to have a significant effect on the resultant near-surface hardness and elastic modulus. Liu et al. [14] have prepared highly oriented CeO$_2$-, Y$_2$O$_3$-, and MgO-doped ZrO$_2$ thin films using a sol–gel process by dip-coating in an ethanol solution consisting of zirconium oxychloride octahydrate and inorganic dopants, as shown in the flow diagram in Figure 7.9. The doped ZrO$_2$ films contained only the zirconia tetragonal phase and exhibited nanoscale morphology (Figure 7.10).

7.2.4
Electrochemical Deposition

Electrochemical deposition is used widely to deposit a thin metallic coating onto the surface of another metal, by the simple electrolysis of an aqueous solution containing the desired metal ion or its complex (Figure 7.11). Electrochemical deposits possess a fine structure and valuable physical properties, such as a high level of hardness and a high reflectivity [15].

In the electrochemical deposition process, reduction takes place when a current is supplied externally, and when the sites for the anodic and cathodic reactions are separate. The reduction mechanism pertaining to the electrochemical deposition

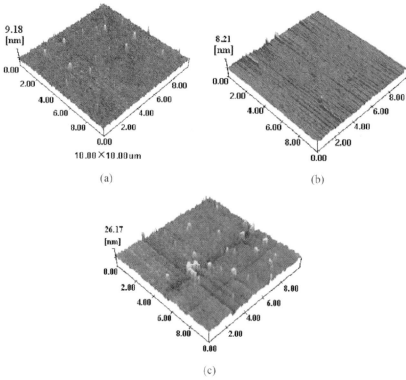

Figure 7.10 Atomic force microscopy images of doped ZrO_2 thin films. (a) 13CSZ; (b) 8YSZ; (c) 8MSZ [14].

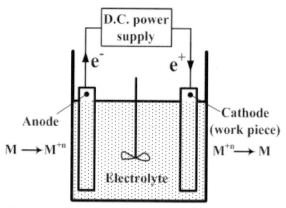

Figure 7.11 Schematic diagram of the cell used for electrochemical deposition [15].

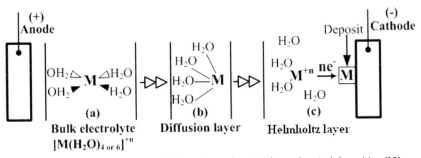

Figure 7.12 Reduction mechanism of solvated metal ion in electrochemical deposition [15].

of a simple solvated metal salt (see Figure 7.12) can also be extended to other ligand-coordinated metal systems. The solvated metal ion present in the electrolyte arrives at the cathode under the influence of the imposed electrical field, as well as by diffusion and convection (Figure 7.12a), and enters the diffusion layer. Although the field strength in the diffusion layer is not sufficiently strong to liberate the free metal ion from the solvated state, the solvated water molecules are aligned by the field (Figure 7.12b); the metal ion then passes through the diffused part of the double layer. As the field strength of the double layer is high (on the order of $10^7\,V\,cm^{-1}$), the solvated water molecules are removed so as to leave the free metal ion (Figure 7.12c), which is then reduced and deposited at the cathode via an ad-atom mechanism [15].

Electrochemical deposition has been used widely to prepare zirconia and composite films. For example, Stefanov et al. [16] investigated the electrochemical deposition of zirconia film on stainless steel, where the films were obtained in an electrolyte of anhydrous ethyl alcohol. With the rising cathodic voltage, the structure and morphology of the zirconia films exhibited essential changes; those films with the most porous structure were formed at 21 V, and those with the highest density at 25 V. The composition in the bulk of the films was close to the stoichiometric value. Shacham et al. [17] have reported a novel electrochemical method for the deposition of ZrO_2 thin films, where films with thicknesses ranging from 50 to 600 nm were obtained by applying moderate positive or negative potentials (from +2.5 V to −1.5 V versus SHE) on the conducting surfaces, which were immersed in a solution of zirconium tetra-n-propoxide [$Zr(OPr)_4$] in *iso*-propanol, in the presence of minute quantities of water (the water/monomer molar ratio ranged from 10^{-5} to 10^{-1}). The magnitude of the applied potential and its duration were thought to provide a convenient means of controlling the film thickness. Thin films of zirconia and an organoceramic composite which consisted of zirconium hydroxide and poly(diallyldimethylammonium chloride) (PDDA) have been produced using cathodic electrodeposition [18], with films of up to 10 μm thickness obtained on Ni and porous Ni–yttria-stabilized zirconia (YSZ) cermet substrates. The amount of material deposited, and its composition, can be controlled by varying the current density and the concentrations of the PDDA and zirconium salt.

7.2.5
Anodic Oxidation and Micro-Arc Oxidation

Anodic oxidation encompasses electrode reactions in combination with electrical field-driven metal and oxygen ion diffusion, leading to the formation of an oxide film at the anode surface. Anodic oxidation is a well-established method for the production of different types of protective oxide films on metals, and different dilute acids (e.g., H_2SO_4, H_3PO_4, acetic acid) can be used as the electrolyte in the process. The structural and chemical properties of the anodic oxides can be varied over a quite wide range by altering the process parameters, such as the anode potential, electrolyte composition, temperature, and current. The anodizing apparatus is shown schematically in Figure 7.13.

Micro-arc oxidation (MAO), which is also referred to as "anodic spark oxidation" or "plasma electrolytic oxidation," is a novel anodic oxidation technique that is used to deposit ceramic coatings on the surfaces of valve metals such as Al, Ti, Mg, Ta, W, Zn, and Zr, as well as their alloys. In MAO processes, the anode which is composed of the valve metal is immersed in an aqueous solution, and an asymmetric alternating voltage applied between the anode and cathode. In the anodic half-circle, the voltage is usually in the range of 150 to 1000 V, while in the cathodic half-circle it is in the range of 0 to 100 V. MAO processes are typically characterized by the phenomenon of electrical discharge on the anode in aqueous solution, while temperatures of up to 10 000 K and local pressures of several hundred bar in the discharge channels have been reported. The quality of the MAO coating is determined by parameters such as the composition and temperature of the electrolyte, the alloy composition, the voltage, current density, and time; high-quality coatings can be created by employing properly selected deposition parameters.

Zhao et al. [20] have fabricated zirconia nanotube arrays with a diameter of approximately 130 nm, lengths of up to 190 μm, and aspect ratios of more than 1400, by anodizing the zirconium foil in a mixture of formamide and glycerol (volume ratio = 1:1) containing 1 wt% NH_4F and 3 wt% H_2O (Figure 7.14). The

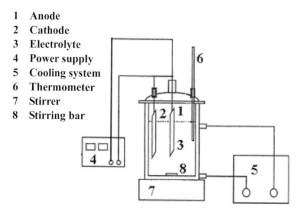

1 Anode
2 Cathode
3 Electrolyte
4 Power supply
5 Cooling system
6 Thermometer
7 Stirrer
8 Stirring bar

Figure 7.13 Schematic diagram of the anodizing apparatus [19].

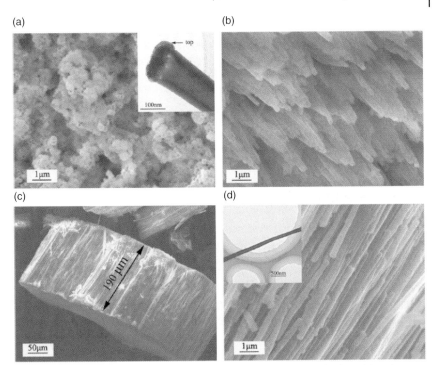

Figure 7.14 Scanning electron microscopy images of the as-prepared samples after anodization of Zr at 50 V for 24 h. (a) Surface structure (the inset shows a transmission electron microscopy image of the top part); (b) Surface structure after being rinsed ultrasonically; (c) Cross-section; (d) Cross-section at higher magnification (the inset shows a transmission electron microscopy image of the zirconia nanotube) [20].

as-prepared nanotube arrays contained amorphous zirconia. Both, monoclinic and tetragonal zirconia coexisted when annealed at 400 °C and 600 °C, whereas monoclinic zirconia was obtained at 800 °C; the ZrO_2 nanotubes retained their shape after heating to up to 800 °C. The lower dissolution rate of zirconia in organic electrolytes was considered to be the main reason for fabricating zirconia nanotube arrays with high aspect ratios.

Tsuchiya et al. [21] have reported the formation of self-organized porous layer of ZrO_2 by the anodization of zirconium in H_2SO_4 electrolytes containing low concentrations of NH_4F. Under optimized electrolyte conditions, and with polarization in the range of several tens of volts, an orderly sponge-like porous ZrO_2 was obtained, the layer thickness of which was on the order of several tens of micrometers, while the pore diameter typically varied from 10 to 100 nm. The pore size was almost independent of the applied potential, whereas the amount of well-structured area on the sample surface increased with higher applied potentials.

Yan et al. [22] have reported that zirconia films can also be prepared by the MAO of zirconium in an aqueous solution, using a pulse power supply. The groups

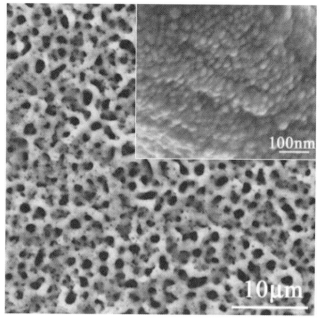

Figure 7.15 Surface view of the MAO zirconia film [22].

results indicated that the MAO-formed films were nanostructured and porous (Figure 7.15), and were composed of zirconia partially stabilized with CaO, together with a small amount of monoclinic ZrO_2.

7.2.6
Magnetron Sputtering

Magnetron sputtering represents one of the most common methods used for depositing thin films, its popularity stemming from the simplicity of the physical processes involved, combined with versatility and flexibility. Magnetron sputtering sources can be classified as diode devices in which the magnetic fields are used in concert with the cathode surface to form electron traps. The charged particles are confined by a closed magnetic field, such that the high-density plasma is produced in the vicinity of the cathode. An enhanced plasma ionization can then be achieved via either additional gas ionization or plasma confinement. This sputtering system can deliver large ion currents to the substrates, and may be operated over a wide range of pressures. When using this process, the coatings can be produced on substrates that may be very large, and also of complex shape.

Huy et al. [23] have deposited zirconia thin films by using RF magnetron sputtering on zircaloy-4 (Zy-4) substrates directly from the ZrO_2 target. These thin films are deposited at different substrate temperatures (from 40 to 800 °C) and over different times (from 10 to 240 min). By comparing the Raman studies on zirconia thin films and on bulk zirconia, it can be concluded that:

- ZrO_2 is not completely dissociated during the deposition process.
- The zirconia films deposited on different substrates are polycrystalline.
- The structure and phase composition of the films depend on the substrate temperature and on the deposition time, and therefore vary as a function of the distance from the film surface.

7.3
Bioactivity of Nano-ZrO$_2$ Coatings and Films

Zirconia is generally considered to be a bioinert ceramic because, when implanted, it shows morphological fixation only to the surrounding tissues, without producing any chemical or biological bonding. However, in the past few years many attempts have been made to prepare bioactive ZrO_2 coatings/films, or to improve their bioactivity by means of post-treatment.

The ability to induce apatite formation on a nanocomposite of cerium-stabilized tetragonal zirconia polycrystals (Ce-TZP) and alumina (Al_2O_3) polycrystals via chemical treatment in aqueous solutions of H_3PO_4, H_2SO_4, HCl, or NaOH has been investigated [24]. When this type of nanocomposite is subjected to such treatment at 95 °C for 4 h, Zr–OH groups that enhance the formation of apatite in simulated body fluids (SBFs) are first formed on the surface. The apatite-forming ability of zirconium metal pretreated in ≥5 M aqueous NaOH solution has also been demonstrated in SBF immersion tests [25]. Apatite nucleation is believed to be induced by Zr–OH groups in a zirconia hydrogel layer that forms on the metal upon exposure to NaOH.

Both, macroporous and nanocrystalline zirconia films have been prepared by the MAO of zirconium, and the effects of chemical treatments in aqueous H_2SO_4 or NaOH solutions on the microstructure and apatite-forming ability of the films have been investigated [26]. Compared to the MAO film, the chemically treated films did not exhibit any apparent changes in their phase component, morphology, and grain size; however, there were more abundant basic Zr–OH groups. The films treated with H_2SO_4 and NaOH solutions could induce apatite formation on their surfaces in SBFs within 1 day, whereas no apatite was detected on the untreated ZrO_2 surface even after 30 days. Thus, it was considered that the enhanced apatite-forming ability of the chemically treated ZrO_2 films was related to the abundant basic Zr–OH groups on the surface.

Plasma-sprayed nanostructured zirconia coatings stabilized with 3 mol% yttria (3Y-TZP) also exhibited good bioactivity [11]. After immersion in the SBF solution for 28 days, a bone-like apatite was formed on the surface of the plasma-sprayed nanostructured zirconia coatings (Figure 7.16), but not on the surface of the polished coating. The Zr–OH groups formed in the aging process, and the

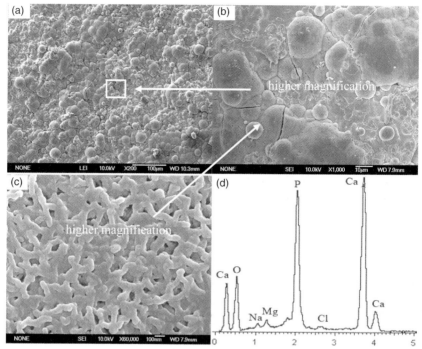

Figure 7.16 (a–c) Surface micrographs of the plasma-sprayed zirconia coating immersed in simulated body fluid solution for 28 days (magnifications as indicated); (d) Energy-dispersive X-ray spectroscopy of the particles on the coating surface [11].

nanostructured surface of the plasma-sprayed 3Y-TZP coating, were believed to be the keys to the bioactivity of the coating.

The ZrO_2 thin films prepared by cathodic arc plasma deposition were immersed in SBF to evaluate their bioactivity; the surface views of the as-deposited and thermally treated ZrO_2 thin films, after soaking in SBF for 28 days, are shown in Figure 7.17. After immersion in SBF for 28 days, the surface of the as-deposited ZrO_2 thin film was completely covered by the apatite layer, although relatively few apatite structures appeared on the surface of the ZrO_2 thin film annealed at 1000 °C for 2 h. These results indicated that the bioactivity of the ZrO_2 thin film might be degraded after thermal treatment.

A high-resolution transmission electron microscopy (HRTEM) image, which was recorded near the apatite and ZrO_2 thin film (Figure 7.18), showed the apatite layer to be in direct contact with the ZrO_2 thin film. In this image, the many disordered areas visible around the crystalline apatite indicated that the apatite layer on the ZrO_2 thin film had only partially crystallized. The (211) planes of the apatite, with a spacing of approximately 0.28 nm, were well resolved in this region. The HRTEM image of the ZrO_2 thin film showed that it consisted of nanosized

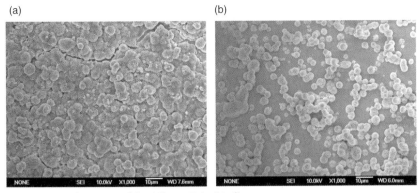

Figure 7.17 Surface views of the different ZrO_2 thin films soaked in simulated body fluid for 28 days. (a) As-deposited thin film; (b) Thermally treated thin film (1000 °C for 2 h) [4].

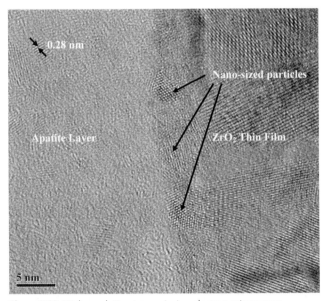

Figure 7.18 High-resolution transmission electron microscopy (HRTEM) image taken near the apatite and ZrO_2 thin film [5].

crystalline ZrO_2, with the outermost ZrO_2 crystal smaller than that in the bulk of the ZrO_2 thin film (as denoted by the arrows in Figure 7.18). The several-nanometer size of the particles in the outermost layer was believed to be one of the reasons why the ZrO_2 thin film was bioactive.

The results obtained with the nano-ZrO_2 thin films and coatings may indicate that nanosized surfaces bode well for bioactivity and biocompatibility, when compared to conventional surfaces. The deposition of calcium ions represents the first, and most crucial, step of carbonate-containing hydroxyapatite nucleation from an ionic solution; indeed, this process is believed to initiate the growth of bone-like apatite on the surface of biocompatible implants [27]. The formation of a negatively charged surface gives rise to apatite precipitation because positive calcium ions are attracted from the solution [28]. The charge densities of the particles are determined by their size; in fact, Vayssieres et al. [29] have suggested that finer nanocrystalline particles might have higher surface charge densities than larger particles. It has also been demonstrated, by using thermodynamic analysis, that the surface or interfacial tension diminishes with decreasing particle size as a result of the increased potential energy of the bulk atoms in the particles [30]. Smaller particles with an increased molar free energy are more likely to adsorb molecules or ions onto the surfaces in order to decrease the total free energy, and thus become more stable. Therefore, the nanosized particles in the outermost layer of the ZrO_2 thin films and coating might represent a key factor for inducing the precipitation of bone-like apatite on surfaces during immersion in SBFs.

Uchida et al. [31] have also reported that pure zirconia gel and zirconia gels containing sodium or calcium induce the formation of apatite in SBFs only when they possess a tetragonal and/or a monoclinic structure. It has been shown previously, that a plasma-sprayed calcium-stabilized zirconia coating is bioactive, with the bioactivity depending on the content of the monoclinic phase in the coating [32].

The bioactivity of zirconia coatings and films is thought to be related to the crystalline structure of zirconia, as well as to the nanostructural surface. When the coatings and films are immersed in a SBF, the water molecules react with zirconia and dissociate to form surface hydroxyl groups. The quantity and nature of the surface hydroxyl groups depend on the crystalline structure of zirconia [33, 34]. Although, the monoclinic phase has the lowest bulk energy among the three zirconia polymorphs [35], its surface energy is higher [36]. However, upon exposure to water molecules there is a greater tendency for this energy to be lowered by the molecular and dissociative adsorption of water, which is exothermic in nature. The adsorption enthalpy of half-monolayer H_2O coverage on the monoclinic zirconia surface was shown to be $-142\,KJ\,mol^{-1}$, whilst that on the tetragonal surface was $-90\,KJ\,mol^{-1}$ [37, 38], indicating the higher reactivity of the monoclinic zirconia surface. Accordingly, the surface concentration of the hydroxyl groups on the monoclinic ZrO_2 surface was higher than that on tetragonal ZrO_2 [34]. In fact, the OH concentration of 6.2 molecules per nm^2 for monoclinic zirconia with a BET surface area of $19\,m^2\,g^{-1}$ was much higher than the value of 3.5 molecules per nm^2 for tetragonal zirconia with a BET surface area of $20\,m^2\,g^{-1}$. Moreover, in the former case the dominant species were tribridged hydroxyl groups that were shown to be more acidic, and in the latter case a combination of bibridged and terminal hydroxyl groups. The higher acidity of the $(Zr)_3OH$ on the monoclinic zirconia surface signified that it would more easily donate protons to form negatively charged $(Zr)_3O^-$. Pettersson et al. [39] have reported that the isoelectric point of the monoclinic zirconia without dopant was 6.4; this indicated that the monoclinic zirconia surface should be negatively charged in an SBF solution with a pH value of 7.4. The negatively charged surface can attract calcium ions from the SBF solution to its surface.

In conclusion, nanosized ZrO_2 coatings and films composed of tetragonal or monoclinic phases can be prepared using a variety of technologies. Moreover, such coatings and films are generally bioactive and can induce the precipitation of apatite onto the surface when they are soaked in SBFs for a certain period of time.

7.4
Cell Behavior on Nano-ZrO$_2$ Coatings and Films

It has been observed that nano-ZrO_2 had no cytotoxic properties when cells were cocultured on its surface. Likewise, it has been shown that nano-ZrO_2 is unable to generate mutations of the cellular genome, and no adverse responses have been reported following the insertion of nano-ZrO_2 into bone or muscle in *"in vivo"* models. This lack of cytotoxicity is illustrated in Figure 7.19, which shows bone marrow mesenchymal stem cells (BMMSCs) seeded onto ZrO_2 films after different culture times. At one day, the BMMSCs were seen to grow and proliferate very well on the film surface (Figure 7.19a), which became partially covered by the cells

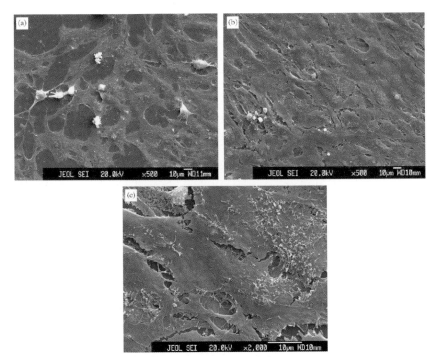

Figure 7.19 Bone marrow mesenchymal stem cells seeded onto the ZrO_2 films after different times. (a) 1 day; (b) 4 days; (c) Higher magnification of 4-day sample [5].

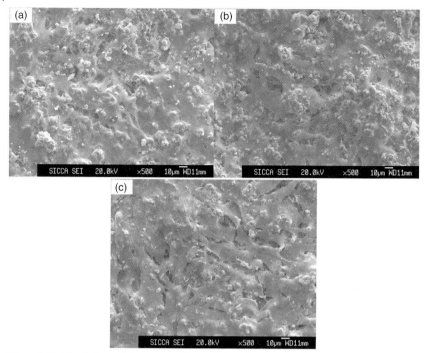

Figure 7.20 Morphology of the MG63 cells seeded on the surface of the plasma-sprayed nanostructural 3Y-TZP coating for different periods of time. (a) 1 day; (b) 3 days; (c) 7 days. [11].

and the extracellular matrix. After a longer seeding time of four days the cells had fused to form a complete layer on the thin film surface (Figure 7.19b). An examination at higher magnification (Figure 7.19c) showed the cells to possess a good morphology, with abundant dorsal ruffles and filapodia

The morphology of MG63 cells seeded onto the surface of a plasma-sprayed nanostructured zirconia coating after one, three, and seven days, is shown in Figure 7.20. The cells exhibited good adhesion and spreading after one day, with dorsal surface ruffles and filopodia seen clearly after three days of culture. After seven days, the coating surface was completely covered by MG63 cells, which indicated that this surface favored their adhesion and growth. Cell attachment, adhesion and spreading represent the first phase of cell–implant interaction, and the quality of this phase is known to influence the capacity of cells subsequently to proliferate and to differentiate. When the proliferation and vitality of MG63 cells cultured on the 3Y-TZP coating and the polystyrene (PS) control were monitored using the alamarBlue™ assay, the results (Figure 7.21) confirmed cell proliferation on both surfaces (albeit with a longer culture time for PS). The fact that no significant differences were identified between the 3Y-TZP coating and PS control indicated that the cytocompatibility of both materials was similar.

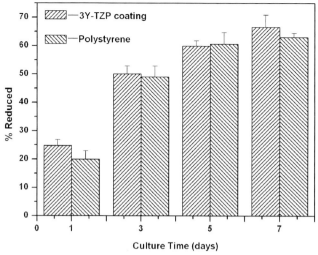

Figure 7.21 Percentage alamarBlue™ reduction for MG63 cells seeded on the plasma-sprayed nanostructural 3Y-TZP coatings and the polystyrene control, over different periods [11].

Vincenzo et al. [40] attempted to address the genetic effects of zirconia coatings on osteoblast-like cells by using a DNA microarray technique. The study results showed clearly that ZrO_2 could upregulate/downregulate certain functional activities of the osteoblast-like cells, such as cell cycle regulation, signal transduction, immunity, and the cytoskeleton components. The surface morphology of the sample was shown to have a major influence on the gene expression of proteins for cellular binding, with cell adhesion/anchoring being improved by the presence of surface canyons (~1 μm deep) placed between the large crystals that were distributed homogeneously over the surface.

7.5
Applications of Nano-ZrO$_2$ Films to Biosensors

Nanostructural materials have been used extensively for both the efficient transport of electrons and optical excitation, based on their high surface-to-volume ratio and tunable electron-transport properties — two factors which are crucial to the function and integration of nanoscale devices. Since the sizes of biological macromolecules such as proteins, enzymes, and nucleic acids are comparable to those of nanoscale building blocks, any interaction between these biomolecules should induce significant changes in the electrical properties of the nanostructured biointerfaces. A range of nanostructured materials, such as nanoparticles, nanotubes, and nanowires — all of which have been produced from metals, semiconductors, carbon, or polymeric species — have been investigated intensely for their ability to fabricate the functional biointerfaces and to enhance the response of biosensors. Nanoscale biointerfaces on electrode surfaces can be obtained in a variety of ways, including modification of the electrode surfaces using biological receptor molecules such as enzymes, antibodies, or oligonucleotides or modification with nanostructured materials. As a result, nanostructured biointerfaces have been recognized as having impressive functional properties that point naturally towards bioelectrochemical applications.

Methods for preparing nanostructured biointerfaces on electrodes include physical or chemical adsorption, self-assembly, sol–gel processes, electrochemical deposition, and electrochemical polymerization. Nanoscale inorganic matrices such as nano Au, Ag, ZrO_2, SiO_2, TiO_2, ZnO, MnO_2, aluminum silicate, magnesium silicate, phyllosilicate sol–gel matrices, and magnetic iron oxide nanoparticles have good mechanical properties and thermal stability, and are also resistant to microbial attack and organic solvents. Therefore, most of these are used as carriers and supports (especially the nanostructured biocomposites), and provide a well-defined recognition interface. For these reasons they have become the ideal materials for the immobilization of biomolecules.

Nanostructural zirconia particles and films have recently been considered as a potential solid support for the immobilization of bioactive molecules in biosensor applications, due to their large specific surface area, high thermal/mechanical/chemical resistance, excellent biocompatibility, and their affinity towards groups containing oxygen. During the past few years, both functionalized and unfunctionalized zirconia nanoparticles have been developed as solid supports for enzyme immobilization and biocatalysis. For example, nanosized zirconia gel has been used to form a reproducible and reversible adsorption–desorption interface for DNA, to immobilize hemoglobin for a novel hydrogen peroxide biosensor, to immobilize DNA for investigating the effects of lanthanide on its electron transfer behavior, for the detection of DNA hybridization, and to be grafted with phosphoric and benzenephosphonic acids for the immobilization of myoglobin. Zirconia films are also able to bond to nonbiomolecules such as PDDA by coelectrodeposition.

Tong et al. have reported a novel and simple immobilization method for the fabrication of hydrogen peroxide (H_2O_2) biosensors using horseradish peroxidase (HRP)–ZrO_2 composite films [41]. Such composite films are synthesized on a gold electrode surface based on the electrodeposition of zirconia doped with HRP by cyclic voltammetric scanning in a KCl solution containing ZrO_2 and HRP. Within the HRP–ZrO_2 film, the HRP retain its bioactivity and exhibits an excellent electrocatalytical response to the reduction of H_2O_2. The results have indicated that ZrO_2 is a good biomaterial capable of retaining the bioactivity of biomolecules, while the HRP doped in biological HRP–ZrO_2 films exhibit a good electrocatalytical response to the reduction of H_2O_2.

A highly reproducible and reversible adsorption–desorption interface of DNA based on the nanosized zirconia film in solutions of different pH-values has been successfully fabricated by Liu et al. [42]. Their results showed that DNA could adsorb onto the nanosized zirconia film from its solution, yet desorb from the nanoparticles in 0.1 M KOH solution. The ZrO_2 film serves as a bridge for the immobilization of DNA on the surface of the glassy carbon (GC) electrode. This reversible adsorption–desorption performance of the nanosized zirconia film gives rise to potential applications in the preparation of removable and reproducible biochips and information storage devices.

The direct electrochemistry and thermal stability of hemoglobin (Hb), when immobilized on a nanometer-sized, zirconium dioxide (ZrO_2)/dimethyl sulfoxide (DMSO) -modified pyrolytic graphite (PG) electrode, have been studied by Liu et al. [43]. Here, Hb showed a high affinity to H_2O_2, and both nanosized ZrO_2 and DMSO could accelerate electron transfer between Hb and the electrode. The ZrO_2 nanoparticles were more important in facilitating the electron exchange than was DMSO, as they provided a three-dimensional stage, while some of the restricted orientations also favored a direct electron transfer between the protein molecules and the conductor surface. The particles were able to stabilize the oxidized form of Hb, to decrease the polarization impedance, and to enhance the thermal stability of the biosensor. These findings may lead to a new approach for the construction of mediator-free sensors by immobilizing proteins or enzymes on the ZrO_2 nanoparticles, enabling the determination of different substrates such as glucose, by using glucose oxidase.

The detection of DNA hybridization is of paramount importance for the diagnosis and treatment of genetic diseases, the detection of infectious agents, and for reliable forensic analyses. A novel electrochemical DNA biosensor based on methylene blue (MB) and zirconia (ZrO_2) film-modified gold electrode for DNA hybridization detection was described by Zhu et al. [44], in which the ZrO_2 film was electrodeposited onto a bare Au electrode surface. This was considered to be a simple, yet practical, means of creating inorganic material microstructures, notably because it overcomes the inherent drawback of sol–gel materials, namely brittleness. The DNA probe, with a phosphate group at its 5′ end, can be attached onto the ZrO_2 film surface, which has a high affinity towards phosphate groups. As the electroactive MB is able to bind specifically to the guanine bases on the DNA molecules, it can be used as an indicator in the electrochemical DNA hybridization assay. The steps involved in the fabrication of the probe DNA-modified electrode, and the detection of a target sequence, are illustrated schematically in Figure 7.22.

Liu et al. [45] have also developed a new method of immobilizing DNA, based on the sol–gel technique. In this experiment, a ZrO_2 gel-derived, DNA-modified electrode was used as the working electrode, with studies being conducted of the electron transfer of DNA in 1.0 mM potassium ferricyanide system in different concentrations of lanthanum(III), europium(III), and calcium(II). The results showed that lanthanide ions could greatly expedite the electron transfer rate of DNA in the $Fe(CN)_6^{3+}$ solution, the order of effect being $Eu^{3+} > La^{3+} \gg Ca^{2+}$; this indicated that the lanthanide ions had a stronger interaction with the immobilized DNA than did the calcium ions, with the order being consistent with the charge:ionic radius ratio. The suggested reasons for the effect that lanthanide ions had on current enhancement included an electrostatic effect and a weak

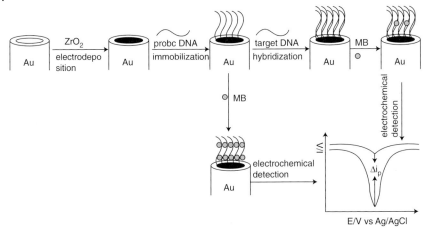

Figure 7.22 Steps involved in the fabrication of the probe DNA-modified electrode, and the detection of a target sequence.

coordination of the La^{3+} ions to DNA. The negative charge on the phosphate backbones of DNA was seen to be neutralized by the metals with positive charges, which in turn decreased the repulsion of the DNA-modified electrode surface to $Fe(CN)_6^{3+}$, and led to an increase in the current.

Four different types of layer-by-layer (LBL) film have been assembled on solid surfaces. These have been designated as PDDA/{ZrO_2}n, {PDDA/ZrO_2}n, {PDDA/NPZrO_2}n, and {PDDA/PSS}n, where ZrO_2 represents the zirconia sol–gel formed by a vapor–surface, sol–gel deposition [46]. Among the four film types, PDDA/{ZrO_2}n demonstrated a better porosity and permeability, and was capable of loading greater quantities of myoglobin (Mb) from solution. Because of the better porosity of this film, the small counterions in the buffer solution could enter and exit it more easily, and this resulted in the film having a better cyclic voltammetry response, according to the mechanism of "electron hopping." The fact that the PDDA/{ZrO_2}n-Mb films showed a larger surface concentration of electroactive Mb (Γ*) than did the {PDDA/ZrO_2}n-Mb films, was probably because the "soft" polyelectrolyte PDDA in the PDDA/{ZrO_2}n films "blocked" some pores or channels in the ZrO_2 layers. Compared to the {ZrO_2/Mb}n films, which were directly assembled LBL by Mb and ZrO_2 sol–gel, the advantage of the PDDA/{ZrO_2}n-Mb films is that any possible denaturation of Mb could be prevented as its loading into the films was completely spontaneous. Moreover, any possible direct contact of Mb with the organic solvent used in the vapor–surface, sol–gel deposition of ZrO_2 could be avoided. In this way, Mb was able to retain its original structure and bioactivity.

Today, the interfaces between protein molecules and inorganic materials are among the "hottest" research topics in biomedicine, biochemistry, biophysics, and even in industrial applications. If there is a need to modify a substrate surface with protein molecules, the first step is to acquire a complete understanding of the mechanism that underlies any specific binding between the target-specific protein and the target in a direct way. Tomohiro Hayashi et al. [47] have analyzed the adhesion force using AFM, and revealed for the first time the mechanism that underlies the specific binding between a titanium surface and ferritin that possesses the sequence of a Ti-binding peptide in its N-terminal domain. Based on such experimental data, it can be concluded that the electrostatic interaction between charged groups, the hydrophobicity or hydrophilicity of the surface, and the surfactant addition, each play important roles in the specific binding between the Ti-binding peptide (minTBP-1) and Ti. Although the adhesion force decreases dramatically in solution with the surfactant, the nonspecific interactions are suppressed and the specificity and selectivity of the minT1-LF are enhanced due to accretion of the surfactant.

In conclusion, nano-ZrO_2 films have been utilized as a solid support for the immobilization of bioactive molecules in biosensor applications due to their high thermal/mechanical/chemical resistance, excellent biocompatibility, and their affinity to certain functional groups. It is possible that the phase (whether tetragonal or monoclinic), morphology, surface microstructure, and crystallite size of ZrO_2 thin films could vary with under different synthesis processes, and this would result in differences between the electrochemical properties and the biocompatibility of the ZrO_2 films. Although nano-ZrO_2 exhibits many excellent characteristics, the electrochemical properties and biocompatibility of nano-ZrO_2 in biosensors remain inadequate. It is important, therefore, to explore new methods that can be used to improve the electrochemical properties and biocompatibility of the nano-ZrO_2 surfaces.

References

1 Piconi, C. and Maccauro, G. (1999) Zirconia as a ceramic biomaterial. *Biomaterials*, **20**, 1–25.

2 Chevalier, J. (2006) What future for zirconia as a biomaterial? *Biomaterials*, **27**, 535–43.

3 Zhang, T., Chu, P.K. and Brown, I.G. (2002) Effects of cathode materials and arc current on optimal bias of a cathodic arc through a magnetic duct. *Applied Physics Letters*, **80**, 3700–2.

4 Liu, X., Chu, P.K. and Ding, C. (2007) Nano-film and coating for biomedical application prepared by plasma-based technologies, in *Ion-Beam-Based-Nanofabrication, Materials Research Society Symposium, 10–12 April 2007, Proceedings Volume 1020*, (eds D. Ila, J. Baglin, N. Kishimoto and P.K. Chu), Materials Research Society, Warrendale, Pennsylvania, USA.

5 Liu, X., Zhao, X., Fu, R.K.Y., Ho, J.P.Y., Ding, C. and Chu, P.K. (2005) Plasma-treated nanostructured TiO_2 surface supporting biomimetic growth of apatite. *Biomaterials*, **26**, 6143–50.

6 Bouos, M., Fauchais, P. and Vardelle, A. (1993) *Plasma Spraying: Theory and Application* (ed. R. Suryanarayanan), World Scientific, Singapore, p. 3.

7 Zeng, Y., Lee, S.W., Gao, L. and Ding, C. (2002) Atmospheric plasma sprayed coatings of nanostructured zirconia. *Journal of the European Ceramic Society*, **22**, 347–51.

8 Liang, B., Ding, C., Liao, H. and Coddet, C. (2006) Phase composition and stability of nanostructured 4.7 wt% yttria-stabilized

zirconia coatings deposited by atmospheric plasma spraying. *Surface and Coatings Technology*, **200**, 4549–56.

9 Liang, B. and Ding, C. (2005) Phase composition of nanostructured zirconia coatings deposited by air plasma spraying. *Surface and Coatings Technology*, **191**, 267–73.

10 Liang, B., Ding, C., Liao, H. and Coddet, C. (2009) Study on structural evolution of nanostructured 3 mol% yttria stabilized zirconia coatings during low temperature ageing. *Journal of the European Ceramic Society*, **29**, 2267–73.

11 Wang, G., Liu, X., Gao, J. and Ding, C. (2009) In vitro bioactivity and phase stability of plasma-sprayed nanostructured 3Y-TZP coatings. *Acta Biomaterialia*, **5**, 2270–8.

12 Chang, S. and Doong, R. (2005) ZrO_2 thin films with controllable morphology and thickness by spin-coated sol–gel method. *Thin Solid Films*, **489**, 17–22.

13 Lucca, D.A., Klopfstein, M.J., Ghisleni, R., Gude, A., Mehne, A. and Datchary, W. (2004) Investigation of sol-gel derived ZrO_2 thin films by nanoindentation. *CIRP Annals Manufacturing Technology*, **53**, 475–8.

14 Liu, W., Chen, Y., Ye, C. and Zhang, P. (2002) Preparation and characterization of doped sol–gel zirconia films. *Ceramics International*, **28**, 349–54.

15 Rao, C.R.K. and Trivedi, D.C. (2005) Chemical and electrochemical depositions of platinum group metals and their applications. *Coordination Chemistry Reviews*, **249**, 613–31.

16 Stefanov, P., Stoychev, D., Stoycheva, M., Ikonomov, J. and Marinova, T. (2000) XPS and SEM characterization of zirconia thin films prepared by electrochemical deposition. *Surface and Interface Analysis*, **30**, 628–31.

17 Shacham, R., Mandler, D. and Avnir, D. (2004) Electrochemically induced sol-gel deposition of zirconia thin films. *Chemistry–A European Journal*, **10**, 1936–43.

18 Zhitomirsky, I. and Petric, A. (2000) Electrolytic deposition of zirconia and zirconia organoceramic composites. *Materials Letters*, **46**, 1–6.

19 Sul, Y.T., Johansson, C.B., Jeong, Y. and Albrektsson, T. (2001) The electrochemical oxide growth behaviour on titanium in acid and alkaline electrolytes. *Medical Engineering & Physics*, **23**, 329.

20 Zhao, J., Wang, X., Xu, R., Meng, F., Guo, L. and Li, Y. (2008) Fabrication of high aspect ratio zirconia nanotube arrays by anodization of zirconium foils. *Materials Letters*, **62**, 4428–30.

21 Tsuchiya, H. and Schmuki, P. (2004) Thick self-organized porous zirconium oxide formed in H_2SO_4/NH_4F electrolytes. *Electrochemistry Communications*, **6**, 1131–4.

22 Yan, Y. and Han, Y. (2007) Structure and bioactivity of micro-arc oxidized zirconia films. *Surface and Coatings Technology*, **201**, 5692–5.

23 Huy, L.D., Laffez, P., Daniel, P., Jouanneaux, A., Khoi, N.T. and Siméone, D. (2003) Structure and phase component of ZrO_2 thin films studied by Raman spectroscopy and X-ray diffraction. *Materials Science and Engineering: B*, **104**, 163–8.

24 Uchida, M., Kim, H.M., Kokubo, T., Nawa, M., Asano, T., Tanaka, K. and Nakamura, T. (2002) Apatite-forming ability of a zirconia/alumina nano-composite induced by chemical treatment. *Journal of Biomedical Materials Research*, **60**, 277–82.

25 Uchida, M., Kim, H.M., Kokubo, T., Miyaji, F. and Nakamura, T. (2002) Apatite formation on zirconium metal treated with aqueous NaOH. *Biomaterials*, **23**, 313–17.

26 Yan, Y., Han, Y. and Lu, C. (2008) The effect of chemical treatment on apatite-forming ability of the macroporous zirconia films formed by micro-arc oxidation. *Applied Surface Science*, **254**, 4833–9.

27 Svetina, M., Ciacchi, L.C., Sbaizero, O., Meriani, S. and De Vita, A. (2001) Deposition of calcium ions on rutile (110): a first principles investigation. *Acta Materialia*, **49**, 2169–77.

28 Li, P., Ohtsuki, C., Kokubo, T., Nakanishi, K., Soga, N. and Groot, K. (1994) The role of hydrated silica, titania and alumina in inducing apatite on implants. *Journal of Biomedical Materials Research*, **28**, 7–15.

29 Vayssières, L., Chanéac, C., Trone, E. and Joliver, J.P. (1998) Size Tailoring of magnetite particles formed by aqueous precipitation: an example of thermodynamic stability of nanometric oxide particles. *Journal of Colloid and Interface Science*, **205**, 205–12.

30 Zhang, H., Penn, R.L., Hamers, R.J. and Banfield, J.F. (1999) Enhanced adsorption of molecules on surfaces of nanocrystalline particles. *Journal of Physical Chemistry*, **B103**, 4656–62.

31 Uchida, M., Kim, H.M., Kokubo, T., Tanaka, K. and Nakamura, T. (2002) Structural dependence of apatite formation on zirconia gel in simulated body fluid. *Journal of the Ceramic Society of Japan*, **110**, 710–15.

32 Wang, G., Liu, X. and Ding, C. (2008) Phase composition and in-vitro bioactivity of plasma sprayed calcia stabilized zirconia coatings. *Surface and Coatings Technology*, **202**, 5824–31.

33 Korhonen, S.T., Calatayud, M. and Krause, A.O.I. (2008) Stability of hydroxylated (-111) and (-101) surfaces of monoclinic zirconia: a combined study by DFT and infrared spectroscopy. *Journal of Physical Chemistry C*, **112**, 6469–76.

34 Jung, K.T. and Bell, A.T. (2000) The effects of synthesis and pretreatment conditions on the bulk structure and surface properties of zirconia. *Journal of Molecular Catalysis A: Chemistry*, **163**, 27–42.

35 Christensen, A. and Carter, E.A. (1998) First-principles study of the surfaces of zirconia. *Physical Review, B*, **58**, 8050–64.

36 Iskandarova, I.M., Knizhnik, A.A., Rykova, E.A., Bagaturyants, A.A., Potapkin, B.V. and Korkin, A.A. (2003) First-principle investigation of the hydroxylation of zirconia and hafnia surfaces. *Microelectronic Engineering*, **69**, 587–93.

37 Ushakov, S.V. and Navrotsky, A. (2005) Direct measurements of water adsorption enthalpy on hafnia and zirconia. Surface enthalpy. *Applied Physics Letters*, **87**, 164103.

38 Radha, A.V., Bomati-Miguel, O., Ushakov, S.V., Navrotsky, A. and Tartaj, P. (2009) Surface enthalpy, enthalpy of water adsorption, and phase stability in nanocrystalline monoclinic zirconia. *Journal of the American Ceramic Society*, **92**, 133–40.

39 Petterson, A., Marino, G., Pursiheimo, A. and Rosenholm, J.B. (2000) Electrosteric stabilization of Al_2O_3, ZrO_2, and $3Y-ZrO_2$ suspensions: effect of dissociation and type of polyelectrolyte. *Journal of Colloid and Interface Science*, **228**, 73–81.

40 Sollazzo, V., Palmieri, A., Pezzetti, F., Bignozzi, C.A., Argazzi, R., Massari, L., Brunelli, G. and Carinci, F. (2008) Genetic effect of zirconium oxide coating on osteoblast-like cells. *Journal of Biomedical Materials Research. Part B, Applied Biomaterials*, **84B**, 550–8.

41 Tong, Z., Yuan, R., Chai, Y., Xie, Y. and Chen, S. (2007) A novel and simple biomolecules immobilization method: electro-deposition ZrO_2 doped with HRP for fabrication of hydrogen peroxide biosensor. *Biotechnology*, **128**, 567–75.

42 Liu, S., Xu, J. and Chen, H.Y. (2004) A reversible adsorption–desorption interface of DNA based on nano-sized zirconia and its application. *Colloid and Surface B. Biointerfaces*, **36**, 155–9.

43 Liu, S., Dai, Z., Chen, H. and Ju, H. (2004) Immobilization of hemoglobin on zirconium dioxide nanoparticles for preparation of a novel hydrogen peroxide biosensor. *Biosensors and Bioelectronics*, **19**, 963–9.

44 Zhu, N., Zhang, A., Wang, Q., He, P. and Fang, Y. (2004) Electrochemical detection of DNA hybridization using methylene blue and electro-deposited zirconia thin films on gold electrodes. *Analytica Chimica Acta*, **510**, 163–8.

45 Liu, S., Xu, J. and Chen, H. (2002) ZrO_2 gel-derived DNA-modified electrode and the effect of lanthanide on its electron transfer behavior. *Bioelectrochemistry*, **57**, 149–54.

46 Wang, G, Liu, Yi, and Hu, N. (2007) Comparative electrochemical study of myoglobin loaded in different types of layer-by-layer assembly films. *Electrochimica Acta*, **53**, 2071–9.

47 Hayashi, T., Sano, K., Shiba, K., Kumashiro, Y., Iwahori, K., Yamashita, I. and Hara, M. (2006) Mechanism underlying specificity of proteins targeting inorganic materials. *Nano Letters*, **6**, 515–19.

8
Free-Standing Nanostructured Thin Films

Izumi Ichinose

8.1
Introduction

A free-standing thin film is defined as a thin film in which at least a part of the film is not in contact with a support material. Although, in chemistry, the term "self-standing" is often used for molecular assemblies with the shape of a thin film, in this chapter we will use only the engineering term "free-standing thin film," a wide range of which is currently available. A metallic or inorganic film with a thickness of several hundreds of nanometers (or more) can be handled as a single sheet because of its intrinsic mechanical strength; these films are naturally free-standing thin films. In fact, even polymers will become mechanically strong films if they are a few micrometers thick. The handling of extremely thin films is difficult, because they are often formed on a substrate with pores. They are also often referred to as "free-standing thin films." This is because the parts of the thin film that cover the pores play a significant role in practical applications. Free-standing thin films are not always solid; for example, a foam film can be formed if a metal frame is immersed in an aqueous solution of surfactant, which then makes it a liquid free-standing thin film. In this chapter, nanometer-thick, free-standing thin films – or "free-standing nanostructured thin films" with nanopores or nanofibrous structures – are described. The various relationships between nanostructured free-standing thin films and the life sciences are also discussed.

8.2
The Roles of Free-Standing Thin Films

8.2.1
Films as Partitions

A free-standing thin film can be used to separate two phases and to control material transfer between those phases which, in general, are either gas or liquid. The

Nanomaterials for the Life Sciences Vol.5: Nanostructured Thin Films and Surfaces.
Edited by Challa S. S. R. Kumar
Copyright © 2010 WILEY-VCH Verlag GmbH & Co. KGaA, Weinheim
ISBN: 978-3-527-32155-1

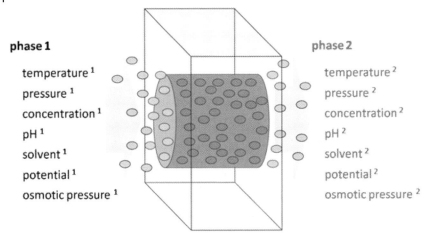

Figure 8.1 Differences in parameters of two phases separated by a membrane, and permeation of molecules through the membrane.

two phases, when separated by the film, have different composition and/or physical states. In other words, one phase may exist under different conditions in comparison with the other phase with regards to temperature, pressure, solute concentration, pH, and so on (Figure 8.1). In the case of a semipermeable membrane, water and small ions can each pass through (permeate) the membranes, but biopolymers cannot. The diffuse transmission of solutes due to concentration differences is employed in dialysis.

Transfer through free-standing thin films is not limited to molecular transfer, however. For example, in solid polymer electrolyte fuel cells, the two electrodes are placed in position but are separated by a polymer thin film with a high-proton conductivity. A potential difference is then generated by a series of electrochemical reactions, with subsequent proton transfer through the film. When producing drinking water from seawater, reverse osmosis membranes are used [1] which consist of a dense, ultrathin polymer layer formed on a porous support. These membranes have a separation function that allows water to permeate, while blocking dissolved ions. The osmotic pressure between seawater and pure water is approximately 2.5 GPa; hence, if a pressure in excess of the osmotic pressure is applied, water will be forced out from the seawater through the dense polymer layer. Although, the roles of free-standing thin films depend mainly on the conditions of the two separated phases, their performance in contrast, is largely affected by the materials and the inner nanostructures of the free-standing thin films. With much recent research activity having been expended in the area of nanostructured free-standing thin films, significant improvements in their performance – notably including material separation – are to be expected in the near future.

8.2.2
Nanoseparation Membranes

One example of improved separation performance is that of nanoseparation membranes. In generally, liquid permeation through a porous thin film can be evaluated using the Hagen–Poiseuille equation; this states that the liquid flux (J) is given by Equation 8.1:

$$J = \varepsilon \pi r_p^2 \Delta p / 8 \mu L \tag{8.1}$$

where L is the film thickness, μ is the liquid viscosity, Δp is the pressure difference between both sides of the film, r_p is the pore radius, and ε is the surface porosity (the ratio of total pore area to the total film area) [2].

As is apparent from Equation 8.1, the liquid flux increases in inverse proportion to the thickness; as compared to a 10 μm-thick film, a 10 nm-thick film gives a 1000-fold greater flux. In the case of molecular diffusion in a dense polymer layer, the permeation rate increases in inverse proportion to the film thickness. Although it is not completely clear to what degree of film thickness this relationship holds true, nanometer-thick separation membranes have attracted interest for the development of a membrane separation system with high energy efficiency. Another important point for nanoseparation membranes is to selectively dissolve a target molecule at high concentration. For example, natural rubber will absorb carbon dioxide selectively compared to carbon monoxide, so that the films have a high permeation rate for carbon dioxide [3]. If carbon dioxide can be absorbed selectively at high concentration by designing the inner structure of the film, then a significant improvement in separation performance can be achieved.

Ultrafiltration membranes are defined as pressure-driven separation membranes that can remove either particles or macromolecules with a size ranging from 2 to 100 nm. These membranes are widely used for water purification, food processing, the recovery of paint, and so on [2–4]. For example, Fauchet and coworkers produced free-standing silicon films with a thickness of 15 nm and pores in a range of 5 to 25 nm, and used these to separate biomacromolecules [5]. As the thickness of these films was approximately one-thousandth that of a typical ultrafiltration membrane, they would be expected to be used as membranes for high-speed separations. In fact, when the diffusion rate of biomacromolecules was measured under a certain concentration gradient, the flux through the ultrathin porous silicon films was approximately tenfold faster than that through conventional ultrafiltration membranes with similar permeation properties (Figure 8.2). One nanofiltration membrane which recently has attracted much attention is that of free-standing films of carbon nanotubes (CNTs) embedded in silicon nitride or polymer matrices [7, 8]. As an example, Holt *et al.* reported that, in films made from double-walled nanotubes with an inner diameter of 1.3–2.0 nm, the flux of gases was more than one order of magnitude faster than the value calculated by using the Knudsen diffusion model (Figure 8.3). Furthermore, the flux of water through these films was more than three orders of magnitude faster than the value

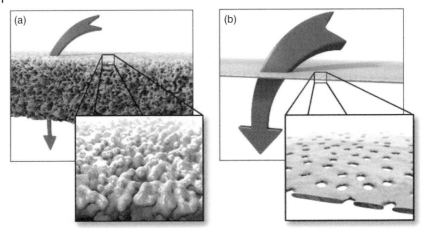

Figure 8.2 (a) Conventional ultrafiltration membrane; (b) Ultrathin porous nanocrystalline silicon membrane. Reproduced with permission from Ref. [6]; © 2007, Nature Publishing Group, London.

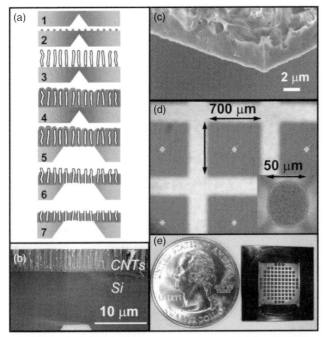

Figure 8.3 (a) Seven-step fabrication process of double-walled carbon nanotube (DWNT) membrane; (b) Scanning electron microscopy (SEM) cross-section of DWNTs on silicon substrate (step 3); (c) SEM cross-section of DWNTs fixed with silicon nitride (step 5); (d) Photographic image of the open membrane area (the inset shows a close-up of one membrane); (e) Photographic image of a membrane chip of 89 open windows. Reproduced with permission from Ref. [7]; © 2006, American Association for the Advancement of Science.

calculated by using a continuous hydrodynamic model [7]. It is supposed that the high-speed permeation of water is caused by a remarkably low frictional resistance of the inner wall of the nanotubes. Whilst, at present the detailed mechanism of water permeation remains unclear, robust CNT films will undoubtedly lead to applications such as water purification and the production of drinking water from seawater.

8.2.3
Biomembranes

The role of free-standing thin films also includes the control of electron and energy transfer between liquid phases or gas phases. Biomembranes are ultrathin films which have superior chemical systems. As an example, various proteins are embedded in the thylakoid membranes of the chloroplast, where chlorophyll (the photosynthetic pigment) degrades water using light energy, after which protons, oxygen molecules and electrons are produced. Subsequently, these electrons lead to the production of NADPH (the reduced form) from $NADP^+$ (the oxidized form). In similar fashion, ATP synthase is able to synthesize ATP by employing the proton concentration gradient inside and outside of the thylakoid membrane.

Aquaporins are membrane proteins that are capable of blocking a solute and ions, allowing the permeation of water [9]. The aquaporins have one or two internal gates each with a width of approximately 0.3 nm, which is of a similar size to a water molecule (of about 0.25 nm). Inside the gates, the hydrogen bindings between water molecules are blocked such that only water, but not protons, are allowed to pass. Aquaporins exist universally in various organism, ranging from bacteria to plants and mammals, where they contribute towards water transportation in biomembranes. One surprising point here is the ultrafast permeation of water through the gates; water molecules may pass through a channel of approximately 0.3 nm at a rate of 4×10^9 molecules per second, which is equal to a flux of $6\,000\,000\,l\,h^{-1} \cdot m^2$. Molecular-sized gates can generate a remarkably large flux [10].

One other interesting biological membrane transport system is that of RNA transfer via nuclear membranes [11]. The nuclear membrane of a eukaryotic cell contains pores that are several tens of nanometers in size, and together form the so-called "nuclear pore complex" [12]. It is said that the pores can block proteins and similar components with a molecular weight of 60 kDa or more, while they pass ribosome precursors or large mRNA molecules at high speed. These molecules are released from the nuclear pore complex by active transport, with the recognition of (and binding to) specific transport signals providing selectivity to the transportation process.

The functions of biomembranes have been substantially clarified in biological terms. Whilst a realization of the functions in artificial systems is not straightforward, it does represent a worthy challenge for thin film research, the mainstream of which is to create innovative separation membranes by the fabrication of free-standing films that are several nanometers thick. In addition, it is vital to design an internal structure that will exceed the superior performance of biomembranes.

8.3
Free-Standing Thin Films with Bilayer Structures

8.3.1
Supported Lipid Bilayers and "Black Lipid Membranes"

A biomembrane is a free-standing thin film in water, and which is composed of a lipid bilayer with a thickness of several nanometers. The lipid bilayer is a basic component of cell membrane, while the cell nucleus and organelles such as mitochondria are also mainly composed of lipid bilayers. The dispersion of phospholipids in water leads to the spontaneous formation of a vesicle structure termed a "liposome." The liposome is also composed of a lipid bilayer, and is therefore useful for examining the functions of biomembranes. Because they exist in water, liposomes are not always suitable for engineering experiments; hence, to solve this problem a method of transferring a lipid bilayer onto a solid substrate has been developed. These membranous transferred bilayers, which are known as supported lipid bilayers (SLBs) [13], are stronger and smoother than the lipid bilayers in water, and consequently are suitable for atomic force microscopy (AFM) and other spectroscopic measurements. If hydrophobic proteins can be embedded into SLBs, then the evaluation of such functions becomes straightforward. Furthermore, when SLBs are used as substrates to assemble biomacromolecules (such as proteins and sugar chains), applications in the area of medicine, such as biosensors or diagnostic sheets become possible.

The oldest model of bilayer architectures appears to be the so-called "black lipid membrane" (BLM), which is a free-standing thin film in water [14]. These membranes may be fabricated by immersing a hydrophobic substrate with a micrometer-sized pore in its center into water, and then coating the inside of the pore with an organic solvent containing phospholipids. Because the hydrophobic substrate is wet with the organic solvent, the phospholipid solution coating the pores will gradually become a thin liquid film, while the outermost surfaces of the liquid film will be covered with a phospholipid monolayer. Finally, the micrometer-sized pore becomes covered with a single phospholipid bilayer that is several nanometers thick. One drawback of the BLMs is their weak mechanical strength, although because two aqueous solutions are separated by the ultrathin free-standing film, BLMs are generally suitable for evaluating the electrical characteristics of lipid bilayers. The large electrical resistance and capacitance of lipid bilayers have each been monitored quantitatively by using BLMs.

8.3.2
Foam Films and Newton Black Films

The aforementioned BLM is a free-standing thin film in water in which the hydrophilic groups of the phospholipids face outwards. In contrast, free-standing thin films containing hydrophobic groups that face outward have been known as "foam films" for quite some time. The foam films are formed by, for example, immersing a metal frame into an aqueous soap solution and pulling the frame out (Figure 8.4). As with soap bubbles, the foam films are thin water films with the surfaces covered by surfactant molecules. Whilst the above-mentioned biomembranes are generally formed from phospholipids with two alkyl chains, foam films can also be formed from a surfactant with only one alkyl chain. The hydrophilic moieties may be anionic, cationic, nonionic, or zwitterionic, and each surfactant will provide stable foam films.

The foam film that is formed in a wire frame takes on a shape which has the minimum surface area; this occurs because the liquid composing the foam film is subjected to pressure derived from the surface tension. This pressure acts positively on a convex area, and negatively on a concave area, according to the Young–Laplace equation. As a result, the foam film formed in a square frame will become flat. On the other hand, at the points where the foam film is in contact with the frame and water surface, a concave will be formed on the liquid surface, and a similar pressure (the capillary pressure) is generated. Therefore, the liquid in the foam film will drain from the flat center part to the concave part on the edge. When the metal frame is held vertically, the thickness of the foam film gradually

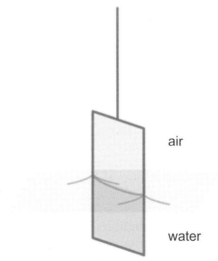

Figure 8.4 A foam film formed by lifting a square wire frame from an aqueous solution of surfactant. The concave surface at the boundary area between the film and water surface induces capillary force.

decreases, and the film becomes iridescent due to a light interference phenomenon. At this moment, the thickness will range from approximately 1 µm to 200 nm. If the foam film is left for longer, however, the film will become increasingly thinner at the top and the color will change, from pale-white to black. The thickness of the black region (known as the "black film") is approximately 100 nm. This phenomenon occurs because, when the thicker film becomes thinned, a repulsion (which is caused mainly by electrostatic interactions) is generated between the two surfactant layers, such that the repulsive force (which is referred to as the "disjoining pressure") and the capillary pressure will balance out. At this point, the black film is termed a "common black film" [15–17].

In contrast, when an aqueous solution of surfactant contains salt such as sodium chloride, the electrostatic repulsion generated between the two surfaces becomes weak, and the black film will become thinner [18]. When the average thickness of the liquid between two surfactant layers becomes several tens of nanometers, the surface tension will gradually weaken. In other words, the van der Waals forces between two surfactant layers becomes stronger and, as a result, the surfactant layers move closer to each other so as to create a "Newton black film" with a thickness of <7 nm. These two types of black film are both equilibrium films; altering a condition such as the capillary pressure will cause a transition from one type to the other [16]. Even in a Newton black film (which is an ultrathin film), a water layer containing a salt exists between the two surfactant layers; it is believed, however, that this water layer provides a steric repulsion toward the two surfactant layers so that the film thickness remains constant.

The stability of foam films is explained by the Gibbs–Marangoni effect. Although the thickness of black films may be ≤100 nm, they do not become dramatically unstable in comparison with foam films, which may be approximately 1 µm thick. In fact, it has been reported that thin foam films are less affected by liquid convection and thus are more stable. Although it is generally considered that Robert Hooke, an English seventeenth century scientist, was the first to report the existence of black films, the realization a black film is actually a thin film first appeared in *Opticks*, a writing by Isaac Newton. Research into black films was subsequently continued by great scholars such as Dewar and Perrin, and later had a major influence on the research into oil films on water that was carried out by Rayleigh, Langmuir, and others, and which helped to create the foundations of the molecular sciences.

8.3.3
Dried Foam Film

Recently, it was found that free-standing, water-free, surfactant bilayers could be obtained if the foam films were prepared in micrometer-sized pores and then dried in air. These films, which are known as "dried foam films" [19], may be prepared by immersing a copper transmission electron microscopy (TEM) grid in an aqueous solution of 10 mM dodecylphosphocholine (DPC), and then lifting it up vertically. At the grid is lifted, a very small volume of the solution is captured in

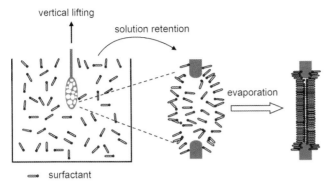

Figure 8.5 Formation of dried foam films from a dilute aqueous solution of surfactant. The free-standing surfactant bilayers form in micrometer-sized pores of a substrate [19].

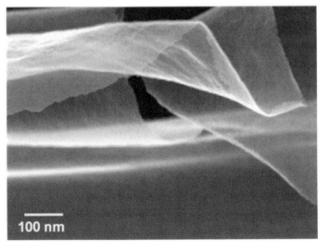

Figure 8.6 Scanning electron microscopy cross-section of a dried foam film prepared in a pore of a copper transmission electron microscopy grid. The surfactant used was dodecyltrimethylammonium bromide [19].

each of its pores. Then, as the concentration of the surfactant (DPC) increases due to evaporation of the water, a Gibbs monolayer will form at the air–water interface. As the two Gibbs monolayers gradually move closer to each other, a free-standing film with the thickness of two surfactant monolayers will be formed (Figure 8.5). An excess amount of surfactant molecules will accumulate around the periphery of the pores during the film-formation process, and accordingly the dried foam film will be an ultrathin free-standing film with a uniform thickness. A scanning electron microscopy image of the dried foam film formed from dodecyltrimethylammonium bromide (DTAB) is shown in Figure 8.6. This image was recorded after the deposition of a 2 nm-thick platinum layer on the surface of the film which, because of its very thin structure, appears quasi-transparent to the electron beam. Based on high-resolution observations of the cross-section, the film thickness was estimated as 2.5 nm after having subtracted the thickness of the platinum layers, while the thickness of the dried foam film was a little less than twice the molecular length of DTAB (1.6 nm). A dried foam film has a structure similar to that of a Newton black film. However, whilst the latter has a very thin water layer containing salt between the two layers of surfactant, the former does not even contain hydration water. In fact, dried foam films can exist stably even under ultrahigh-vacuum conditions, with some demonstrating thermal stability above 150 °C. The dried foam films can be obtained from either cationic or zwitterionic surfactants, and may also be prepared from long-chain silane alkoxides, such as octadecyltrimethoxysilane. Furthermore, not only water but also polar organic solvents (e.g., ethanol) can be used as the solvent. In contrast, homogeneous films have not been prepared from anionic surfactants such as sodium dodecylsulfate (SDS) and nonionic surfactants which have a poly(ethyleneglycol) (PEG) group as the hydrophilic moiety. Although the reason for this is not yet completely clear, it can be assumed that the strong interaction among the hydrophilic groups is essential to stabilize the dried foam films.

It has been reported that various inorganic materials can be deposited on the surfaces of the dried foam films [20], and this may represent an interesting and novel approach for the fabrication of inorganic free-standing thin films by using liquid foam films. When DPC was used, dried foam films with a thickness of only 3 nm could not be broken after depositing carbon or silicon layer with a thickness of 100 nm or more by using either thermal or electron beam deposition (Figure 8.7). The reason why these inorganic materials could be deposited was most likely related to the mechanical stability of the dried foam films. Based on nanoindentation measurements using AFM, the Young's modulus of DPC films was estimated as about 15 MPa [21], a value which ranged from one-tenth to one-hundredth of that of elastic polymers such as polyethylene, and three to four orders of magnitude greater than that of biomembranes. Whilst the above-mentioned deposition seems to be limited to amorphous materials, in the case of platinum or indium (both of which cause aggregation of the corresponding nanocrystals during the deposition process) the maximum thickness of the deposited layers was approximately 10 nm.

8.3.4
Foam Films of Ionic Liquids

Whilst, in general, a foam film is considered to be a water layer covered with two surfactant layers, it has been found that foam films can also be obtained from ionic liquids, if the correct surfactant is used [22]. One example of this is the combination of an ionic liquid, 1-ethyl-3-methyl-imidazolium, and a nonionic surfactant. Figure 8.8 shows the structural changes of foam films prepared by using the above-described ionic liquid and nonionic surfactant with a polyethylene

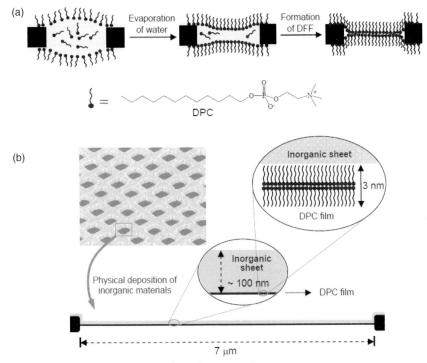

Figure 8.7 (a) Formation of a dried foam film (DFF) of dodecylphosphocholine (DPC); (b) Schematic diagram of an inorganic sheet deposited on a free-standing surfactant bilayer [20].

oxide chain (Brij-35) as the hydrophilic moiety (Figure 8.8). Initially, the foam film shows a rainbow color, but the interval of interference fringes then becomes larger with decreasing thickness and, after 50 min, a white region appears at the top of the film. The film thickness of this white region is approximately 200 nm. A foam film made from an ionic liquid has a high viscosity, so that the film can be transferred onto a porous substrate. On transfer, however, the film becomes increasingly thinner due to capillary forces from the inner wall of the pore, and finally the thickness becomes approximately 9 nm. Such a thin film may be considered as a Newton black film solvated with an ionic liquid.

Although foam films made from ionic liquids are liquid films, they are able to maintain their film structure even under ultrahigh-vacuum conditions. Furthermore, whilst the above-mentioned dried foam films are formed in micrometer-sized pores, foam films of ionic liquid may occur in sizes of at least several hundreds of micrometers. An important point here is that such films can be prepared by using a polymer surfactant. Figure 8.9 shows a Newton black film prepared from a triblock polymer, Pluronic F127, and an ionic liquid, [EMIM][BF$_4$]. This surfactant polymer has a polypropylene glycol chain at the center, which acts

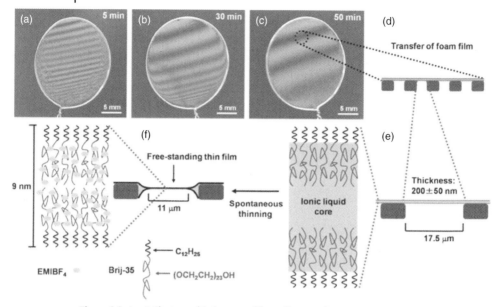

Figure 8.8 (a–c) Photographic images of foam films made from ionic liquid ([EMIM][BF$_4$]) and nonionic surfactant (Brij-35); (d,e) The film is transferred onto a porous substrate; (f) Spontaneous thinning to ultrathin free-standing films [22].

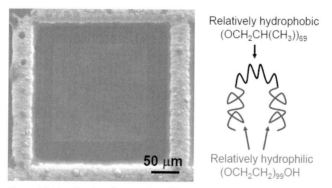

Figure 8.9 An ultrathin free-standing film made from a surfactant polymer (Pluronic F127) and an ionic liquid ([EMIM][BF$_4$]) [22].

as a hydrophobic moiety, while the PEG chains at both ends are hydrophilic and interact strongly with the ionic liquid. The obtained film covers (homogeneously) a pore of $250 \times 250\,\mu m^2$, and in this case the film thickness was 19 nm. The excess polymers tended to accumulate around the periphery of the film, where they increased the film thickness.

As is normally associated with soap bubbles, general foam films disappear as they are dried in air; however, this problem can be solved by using a nonvolatile ionic liquid. Furthermore, comparatively large ultrathin free-standing films can be obtained by using surfactant polymers and, when in time their mechanical strength is improved, these films will have great potential as separation membranes. The formation of foam films represents an attractive fabrication process, in which nanometer-thick free-standing films can be obtained from solution. Yet, in the future it will most likely be possible to design a wide range of nanostructures by using chemical methods.

8.4
Free-Standing Thin Films Prepared with Solid Surfaces

Free-standing films with a thickness of several nanometers may be fabricated using lipid bilayer techniques. Miyashita and coworkers successfully fabricated free-standing films of bilayer thickness (3.3 nm) by using vinyl polymers with long alkyl side chains [23]. These ultrathin polymer films were designed to be stabilized by the hydrogen bonds of hydrophilic polymer chains. The free-standing films were prepared by transferring the polymeric monolayer formed at the air–water interface onto a solid substrate that had been coated with a sacrificial cellulose acetate layer, and then dissolving the sacrificial layer with acetone. The films which were peeled from the substrates were readily transferred onto a TEM grid with 100 μm pores. The air–water interface was also used to prepare inorganic free-standing films. For example, Ozin *et al.* synthesized mesoporous silica films by using an acidic aqueous solution of tetraethoxysilane and a cationic surfactant [24]. After aging the solution at 80 °C over a reaction time of minutes to days, both free-standing and oriented mesoporous films were formed spontaneously at the air–water interface.

Free-standing films of several tens of nanometers thickness can also be obtained by means of a layer-by-layer (LBL) adsorption of oppositely charged polyelectrolytes [25]. This technique requires films to be produced on a smooth substrate and then to be peeled from the substrate, although on occasion the peeling procedure may cause damage to the ultrathin films. Consequently, rigid components are often used as a component of free-standing thin films. For example, Kotov *et al.* fabricated robust free-standing thin films from a combination of chemically modified water-soluble CNTs and various polyelectrolytes. The maximum tensile strength of the films produced was 200–300 MPa, this being approximately 10-fold higher than that of ordinary polymer/CNT composites [26]. A method for covering pores that were approximately 100 μm in diameter with free-standing thin films by using an LBL technique was reported by Mallwitz and Laschewsky [27]. These authors alternately immersed a porous substrate into aqueous solutions of polyacrylic acid and polyallylamine but, for the first three cycles, omitted the rinse and drying procedures usually required for LBL adsorption. In this way, pores with a size of $100 \times 100\,\mu m^2$ could be covered with a polymer film of about 40 nm thickness. It was also possible to chemically stabilize the polymer films by heat treatment at 120 °C, as the polyacrylic acid and polyallylamine were crosslinked with each other.

This technique should be especially interesting because it allows free-standing thin films to be obtained without the use of sacrificial layers.

In a related investigation, Kunitake et al. reported the preparation of robust free-standing thin films by using a spin-coating process. Here, zirconium butoxide ($Zn(O^nBu)_4$) and two types of acrylic monomer were spin-coated onto a substrate with a sacrificial layer, followed by irradiation with ultraviolet light and the recovery of a polymerized layer in ethanol. In this way, robust free-standing films with a thickness of 35 nm and a width of several centimeters were obtainable [28]. Moreover, the films were very soft in texture, in spite of a maximum tensile strength of 100 MPa.

8.5
Free-Standing Thin Films of Nanoparticles

Free-standing thin films of polyelectrolytes prepared using LBL techniques are relatively unstable against strong acids and bases. In addition, because the films are assembled by electrostatic interactions, they often break when other strong polyelectrolytes such as sodium polystyrene sulfonate are added. Consequently, these films generally need to be stabilized by crosslinking. However, it seems that comparatively stable free-standing thin films can be obtained by using colloid particles as a component of the alternate LBL assembly. Tsukruk et al. prepared free-standing thin films with a thickness of 25 to 70 nm by alternately assembling gold nanoparticles and polyelectrolyte multilayers on a sacrificial layer and subsequently peeling them from the substrate (Figure 8.10) [29]. Despite a thickness of several tens of nanometers, the films were obtainable in a size range of millimeters to centimeters; moreover, the films had a Young's modulus of 8 ± 3.5 GPa – an unprecedented high value for nanocomposite thin films. When Tsukruk et al. investigated the deformation of free-standing thin films under applied pressure (4 kPa), they found that a free-standing film covering a pore of 600 μm diameter possessed remarkable self-restoration properties. Specifically, when the films were expanded under pressure, they returned to their original flat form in approximately 10 s.

The free-standing thin films of nanoparticles reported by Jaeger et al. also showed interesting mechanical properties [30]. Here, the surfaces of gold nano-

Figure 8.10 Schematic illustration of a free-standing film composed of gold nanoparticles and polyelectrolyte multilayers. Reproduced with permission from Ref. [29]; © 2004, Nature Publishing Group, London.

(a)

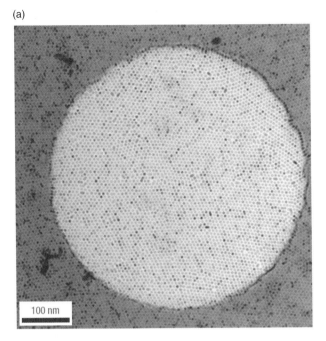

(b)

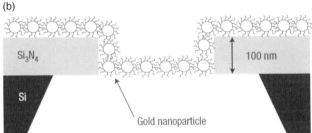

Figure 8.11 (a) Free-standing array of surfactant-modified gold nanoparticles covering a 500 nm pore in diameter; (b) Schematic diagram of the array configuration inside a single pore. Reproduced with permission from Ref. [30].; © 2007, Nature Publishing Group, London.

particles of 6 nm diameter were first modified with dodecanethiol, after which a small volume of the gold nanoparticle solution in toluene was spread on a water droplet. During evaporation of the toluene, the nanoparticles self-assembled at the air–water interface to produce a densely packed, two-dimensional array (Figure 8.11). A nanoparticle sheet prepared in this way could be transferred onto a mesh of TEM microgrid. This single nanoparticle sheet was approximately 9.4 nm thick, such that nanoparticles of 6 nm diameter could be coated with a dodecanethiol layer of 1.7 nm thickness. Such films show a Young's modulus of several GPa, a rubber-like flexibility, and a thermal stability of approximately 100 °C. These robust and elastic films may be used as thin-film sensors.

Recently, it has become clear that nanoparticles have similar functions to those of the surfactant. The nanoparticles stabilize an oil–water interface or an air–water interface to form an emulsion, foam, or powder containing water [31, 32], while the characteristics of the surfactant appear in both hydrophilic and hydrophobic nanoparticles. Free-standing thin films of nanoparticles are mechanically more stable than thin films of the surfactant; for example, the above-mentioned dried foam films (free-standing bilayers of surfactant) have an elastic modulus of several tens of MPa. Rigid bilayers may be obtained from the salts of anionic and cationic surfactant molecules (a so-called "catanionic" surfactant), and these have elastic moduli in the region of 100 MPa [33]. In contrast, free-standing thin films of nanoparticles have moduli that are 10- to 100-fold larger than those of surfactant-based, free-standing thin films. The production of large-area, free-standing thin films of nanoparticles was reported by Xia and Wang [34]. Although, in general, rigid nanoparticle films are not suitable for scaling-up, it seems possible to form free-standing thin films with areas of up to several square centimeters, and a thickness of approximately 12 nm.

8.6
Nanofibrous Free-Standing Thin Films

8.6.1
Electrospinning and Filtration Methods

Nanofibrous free-standing thin films are usually fabricated by using an electrospinning technique where a polymer solution is poured into a syringe-like container and a high voltage is applied. When the polymer solution is ejected through container nozzle to a grounded electrode (usually a metallic plate), the solution flow becomes extra fine, the solvent evaporates simultaneously, and a nanofibrous sheet is formed on the electrode [35]. One advantage of this technique is that various nanofibrous free-standing thin films can be formed by mixing different components into polymer solutions.

Free-standing thin films can also be fabricated by the filtration of nanofibrous materials, a method that is similar to the preparation of paper using cellulose microfibers. Wu et al. fabricated nanofibrous free-standing thin films from aqueous dispersions of single-walled carbon nanotubes (SWNTs) (Figure 8.12) [36]. When prepared using a pulsed laser, the SWNTs may contain many impurities, but can be purified by refluxing in aqueous nitric acid for a few days, followed by repeated filtration. The surfaces of the purified nanotubes become partially oxidized, which allows them to be dispersed in water, using a surfactant. The nanotubes were filtered on an esterified cellulose filter, washed thoroughly, and immersed in acetone to dissolve the cellulose filter. The result was a transparent free-standing thin film with electrical conductivity, a thickness of ≥ 50 nm, and a diameter in

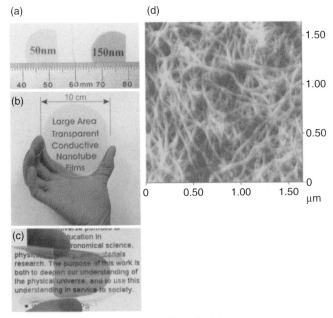

Figure 8.12 (a) Transparent SWNT films; (b) A large 80 nm-thick SWNT film on a sapphire substrate; (c) The same film on a flexible sheet; (d) AFM image of a SWNT film surface. Reproduced with permission from Ref. [36]; © 2004, American Association for the Advancement of Science.

excess of 10 cm. Whilst the transmittance of visible light was similar to that of indium tin oxide (ITO; a conductive metal oxide thin film), that of infrared light was considerably greater. The free-standing thin films of CNTs, which are known as "buckypaper," are said to be used in spacecraft and next-generation electronics.

Filtration represents a simple and excellent method for fabricating free-standing thin films with a large area. Although the film thickness is well controlled by the volume of the nanofiber dispersions to be filtered, the filtration method requires nanofibrous materials with a high aspect ratio, such as CNTs. Notably, in order to prepare ultrathin nanofibrous films, extremely fine nanofibers and a high dispersibility in solvent are required. In addition, the nanofibers must be rigid and sufficiently tough (in mechanical terms) to undergo treated with suction filtration. The SWNTs satisfy all of these requirements.

8.6.2
Metal Hydroxide Nanostrands

Recently, extremely fine nanofibers that were very similar to SWNTs were identified and referred to as "metal hydroxide nanostrands" [37]. For example,

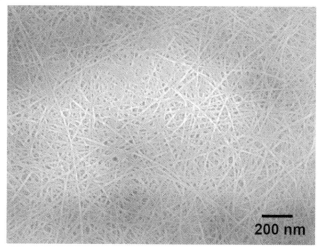

Figure 8.13 Scanning electron microscopy image of copper hydroxide nanostrands. This image was obtained after depositing 2 nm-thick platinum [38].

when a dilute copper nitrate solution ($Cu(NO_3)_2$; 1.6 mM) is mixed with aqueous ethanolamine to create a neutral pH and aged for one day, remarkably long and thin nanofibers of copper hydroxide are formed. Indeed, their length exceeds several tens of micrometers despite a diameter of only approximately 2.5 nm [38] that is only slightly larger than that of SWNTs (0.7–2.0 nm). A scanning electron microscopy (SEM) image of the nanofibers of copper hydroxide is shown in Figure 8.13. Similar nanofibers were obtained by mixing dilute aqueous solutions of sodium hydroxide and cadmium nitrate; the resultant nanofibers of cadmium hydroxide grew rapidly, producing a length of several micrometers after only 10 min. Therefore, the nanofibers were discovered at an earlier time [37]. The diameter of the cadmium hydroxide nanofibers, at 1.9 nm, was very similar to the 2.0 nm diameter of double-stranded DNA (ds-DNA). These nanofibers could also be obtained as an aqueous dispersion, and behaved in similar fashion to an organic polyelectrolyte. Because their colloidal properties were very close to those of ds-DNA, these nanofibers were named "cadmium hydroxide nanostrands." Subsequently, similar nanostrands of zinc hydroxide were prepared, with widths ranging from 2 to 2.5 nm (these varied according to the viewing angle) [39]. Zinc is a biologically essential element, the recommended daily intake of which is 10 mg for an adult male human. The zinc hydroxide nanostrands were shown to dissolve readily in water, and to form complexes with amino acids and other ligands. For a combination of these reasons, they may be referred to as "nontoxic" nanomaterials.

The most important feature of metal hydroxide nanostrands is the fact that they are extremely positively charged. Based on the results of adsorption experiments with negatively charged dyes, approximately one-sixth of the cadmium atoms that comprise cadmium hydroxide nanostrands are thought to be positively charged; this corresponds to about one-third of the surface cadmium atoms. Similarly, approximately half of the copper atoms at the surfaces of copper hydroxide nanostrands appear to be positively charged. The structure of a zinc hydroxide nanostrand is shown in Figure 8.14; this nanostrand is constructed from hexagonal clusters that are stacked in the direction of the c-axis, and approximately one-fourth of the zinc atoms at the surfaces are positively charged. Because of these abundant positive charges, the metal hydroxide nanostrands can maintain a distance from each other, and can be uniformly dispersed in water.

The above-mentioned nanostrands are formed in dilute aqueous solutions of copper (also cadmium and zinc) nitrate at near-neutral pH. The typical concentrations are 2.0 mM for the copper ion and 0.8 mM for the base, under which conditions the nanostrands appear to be obtained almost stoichiometrically with respect to the added base. In other words, if the base concentration is 0.8 mM, approximately 0.4 mM of copper ions will be converted to nanostrands in water. The coexistence of many copper ions that remained unchanged might be related to the fact that the nanostrands carry remarkable positive charges. Whilst at present this is speculation, these nanostrands are the thinnest structures known to be capable of existing in water; in fact, if they were to become much thinner they might decompose to hydrated metal ions. Such speculation would, however, explain why these nanostrands have a constant width.

8.6.3
Nanofibrous Composite Films

Metal hydroxide nanostrands may be readily filtered on a membrane filter with pores of a few hundred nanometers, to produce ultrathin, nanofibrous, free-standing films. Unlike CNTs, the nanostrands do not form bundle-like assemblies, and consequently the nanofibrous films are much thinner and more dense than films formed from CNTs. It is also possible that the free-standing films of metal hydroxide nanostrands are composed of ultrafine fibers that are 1000- to 10 000-fold thinner than the cellulose fibers that constitute normal filter paper, the fibers of which have widths of several tens of micrometers. As many pores of several tens of nanometers in size are formed in the gaps of the nanostrands, the latter might have a potential application in air filters for the removal of viral particles. Likewise, because the nanostrands have considerable positive charges on their surfaces they would adsorb anionic dyes, proteins, nanoparticles, and many other compounds with negative charges in water, to form the corresponding nanocomposite fibers. Free-standing thin films can be also obtained from these fibers by filtration. A TEM image of nanocomposite fibers of cadmium hydroxide nanostrands and gold nanoparticles is shown in Figure 8.15b. Here, gold nanoparticles of 20 nm diameter were adsorbed electrostatically onto the nanostrands to produce a weakly gelled solution that could be filtered to yield free-standing thin films of the nanoparticles. An SEM image of the nanocomposite fibers is shown in Figure 8.15a. Similar free-standing thin films may also

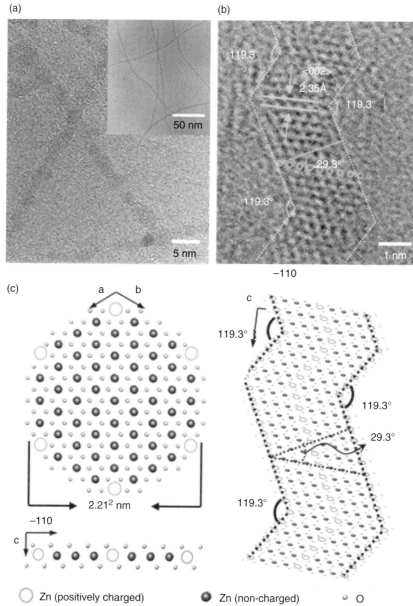

Figure 8.14 (a,b) High-resolution transmission electron microscopy images of zinc hydroxide nanostrands; (c) The proposed crystal structure [39].

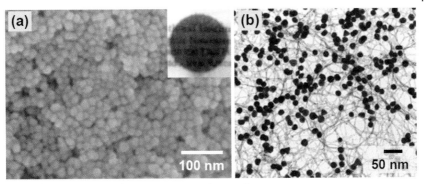

Figure 8.15 (a) Scanning electron microscopy image of free-standing nanocomposite films prepared with 20 nm gold nanoparticles and cadmium hydroxide nanostrands; (b) TEM image of the corresponding nanocomposite fibers. The inset in panel (a) shows a photographic image of the free-standing nanocomposite film [40].

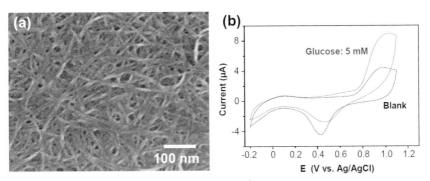

Figure 8.16 (a) Scanning electron microscopy image of a free-standing glucose oxidase/nanostrand composite film; (b) The electrochemical properties of the composite film [40].

be fabricated from a combination of nanostrands and proteins such as cytochrome c, ferritin, and glucose oxidase (GOx). It has been revealed from electrochemical experiments that GOx, when embedded in a free-standing nanocomposite film, has a similar enzymic activity to the native enzyme. An SEM image of a nanocomposite film formed from cadmium hydroxide–GOx composite nanofibers is shown in Figure 8.16a. Here, the free-standing thin film was transferred onto a gold electrode and the redox properties were measured. The current–voltage curves are shown in Figure 8.16b. The current response of the proteins, when embedded into the film, was significantly amplified by the presence of 5 mM glucose, which shows that glucose was oxidized and hydrogen peroxide generated at the electrode.

8.6.4
Nanoseparation Membranes

Although, in general metal hydroxide nanostrands are not thermally stable, this problem was recently overcome by uniformly coating the surfaces with conjugated polymers [41]. As an example, aminoethanol was added to an aqueous solution of copper nitrate, after which the mixture was aged for several days to form the developed nanostrands. A very small amount of sodium polystyrene sulfonate was then added, after which pyrrole and ammonium peroxydisulfate (as oxidant) were added, and the mixture again aged for several days. By adopting these procedures the copper hydroxide nanostrands were coated with an extremely thin layer of polypyrrole. As shown in Figure 8.17b, the thickness of the coating layer was approximately 3 nm, which was very similar to the width of the nanostrand (2.5 nm). Because the core–shell nanocomposite fibers are obtained as an aqueous dispersion, it is possible to prepare the nanofibrous free-standing thin films by filtration; such a film is shown in Figure 8.17a. As polypyrrole is chemically stable, the free-standing thin films of the nanocomposite fibers were also stable over a wide range of pH-values, and at temperatures up to at least 300 °C. As seen in the SEM image (Figure 8.17a), the surfaces of the free-standing films included pores that were several tens of nanometers in diameter, although because the nanocomposite fibers overlapped each other the films were able to block water-soluble proteins of several nanometers diameter. The UV-visible spectra of a dilute myoglobin solution, before and after passing through this free-standing thin film, are shown in Figure 8.18. Based on the peak intensities, approximately 90% of the myoglobin (particle size $2.5 \times 3.5 \times 4.5 \, nm^3$) was filtered out. One surprising point was the ultrafast permeation of water through the nanofibrous membrane. Although in the above-mentioned myoglobin filtration the rejection was 90%, the

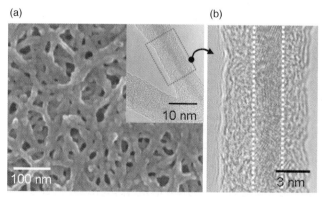

Figure 8.17 (a) Scanning electron microscopy image of nanofibrous free-standing film of copper hydroxide nanostrand coated with polypyrrole; (b) High-resolution transmission electron microscopy images of the nanocomposite fibers [also inset to panel (a)] [41].

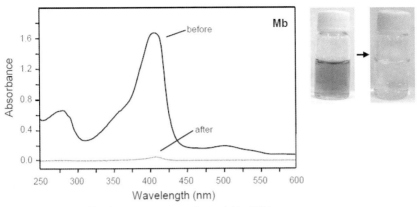

Figure 8.18 UV-visible absorption spectra of myoglobin (Mb) solution before and after filtration [41].

filtration rate was $5900 \, l \, h^{-1} \, m^{-2} \, bar^{-1}$, which was approximately two orders of magnitude greater than that of normal ultrafiltration membranes. This effect was achieved because the thickness of the free-standing thin films was only 40 nm. The free-standing films of polypyrrole-coated nanostrands were very tough, and the applied pressure was not less than 100 kPa if the film was placed on a porous aluminum membrane with 0.2 μm pores. As a result, a very rapid ultrafiltration could be achieved.

8.7
Conclusions

In this chapter, we have introduced new fabrication methods for nanostructured free-standing thin films. In the past, free-standing films have been produced conventionally by either the casting or spin-coating of polymer solutions, or by the spreading of a polymer solution onto an air–water interface. Problems were encountered, however, when attempting to obtain free-standing thin films with thicknesses of only several hundreds of nanometers, or less. Previously, inorganic free-standing thin films were formed by thermal deposition on a substrate, after which the substrate had to be etched. Unfortunately, this method proved to be very expensive and difficult to carry out when fabricating free-standing thin films with a large surface area. In order to overcome such problems, mechanically stable free-standing thin films have been designed and new processes developed. At the same time, the fabrication of nanothickness, nonwoven fabrics became possible by using long and thin fibers such as CNTs. This advance was mainly due to the discovery of new inorganic nanofibers and to improvements in the fabrication process. In particular, because metal hydroxide nanostrands can be produced only by mixing metal nitrate and alkali in water, this has led to a general method for the production of nanothickness, nanofibrous, free-standing films.

These free-standing thin films may be rapidly incorporated into applications such as water purification and, because the mechanical strength has been improved and the films have become very thin, such that filtration rates have been improved by two to three orders of magnitude. In future, the separation performance of nanometer-thick free-standing films will be further improved by controlling not only the film's pore size but also its affinity towards molecules and macromolecules. In particular, the high-speed separation of biomacromolecules such as proteins and polysaccharides might prove to be a major innovation in the fields of medicine and food production. It is also clear that nanostructured free-standing thin films will make significant contributions to the separation of materials such as water, ethanol, methane, and carbon dioxide, which are considered to be the "basic components" of the 21st century. Clearly, the main future target is to develop the excellent performance of nanoseparation membranes for small molecules, as this will undoubtedly have a major social impact in terms of energy-related and environmental problems.

References

1 Baker, R.W. (2004) *Membrane Technology and Applications*, 2nd edn, John Wiley & Sons, Inc., Chichester.

2 Li, K. (2007) *Ceramic Membranes for Separation and Reaction*, John Wiley & Sons, Ltd, Chichester.

3 Schäfer, A.I., Fane, A.G. and Waite, T.D. (2005) *Nanofiltration: Principles and Applications*, Elsevier, Amsterdam.

4 Mulder, M. (1996) *Basic Principles of Membrane Technology*, 2nd edn, Kluwer Academic Publishers, Dordrecht.

5 Striemer, C.C., Gaborski, T.R., McGrath, J.L. and Fauchet, P.M. (2007) Charge- and size-based separation of macromolecules using ultrathin silicon membranes. *Nature*, **445**, 749–53.

6 Van den Berg, A. and Wessling, M. (2007) Silicon for the perfect membrane. *Nature*, **445**, 726.

7 Holt, J.K., Park, H.G., Wang, Y., Stadermann, M., Artyukhin, A.B., Grigoropoulos, C.P., Noy, A. and Bakajin, O. (2006) Fast mass transport through sub-2-nanometer carbon nanotubes. *Science*, **312**, 1034–7.

8 Hinds, B.J., Chopra, N., Rantell, T., Andrews, R., Gavalas, V. and Bachas, L.G. (2004) Aligned multiwalled carbon nanotube membranes. *Science*, **303**, 62–5.

9 Preston, G.M., Carroll, T.P., Guggino, W.B. and Agre, P. (1992) Appearance of water channels in *Xenopus* oocytes expressing red cell CHIP28 protein. *Science*, **256**, 385–7.

10 Zeidel, M.L., Ambudkar, S.V., Smith, B.L. and Agre, P. (1992) Reconstitution of functional water channels in liposomes containing purified red cell CHIP28 protein. *Biochemistry*, **31**, 7436–40.

11 Rodriguez, M.S., Dargemont, C. and Stutz, F. (2004) Nuclear export of RNA. *Biology of the Cell*, **96**, 639–55.

12 Alber, F., Dokudovskaya, S., Veenhoff, L.M., Zhang, W., Kipper, J., Devos, D., Suprapto, A., Karni-Schmidt, O., Williams, R., Chait, B.T., Sali, A. and Rout, M.P. (2007) The molecular architecture of the nuclear pore complex. *Nature*, **450**, 695–701.

13 Castellana, E.T. and Cremer, P.S. (2006) Solid supported lipid bilayers: from biophysical studies to sensor design. *Surface Science Reports*, **61**, 429–44.

14 Mueller, P., Rudin, D.O., Tien, H.T. and Wescott, W.C. (1962) Recognition of cell membrane structure *in vitro* and its transformation into excitable system. *Nature*, **194**, 979–80.

15 Tien, H.T. and Ottova-Leitmannove, A. (2000) *Membrane Biophysics: As Viewed from Experimental Bilayer Lipid Membranes (Planar Lipid Bilayers and Spherical Liposomes)*, Elsevier, Amsterdam.

16 Exerowa, D. and Kruglyakov, P.M. (2000) *Foam and Foam Films: Theory, Experiment, Application*, Elsevier, Amsterdam.

17 Lyklema, J. (2005) *Fundamentals of Interface and Colloid Science*, Vol. 5, Elsevier, Amsterdam.

18 Bélorgey, O. and Benattar, J.J. (1991) Structural properties of soap black films investigated by X-ray reflectivity. *Physical Review Letters*, **66**, 313–16.

19 Jin, J., Huang, J. and Ichinose, I. (2005) Dried foam films: self-standing, water-free, reversed bilayers of amphiphilic compounds. *Angewandte Chemie, International Edition*, **44**, 4532–5.

20 Jin, J., Wakayama, Y., Peng, X. and Ichinose, I. (2007) Surfactant-assisted fabrication of free-standing inorganic sheets covering an array of micrometre-sized holes. *Nature Materials*, **6**, 686–91.

21 Jin, J., Sugiyama, Y., Mitsui, K., Arakawa, H. and Ichinose, I. (2008) Nanomechanical properties of reversed surfactant bilayers formed in micrometre-sized holes. *Chemical Communications*, 954–6.

22 Bu, W., Jin, J. and Ichinose, I. (2008) Foam films obtained with ionic liquid. *Angewandte Chemie, International Edition*, **44**, 902–5.

23 Endo, H., Mitsuishi, M. and Miyashita, T. (2008) Free-standing ultrathin films with universal thickness from nanometer to micrometer by polymer nanosheets assembly. *Journal of Materials Chemistry*, **18**, 1302–8.

24 Yang, H., Coombs, N., Sokolov, I. and Ozin, G.A. (1996) Free-standing and oriented mesoporous silica films grown at the air-water interface. *Nature*, **381**, 589–92.

25 Decher, G. and Schlenoff, J.B. (2003) *Multilayer Thin Films: Sequential Assembly of Nanocomposite Materials*, Wiley-VCH Verlag GmbH, Weinheim.

26 Mamedov, A.A., Kotov, N.A., Prato, M., Guldi, D.M., Wicksted, J.P. and Hirsch, A. (2002) Molecular design of strong single-wall carbon nanotube/polyelectrolyte multilayer composite. *Nature Materials*, **1**, 190–4.

27 Mallwitz, F. and Laschewsky, A. (2005) Direct access to stable, freestanding, polymer membranes by layer-by-layer assembly of polyelectrolytes. *Advanced Materials*, **17**, 1296–9.

28 Vendamme, R., Onoue, S.-Y., Nakao, A. and Kunitake, T. (2006) Robust free-standing nanomembranes of organic/inorganic interpenetrating networks. *Nature Materials*, **5**, 494–501.

29 Jiang, C., Markutsya, S., Pikus, Y. and Tsukruk, V.V. (2004) Freely suspended nanocomposite membranes as highly sensitive sensors. *Nature Materials*, **3**, 721–8.

30 Mueggenburg, K.E., Lin, X.-M., Goldsmith, R.H. and Jaeger, H.M. (2007) Elastic membranes of close-packed nanoparticles arrays. *Nature Materials*, **6**, 656–60.

31 Herzig, E.M., White, K.A., Schofield, A.B., Poon, W.C.K. and Clegg, P.S. (2007) Bicontinuous emulsions stabilized solely by colloidal particles. *Nature Materials*, **6**, 966–71.

32 Binks, B.P. and Murakami, R. (2006) Phase inversion of particle-stabilized materials from foams to dry water. *Nature Materials*, **5**, 865–9.

33 Dubois, M., Demé, B., Gulik-Krzywicki, T., Dedieu, J.-C., Vautrin, C., Désert, S., Perez, E. and Zemb, T. (2001) Self-assembly of regular hollow icosahedra in salt-free catanionic solutions. *Nature*, **411**, 672–5.

34 Xia, H. and Wang, D. (2008) Fabrication of macroscopic freestanding films of metallic nanoparticle monolayers by interfacial self-assembly. *Advanced Materials*, **20**, 4253–6.

35 Greiner, A., Wendorff, J.H. and Ramakrishna, S. (2009) *Electrospinning: A Versatile Route to Nanofibers*, Wiley-VCH Verlag GmbH, Weinheim.

36 Wu, Z., Chen, Z., Du, X., Logan, J.M., Sippel, J., Nikolou, M., Kamaras, K., Reynolds, J.R., Tanner, D.B., Hebard, A.F. and Rinzler, A.G. (2004) Transparent, conductive carbon nanotube films. *Science*, **305**, 1273–6.

37 Ichinose, I., Kurashima, K. and Kunitake, T. (2004) Spontaneous formation of cadmium hydroxide nanostrands in water. *Journal of the American Chemical Society*, **126**, 7162–3.

38 Luo, Y.-H., Huang, J., Jin, J., Peng, X., Schmitt, W. and Ichinose, I. (2006) Formation of positively charged copper hydroxide nanostrands and their structural characterization. *Chemistry Materials*, **18**, 1795–802.

39 Peng, X., Jin, J., Kobayashi, N., Schmitt, W. and Ichinose, I. (2008) Time-dependent growth of zinc hydroxide nanostrands and their crystal structure. *Chemical Communications*, 1904–6.

40 Peng, X., Jin, J., Ericsson, E.M. and Ichinose, I. (2007) General method for ultrathin free-standing films of nanofibrous composite materials. *Journal of the American Chemical Society*, **129**, 8625–33.

41 Peng, X., Jin, J. and Ichinose, I. (2007) Mesoporous separation membranes of polymer-coated copper hydroxide nanostrands. *Advanced Functional Materials*, **17**, 1849–55.

9
Dip-Pen Nanolithography of Nanostructured Thin Films for the Life Sciences

Euiseok Kim, Yuan-Shin Lee, Ravi Aggarwal, and Roger J. Narayan

9.1
Introduction

The growth of nanotechnology during the past decade has led to the development of different methods for the fabrication of nanostructures. In particular, nanolithographic methods such as electron beam lithography, nanoimprinting and tip-based lithography have enabled the fabrication of features as small as 10 nm. This, in turn, has opened up the possibility of developing both microbioarrays and nanobioarrays, which can revolutionize not only medical diagnostics but also areas such as proteomics [1]. In general, small-sized biopatterning provides more sensitivity and a higher throughput for protein analysis and medical diagnostics, because a large number of targets can be evaluated in a more rapid manner. In addition, it can provide significantly lower limits of detection [2]. Whilst most currently available protein arrays are generated using spotting methods (see Figure 9.1), these methods cannot meet the demand for smaller feature sizes. Hence, there has been increasing interest in adopting nanolithography methods for micro- and nanobiopatterning because of the improved resolution that these methods provide over current techniques, including ink-jet printing and the ring and pin method.

A variety of substrates such as metals, silicon and glass, which often are coated with thin films, have been used for biopatterning using nanolithography. The surface of these substrates is sometimes modified in order to improve or facilitate the binding and immobilization of the biomaterials for subsequent applications [3].

Each nanolithography method has its own advantages and disadvantages for life sciences applications. For example, whereas some methods can be used to deposit biomaterials directly onto thin substrates, others can deposit biomaterials only indirectly by using a prepatterning for self-assembly. In addition, different process conditions may be required for the different methods, and the resolution of the patterning may also differ. Consequently, the selection of a nanolithographic method for biopatterning requires careful consideration of all these aspects. Today, several reviews are available that describe nanolithography and biopatterning, and

Nanomaterials for the Life Sciences Vol.5: Nanostructured Thin Films and Surfaces.
Edited by Challa S. S. R. Kumar
Copyright © 2010 WILEY-VCH Verlag GmbH & Co. KGaA, Weinheim
ISBN: 978-3-527-32155-1

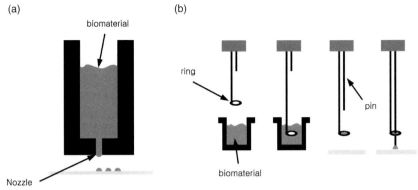

Figure 9.1 A description of the current methods used to generate microarrays. (a) Inkjet printing method; (b) Ring and pin method.

which discuss the advantages and disadvantages of nanolithographic methods for applications in the life sciences [4, 5].

Dip-pen nanolithography (DPN) is a relatively new, tip-based nanolithography method, which was first demonstrated in 1999 [6]. Since its discovery, DPN has been investigated for its applications in nanofabrication, for semiconductors, and also in the life sciences because of its simplicity and flexibility as a nanolithographic method. In fact, it has been shown that DPN can be used either directly or indirectly to fabricate bioarrays, and with a high process flexibility. In this chapter, attention will be focused on DPN processes and their application in the life sciences. Various approaches towards the fabrication of bioarrays, including direct and indirect DPN, together with details of the most important process parameters, and of several applications (with examples) will also be discussed.

9.2
Dip-Pen Nanolithography

Dip-pen nanolithography is a direct writing method in which the tip of an atomic force microscope is used to transfer molecules or nanoparticles to the surface of a substrate so as to create a desired pattern. In DPN, the material to be patterned is coated onto the microscope tip, and then transferred from the tip to the substrate surface through a water meniscus. This process can be used to create self-assembled monolayers (SAMs) on the substrate, with great selectivity. Although the DPN concept is similar to that of micro/nano-contact printing, it provides a greater flexibility because it allows control over the size of the pattern by manipulating both the tip speed and dwell time (see below). Indeed, this process can be used to pattern features with sub-100 nm resolution.

DPN has several advantages over other nanolithography methods. Notably, it involves neither expensive equipment nor a restricted environment (as does electron beam lithography), and can be carried out under ambient conditions. In contrast to micro-contact printing and nano-imprinting methods, it is not necessary to fabricate a pattern mold for DPN because the process itself can be used to control the size and shape of the pattern, via the tip movement. Whilst the tip movement speed will control the volume of ink that is transferred to the substrate, using the tip to transfer material does place certain limitations on the process. For example, the possible geometries generated by DPN are very limited because the diffusion rate of the ink coated onto the tip is the same in all directions on the substrate. This means that any polygonal pattern which includes sharp edges cannot be created using DPN. When the tip moves around in a polygonal area on the substrate, the resulting patterned shape may be a dot rather than a polygon, because of the diffusion. Hence, DPN is suitable for the patterning only of dots and lines which do not have sharp edges. The patterning of particle-based materials does not have this limitation, as it is not affected to any great degree by diffusion.

A second disadvantage of DPN is that it cannot support high process speeds because of its serial nature. For example, if a dot pattern has to be created, the microscope tip must move to every single dot and stay there for the same dwell time. As an example, if DPN was used generate a 50×50 dot array with a 10 s dwell time in an $80 \times 80\,\mu m^2$ substrate, it would take approximately 7 h; thus, the duplication of 10 patterns would take 70 h. To overcome this problem, a significant amount of research is currently being conducted in the development of a parallel DPN process. In one such effort, an AFM probe fitted with 55 000 tips over a 1 cm^2 area was developed for parallel writing, and achieved a very high throughput (writing rate $3 \times 10^7\,\mu m^2 h^{-1}$ [7]) was achieved. A parallel polymer pen probe, instead of a conventional silicon nitride pen, has also been developed for parallel DPN [8].

9.2.1
Important Parameters

The quality and resolution of patterns fabricated by DPN depends on many factors:

- **Chemical interaction:** The most important factor is the chemical interaction between the ink material and the substrates. Only if there is good ink–substrate interaction can DPN create a stable pattern. Although it is possible to transfer the ink from tip to substrate when the chemical interaction is poor, the pattern will not be stable as the ink will simply be placed on the substrate, without any bonding. Based on these chemical interactions, several combinations of inking material and substrate may be suitable for DPN [9].

- **Diffusion:** The diffusion of the ink represents another critical parameter that determines the resolution and reproducibility of the DPN patterns. Ink diffusion during DPN is controlled by many factors, including humidity, temperature, and substrate roughness. Among these factors humidity is dominant, as the formation and characteristics of the water meniscus between the microscope tip and substrate are dependent on humidity. However, the influence of humidity also depends on the ink material, and must be examined for each ink–substrate combination.

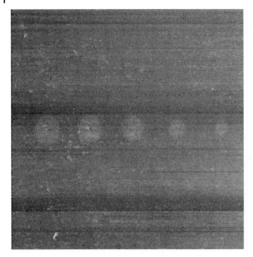

—— 2 μm

Figure 9.2 The effect of dwell time on the size of MHA dots on a gold-coated mica substrate. Dwell time from right to left: 40 s; 30 s; 20 s; 10 s; 5 s.

- **Tip size:** In order to achieve a high resolution of patterns, the tip size is an important factor, as a larger tip radius will mean a greater contact area being generated between the tip and the substrate, and hence poor resolution.
- **Surface roughness and grain size of the substrate:** Both of these factors may influence the resolution of the pattern, as well as its reproducibility.

Consequently, the best DPN patterns are achieved with optimum humidity, a small tip radius, and a smooth substrate surface, and with correct control of these parameters DPN can easily create sub-100 nm structures. Feature sizes generated by DPN can also be controlled by the dwell time and tip speed (see Figure 9.2):

- **Dwell time:** In order to create a dot, the tip is held at a desired point on the substrate for a certain time; this is known as the "dwell time." A long dwell time means a longer diffusion length and a larger diameter of the patterned dot. On the other hand, lines can be created by moving the tip at a certain speed; again, a slower tip speed will result in more time for diffusion and hence a higher width of the patterned lines. The best resolution achieved to date for DPN patterns has been ~15 nm, for the patterning of mercaptohexadecanoic acid (MHA) on gold substrates [10].

9.2.2
Applications of DPN

As discussed above, biomaterials such as proteins and viruses can be patterned using DPN for medical diagnostic and biochip/sensor applications [11, 12]. Recently,

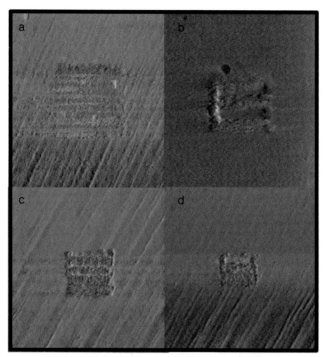

Figure 9.3 Atomic force microscopy images of deposited gold particle on modified silicon substrates using DPN. The sizes of the gold squares were: (a) $1.150 \times 1.150\,\mu m^2$; (b) $0.950 \times 0.950\,\mu m^2$; (c) $0.760 \times 0.760\,\mu m^2$; (d) $0.570 \times 0.570\,\mu m^2$.

DPN has also been used both directly and indirectly to pattern gold nanoparticles, magnetic nanoparticles, and polymers [13–15]. Figure 9.3 shows the atomic force microscopy (AFM) images of gold particles that have been directly patterned on a modified silicon substrate. Although nanoparticles of several materials can be patterned on suitable substrates, the binding strength between the nanoparticles and the substrates is not sufficiently strong for the DPN method to be adopted for applications such as the repair of masks or the creation of mask resists for nanofabrication. There are also certain issues associated with dimensional control in both horizontal and vertical directions. Notably, for the direct DPN patterning of nanoparticles it is almost impossible to control the pattern thickness; however, an indirect patterning based on DPN may be used to generate SAMs of nanoparticles.

Polymers and molten metals can also be patterned using electrochemical and thermal variants of DPN:

- In *electrochemical DPN*, a voltage is applied to the substrate through a cantilever tip during the DPN process, and this causes the surface of the substrate to be chemically modified. A pattern of polymer structures on silicon substrates has been successfully generated using electrochemical DPN [16]. This was achieved by spin-coating the substrate with a thin monomer layer, which was then selectively polymerized using electrochemical DPN.

- *Thermal DPN* uses heat rather than an electric voltage to modify the surface of the substrates. Thermal DPN has been used to pattern polymers [17] and metals [18], in which a thermally heated AFM cantilever tip will enable the modification of a precoated polymer layer on the substrates. Thermal DPN can also be used to deposit molten metals transferred through the heated cantilever to the substrate.

Typical structures created by the electrochemical and thermal DPN processes are shown in Figure 9.4. The primary advantage of these DPN methods is that the structures that they pattern are very stable.

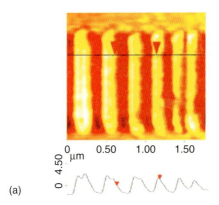

(a)

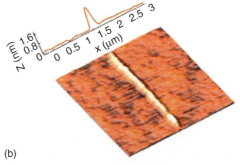

(b)

Figure 9.4 Nanostructures generated by electrochemical and thermal DPN. (a) Polymer structures on the silicon substrate generated by electrochemical DPN. Reproduced with permission from Ref. [16]; © 2006, American Chemical Society; (b) The metal structure on the glass substrate generated by thermal DPN. Reproduced with permission from Ref. [18].; © 2006, American Institute of Physics.

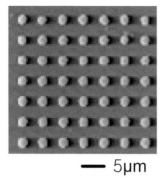

(a) Gold dot array on a mica substrate. Permission for this figure is being obtained.

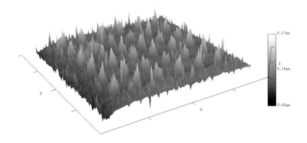

(b) Gold dot array on a silicon substrate. Permission for this figure is being obtained.

Figure 9.5 Atomic force microscopy images of a gold dot array on (a) a mica substrate and (b) a silicon substrate after wet etching.

Although, DPN cannot be used directly to fabricate three-dimensional (3-D) nanostructures, if it is combined with processes such as wet/dry etching, the two-dimensional (2-D) DPN patterns can easily be transferred into three dimensions. Hence, indirect DPN may represent a substitute for current nanolithography methods. The SAMs on substrates deposited by DPN can also serve as a mask for wet chemical etching. For example, monolayers of thiols patterned on gold substrates can protect a patterned area against a gold etchant, such that the thiols patterned on a gold substrate may serve as a positive mask. Figure 9.5 shows gold patterns on mica and silicon substrates after chemical etching; these were generated using DPN by patterning 16-MHA onto gold-coated substrates, followed by an iron-nitrate/thiourea etching. Both, positive [19] and negative [20] patterns created by DPN and chemical etching have been reported. Regardless of the area of application, the most important issue for DPN is to generate repeatable and high-resolution patterns; however, in order to achieve such a feat, all of the parameters that affect DPN must be carefully controlled.

9.3
Direct and Indirect Patterning of Biomaterials Using DPN

9.3.1
Background

When conducting DPN of biomaterials, there are two important issues: (i) that the biomaterials should maintain their native activity and properties after DPN patterning; and (ii) that the target materials are immobilized on the pattern area of substrate, using appropriate methods. In general, four methods can be used to create such immobilization, namely *electrostatic/hydrophobic interaction, physical entrapment, covalent binding,* and *biorecognition* [21] (for a visual description of these, see Figure 9.6). Electrostatic immobilization utilizes the positive and negative charge between substrates and target materials, while covalent bonding uses a common ion between the predeposited materials on a substrate and the target material. In contrast, biorecognition utilizes the protein–protein interaction of antibody and antigen. Among these methods, covalent binding and biorecognition are the most frequently used for biomaterial deposition, while the nonspecific adsorption of a material onto the substrate is used occasionally for physical entrapment. For example, hydrophobic interactions are commonly used in materials for microarray applications. Nonspecific interactions between the substrate materials and patterned materials may occur during the patterning process; however, the hydrophobic forces may sometimes cause denaturation of the proteins.

Two approaches, namely direct and indirect, are available for patterning biomaterials with DPN on the substrate:

- In *direct patterning,* a coated material on the microscope tip is deposited directly onto the substrate, without any post processing, with the tip carrying the target

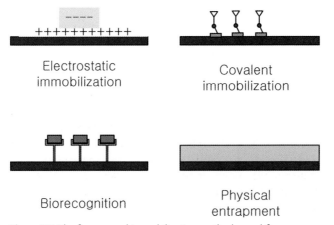

Figure 9.6 The four general immobilization methods used for bioarrays. Reproduced with permission from Ref. [21]; © 2006, Elsevier.

material(s) to the substrate. Whilst it is possible that biomaterials are attached via a surface modification of the tip, it is also necessary to generate a binding force between the deposited material and the substrate, usually with the aid of organic and chemical linkers for biomaterials. Following deposition of the linker layer, which can bind with both the substrate and target material, the biomaterial can be patterned onto the substrate using DPN.

- In *indirect patterning*, the biomaterial is not directly patterned using the microscope tip. Instead an organic or chemical linker – but not the target material – can be patterned, which in turn means that post-processing for the deposition of target materials is essential. When the substrate has been patterned with a binding material, the unpatterned area will be passivated against the target material, after which the substrate is either immersed in a target material (which is in a liquid state) or is incubated with a target material. The latter will then be deposited selectively on the patterned area of the substrate.

For both approaches, purification is an important issue when immobilizing a target material onto a substrate. In order to accomplish a high density of arrays within the restricted area, it is vital that the size of a dot should be minimized, and sub-100 nm-diameter protein dots of <10 nm thickness have been successfully generated in this respect [11, 22]. Several different types of substrate can be used to pattern biomaterials, including metals, glasses, and polymers. The direct and indirect patterning approaches for depositing biomaterials are discussed, and examples provided, in the following sections.

9.3.2
Direct Patterning

The direct patterning of biomaterials using DPN has opened up the possibility of fabricating complex biomaterial arrays for functional applications. The challenge remains during the DPN process, however, to transfer the biomaterial that has been coated onto the microscope tip to a substrate, without impairing biological activity [5]. Several different methods have been devised to overcome this problem. First, the target material can be supplied directly to, rather than coated on, the tip, using a process that resembles conventional ink-jet printing. In this respect, *nanofountain probes* (see Figure 9.7) have been created to deliver the target materials to the substrate during DPN [23]; these probes consist of a microfluidics set-up, a volcano tip, a microchannel network, and a reservoir. The reservoir holds a large amount of the target material, which it supplies to the tip through the microchannel network inside the probe cantilever. When compared to a conventional DPN process, where the microscope tip is coated with the target material, a large volume of material can be stored using this technique. By using the nanofountain probe method, DNA dot patterns with a minimum feature size of ~200 nm have been successfully created, without impairing the biological activity of the DNA. In order to improve the patterning rates, probes with multiple tips (up to 12) have been used to produce DNA arrays, with the feature size being controlled by varying the

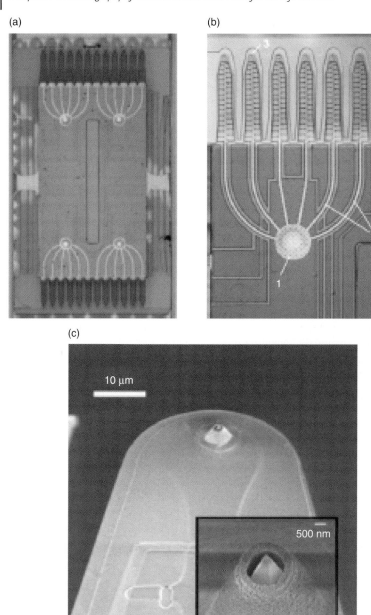

Figure 9.7 Description of the nanofountain probe. (a) The overall system; (b) The reservoir and microchannel network for multiple tips; (c) The volcano tip. Reproduced with permission from Ref. [23]; © Wiley-VCH Verlag GmbH & Co. KGaA.

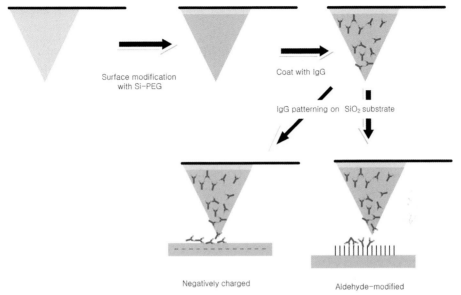

Figure 9.8 A description of the direct deposition of protein on SiO$_2$. For details, see the text. Reproduced with permission from Ref. [22]; © Wiley-VCH Verlag GmbH & Co. KGaA.

dwell time. When using this method it is very important to control the microfluidics within the probe in order to maintain a constant patterning rate.

A second approach is to chemically modify the surface of the microscope tip and/or the substrate surface, in order to deliver and pattern the target materials and achieve consistent and high-density bioarrays. In general, a silicon nitride tip cannot be wetted by biomaterials such as proteins and DNA; consequently, the tip must be modified by an appropriate chemical treatment to render it biocompatible and hydrophilic. Several methods have been introduced to coat the target materials onto microscope tips. In the case of a silicon nitride tip coated with 3-aminopropyltrimethoxysilane (MPTMS), DNA was successfully deposited directly onto a gold surface using DPN [24]. The deposition of sub-100 nm protein features has also been reported on silicon oxide surfaces [22]; this was achieved by modifying the cantilever with a coating of 2-[methoxypoly(ethyleneoxy)propyl] trimethoxysilane (Si-PEG), which easily adsorbs the protein.

As shown in Figure 9.8, two different approaches – namely, electrostatic interaction or covalent bonding between the protein and substrate – were used to create a stable pattern on the substrate. Protein dots with minimum size of ~55 nm have been patterned using this approach. Lee *et al.* used a coating of gold and mercaptoundecanoic acid (MUDA) on the microscope tip to increase the hydrophilicity, which in turn helped to maintain protein stability during the DPN process [25]. In this study, a gold-coated silicon substrate was coated with a commercial linker solution (Prolinker™) to facilitate linking of the proteins to the substrates. Following modification of the tip and substrate, the protein was successfully patterned onto the substrate using direct DPN. In general, the immobilization of biomaterials onto the substrate represents a common problem, regardless of the patterning method used. On the other hand, the adsorption of biomaterials onto the tip is an important issue only for the direct DPN method.

9.3.3
Indirect Patterning

As noted above, the indirect patterning of biomaterials by DPN involves the tip being used indirectly to pattern a biomaterial onto a substrate. In this case, DPN is used to create a pre-pattern for binding a target material onto the substrate, which in turn acts as a linker for a given biomaterial. Following this pre-patterning by DPN, the biomaterial is applied to the entire substrate, including the unpatterned region; thus, there is a possibility that the target material might also bind with the unpatterned zone of substrate. It follows that the passivation of unpatterned areas is necessary in order to prevent nonspecific binding. Many reports exist of the patterning of biomaterials using DPN in indirect mode, with one of earliest and best-known approaches being to create MHA patterns on the substrate and then to bind proteins using those MHA patterns [26]. In order to prevent any nonspecific binding of protein, 11-mercaptoundecyl-tri(ethylene glycol) was used for passivation. As illustrated schematically in Figure 9.9, anti-IgG and lysozyme have each been patterned successfully using this approach. Baserg *et al.* have also demonstrated the patterning of a DNA array on a gold substrate [27] by first creating an MHA pattern on the gold substrate, and then passivating the unpatterned area with 1-octadecanthiol (ODT). The final stage was to bind the DNA with the MHA pattern on the substrate. In these studies, a silver and gold template was fabricated to enable an easy retrieval of the DNA array; for this, the substrate consisted of a square gold pattern on a silver layer, with the DNA pattern being created inside the square area.

Other materials may also be patterned on gold substrates that can serve as templates for biomaterial arrays [11]. For example, 11-mercaptoundecanoyl-*N*-hydroxysuccinimide ester (NHSC11SH) was used to pattern template dots instead of MHA on a gold substrate. The 11-mercaptoundecyl-tril was used to passivate the unpatterned zone, and this was followed by an incubation of the substrate in protein A/G (a protein created using genetic engineering to enhance binding affinity) and antibody . For these studies, a multipen probe with 26 tips was used to achieve a high throughput, such that a high-density bioarray was fabricated. In most cases of indirect patterning, the unpatterned zone of the substrate should be passivated by chemical or other methods, following molecular patterning to create linkers for the biomaterial. In some cases, however, a physical change in the substrate surface can be used for fabricating bioarrays. For example, layers of heptylamine plasma-polymer (HApp) and diethylene glycol dimethyl ether plasma-polymer (DGpp) were deposited on a mica substrate [28]. Whilst HApp is known generally as a protein-adsorbing surface, DGpp is not; consequently,

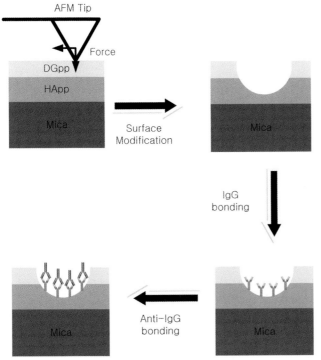

Figure 9.9 Protein patterning process using a modification of a multilayer substrate. Reproduced with permission from Ref. [28]; © Wiley-VCH Verlag GmbH & Co. KGaA.

HApp could act as a linker for the proteins and DGpp as a passivation layer to prevent any nonspecific binding of protein to the substrate. The top DGpp layer was removed physically at selected spots by mechanical plowing with the microscope tip. Following this physical change to the surface of the substrate, rabbit IgG protein was patterned in those areas which had been physically modified; finally, anti-IgG was selectively bound with rabbit IgG inside the patterned area. One important issue in the fabrication of biopatterns is the isolation of single proteins from complex proteins, which may be a time-consuming and labor-intensive process. Clearly, it would highly beneficial if such patterning could be achieved without a need for purification. Currently, most immobilization methods are incapable of linking certain biomaterials in selective fashion; however, one possible means to overcome this problem would be to carry out a selective immobilization during the patterning process, using a nitrilotriacetic acid (NTA)/Ni^{2+} pattern in order to bind His-tagged target protein [29]. For this, NTA was patterned using DPN onto a glass substrate coated with a MPTMS monolayer. The unpatterned area was subsequently passivated by polyethylene glycol (PEG)-maleimide to prevent nonspecific binding. With the NTA/Ni^{2+}-patterned substrate, the final linker pattern on the glass substrate could be generated. Finally, the patterned substrate was shown to bind target proteins from a protein complex. The patterning method, using NTA/Ni^{2+} pattern as linkers for His-tagged protein, is well known and had been reported previously [30]; however, the protein array had earlier been patterned on a silicon oxide (SiO$_2$) substrate using nanoimprint lithography.

In general, the indirect patterning process is simpler than its direct counterpart. Moreover, the indirect method does not necessitate the delivery of biomaterial using the tip because DPN is involved only in patterning the organic linkers for the final pattern. Therefore, a high density of bioactive biomaterials can be easily maintained during and after the patterning process. Despite these advantages, it is difficult to create patterns with multiple biomaterials on the same substrate so as to achieve a better functionality for advanced applications.

Although, in this chapter we have discussed only the basic methodology used to fabricate bioarrays via direct and indirect DPN methods, several other tip-based nanolithographic methods, including *AFM scratching* and *nanografting*, have been reported. *Nanografting* can be used to selectively replace the prepatterned bioarrays with different biomaterials [31], by using not only a direct contact between the tip and the substrates, but also self-assembly. Based on the basic methodology of patterning bioarrays, applications such as biochips, medical diagnostics and biosensors are discussed in the following sections.

9.4
Applications of DPN for Medical Diagnostics and Drug Development

Currently, the most promising application areas of biopatterning are those of proteomics and genomics, including protein and DNA structural/functional analysis. Yet, biopatterning can also be applied to medical diagnostics and drug development, to achieve less complexity and a high throughput. On examining the current blood test algorithms and microarray-based algorithms (see Figure 9.10), the simplicity of the latter over the former is very clear. For example, the microarray-based methods are cost-effective, based on the fact that the use of multiple instruments with dedicated operators and specific reagents can be largely eliminated. The microarray-based approach also provides a much greater flexibility, as the incremental cost of additional probes is low. Moreover, the need for repeat testing may be eliminated by using reaction patterns based on multiple probes per target [32]. As mentioned above, nano/micro bioarrays provide several unique advantages over the current technology. For example, a high throughput can be achieved with nano/micro bioarrays because high-density patterning within a relatively small area provides parallel analyses and a larger number of results. In addition, multiple tests can be substituted with a single micro/nano bioarray test. Moreover, each bioarray of proteins, DNA, or other small molecule, has its own unique applications [33]. Nano/micro bioarray testing is also suitable for investigating pharmacogenomics when testing new drug candidates for their efficacy, adverse side effects, and toxicity. In fact, the information obtained through medical screening

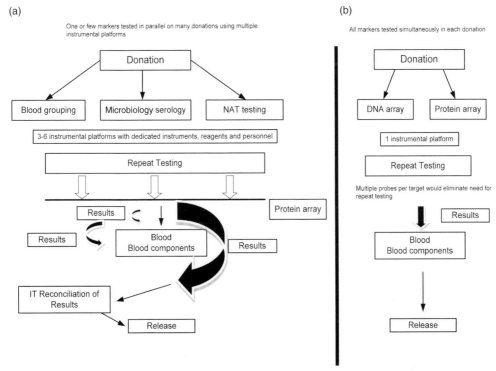

Figure 9.10 Blood testing algorithms. Reproduced with permission from Ref. [32].; © Wiley-VCH Verlag GmbH & Co. KGaA.

and diagnostics may lead to the discovery of a new drug for a certain type of disease.

9.4.1
General Methods of Nano/Micro Bioarray Patterning

Today, there is a strong move towards the miniaturization of diagnostic testing for diseases such as cancer and various allergies. Such miniaturization is driven not only by economic reasons but also by functional reasons such as simplicity, low sample volume, high throughput, and flexibility [34]. A variety of applications for micro/nano bioarrays have been developed to date, and some current applications are listed in Table 9.1. Overall, while microarrays are currently widely used for many medical applications, nanoarrays are still very much at the developmental stage.

In medical diagnostics with micro/nano arrays, an antibody can be used to detect a virus, a cancer, or an allergen [50]. In contrast, for drug testing and/or screening, the target virus, disease or cancer cells may be patterned and grown

Table 9.1 Applications of micro/nano bioarrays for medical diagnostic and drug development.

Application	Reference articles	
	Microarray	Nanoarray (using DPN)
Cancer diagnosis	Ovarian cancer [35] Breast cancer [36] Pancreatic cancer [37] Prostate cancer [38]	Prostate cancer [38]
Virus patterning/detection	Foot-and-mouth disease virus [39] Influenza virus [40] Plum pox virus [41]	rSV5-EGFP [42] CV1 [43] HRV [44]
Allergy detection	Glass/Tree pollen and latex allergen [45] Orchard grass, cow milk, egg white [46]	–
Drug development	Drug toxicity [47, 48] Drug testing [49]	–

with the candidate drugs. Protein–protein and protein–drug interactions may then be employed for drug testing and screening. In the case of medical diagnostics, each type of antibody has its own functionality; for example, immunoglobulin E (IgE) is the correct selection for allergy tests, whilst IgG can be used in the detection of viruses. The information acquired from medical diagnostics can subsequently be used (as intended) for diagnostic purposes, and perhaps also in the first step(s) of drug development. There are, however, several critical issues related to the fabrication of micro/nano arrays for medical purposes. The primary problem is the immobilization of protein or DNA arrays on the substrate, followed closely by the need to maintain the bioactivity of bioarrays during and after patterning. This may be a less-serious problem in the fabrication of microarrays due to: (i) the relatively large volume of each single drop; and (ii) the processing conditions that are typically used in microarray fabrication. Nonetheless, these issues cannot be ignored, as the fabrication of nanoarrays for nanolithography involves very small volumes of biomaterial. Some of these processes may also require restricted conditions, such as a high vacuum. The mechanism by which interactions between the patterned materials and target materials is detected during experiments remains another important issue; with quantum dots (QDs) having been used by others to detect such interactions and to obtain high-intensity signals.

9.4.2
Virus Array Generation and Detection Tests

Micro/nano bioarrays can each be applied to study the properties and infective mechanisms of viruses. The detection of viruses by using microarrays is currently

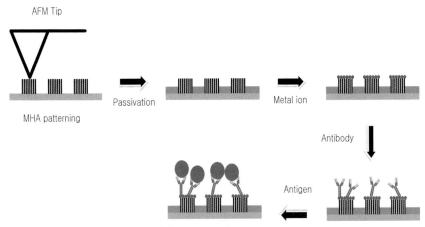

Figure 9.11 Schematic diagrams of the process of virus patterning generation. Reproduced with permission from Ref. [43]; © Wiley-VCH Verlag GmbH & Co. KGaA.

a highly active research area, due mainly to the importance for human medical diagnostics and for livestock and agricultural products [39]. A detailed understanding of the pathogenic progress between a virus and an infected host cell is not only of fundamental importance, but also could (potentially) lead to new approaches for antiviral drug development [42]. Today, several approaches are available for patterning viruses while maintaining bioactivity and immobilization. Recently, Vega *et al.* fabricated tobacco mosaic virus (TMV) patterns on metal templates [51], and were also successful in patterning Parainfluenza virus SV5 (rSV5-EGFP) on a glass substrate [42]; in this case, an MHA pattern was created on a glass substrate using DPN. Here, 11-mercaptoundecylpenta(ethylene glycol) (PEG-SH) was used to passivate the unpatterned region in order to prevent any nonspecific bonding between the substrate and the virus particles. Following its passivation, the substrate was immersed in an ethanolic solution of $Zn(NO_3)_2 \cdot 6H_2O$ to immobilize the Zn^{2+} ions and to bind the antibodies (polyclonal rabbit antibody) to the MHA/Zn^{2+} pattern. Following an incubation for 2 h, the antibodies were bound with the target virus (SV5), after which an infection test on the virus pattern was conducted using CV1 (African green monkey kidney) cells. The process for generating the virus pattern is shown schematically in Figure 9.11.

The differential interference contrast microscopy (DIC) and fluorescence microscopy images obtained between 74 and 89 h post infection are shown in Figure 9.12. The fluorescence microscopy images indicate that the infection had intensified over time, with 14 of the CV1 cells having been infected at 95 h post infection. Another virus patterning was reported by the same group [43], using an identical method of virus patterning with CV1 cells. The application of micro/nano arrays has been also demonstrated for the detection of human rhinovirus (HRV) particles [44], using AFM scanning. Clearly, although recent developments into the detection of viruses using micro/nano protein arrays has been highly successful, there

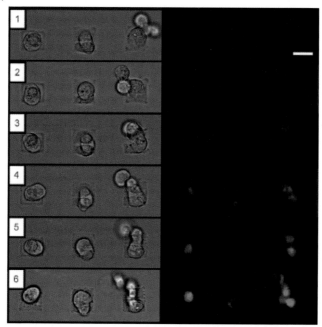

Figure 9.12 Differential interference contrast (DIC) microscopy images (left) and fluorescence microscopy images (right) during an infection test. Each panel of the figure relates to a 3 h time interval during the 74–89 h post infection period. The intensity of green fluorescence is seen to increase over time. Reproduced with permission from Ref. [42]; © Wiley-VCH Verlag GmbH & Co. KGaA.

remains a need to develop patterns of a smaller size, as well as new detection methods.

9.4.3
Diagnosis of Allergic Disease

Allergy testing represents another important example of medical diagnostics using micro/nano bioarrays. Allergic diseases include a heterogeneous group of type I IgE-mediated hypersensitivity reactions that affect more than 25% of the world's population in developed countries [52]. The basic approach to allergy patterning is very similar to that of virus patterning (see Section 9.4.2), except that the IgE antibody rather than the IgG antibody should be immobilized. When an allergen is trapped by IgE, the antigen–antibody interaction can be detected and visualized using appropriate methods that include fluorescence and chemiluminescence. A comparison of allergy tests (type I) conducted using a microarray with conventional allergy tests, such as enzyme-linked immunosorbent assay (ELISA) and the radioallergosorbent test (CAP/RAST) [53], showed the microarray test to provide a greater sensitivity than the CAP/RAST. Moreover, the dynamic range of the microarray was similar to that of the CAP/RAST, but much higher than the ELISA; the microarray was also able to accelerate the test considerably so that the results were obtained more rapidly. Although microarrays are unable to provide high detection rates (due to their small volume) compared to conventional methods, they will provide sufficient data in the majority of cases. Recently, microarrays have been used successfully in a variety of allergen tests for pollen, foods, and latex [45, 46], although a DPN-based device for allergy testing has not yet been described. Nonetheless, the nanopatterning of antibodies to detect antigens is very similar to the patterning of IgG, the details of which have been reported previously.

9.4.4
Cancer Detection Using Nano/Micro Protein Arrays

In recent years, nanotechnology has opened new opportunities for cancer diagnostics, notably with regards to the early detection of cancer, tumor imaging, and antitumor drug delivery [54, 55]. Indeed, today micro/nano protein arrays are used widely in the study and diagnosis of many different types of cancer. Over the years, although the "conventional" methods of cancer research have been slow, they have nonetheless provided a limited insight into the global gene expression patterns that occur during the different stages of tumor formation. According to one report [56], within the United States 72% of lung cancer patients, 57% of colorectal cancer patients and 34% of breast cancer patients are diagnosed at a late stage. In order to effect a cure for cancer, it is very important that cancerous cells are detected at as early a stage as possible, and micro/nano array technology has shown great promise in achieving this goal. Today, micro/nano protein arrays represent an alternative to the current cancer detection methods. As with other micro/nano protein arrays, patterning can be achieved using currently available microarray methods, as well as nanolithography, the key problem being the detection of the "clues" of cancer in the relatively small volumes of human serum and blood that are available. Previously, protein microarrays, in conjunction with QD probes, have been used for the early detection of cancer [57]. In this case the microarrays were termed "sandwich arrays" because they consisted of an antigen located between two antibodies. Each antibody acted as a linker on the substrate and as a detector for the cancer test, with the QD probes being used to visualize detection during the experiments. In general, cancer detection on micro/nano protein arrays is a critical issue when acquiring the test results. Following fabrication of the protein arrays, cytokine detection tests [e.g., for tumor necrosis factor-α (TNF-α), interleukin (IL)-8, -6, -13 and -1β, and macrophage inflammatory protein-1β (MIP-1β)] were conducted, based on the knowledge that cytokine levels in human blood are altered in the presence of certain types of cancer. The study results indicated that this method could be used to detect each cytokine above a certain concentration, such that ovarian cancer [35], breast cancer [36], and pancreatic cancer [37] could be detected using protein/DNA microarrays. In recent

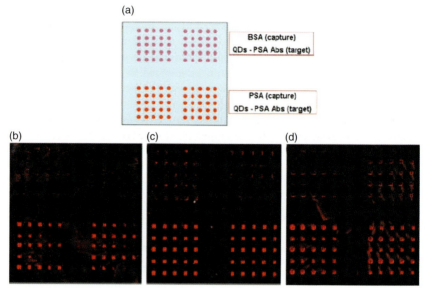

Figure 9.13 (a) Schematic illustration of a microarray; (b–d) Images of protein–protein interaction obtained at: (b) 1 h; (c) 2 h; and (d) 5 h after incubation had commenced. Reproduced with permission from Ref. [38]; © Wiley-VCH Verlag GmbH & Co. KGaA.

years the "battery" of cancer-detection tests has expanded with the development of denser and smaller-sized protein patterns using DPN [38]. Among these developments, QDs conjugated to antibodies for prostate-specific antigen (PSA) were patterned on aldehyde-functionalized silicon substrates, using DPN; in this way, prostate cancer could be detected using the protein arrays, with few control factors. The bovine serum albumin (BSA) blocking layer used to pattern the protein arrays caused strong, nonspecific interactions between the antibodies and normal proteins that might have prevented the protein array from detecting prostate cancer. However, in the absence of these blocking layers the nonspecific interaction was no longer observed. The detection of such an interaction, and a comparison of detection test results between normal protein and prostate-specific antigens, are shown in Figure 9.13. The extent of nonspecific interaction was clearly negligible.

Although, within this chapter, bioarrays have been discussed only for use in medical diagnostics, there is no doubt that they offer immense advantages over conventional diagnostics. However, a number of issues remain that must be addressed before bioarrays can be applied to diagnostics, notably the efficiency of the targeting, the quality of the bioarrays, and regulatory issues. Fortunately, none of these points is expected to hinder the growth of the new technology, which clearly has the potential to revolutionize health care in the near future.

9.4.5
Drug Development

The process of drug development incorporates a sequence of stages that include the screening and classification of disease, and the selection of candidates for drug testing. Strictly speaking, medical diagnostics (as discussed above) may also represent one part of the drug development procedure. Today, bioarrays are able to play an important role in drug development, because they can offer major advantages over conventional methods, including lower costs, time savings, and simplicity of the process. The information gathered from medical diagnostics and screening can be used to discover a new drug for a particular disease [58]. The first stage of drug development is to screen a target so as to identify a disease pathway. Then, based on the results of the first stage, the drug candidate is selected and tested to verify its efficacy and mechanism of action. The final stage incorporates the determination of any adverse side effects that the drug candidate might cause. Today, each stage of development can be started using microarrays containing diseased cells; in this way a disease pathway might be identified using disease microarrays, whereas at a later stage the drug candidate may be tested based on the disease arrays. In the drug-testing scheme the target cells (e.g., tumors or viruses) are patterned, after which the drug candidates are added to the target cell arrays and incubated with the target cells. Any interaction that occurs between the target cells and a drug candidate may be detected by fluorescence. The major advantage of microarrays is that they provide an opportunity to measure parameters on hundreds of individual cells and to generate an average value, rather than measuring the same parameters on the entire cell population [59].

Recently, Lee et al. have developed a drug-metabolizing enzyme–toxicology assay chip, termed the MetaChip, which is based on microarray technology [60]. The primary objective of this investigation was to estimate the cytotoxicity of a prodrug during the development of a novel pharmacologic agent. A schematic diagram of the experimental set-up, together with the study results, is shown in Figure 9.14. Here, a sol–gel solution with cytochrome P450 was patterned on a glass slide, which was coated with methyltrimethoxysilane (MTMOS), using a microarrayer. The anticancer therapeutic agent cyclophosphamide was then patterned over the sol–gel solution, and MCF7 human breast cancer cells subsequently cultured. The cytotoxicity of the prodrug was assessed by monitoring the proportion of dead cells present.

Although nanoarrays have not yet been applied to drug screening (as discussed above), if they were to be created using DPN they could be applied to achieve a higher sensitivity and throughput.

9.4.6
Lab-on-a-Chip Using Microarrays

The term "Lab-on-a-chip" (LOC) refers to the technology in which several functions that are typically performed in a traditional laboratory are, instead, placed on

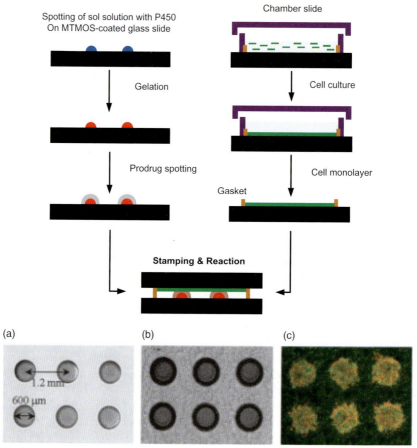

Figure 9.14 Schematic diagram of the MetaChip and images of the results obtained. (a) The P450 sol–gel spots; (b) Sol–gel spots with the prodrug solution; (c) Cell monolayer after removing the sol–gel array. The red coloration indicates dead cells. Reproduced with permission from Ref. [60]; © 2005, National Academy of Sciences, USA.

a small chip. Although LOC technology is currently being used for medical diagnostics, it has plenty of scope for further development and applications. One possible area of use might be in environmental detection. Whilst medical treatment is absolutely essential for every patient, the early detection of some types of disease is more important than the medical treatment. In this respect, the LOC represents a very promising means of achieving the early detection of a disease, especially for those individuals who live in regions that lack comprehensive medical facilities. This technology has a tremendous (though as yet unproven) potential to improve the health of people in developing countries [61]. Compared to conventional medical diagnostics, which may be both expensive and time-consuming, the LOC can be cheaper and much more accessible to the population. It also facilitates the early detection of diseases and viruses because of the high sensitivity of the chip, as well as an easy accessibility. Today, protein microarrays may be integrated with microfluidics and sensing devices for medical use, with many companies producing LOCs for applications such as allergic diagnostics and other types of disease. In one commercially available LOC (produced by STMicroelectronics) [62], the chip consists of four parts: the microfluidics input ports; the polymerase chain reaction (PCR) region; the microfluidics connection; and the microarray. In operational mode, the sample is loaded through the microfluidics input ports and amplified by the PCR, after which the microarray is used to detect the virus/disease. In future developments the microarray might be replaced by nanoarrays patterned by DPN, to create high-sensitivity assays.

9.5
Summary and Future Directions

In this chapter, we have discussed the DPN process and highlighted its applications in the life sciences, notably in the fabrication of bioarrays. The most important issue when patterning of biomaterials is to immobilize arrays on substrates without impairing their bioactivity. With regards to medical diagnostics and drug development, bioarrays offer advantages that include high throughout and sensitivity that will most likely revolutionize these areas. The LOC concept, which involves the incorporation of micro bioarrays for medical diagnostics, was also discussed. In terms of a high functionality, the bioarray is the primary and most important component of the LOC, but to achieve high sensitivity and throughput, the bioarray size must be minimized. Reductions in the size of protein arrays can be achieved using DPN; typically, it can be used to pattern small features with sizes in the region of 50 nm, and to provide flexibility in terms of bioarray size. Many issues remain to be addressed, however. Notably, the repeatability of the pattern size is not good, and the process also requires pre- and post-processing steps. In this regard, the development of parallel DPN which, potentially, may improve the patterning rate, has shown great promise.

References

1 Gulmann, G., Sheehan, K.M., Kay, E.W., Liotta, L.A. and Petricoin, E.F. III. (2006) Array-based proteomics: mapping of protein circuitries for diagnostics, prognostics, and therapy guidance in cancer. *Journal of Pathology*, **208**, 595–606.

2 Salaita, K., Wang, Y. and Mirkin, C.A. (2007) Applications of dip-pen nanolithography. *Nature Nanotechnology*, **2**, 145–55.

3 Kricka, L.J., Master, S.R., Joos, T.O. and Fortina, P. (2006) Current perspectives in protein array technology. *Annals of Clinical Biochemistry*, **43**, 457–67.

4 Christman, K.L., Enriquez-Rios, V.D. and Maynard, H.D. (2006) Nanopatterning

proteins and peptides. *Soft Matter*, **2**, 928–39.
5 Mendes, P.M., Yeung, C.L. and Preece, J.A. (2007) Bio-nanopatterning of surfaces. *Nanoscale Research Letters*, **2**, 373–84.
6 Piner, R.D., Zhu, J., Xu, F., Hong, S.H. and Mirkin, C.A. (1999) "Dip-pen" nanolithography. *Science*, **283**, 661–3.
7 Haaheim, J.R., Tevaarwerk, E.R., Fragala, J. and Shile, R. (2008) Commercially available high-throughput dip pen nanolithography. *Proceedings of the SPIE*, **6959**, 695901-1.
8 Huo, F., Zheng, Z., Zheng, G., Giam, L.R., Zhang, H. and Mirkin, C.A. (2008) Polymer pen lithography. *Science*, **321**, 1658–60.
9 Ginger, D.S., Zhang, H. and Mirkin, C.A. (2004) The evolution of dip-pen nanolithography. *Angewandte Chemie, International Edition*, **43**, 30–45.
10 Hong, S., Zhu, J. and Mirkin, C.A. (1999) Multiple ink nanolithography: toward a multiple-pen nano-plotter. *Science*, **286**, 523–5.
11 Lee, S.W., Oh, B.K., Sanedrin, R.G., Salaita, K., Fujigaya, T. and Mirkin, C.A. (2006) Biologically active protein nanoarrays generated using parallel dip-pen nanolithography. *Advanced Materials*, **18**, 1133–6.
12 Kim, J.D., Ahn, D.G., Oh, J.W., Park, W.J. and Jung, H.G. (2008) Ribosome display and dip-pen nanolithography for the fabrication of protein nanoarrays. *Advanced Materials*, **20**, 3349–53.
13 Roy, D., Munz, M., Colombi, P., Bhattacharyya, S., Salvetat, J.P., Cumpson, P.J. and Saboungi, M.L. (2007) Directly writing with nanoparticles at the nanoscale using dip-pen nanolithography. *Applied Surface Science*, **254**, 1394–8.
14 Sheu, J.T., Wu, C.H. and Chao, T.S. (2006) Selective deposition of gold particles on dip-pen nanolithography patterns on silicon dioxide surfaces. *Japanese Journal of Applied Physics*, **45** (4B), 3693–7.
15 Wei, J.H., Coffey, D.C. and Ginger, D.S. (2006) Nucleating pattern formation in spin-coated polymer blend films with nanoscale surface templates. *Journal of Physical Chemistry B*, **110**, 24324–30.
16 Jegadesan, S., Sindhu, S., Advincula, R.C. and Valiyaveettil, S. (2006) Direct electrochemical nanopatterning of polycarbazole monomer and precursor polymer films: ambient formation of thermally stable conducting nanopatterns. *Langmuir*, **22**, 780–6.
17 Yang, M.Y., Sheehan, P.E., King, W.P. and Whitman, L.J. (2006) Direct writing of a conducting polymer with molecular-level control of physical dimensions and orientation. *Journal of the American Chemical Society*, **128**, 6774–5.
18 Nelson, B.A., King, W.P., Laracuente, A.R., Sheehan, P.E. and Whitman, L.J. (2006) Direct deposition of continuous metal nanostructures by thermal dip-pen nanolithography. *Applied Physics Letters*, **88**, 033104-1–3.
19 Wei, J.H. and Ginger, D.S. (2007) A direct-write single-step positive etch resist for dip-pen nanolithography. *Small*, **3** (12), 2034–7.
20 Zhang, H., Amro, N.A., Disawal, S., Elghanian, R., Shile, R. and Fragala, J. (2007) High-throughput dip-pen-nanolithography-based fabrication of Si nanostrucutres. *Small*, **3** (1), 81–5.
21 Cretich, M., Damin, F., Pirri, G. and Chiari, M. (2006) Protein and peptide arrays: recent trends and new directions. *Biomolecular Engineering*, **23**, 77–88.
22 Lim, J.H., Ginger, D.S., Lee, K.B., Heo, J.S., Nam, J.M. and Mirkin, C.A. (2003) Direct-write dip-pen nanolithography of proteins on modified silicon oxide surfaces. *Angewandte Chemie, International Edition*, **115**, 2411–14.
23 Kim, K.H., Sanedrin, R.G., Ho, A.M., Lee, S.W., Moldovan, N., Mirkin, C.A. and Espinosa, H.D. (2008) Direct delivery and submicrometer patterning of DNA by a nanofountain probe. *Advanced Materials*, **20**, 330–4.
24 Demers, L.M., Ginger, D.S., Park, S.J., Li, Z., Chung, S.W. and Mirkin, C.A. (2002) Direct patterning of modified oligonucleotides on metals and insulators by dip-pen nanolithography. *Science*, **296**, 1836–8.

25 Lee, M.S., Kang, D.K., Yang, H.K., Park, K.H., Choe, S.Y., Kang, C.S., Chang, S.I., Han, M.H. and Kang, I.C. (2006) Protein nanoarray on Prolinker™ surface constructed by atomic force microscopy dip-pen nanolithography for analysis of protein interaction. *Proteomics*, **6**, 1094–103.

26 Lee, K.B., Park, S.J., Mirkin, C.A., Smith, J.C. and Mrksich, M. (2002) Protein nanoarrays generated by dip-pen nanolithography. *Science*, **295**, 1702–5.

27 Baserga, A., Vigano, M., Casari, C.S., Turri, S., Bassi, A.L., Levi, M. and Bottani, C.E. (2008) Au-Ag template stripped pattern for scanning probe investigations of DNA arrays produced by dip pen nanolithography. *Langmuir*, **24**, 13212–17.

28 Muir, B.W., Fairbrother, A., Gengenbach, T.R., Rovere, F., Addo, M.A., McLean, K.M. and Hartley, P.G. (2006) Scanning probe nanolithography and protein patterning of low-fouling plasma polymer multilayer films. *Advanced Materials*, **18**, 3079–82.

29 Kim, K.H., Kim, J.D., Kim, Y.J., Kang, S.H., Jung, S.Y. and Jung, H.G. (2008) Protein immobilization without purification via dip-pen nanolithography. *Small*, **4** (8), 1089–94.

30 Maury, P., Escalante, M., Péter, M., Reinhoudt, D.N., Subramaniam, V. and Huskens, J. (2007) Creating nanopatterns of His-tagged proteins on surfaces by nanoimprint lithography using specific NiNTA-histidine interactions. *Small*, **9**, 1584–92.

31 Tinazli, A., Piehler, J., Beuttler, M., Guckenberger, R. and Tampé, R. (2007) Native protein nanolithography that can write, read and erase. *Nature Nanotechnology*, **2**, 220–5.

32 Petrik, J. (2006) Diagnostic applications of microarrays. *Transfusion Medicine*, **16**, 233–47.

33 Uttamchandani, M., Wang, J. and Yao, Q.S. (2006) Protein and small molecule microarrays: powerful tools for high-throughput proteomics. *Molecular Biosystems*, **2**, 58–68.

34 Renault, N.K. and Mirotti, L. (2007) Biotechnologies in new high-throughput food allergy test: why we need them. *Biotechnology Letters*, **29**, 333–9.

35 Chatterjee, M., Mohapatra, S., Ionan, A., Bawa, G., Ail-Fehmi, R., Wang, X., Nowak, J., Ye, B., Nahhas, F.A., Lu, K., Witkin, S.S., Fishman, D., Munkarah, A., Morris, R., Levin, N.K., Shirley, N.N., Tromp, G., Abrams, J., Draghich, S. and Tainsky, M.A. (2006) Diagnostic markers of ovarian cancer by high-throughput antigen cloning and detection on arrays. *Cancer Research*, **66** (2), 1181–90.

36 Anderson, K.S., Ramachandran, N., Wong, J., Raphael, J.V., Hainsworth, E., Demirkan, G., Cramer, D., Aronzon, D., Hodi, F.S., Harris, L., Logvinenko, T. and Labaer, J. (2008) Application of protein microarrays for multiplexed detection of antibodies to tumor antigens in breast cancer. *Journal of Proteome Research*, **7**, 1490–9.

37 Ingvarsson, J., Wingren, C., Carlsson, A., Ellmark, P., Wahren, B., Engström, G., Harmenberg, U., Krogh, M., Peterson, C. and Borrebaeck, A.K. (2008) Detection of pancreatic cancer using antibody microarray-based serum protein profiling. *Proteomics*, **8**, 2211–19.

38 Gokarna, A., Jin, L.H., Hwang, J.S., Cho, Y.H., Lim, Y.T., Chung, B.Y., Youn, S.H., Choi, D.S. and Lim, J.H. (2008) Quantum dot-based protein micro- and nanoarrays for detection of prostate cancer biomarkers. *Proteomics*, **8**, 1809–18.

39 Baxi, M.K., Baxi, S., Clavijo, A., Burton, K.M. and Deregt, D. (2006) Microarray-based detection and typing of foot-and mouth disease virus. *The Veterinary Journal*, **172**, 473–81.

40 Townsend, M.B., Dawson, E.D., Mehlmann, M., Smagala, J.A., Dankbar, D.M., Moore, C.L., Smith, C.B., Cox, N.J., Kuchta, R.D. and Rowlen, K.L. (2006) Experimental evaluation of the fluchip diagnostic microarray for influenza virus surveillance. *Journal of Clinical Microbiology*, **44** (8), 2863–71.

41 Pasquini, G., Barba, M., Hadidi, A., Faggioli, F., Negri, R., Sobol, I., Tiberini, A., Caglayan, K., Mazyad, H., Anfoka, G., Ghanim, M., Zeidan, M. and Czosnek, H. (2008) Oligonucleotide microarray-based detection and genotyping of Plum pox

virus. *Journal of Virological Methods*, **147** (1), 118–26.

42 Vega, R.A., Shen, C.K.F., Maspoch, D., Robach, J.G., Lamb, R.A. and Mirkin, C.A. (2007) Monitoring single-cell infectivity from virus-particle nanoarrays fabricated by parallel dip-pen nanolithography. *Small*, **3** (9), 1482–5.

43 Vega, R.A., Maspoch, D., Shen, C.K.F., Kakkassery, J.J., Chen, B.J., Lamb, R.A. and Mirkin, C.A. (2006) Functional antibody arrays through metal ion-affinity templates. *Chembiochem*, **7**, 1653–7.

44 Artelsmair, H., Kienberger, F., Tinazli, A., Schlapak, R., Zhu, R., Preiner, J., Wruss, J., Kastner, M., Saucedo-Zeni, N., Hoelzl, M., Rankl, C., Baumgartner, W., Howorka, S., Blass, D., Gruber, H.J., Tampé, R. and Hinterdorfer, P. (2008) Atomic force microscopy-derived nanoscale chip for the detection of human pathogenic viruses. *Small*, **6**, 847–54.

45 Deinhofer, K., Sevcik, H., Balic, N., Harwanegg, C., Hiller, R., Rumpold, H., Mueller, M.W. and Spitzauer, S. (2004) Microarrayed allergens for IgE profiling. *Methods*, **32**, 249–54.

46 Ohyama, K., Omura, K. and Ito, Y. (2005) A photo-immobilized allergen microarray for screening of allergen-specific IgE. *Allergology International*, **54**, 627–31.

47 Liguori, M.J., Anderon, M.G., Bukofzer, S., McKim, J., Pregenzer, J.F., Retief, J., Spear, B.B. and Waring, J.F. (2005) Microarray analysis in human hepatocytes suggests a mechanism for hepatotoxicity induced by trovafloxacin. *Hepatology*, **41** (1), 177–86.

48 Tan, Y., Shi, L., Hussain, S.M., Xu, J., Tong, W., Frazier, J.M. and Wang, C. (2006) Intergating time-courses microarray gene expression profiles with cytotoxicity for identification of biomarkers in primary rat hepatocytes exposed to cadmium. *Bioinformatics*, **22** (1), 77–87.

49 Zhao, H. and Yan, H. (2007) HoughFeature, a novel method for assessing drug effects in three-color cDNA microarray experiments. *BMC Bioinformatics*, **8**, 256-1–10.

50 Wingren, C. and Borrebaeck, C.A.K. (2006) Antibody microarrays: current status and key technological advances. *Journal of Integrative Biology*, **10** (3), 411–27.

51 Vega, R.A., Maspoch, D., Salaita, K. and Mirkin, C.A. (2005) Nanoarrays of single virus particles. *Angewandte Chemie, International Edition*, **117**, 6167–9.

52 González-Buitrago, J.M., Ferreira, L., Isidoro-García, M., Sanz, C., Lorente, F. and Dávila, I. (2007) Proteomic approaches for identifying new allergens and diagnosing allergic diseases. *Clinica Chimica Acta*, **385**, 21–7.

53 Jahn-Schmid, B., Harwanegg, C., Hiller, R., Bohle, B., Ebner, C., Scheiner, O. and Muller, M.W. (2003) Allergen microarray: comparison of microarray using recombinant allergens with conventional diagnostic methods to detect allergen-specific serum immunoglobulin E. *Clinical and Experimental Allergy*, **33**, 1433–49.

54 Ferrari, M. (2005) Cancer nanotechnology: opportunities and challenges. *Nature Reviews Cancer*, **5**, 161–71.

55 Sengupta, S. and Sasisekharan, R. (2007) Exploiting nanotechnology to target cancer. *British Journal of Cancer*, **96**, 1315–19.

56 Zhang, X., Li, L., Wei, D., Yap, Y. and Chen, F. (2007) Moving cancer diagnostics from bench to bedside. *Trends in Biotechnology*, **25** (4), 666–73.

57 Zajac, A., Song, D., Qian, W. and Zhukov, T. (2007) Protein microarrays and quantum dot probes for early cancer detection. *Colloids and Surfaces B: Biointerfaces*, **58**, 309–14.

58 Ng, J.H. and Hag, L.L. (2002) Biomedical applications of protein chips. *Journal of Cellular and Molecular Medicine*, **6** (3), 329–40.

59 Castel, D., Pitaval, A., Debily, M.A. and Gidrol, X. (2006) Cell microarrays in drug discovery. *Drug Discovery Today*, **11**, 616–22.

60 Lee, M.Y., Park, C.B., Dordick, J.S. and Clark, D.S. (2005) Metabolizing enzyme toxicology assay chip (MetaChip) for high-throughput microscale toxicity

analyses. *Proceedings of the National Academy of Sciences of the United States of America*, **102** (4), 983–7.

61 Chin, C.D., Linder, V. and Sia, S.K. (2007) Lab-on-a-chip devices for global health: past studies and future opportunities. *Lab on a Chip*, **7**, 41–5.

62 Palmieri, M., Alessi, E., Conoci, S., Marchi, M., and Panvini, G. (2008) Develop the "In-Check" platform for diagnostic applications. *Proceedings of the SPIE*, **6886**, 688602-1–14.

10
Understanding and Controlling Wetting Phenomena at the Micro- and Nanoscales

Zuankai Wang and Nikhil Koratkar

10.1
Introduction

Wetting phenomena are not only ubiquitous in our everyday lives, but also play an important role in many technological processes, as well as in many biological systems [1, 2]. Controlling the wetting property of solid materials is a classical and key issue in surface engineering. As material systems and devices are increasingly shrinking to ever-smaller length scales (i.e., from macro to micro to nano), there is an increasing interest in the study of wetting phenomena at the micro and nano scales [3, 4]. With the increased surface area to volume ratio at the micro/nanoscale, some natural questions with regards to wetting phenomena would be: How do micro/nanostructured surfaces affect wetting phenomena? And is it possible to engineer the solid–liquid–gas interactions in a clever and well-controlled way so as to achieve specific functionalities? Thanks to rapid advances in microfabrication and nanofabrication technologies, it is possible to study systematically, for the first time, wetting phenomena at these scales. By designing and synthesizing microstructures and nanostructures with desired sizes and dimensions, various research groups have been able to uncover the fundamental mechanisms that govern wettability at the micro/nano scales, and to apply this knowledge to important applications such as surface self-cleaning, microfluidics, and energy management. Given the dynamic and rapidly evolving nature of this field, there is a need to review several recent developments in this area, and this is the objective of this chapter.

At the outset, it is worthwhile reviewing some examples of natural or biological systems that exploit wetting phenomenon at very small scales. During the course of evolution, Nature has developed unique strategies to take advantage of wetting phenomena for complex functions, by producing superior materials and structures. Below are presented three very interesting observations from Nature:

- **Observation 1: Why is the Lotus leaf water-repellent?:** The lotus leaf, with its self-cleaning surface, has attracted the attentions of biophysical scientists and engineers for many years [5–8]. The remarkable self-cleaning ability of the leaf is enabled by the combined physical morphology (two-scale roughness) and repulsive chemical properties (nanoscale wax is hydrophobic). This imparts extreme water repellency (or super-hydrophobicity) to the leaf, enabling water droplets simply to roll off the surface of the leaf at very low inclination (or sliding) angles.

- **Observation 2: How does the water strider stand on water?**: Water striders can stand effortlessly, but also move very quickly, on the water surface. In the past, scientists have discovered that the secret behind this lies in the superhydrophobic surfaces of the insect's legs, which are covered by large numbers of oriented tiny hairs (microsetae) with fine nanogrooves. Entrapped air in these two-scale roughness structures imparts a superhydrophobic wetting property, which results in an upward curvature force that is much larger than the insect's weight [9–12].

- **Observation 3: How does the desert beetle harvest and control water?**: In the Namib Desert in southern Africa, rainfall is scarce. However, the desert beetle has evolved to take perfect advantage of the tiny amount of water available in the desert by its unique droplet-producing and -controlling mechanism [13, 14]. The back and legs of the beetle are textured, and exhibit an extreme hydrophobic-to-hydrophilic spatial contrast. The water-harvesting system operates by collecting tiny water droplets on the hydrophilic seeding points on the beetle's back; these are surrounded by hydrophobic areas that drive the water towards the hydrophilic collection points. When the droplet grows to a critical size, and when the pinning force can be easily overcome with the help of the wind, the droplet becomes detached and rolls downwards to the beetle's mouth.

Based on the above observations from Nature, it is obvious that nano- and microtexturing has a major impact on wetting phenomena. Without its special two-scale roughness features, it would be impossible for the Lotus leaf to maintain its self-cleaning property; without the dominance of capillarity and superhydrophobicity enabled by its two-scale roughness, it would be impossible for the water strider to walk freely on water; and finally the extreme wetting contrast enabled by nano/microtexturing enables the desert beetle to harvest and control water. Just as Nature uses nano- and microengineering to control wetting and to achieve desired functionalities, the question arises as to whether the same can be achieved with man-made systems and devices.

The first step to achieving this is to design and create artificial superhydrophobic surfaces with micro/nanoscale surface roughness features. Consequently, an initial review is provided of the fabrication of such surfaces and the characterization of their wettability, under static conditions. The wetting behavior of the superhydrophobic rough surfaces under dynamic and droplet impact conditions is then be reviewed, as are the strategies employed to control and manipulate the wettability of surfaces using electrical (e.g., electrostatic and electro-wetting) methods. Following a description of the emerging electrochemical techniques that enable wettability switching in carbon nanotube (CNT) membranes, a summary is provided of the present field, together with some future perspectives for continued research. The aim of this chapter is to cover the fundamental physical mechanisms of wetting, contact angle, wetting transition, dynamic wetting and impact, the design/creation of superhydrophobic surfaces, and the active control of wettability and wettability switching. It should be noted that several other excellent articles [1] and reviews [2–4, 8] on similar topics are available, although none of these covers all of the above topics. In particular, the chapter's aim is to review the latest developments and advancements in this rapidly evolving and fascinating field.

10.2
Wetting and Contact Angle

"Wetting" refers to the study of how a liquid, when deposited onto a solid (or another liquid) substrate, spreads out. Roughly speaking, two extreme wetting limits are often desired: (i) *complete wetting*, where a liquid, when brought into contact with a solid, forms a film spontaneously [8]; and (ii) *complete drying*, where the liquid drops remain spherical without developing any contact with the surface. The specific wettability can be characterized by measuring the *contact angle* of a liquid placed on the surface of a solid. The contact angle is the angle formed by the solid–liquid interface and the liquid–vapor interface, measured from the side of the liquid. On a flat, smooth, and chemically homogeneous surface, the contact angle can be related to the interfacial energy by the Young's equation [1]:

$$\cos\theta = \frac{\gamma_{SG} - \gamma_{SL}}{\gamma_{LG}} \tag{10.1}$$

where γ_{SG}, γ_{SL}, and γ_{LG} are the surface tensions at the solid–air, solid–liquid, and liquid–air interfaces, respectively. Generally, the surface is termed as "hydrophobic" when the contact angle is higher than 90°, otherwise it is hydrophilic. The smaller the contact angle, the more hydrophilic the surface. Generally, if the contact angle on the solid surface is greater than 150°, the surface is called "superhydrophobic."

From Equation 10.1, it can be seen that the contact angle on the solid surface can be altered by modifying the solid surface energy. For example, a hydrophilic surface could be transformed into a hydrophobic surface by simply depositing on that surface a thin coating of fluorinated molecules. However, the greatest contact angle that can be achieved on flat surfaces by the use of such coatings is typically less than 120°. Therefore, superhydrophobicity cannot be achieved on flat surfaces by modifying the surface energy alone. In order to achieve a higher contact angle, a rough or textured surface is necessary, as in the Lotus leaf.

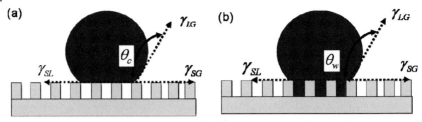

Figure 10.1 Two wetting states on rough surfaces. (a) Cassie state; (b) Wenzel state.

Wetting on rough surfaces may assume either of two regimes (Figure 10.1): (i) *homogeneous wetting*, where the liquid completely penetrates the roughness grooves (as explained by the Wenzel model [15]); or (ii) *heterogeneous wetting*, where air is trapped underneath the liquid inside the roughness grooves (as explained by the Cassie model [16]).

In the Wenzel model, the apparent contact angle on a rough surface in the homogeneous regime, θ_w is expressed as:

$$\cos\theta_w = r(\cos\theta) \tag{10.2}$$

where θ is the contact angle on the flat surface and r is the roughness ratio, defined as the ratio of the true area of the solid surface to its projection area. Since roughness is always larger than 1, from Equation 10.2 it is clear that, θ_w will be greater than θ if the surface is originally hydrophobic ($\theta > 90°$), otherwise, θ_w is less than θ.

The apparent contact angle in the heterogeneous regime, θ_c, is given by the Cassie–Baxter (CB) equation [16]:

$$\cos\theta_c = \phi(\cos\theta+1)-1 \tag{10.3}$$

$\phi = \dfrac{\pi}{2\sqrt{3}}\dfrac{a^2}{b^2}$ where ϕ is the solid fraction, expressed as the projected area of the solid–liquid contacts divided by the total projected area of the solid surface. It is worthwhile to mention that, as the water droplet remains on the top of the rough surface, the apparent contact angle in the Cassie model does not depend on the height of the roughness structures; this is in contrast to the Wenzel model, in which the contact angle shows a dependency of the height of the roughness features. As an example, for a pattern comprised of circle pillars in a hexagonal matrix, the solid fraction (ϕ) and roughness (r) can be expressed as:

$$\phi = \frac{\pi}{2\sqrt{3}}\frac{a^2}{b^2}, \quad r = 1+4\phi\frac{H}{a} \tag{10.4}$$

where b is center-to-center distance between the pillars, while a and H are the diameter and height of the pillars, respectively.

10.3
Design and Creation of Superhydrophobic Surfaces

10.3.1
Design Parameters for a Robust Composite Interface

To mimic the Lotus leaf, a surface is required which has an excellent air-trapping ability, in which the droplet remains in a Cassie state [8, 17–21]. A Cassie droplet is critical in ensuring that the droplet can roll easily from the rough surface; this is an important property for self-cleaning objectives. Thus, geometrically, the asperities in the solid surface must be tall enough to ensure that liquid protruding between them does not contact the underlying solid. Thermodynamically, the droplet must remain in a low-energy state [22–24]. By using the energy approach (demonstrated in Refs [22, 23]), the Gibbs free energy (G) of a droplet in equilibrium can be represented by:

$$G \sim (1-\cos\theta)^{2/3}(2+\cos\theta)^{1/3} \tag{10.5}$$

It can be easily verified that the right-hand side of Equation 10.5 is a monotonically increasing function of θ. As a result, an equilibrium drop shape with a lower value of the apparent contact angle will have a lower energy. Thus, the Cassie wetting mode corresponds to a lower Gibbs free energy than the Wenzel wetting mode if – and only if – the apparent contact angle in the Cassie state is lower than the Wenzel state. Equating the Cassie and Wenzel equations (Equations 10.2 and 10.3) gives the critical equilibrium contact angle (θ_c) as expressed in Equation 10.6. Contact angles above the equilibrium value lead to a lower free energy state for the composite interface than for the fully wetted interface [24, 25]:

$$\cos\theta_c = \frac{1-\phi}{\phi-r} \tag{10.6}$$

10.3.2
Creation of Superhydrophobic Surfaces

A large number of elegant techniques to fabricate rough surfaces that exhibit superhydrophobicity have been reported [26–34]. Generally, these can be divided into two categories: (i) the creation of a rough surface on a hydrophobic material (the one-step approach); or (ii) the modification of a rough surface with hydrophobic coatings (the two-step approach). Specifically, the fabrication methods reported include: template synthesis; electrochemical deposition; photolithography; chemical vapor deposition (CVD), and sol–gel processing [26–34].

The Cassie and Wenzel models described above provide the general guidelines for the design of superhydrophobic rough surfaces. These models indicate that the contact angle on a rough surface is governed by two parameters: the contact angle on the flat surface (θ); and the solid fraction (ϕ) or roughness (r). It should be noted that the solid fraction and roughness are not absolute quantities, but are relative, as this allows significant flexibility in the design and creation of superhydrophobic surfaces. For example, by retaining the same solid fraction the same static contact angle can achieved by using either nanoscale or microscale roughness, or a combination of these.

10.3.3
Superhydrophobic Surfaces with Unitary Roughness

In Nature, there exist many superhydrophobic surfaces with unitary features (i.e., microscale or nanoscale roughness only) [35]. For example, the Ramee leaf is uniformly covered by tiny fibers with diameter of 1–2 μm; this structure effectively entraps air and forms a superhydrophobic surface with a water contact angle of ~164°. Interestingly, this phenomenon is also observed in the Chinese watermelon, indicating that hierarchical dual-scale structures (i.e., combinations of micro- and nanostructures) are not a necessary condition to form effective superhydrophobic surfaces. Superhydrophobic surfaces with nanoscale roughness alone have also been demonstrated on various materials [26, 29, 36–38], such as ZnO nanostructures, silicon carbide nanowires, nanofibers, metal nanorods, and CNTs. Artificial superhydrophobic surfaces with microscale roughness [26, 36, 39–41] have been developed on materials such as polydimethylsiloxane (PDMS), silicon, polymers, and block copolymers.

10.3.4
Superhydrophobic Surfaces with Two-Scale Roughness

Although unitary roughness can exhibit superhydrophobicity, the general consensus within the scientific community is that micro- and nanometer-scale hierarchical structures promote more robust superhydrophobicity, especially from the point of view of dynamic wetting [42–44]. Recent experiments [45] have demonstrated that, although multiwalled carbon nanotube (MWNT) arrays (comprising a nanoscale roughness only) exhibit a high static water contact angle, they suffer from a large sliding angle due to a fragile Cassie state that can easily be broken under dynamic conditions, causing the droplet to become pinned to the surface (Figure 10.2a). Although such a combination of high contact angle and a large adhesion force to the surface is counterintuitive, it has also been reported on other nanostructures, for example, polystyrene nanostructures, TiO_2 nanostructures and PDMS [46–49]. For the TiO_2 nanostructure, the combination of hydrogen bonding between the nitro groups and the hydroxyl groups at the TiO_2 solid–liquid interface is responsible for the increase of adhesion [48]. Since there is no hydrogen bonding available in the MWNT array, the wide tilt angle is ascribed to van der Waal's forces between the liquid and solid film. Recently, various research groups have created stable superhydrophobic MWNT arrays by coating the MWNT with low-surface energy compounds (e.g., fluoropolymers such as polytetrafluoroethylene) [50].

10.3 Design and Creation of Superhydrophobic Surfaces

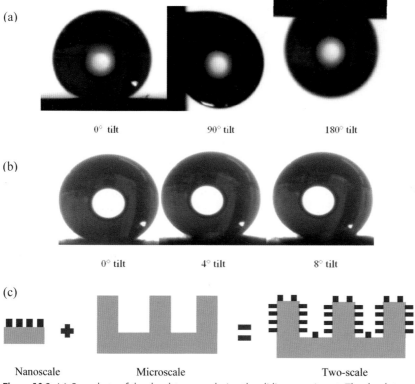

Figure 10.2 (a) Snapshots of the droplet shape for a water droplet resting on multiwalled carbon nanotube (MWNT) arrays during dynamic sliding experiments; (b) Snapshots of the droplet shape with inclination angle for a water droplet resting on a patterned (two-scale) MWNT array during the sliding experiment. The droplet rolls off the two-scale surface at an inclination angle of ~8°; (c) Schematic of surfaces with different roughness features. Reproduced with permission from Ref. [45]; © 2007, American Institute of Physics.

An alternative to improving the dynamic wetting properties without any surface treatment is to adopt a two-scale roughness structure, as in the Lotus leaf (Figure 10.2c) [31, 45, 51]. The presence of two-scale roughness provides a more robust liquid–gas–solid interface, so that the water droplet can slide off the surface at a sliding angle of 8° (Figure 10.2b) [45].

In order to better understand the impact of two-scale roughness on static and dynamic wetting, let us consider the Wenzel and Cassie equations. Assuming that r_m, r_n, ϕ_m, ϕ_n are the roughness and the solid fraction of the microscale and nanoscale features, the effective roughness in the two-scale structure (Figure 10.2c) is $r_m r_n$ and the effective solid fraction is given by $\phi_m \phi_n$. Equations 10.2, 10.3 and 10.6 for the surface with nanoscale, microscale and two-scale roughness can be expressed as follows:

- Nanoscale roughness:
  Cassie equation: $\cos\theta_{nc} = \phi_n(\cos\theta + 1) - 1$
  Wenzel equation: $\cos\theta_{nw} = r_n(\cos\theta)$
  Criterion that Cassie state is more energy-favorable: $\cos\theta < \dfrac{1-\phi_n}{\phi_n - r_n}$

- Microscale roughness:
  Cassie equation: $\cos\theta_{mc} = \phi_m(\cos\theta + 1) - 1$
  Wenzel equation: $\cos\theta_{mw} = r_m(\cos\theta)$
  Criterion that Cassie state is more energy-favorable: $\cos\theta < \dfrac{1-\phi_m}{\phi_m - r_m}$

- Two-scale roughness:
  Cassie equation: $\cos\theta_{mnC} = \phi_m\phi_n(\cos\theta + 1) - 1$
  Wenzel equation: $\cos\theta_{mnW} = r_m r_n(\cos\theta)$
  Criterion that Cassie state is more energy-favorable: $\cos\theta < \dfrac{1-\phi_m\phi_n}{\phi_m\phi_n - r_m r_n}$

For the same solid surface with a contact angle of θ, since $\phi_m\phi_n < \phi_m$, ϕ_n and $r_m r_n > r_m$, r_n, it is clear from the above equations that the required contact angle required to maintain a robust Cassie state is the lowest for the surface with the dual-scale roughness features. Therefore, by engineering a two-scale (micro/nano) roughness, it is relatively easier for the droplet to remain in the Cassie state and not to become pinned to the surface.

It should be pointed out that such two-scale hierarchical architectures have several important applications beyond superhydrophobicicty and self-cleaning properties. Many unique properties in Nature are related to this multiscale structure. For example, the directional adhesion on the superhydrophobic wings of the butterfly *Morpho aega* arises from the direction-dependent arrangement of nanostripes and microstructures being overlapped on the wings' surface [52]. A droplet can easily roll from the wing surface of *M. aega* along the radially outward direction of the central axis of the body, while the droplet becomes tightly pinned perpendicular to this direction.

10.3.5
Superhydrophobic Surfaces with Reentrant Structure

Equation 10.6 also indicates that a robust composite state requires that the flat surface contact angle should be hydrophobic and larger than the critical equilibrium contact angle. This complicates the choice of the surface, given the fact that many wax structures in Nature are weakly hydrophilic. Recently, superhydrophobic surfaces made from hydrophilic materials have been reported, as shown in Figure 10.3 [53, 54]. The presence of overhung or reentrant structures prevents the liquid from entering into the indents between the microtextures, and can result in a more robust composite (Cassie) state. As a result, air can become trapped in the rough structures, even though the roughness features are composed of a purely hydrophilic material. Thus, a hydrophobic surface material is not a prerequisite for the entrapment of gas under a drop, and it is possible that entrapped gas can

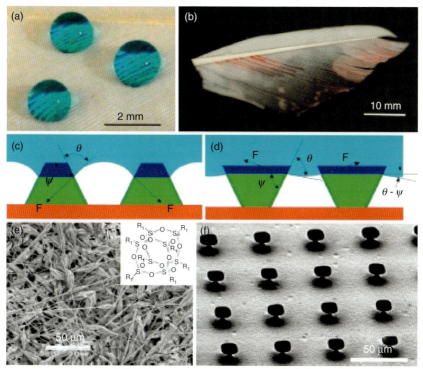

Figure 10.3 Critical role of reentrant texture. (a,b) Droplets of water (colored with blue) and rapeseed oil (colored with red) on a duck feather; (c,d) Schematic diagrams illustrating possible liquid–vapor interfaces on two different surfaces having the same solid surface energy and the same equilibrium contact angle, but different geometric angles; (e) A scanning electron microscopy (SEM) image of an electrospun surface containing 44.4 wt% fluorodecyl polyhedral oligomeric silsesquioxane (POSS), and possessing the beads-on-strings morphology. The inset shows the molecular structure of the fluorodecyl POSS molecules. The alkyl chains (R_f) have the molecular formula $CH_2CH_2(CF_2)_7CF_3$; (f) An SEM image of a microhoodoo surface. The samples are viewed from an oblique angle of 30°. Reproduced with permission from Ref. [54]; © 2008, National Academy of Sciences, USA.

exist on a surface made from hydrophilic materials when the maximum asperity slope is greater than $\pi/2$, as indicated in Figure 10.3.

10.4
Impact Dynamics of Water on Superhydrophobic Surfaces

Previously, the design and creation of superhydrophobic surfaces by using micro/nanoscale roughness has been discussed. In many practical situations, the superhydrophobic nature of the surface must be maintained under dynamic conditions; for example, in de-icing applications, the impact of water droplets on the surface is encountered [55, 56]. The ability of a droplet to rebound from the surface is critical in such situations. Droplet impact is also important in lab-on-chip devices, semi-conductor chips, ink-jet printing, rapid spray cooling of hot surfaces, spray-painting and -coating, precision solder-drop dispensing in microelectronics, as well as in liquid atomization and cleaning [57].

Drop impact phenomena on dry solids are extremely diverse and surprising. The outcome can be deposition, prompt splash, corona splash, receding break-up, partial rebound, and complete rebound, depending on the impact velocity, drop size, the properties of the liquid, the interfacial tension of the surface, the roughness and the wettability of the solid surface [57, 58]. In spite of more than 100 years of research, such phenomena are still far from being fully understood, especially on superhydrophobic rough surfaces.

In the following sections, the results of recent studies are presented [56–61] in which the impact behavior of superhydrophobic surfaces with nanoscale roughness are compared to that of surfaces with microscale roughness features.

10.4.1
Impact Dynamics on Nanostructured MWNT Surfaces

Impact experiments were conducted on two nanostructured MWNT surfaces with different wettabilities [56]; one with a contact angle of 163°, and another of 140°. In the experiment, droplets of deionized water were impacted on the array at velocities ranging from 10 to 60 cm s^{-1}. The snapshots of a water droplet impacting an MWNT array with a static contact angle of 163° are shown in Figure 10.4a; the impact velocity for this case is ~56 cm s^{-1}. As shown in the images, the droplet first

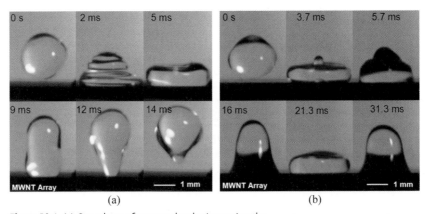

Figure 10.4 (a) Snapshots of a water droplet impacting the surface of a carbon nanotube (CNT) array. The static contact angle of the nanotube array was 163°; (b) Snapshots of a water droplet impacting the surface of a CNT array. The static contact angle of the nanotube array was 140°. Reproduced with permission from Ref. [56]; © 2007, American Institute of Physics.

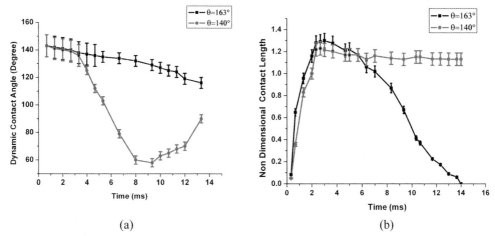

Figure 10.5 (a) Time history of the droplet's dynamic contact angle plotted for the droplet impinging on the 140° and 163° contact angle MWNT arrays; (b) Normalized contact line length plotted versus time for droplets to impinge on 140° and 163° static contact angle MWNT arrays with a velocity of 0.56 m s^{-1}. Reproduced with permission from Ref. [56]; © 2007, American Institute of Physics.

deforms and flattens into a pancake shape, but then retracts and finally rebounds from the surface. The droplet remains completely intact during the collision, and does not splash or fragment into smaller droplets. The total contact time (~14 ms) of the droplet with the MWNT array was found to be independent of the impact velocity, and this was consistent with the results of others [62]. Under the same impact velocity (56 cm s^{-1}), for the MWNT array with a lower static contact angle of ~140°, the droplet spreads but does not rebound from the surface, as indicated in Figure 10.4b.

As shown in Figures 10.4 and 10.5, during the early spreading phase, there was no significant difference between the response of the 140° and 163° contact angle arrays. This can be understood by considering the competition between the inertial force and the viscous force, which governs the spreading process. The competition can be characterized by using the Reynolds number (R_e):

$$R_e = \frac{\rho V R}{\mu} \quad (10.7)$$

where ρ is the density of water, V is the impact velocity, R is the radius of the droplet, and μ is the viscosity. By substituting the impact velocity into the Reynolds equation, the computed Reynolds number is ~560, which is much larger than 1. Therefore, it is understandable that the spreading patterns on the two surfaces are similar, as the inertial force is dominant over the viscous force. The inertial force in the impact is almost identical for the two surfaces, while the MWNT surface with a higher contact angle will offer a much reduced viscous force. When the inertial force is equal to the capillary and viscous forces, the water droplet will begin to decelerate. During the decelerating process, it is expected that the wettability of the solid surface would have a strong influence on the contact line dynamics. At the end of the spreading process, the surface energy available in the droplet on the superhydrophobic surface is much greater than on the hydrophobic surface.

During the receding process, as the droplet on the 163° array begins to retract, its receding contact angle shows a small reduction compared to its advancing contact angle; this is in contrast to the 140° array on which a dramatic reduction was observed in the receding contact angle (see Figure 10.5a). Also, during the retracting process the contact length for the droplet on the 140° array remains essentially constant (it is pinned) and does not decrease with time (see Figure 10.5b). This pinning of the contact line ultimately causes the droplet to remain attached to the surface and prevents it from rebounding. In contrast, the contact length for the 163° array decreases monotonically to zero as the droplet contracts and subsequently lifts from the surface. Similar results were also observed [56] at a higher impact velocity of $1.4\,\mathrm{m\,s^{-1}}$.

10.4.2
Impact Dynamics on Micropatterned Surfaces

As discussed in the previous section, complete rebounding on superhydrophobic surfaces with nanoscale roughness (static contact angle ~163°) was observed under impact velocities of $1.4\,\mathrm{m\,s^{-1}}$ and $0.56\,\mathrm{m\,s^{-1}}$. Now, a very pertinent question is: under the same static contact angle, are the impact behaviors on micro- and nanostructured surfaces identical?

In order to answer this question, a superhydrophobic Si micropillar surface (Figure 10.6b) with porosity (f) of 0.94 ($f = 1 - \phi$) was fabricated using a deep reactive ion-etching process. According to the prediction from the Cassie model, the static contact angle is 162.68°, which is almost the same as the static contact angle on the surface with nanoscale roughness. It was found that no rebounding occurs (see Figure 10.6a) at a low impact velocity of $0.56\,\mathrm{m\,s^{-1}}$, as opposed to complete rebounding on the nanoscale surface. With higher impact velocities, however, the rebounding ability of the droplet on the micropatterned Si surface becomes progressively worse with the water droplet invariably transiting into the Wenzel state and being pinned to the surface. These remarkable differences in dynamic behaviors on the surfaces with nanoscale and microscale roughnesses can be explained by considering the competition between dynamic pressure and critical pressure. The *critical pressure* is a measure of the pressure needed to pump the water into the microgrooves. If the *dynamic pressure* overcomes the critical pressure, then the water will be prone to transit from the Cassie to the Wenzel state, and rebounding is prevented. Otherwise, it is expected that rebounding would occur. The critical pressure that prevents the collapse of Cassie to Wenzel state is expressed as [63]:

10.4 Impact Dynamics of Water on Superhydrophobic Surfaces | 343

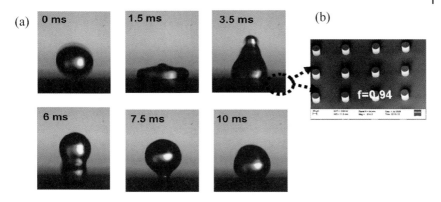

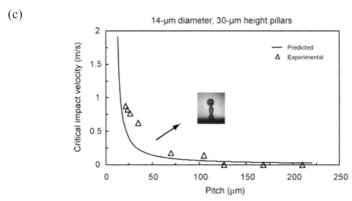

Figure 10.6 Droplet impact behavior on superhydrophobic surfaces with microscale roughness. (a) Snapshots of a water droplet impacting the surface with microscale roughness; (b) Scanning electron microscopy image of the micropatterned surface. The diameter and height of pillars are 40 and 67 μm, respectively; (c) Measured critical impact velocity [59] of a droplet with 1 mm radius as a function of geometric parameters (triangles). Data are compared with the criterion of impact velocity for the pinning of a droplet (solid lines) for two series of patterned Si with different pitch values. Reproduced with permission from Ref. [59]; © 2008, American Chemical Society.

$$P = -\gamma\cos\theta\frac{L}{A} = \gamma\cos\theta\frac{4}{a\left(1-\frac{1}{\phi}\right)} = \gamma\cos\theta\frac{4(f-1)}{af} \tag{10.8}$$

where L is the perimeter of the capillary, a is the diameter of the capillary, A is the cross-sectional area of the capillary, $f\ (= 1 - \phi)$ is the porosity of the surface, and θ is the contact angle on the flat surface.

The impacting dynamic pressure (P_d) of the droplet is related to its impact velocity [53], and is expressed as:

$$P_\text{d} = \frac{1}{2}\rho V^2 \tag{10.9}$$

where ρ and V are the density and impacting velocity, respectively.

At an impact velocity of $0.56\,\text{m s}^{-1}$, the impact pressure is 156 Pa. For the surface with nanoscale roughness, the critical pressure is 62.8 kPa (with $f = 0.96$, $a = 10\,\text{nm}$), which is much higher than the critical pressure of 138 Pa for the Si micropillar surface (with $f = 0.94$, $a = 40\,\mu\text{m}$). This explains why no rebounding occurs on the surface with microscale roughness, as the impacting pressure is greater than the critical pressure. However, this situation is reversed for the nanoscale roughness surface (MWNT array), which enables complete droplet rebound, as described in Section 10.4.1.

Jung and Bhushan [59] have studied impact dynamics on micropatterned surfaces with different porosities. It was found that there exists a critical impacting velocity (see Figure 10.6c); here, droplets that impact the surface with a lower velocity bounce off the surface, whereas those with a higher impact velocity adhered to the surface. Adherence was associated with the droplet being in the Wenzel state with a large solid–liquid contact area, whereas those droplets that bounced off the surface were in the Cassie state, with an air pocket beneath them.

Quéré et al. have investigated droplet impact on superhydrophobic surfaces at a relatively higher impacting velocity ($\sim 4.3\,\text{m s}^{-1}$) [61]. Here, it was observed that the shock produced peripheral satellite droplets that were much smaller in size than the original droplet. It was noted that, on the superhydrophobic surface, the threshold velocity for the ejection of satellite droplets was much lower than on a hydrophobic flat surface. This was because the viscous dissipation on the superhydrophobic surface was relatively lower, due to the presence of air pockets under the droplet.

10.5
Electrically Controlled Wettability Switching on Superhydrophobic Surfaces

Previously, the design and creation superhydrophobic surfaces and the impact dynamics of water droplets on such superhydrophobic surfaces has been reviewed. For many practical applications, the ability to dynamically control wettability is greatly desired [64–67], and today many switching techniques that allow the control of wetting properties have been developed, ranging from magnetic to optical and from chemical to mechanical [68–72]. Recent progress in the electrical control of wetting properties on superhydrophobic surfaces, in real-time mode, is reviewed in the following sections.

10.5.1
Reversible Control of Wettability Using Electrostatic Methods

Chen et al. [73] reported on the modification of surface wetting induced by morphology (SWIM) changes under the influence of electrostatic force. In this study, a free-standing metal/polymer membrane with hydrophobic microposts was sustained by SU8 spacers. Electrostatic force was used to deflect the metal/polymer membrane, thereby changing the surface morphology as well as the relative interfacial fraction of liquid-to-solid phase. The water contact angles under this mechanism could be reversibly manipulated from 131° to 152°, depending on the liquid–solid interface fraction, which is manipulated by the electrostatic force.

10.5.2
Electrowetting on Superhydrophobic Surfaces

Electrowetting is a popular method used to control wetting in microsystems. In this method, an insulating layer is introduced between the electrode and the electrolyte. The application of an electric potential between the electrode and electrolyte results in an accumulation of charges in the electrical double layer (EDL); this lowers the surface tension of the liquid, enabling it to wet the solid surface. Due to its low power consumption and easy integration with integrated chips, during the past 25 years a large number of devices based on electrowetting have been devised [74–78]. In particular, electrowetting on planar surfaces has been used successfully to actuate microdroplets in digital microfluidic devices. On planar surfaces, the electrowetting equation describing the relationship between the contact angle and applied voltage can be expressed as follows:

$$\cos\theta_e(V) = \cos\theta + \frac{\varepsilon\varepsilon_r V^2}{2\gamma d} \qquad (10.10)$$

where θ and θ_e are the initial and spreading contact angles, respectively, d is the thickness of the insulating layer, V is the applied voltage, and γ is the surface tension of the liquid. As the contact angle is proportional to V^2, the electrowetting response on planar surfaces is fully reversible.

On rough surfaces, the apparent contact angle and applied voltage relationship can be expressed as [79]:

$$\cos\theta_e = \cos\theta + \phi\frac{\varepsilon\varepsilon_r}{2d\gamma}V^2 \qquad (10.11)$$

where ϕ is the solid fraction. Since $\phi < 1$, it is clear that the electrowetting effect is much weaker on superhydrophobic (rough) surfaces than on flat surfaces. Moreover, once the water droplet has completely wetted the roughness features, its contact line would be prone to pinning; this would result in a large contact angle hysteresis and poor reversibility of the electrowetting on such rough surfaces [79, 80]. For example, studies with superhydrophobic SU-8 patterned surfaces and nanostructured silicon posts revealed that reversible wetting was not observed during electrowetting experiments [81, 82]. For the Su-8 micropatterned surfaces coated with a thin layer of Teflon, the contact angle decreased from an initial 152° to 90° under an activation voltage of 130 V, and switched back to 114° when the voltage was turned off [81]. On nanostructured silicon posts, the droplet transited from a superhydrophobic state to complete wetting with the application of ~50 V bias. When the liquid has penetrated all the way to the bottom of the nanostructured surface, no reversibility in the wetting was observed when the bias was turned off [82].

The first report of reversible electrowetting on rough surfaces in an air environment was recently demonstrated on heterogeneous Si nanowires networks [83]. This consisted of a two-layered structure: a low density of nanowires with a height of 15 µm; and a high density of nanowires with a height of 20 µm. It would be expected that the presence of such a two-layer structure would inhibit liquid penetration to the bottom of the surface, and the liquid could therefore reverse back when the electrowetting voltage was removed. Although these studies have provided a promising solution for the realization of reversible electrowetting on superhydrophobic rough surfaces, the switching ability was limited to very large contact angles (160–137°).

In order to realize fully reversible electrowetting on superhydrophobic surfaces, it is important to develop theoretical models for wetting and dewetting on rough surfaces. For example, Patankar *et al.* analyzed the energy barrier between the Cassie and Wenzel wetting states in the absence of an electrowetting voltage [23]. Bahadur and coworkers, accounted for the effect of electrowetting voltage by using an energy-minimization based model, and analyzed the influence of roughness and interfacial energy in determining the apparent contact angle of a droplet in the Cassie and Wenzel states [84].

Recently, a systematic experimental study of electrically controlled wetting and dewetting mechanism on rough surfaces was conducted by the authors' group [85]. In this study, superhydrophobic surfaces with different wettabilities were first developed using a two-step process, first creating a rough surface and then modifying it with a hydrophobic coating. Specifically, desired photoresist features were first patterned on a Si substrate using photolithography. The deposited photoresist not only defined the rough surface of interest, but also acted as the mask during the etching process. During etching, passivation gas C_4F_8 and etching gas SF_6 were passed alternatively; by doing so, the passivation layer (fluorocarbon film) generated from the reaction of silicon with the passivation gas prevented the sidewall from etching, and thus a structure with very high aspect ratio was achieved. When the desired structure had been produced, the photoresist layer could be stripped off, and then only the passivation gas could be passed. As a result, the entire surface features will become covered by a uniform layer of a thin fluorocarbon film. Wetting experiments for such Si micropillar arrays with different porosities confirmed that the arrays showed a superhydrophobic wetting property; the Cassie model showed excellent correlation with test data for arrays with different porosities [85].

As expected, under the application of applied bias, water droplets are found to wet the surface (see Figure 10.7a and b). When the applied voltage is removed, the dewetting is found to depend heavily on the porosity of the microarray. For arrays with high porosity (as in Figure 10.7b), the droplet contact angle does not

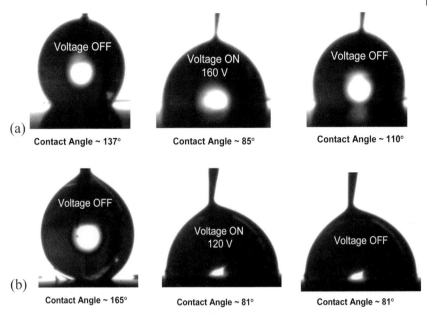

Figure 10.7 Electrowetting experiments for different arrays with different porosity and aspect ratio. (a) Porosity = 0.42; (b) Porosity = 0.96. Reproduced with permission from Ref. [85]; © 2008, American Scientific Publishers.

change significantly and the droplet remains pinned to the substrate (i.e., no significant change in contact line length). Such contact line pinning and contact angle hysteresis is indicative of a sticky Wenzel state, suggesting that the droplet now resides stably in the Wenzel state and cannot be expelled from the microarray. In contrast, the microarray with a porosity of 0.42 expels the droplet from the array (Figure 10.7a) and achieves a hydrophobic state (with contact angle ~110°). Note that the recovered contact angle (~110°) is lower than the initial contact angle (~137°) prior to beginning the electrowetting experiment. Therefore, complete dewetting is not observed even for the array with a porosity of 0.42. This may be caused by energy dissipation and potential energy loss during the motion of the water front along the microgrooves. For arrays with an intermediate porosity (between 0.42 and 0.96), relatively weak dewetting (not shown here) was observed compared to the results shown in Figure 10.7a for the microarray with a porosity of 0.42.

The dewetting difference for arrays with different porosities can be explained by comparing the relative energies of the Cassie and Wenzel states. By equating the energy in the Wenzel state and Cassie states, a critical aspect ratio (P) was found (Equation 10.12), which is governed by the porosity (f) of the array and the flat surface contact angle (θ). Above the critical aspect ratio, the Cassie wetting state is more stable compared to the Wenzel state, and *vice versa*.

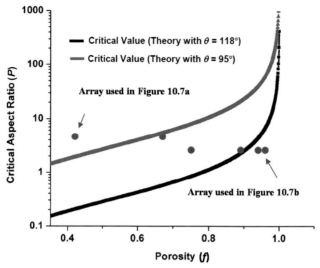

Figure 10.8 The relationship between the critical aspect ratio of the roughness features, porosity, and flat surface contact angle. Above the curve, the Cassie wetting state is more stable compared to the Wenzel state, and *vice versa*.

$$P = -\frac{f}{4(1-f)} \cdot \frac{1+\cos\theta}{\cos\theta} \tag{10.12}$$

As shown in Figure 10.8 it is clear that, for a porosity of 0.42 (i.e., for the pattern tested in Figure 10.7a), the actual aspect ratio (~4.6) is over an order of magnitude larger than the critical aspect ratio. This means that the Wenzel state is metastable for this case, and the Cassie state is the lower energy state. Consequently, as the applied bias is removed, the electrowetted liquid tends to switch back to the Cassie state and achieves a contact angle of ~110°. By contrast, at a porosity of 0.96 (i.e., for the pattern tested in Figure 10.7b), the actual aspect ratio of the pattern (~2.6) is significantly lower than the critical aspect ratio (~6.7). As a consequence, the Wenzel state is much more stable in this case and therefore the droplet will remain in this state, even after the applied bias has been removed. These results indicate that, when the applied voltage is removed, the dewetting process is highly sensitive to the porosity of the substrate, the aspect ratio of the Si micropillars, and to the wetting property of the material used to coat the pillars. With an appropriate selection of these parameters, water can partially dewet the rough surface and revert back to a hydrophobic state when the electrowetting voltage is removed.

10.5.3
Novel Strategies for Reversible Electrowetting on Rough Surfaces

A radically different strategy to achieve reversible electrowetting on superhydrophobic surfaces was developed by Krupenkin *et al.* [86], by using external heating. An embedded microheating element beneath the structure locally heated the

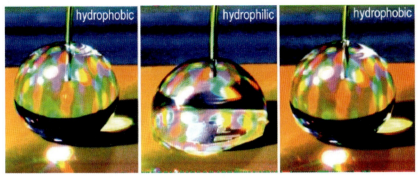

Figure 10.9 Snapshots of water droplet electrowetting response on a rough surface. Reversible wetting is enabled by heating the substrate. Reproduced with permission from Ref. [86]; © 2007, American Chemical Society.

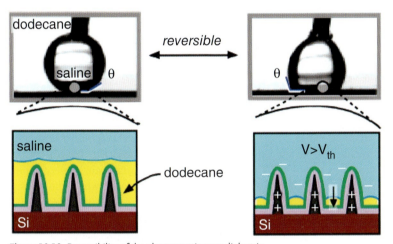

Figure 10.10 Reversibility of the electrowetting-on-dielectric (EWOD) phenomenon on a superhydrophobic surface by immersion of the water drop in dodecane. Reproduced with permission from Ref. [87]; © 2007, American Chemical Society.

surface and caused liquid boiling and droplet expulsion from the surface (Figure 10.9). This technique represented a major step forward to enable reversible electrowetting on superhydrophobic rough surfaces. However, an extra energy supply (i.e., substrate heating) was required.

Another strategy to achieve a reversible wetting behavior over a wide range would be to replace the air environment with a hydrophobic liquid [87]. As reported by Dhindsa *et al.*, reversible electrowetting was achieved on carbon nanofiber scaffolds by the use of a competitive, two-liquid (dodecane/saline) environment. The nanostructured scaffold with a $5 \times 5\,\mu m$ pitch was grown on Si substrates, electrically insulated with a conformal dielectric, and hydrophobized with a fluoropolymer. As indicated in Figure 10.10, the electrowetted liquid with a contact angle of 120° could reversibly switch back to its initial state of 160°.

10.6
Electrochemically Controlled Wetting of Superhydrophobic Surfaces

Aligned MWNT arrays, due to their frictionless surfaces and nanoscale pores, hold great promise as membranes in microfiltration and separation technologies [88–91], and where the well-controlled wettability of MWNT arrays is a major requirement. As noted previously, MWNT arrays (membranes) exhibit a superhydrophobic property by inhibiting the wetting of water on these surfaces. Thus, in order for these arrays to be wetted by water, an energy barrier must be overcome if a transition from a superhydrophobic to a hydrophilic state is to be effected. To achieve this, both Hinds and Holt and their coworkers used plasma etching to convert a superhydrophobic nanotube surface into a hydrophilic surface when investigating the transport of liquid flow through MWNT membranes [92, 93]. Although both chemical and mechanical means can be used to alter the wetting and flow properties of fluids through nanotube membranes, electrical methods offer a powerful, nondestructive, and selective way to achieve this [94].

10.6.1
Polarity-Dependent Wetting of Nanotube Membranes

In order to tailor the wettability of MWNT membranes by using a simple electrochemical method, a Pt wire can be inserted into the water droplet to establish electrical contact (Figure 10.11a). First, the droplet response was studied with the membrane as the anode (positive potential) and the Pt wire as the cathode (negative potential). The droplet shape and contact angle were found to remain unchanged up to a voltage of ~1.7 V; however, when the critical voltage was reached there was an abrupt transition from the superhydrophobic state to the hydrophilic state, and the droplet rapidly sank into the nanotube forest (see Figure 10.11c). The droplet pumping (or sinking) speed was determined by the applied voltage; for example, with a voltage of 1.8 V the droplet took over 140 s to become submerged, but with 2.6 V the time was 90 s. This time was reduced to 60 s for 3.6 V, and to only 40 s for an applied voltage of 4.5 V (Figure 10.11b).

The droplet response with the CNT membrane as cathode was also investigated. Snapshots of the droplet shape at different applied voltages are shown in Figure 10.12a. Here, the droplet shape for up to 60 V of applied bias was almost identical to that at 0 V, but for voltages exceeding 63 V a gradual reduction in contact angle was observed. The measured contact angle compared to the applied voltage is shown in Figure 10.12b; the threshold voltage required to activate sinking of the droplet into the membrane was 95 V, almost 50-fold greater than when the membrane was the anode.

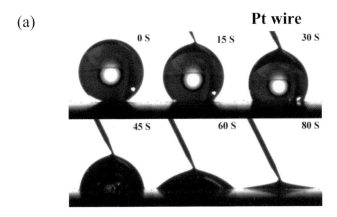

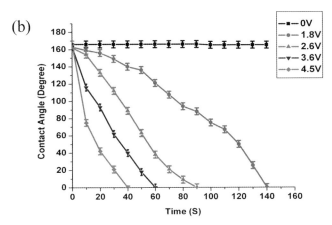

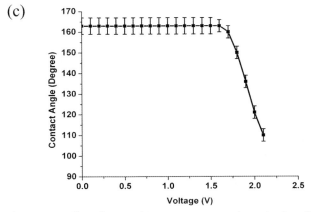

Figure 10.11 Effect of positive bias (nanotubes as anode) on the droplet response. (a) Snapshots of water droplet shape change with +2.6 V potential applied with MWNT film as anode and Pt wire as cathode. The droplet sinks into the nanotube membrane in about 90 s; (b) Apparent contact angle variation with time for different applied voltages; (c) Apparent contact angle variation with positive potential applied to the MWNT membrane. Reproduced with permission from Ref. [94]; © 2007, American Chemical Society.

(a)

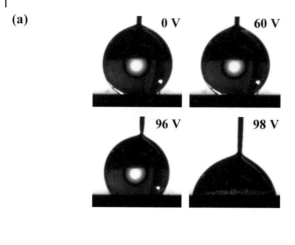

(b)

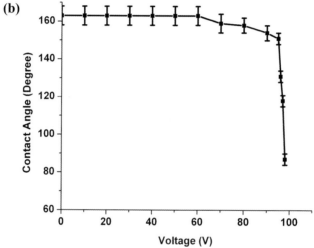

Figure 10.12 Effect of negative bias (nanotubes as cathode) on the droplet response. (a) Snapshots of water droplet shape change with negative bias applied to the nanotube membrane. No significant droplet shape change is observed until ~60 V; (b) Apparent contact angle variation with negative bias applied to the nanotube membrane. Reproduced with permission from Ref. [94]; © 2007, American Chemical Society.

10.6.2
Mechanism of Polarity-Dependent Wetting and Transport

In order to explain the above observations, it is important to consider the classical electrowetting phenomenon. In a recent study, it was shown that the application of an electric potential to a single-walled CNT attached to the tip of an atomic force microscope could be used to wet mercury, although no polarity-dependent wetting was reported [95]. In classical electrowetting, the contact angle change with voltage

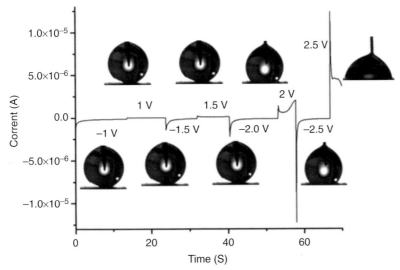

Figure 10.13 Current and droplet shape measurements performed with a nanotube membrane and Pt wire as the electrodes. Reproduced with permission from Ref. [94]; © 2007, American Chemical Society.

is symmetric (Equation 10.11) for both positive and negative bias. Thus, the abrupt transition from nonwetting to wetting state at low voltages cannot be explained by electrowetting, as it occurs only when the nanotube membrane is used as the anode.

In an effort to describe this situation, the role of water electrolysis was considered with regards to the polarity dependence of electrowetting. To examine the extent of electrolysis, current measurements were made (Figure 10.13) between the CNT membrane electrode and the Pt wire during electrowetting experiments. A negative bias (−1.0 V) was first applied to the membrane for approximately 15 s, after which the bias was flipped and +1 V potential applied. This process of polarity reversal was repeated at 1.5 V, 2 V, and 2.5 V. Notably, sinking of the droplet (at +2 V bias with the membrane as anode) was accompanied by a sharp increase in the current. However, when the polarity was flipped to −2 V (with the membrane as cathode) the droplet immediately ceased to sink (see Figure 10.13) and the steady-state current also decreased. Again, when the polarity was switched to +2.5 V, the droplet resumed its sinking and the current increased. The sharp increases in current at +2 V and +2.5 V were indicative of electrolysis and suggested that, in this system, the droplet sinking was linked to the presence of electrolysis. During electrolysis, an electrochemical oxidation of the nanotube anode might have been responsible for the observed abrupt transition (switching) from a nonwetting to a wetting state. The results of a recent study showed that a minimum anode voltage of 1.7 V is necessary to activate the oxidation of a MWNT electrode, and this correlates well with the onset voltage for droplet sinking observed in the above-described experiments [96].

In order to understand the effect of the attachment of oxygen-containing functional groups on the wetting behavior of the nanotube, a theoretical study was carried out of the interaction of H_2O molecules with pristine and oxidized nanotubes, using a first-principles density functional method [97]. For the oxidized nanotube, two cases were considered: (i) where the oxygen is preadsorbed onto a Stone–Wales defect; and (ii) where vacancy defect that is occupied by oxygen is considered (the oxygen molecule dissociates on the vacancy site, and one of the atoms is bonded to two dangling carbon atoms while the other oxygen atom binds to the third dangling carbon and is saturated with hydrogen). In both cases, the binding energy of the H_2O molecule with the oxidized nanotube was calculated, and the results compared to binding of the H_2O molecule with a pristine (or nonoxidized) nanotube. The simulation results indicated that the binding energy of water with the oxidized nanotube (0.16–0.31 eV) was several-fold that of the pristine nanotube (~0.03 eV), and comparable to the hydrogen-binding energy in liquid water and in a water dimer (~0.19 eV); this confirmed that nanotube oxidation would indeed make the membrane strongly hydrophilic. These results also explained why the droplet ceased to sink when the nanotube polarity was flipped (as shown in Figure 10.13). When the nanotube was used as the cathode, no further oxidation occurred and only the oxidized region was wetted; the remaining portion of the nanotube remained nonwetting to the droplet.

10.6.3
Potential Applications of Electrochemically Controlled Wetting and Transport

The electrochemically controlled wetting of nanotube membranes offers many potential applications, due to its simplicity and low-voltage operation. For example, the applied bias can be used to controllably wet the membrane surface; the transport of water through the nanotube membrane can in fact be repeatedly stopped and restarted on demand, depending on the applied voltage (see Figure 10.14). This represents a simple low-voltage technique for gating the flow of aqueous liquids in nanofluidic systems.

Another interesting application of this wetting phenomenon is shown in Figure 10.15. Here, three droplets are placed side-by-side on the nanotube surface (see top panel of figure). No voltage is applied to the droplet on the extreme right, and it shows no contact angle change with time. A bias of +2 V was applied to the center droplet for 20 s, and the bias then switched off. The contact angle of the droplet decreased to 115° in 20 s (middle panel in Figure 10.15). Next, the bias (+2 V) was applied to the center droplet for a further 20 s. When the bias was switched off, the contact angle of the middle droplet decreased to 90° (lower panel of Figure 10.15). The same procedure of biasing the droplet in two 20 s intervals was repeated for the droplet to the extreme left, but at a higher potential (+4 V). The contact angle of the droplet decreased to 93° in 20 s, and then to 69° in 40 s. It should be noted that, after the voltages have been switched off, sinking of the droplets ceased and they located stably on the nanotube surface at the reduced

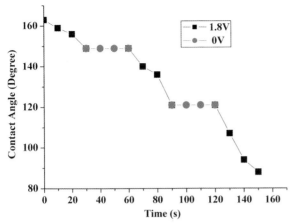

Figure 10.14 Selectively stopping and then restarting water transport through the nanotube membrane is demonstrated by simply cycling the applied voltage from 0 to +1.8 V. Reproduced with permission from Ref. [94]; © 2007, American Chemical Society.

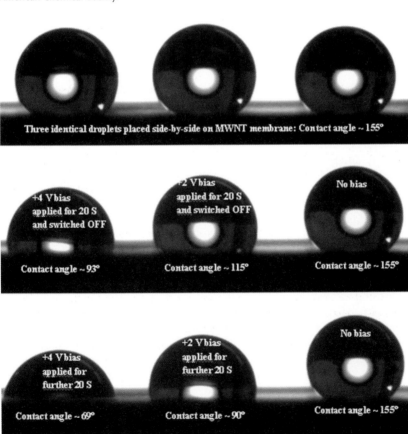

Figure 10.15 Using electrochemical oxidation of nanotubes to set up a controllable wetting gradient on the membrane surface. Reproduced with permission from Ref. [94]; © 2007, American Chemical Society.

contact angles. In this way, it is possible to control precisely the local wettability of any specific point on the nanotube surface, while wettability gradients from superhydrophobic to hydrophilic can be engineered on the nanotube membrane surface. Such extreme hydrophilic/hydrophobic spatial contrast has potential applications that include biomimetic water-harvesting surfaces, drug-release coatings, and microchannel and lab-on-a-chip devices.

10.7
Summary and Future Perspectives

The study of wetting phenomena at both the microscale and nanoscale represents a dynamic and rapidly evolving area of research. The recent important developments in this field that have been reviewed in this chapter permit the following main conclusions to be drawn:

- Roughness has a significant impact on the static and dynamic wetting properties of solid surfaces. Both, unitary roughness (microscale or nanoscale only) and two-scale hierarchical roughness can enable superhydrophobicity. However, the presence of hierarchical structures not only amplifies the local contact angle and decreases contact angle hysteresis, but also (most importantly) results in a more robust "Cassie" state. Such a robust composite liquid–gas–solid interface with excellent air-trapping ability is critical for self-cleaning applications.

- The implementation of reentrant structures adds a new dimension to the design of novel artificial surfaces. Both, superhydrophobic and superoleophobic surfaces can be achieved from hydrophilic substrates.

- The impact dynamics of water droplets on surfaces with microscale roughness is very different from that of surfaces with nanoscale roughness. Under the same static contact angle, the critical velocity above which rebounding does not occur in the nanostructured surface was seen to be much larger compared to the micropatterned surface. Therefore, nanostructured surfaces are superior to micropatterned surfaces in terms of maintaining their hydrophobicity under impact conditions. It is for this reason that microscale surfaces alone are not appropriate for the design of superhydrophobic surfaces.

- The wettability of hydrophobic rough surfaces can be manipulated in real time by using electrowetting, a method that tunes the contact angle by reducing liquid–solid interfacial tension. However, fully reversible electrowetting in air environments without any extra energy supply (e.g., substrate heating) has not yet been realized. Nonetheless, partial dewetting on hydrophobic rough surfaces can be achieved by a careful choice of the aspect ratio, coating property, and porosity of the solid surface.

- The wettability of superhydrophobic rough surfaces (e.g., CNT membranes) can also be manipulated (albeit irreversibly) using electrochemical methods.

By electrochemically oxidizing the nanotube surface, the wettability of the membrane can be switched instantly from superhydrophobic to hydrophilic.

- The electrochemical control of wettability of nanotube membranes has several exciting applications, including the directional control of water flow by reversing the polarity of the applied bias, the ability to controllably switch the water flow ON and OFF by controlling the magnitude of the applied bias, and the ability to generate extreme hydrophobic-to-hydrophilic spatial contrast on the membrane surface. These properties might find various applications in nanofluidic, separation, and water-harvesting technologies.

10.7.1
Future Perspectives

Although significant advances have been made in our fundamental understanding of interfacial phenomena at both the microscale and nanoscale, many interesting questions remain unresolved. Hence, the following topics are proposed for future investigation:

- **Bottom-up design of superhydrophobic multiscale hierarchical structures:** In Nature, the Lotus leaf and the feet of the Gecko share multiscale hierarchical structures. Significant progress has been made on the optimization of the contact and adhesion mechanism of the Gecko. Although it is argued that two-scale hierarchical structures are an ideal choice for a robust composite interface, predictive models and extensive experiments to demonstrate this are missing. Based on wetting models (e.g., Cassie, Wenzel), contact angle hysteresis models and energy-minimization approaches, it should be possible to further optimize the Lotus leaf structure, as has been successfully accomplished in the case of Gecko structures. These studies will provide a unique insight into the creation of stable superhydrophobic materials using emergent nano- and micro-fabrication technologies.

- **Digital electrowetting on superhydrophobic surfaces:** Electrically controlled wetting and dewetting on superhydrophobic surfaces has been demonstrated. By switching the underlying electrodes on and off in a programmed and digital manner, it should be possible to achieve rapid and controllable droplet movement on superhydrophobic surfaces. The successful development of such digital electrowetting techniques on superhydrophobic surfaces (and surfaces with wetting gradients) might have broad applications in microfluidics, lab-on-a-chip, and thermal management systems.

Acknowledgments

These studies were supported in part by NSF Award 0403789 (to N.K.). The authors thank their collaborators at Rice University (Professor Pulickel Ajayan's group) and at Rensselaer Polytechnic Institute (Professor Amir Hirsa's Group) for fruitful collaboration in the fabrication and testing of the wetting properties of nano- and micro-structure arrays.

References

1 Bhushan, B., Nosonovsky, M. and Jung, Y.C. (2008) Lotus effect: roughness-induced superhydrophobic surfaces, in *Nanotribology and Nanomechanics* (ed. B. Bhushan), Springer, Berlin, Heidelberg, pp. 995–1072.
2 Herminghaus, S., Brinkmann, M. and Seemann, R. (2008) Wetting and dewetting of complex surface geometries. *Annual Review of Materials Research*, 38, 101–21.
3 Rauscher, M. and Dietrich, S. (2008) Wetting phenomena in nanofluidics. *Annual Review of Materials Research*, 38, 143–72.
4 Sun, T., Feng, L., Gao, X. and Jiang, L. (2005) Bioinspired surfaces with special wettability. *Accounts of Chemical Research*, 38, 644–52.
5 Neinhuis, C. and Barthlott, W. (1997) Characterization and distribution of water repellent, self-cleaning plant surfaces. *Annals of Botany*, 79, 667–77.
6 Yu, Y., Zhao, Z.-H. and Zheng, Q.-S. (2007) Mechanical and superhydrophobic stabilities of two-scale surface structure of lotus leaves. *Langmuir*, 23, 8212–16.
7 Feng, X.J. and Jiang, L. (2006) Design and creation of superwetting/antiwetting surfaces. *Advanced Materials*, 18, 3063–78.
8 Quéré, D. (2008) Wetting and roughness. *Annual Review of Materials Research*, 38, 71–99.
9 Gao, X.F. and Jiang, L. (2004) Water-repellent legs of water striders. *Nature*, 432, 36.
10 Song, Y.S. and Sitti, M. (2007) Surface-tension-driven biologically inspired water strider robots: theory and experiments. *IEEE Transactions on Robotics*, 23, 578–89.
11 Feng, X.Q., Gao, X.F., Wu, Z.N., Jiang, L. and Zheng, Q.S. (2007) Superior water repellency of water strider legs with hierarchical structures: experiments and analysis. *Langmuir*, 23, 4892–6.
12 Dickinson, M. (2003) Animal locomotion: how to walk on water. *Nature*, 424, 621–2.
13 Parker, A.R. and Lawrence, C.R. (2001) Water capture by a desert beetle. *Nature*, 414, 33–4.
14 Zhai, L., Berg, M.C., Cebeci, F.C., Kim, Y., Milwid, J.M., Rubner, M.F. and Cohen, R.E. (2006) Superhydrophobic carbon nanotube forests. *Nano Letters*, 6, 1213–17.
15 Wenzel, R.N. (1936) Resistance of solid surfaces to wetting by water. *Industrial and Engineering Chemistry*, 28, 988–94.
16 Cassie-Baxter, B.D. and Baxter, S. (1944) Wettability of porous surfaces. *Transactions of the Faraday Society*, 40, 546–51.
17 Lafuma, A. and Quéré, D., (2003) Superhydrophobic states. *Nature Materials*, 2, 457–60.
18 Callies, M. and Quéré, D. (2005) On water repellency. *Soft Matter*, 1, 55–61.
19 Herminghaus, S. (2000) Roughness-induced nonwetting. *Europhysics Letters*, 165–70.52
20 Extrand, C.W. (2004) Criteria for ultrahydrophobic surfaces. *Langmuir*, 20, 5013–18.
21 Chen, W., Fadeev, A.Y., Hsieh, M.C., Öner, D., Youngblood, J.P. and McCarthy, T.J. (1999) Ultrahydrophobic and ultralyophobic surfaces–some comments and some examples. *Langmuir*, 15, 3395–9.
22 He, B., Patankar, N.A. and Lee, J. (2003) Multiple equilibrium droplet shapes and design criterion for rough hydrophobic surfaces. *Langmuir*, 19, 4999–5003.
23 Patankar, N. (2004) Transition between superhydrophobic states on rough surfaces. *Langmuir*, 20, 7097–102.
24 Nosonovsky, M. (2007) Multiscale roughness and stability of superhydrophobic biomimetic interfaces. *Langmuir*, 23, 3157–61.

25 Tuteja, A., Choi, W., McKinley, G.H., Cohen, R.E. and Rubner, M.F. (2008) Design parameters for superhydrophobicity and superoleophobicity. *MRS Bulletin*, **33**, 752–8.

26 Li, X.-M., Reinhoudt, D. and Crego-Calama, M. (2007) What do we need for a superhydrophobic surface? A review on the recent progress in the preparation of superhydrophobic surfaces. *Chemical Society Reviews*, **36**, 1350–68.

27 Zhao, N., Shi, F., Wang, Z. and Zhang, X. (2005) Combining layer-by-layer assembly with electrodeposition of silver aggregates for fabricating superhydrophobic surfaces. *Langmuir*, **21**, 4713–16.

28 Kulinich, S.A. and Farzaneh, M. (2004) Hydrophobic properties of surfaces coated with fluoroalkylsiloxane and alkylsiloxane monolayers. *Surface Science*, **573**, 379–90.

29 Zhang, X.T., Sato, O. and Fujishima, A. (2004) Water ultrarepellency induced by nanocolumnar ZnO surface. *Langmuir*, **20**, 6065–7.

30 Lee, W., Jin, M.K., Yoo, W.C. and Lee, J.K. (2004) Nanostructuring of a polymeric substrate with well-defined nanometer-scale topography and tailored surface wettability. *Langmuir*, **20**, 7665–9.

31 Zhu, L.B., Xiu, Y.H., Xu, J.W., Tamirisa, P.A., Hess, D.W. and Wong, C.P. (2005) Superhydrophobicity on two-tier rough surfaces fabricated by controlled growth of aligned carbon nanotube arrays coated with fluorocarbon. *Langmuir*, **21**, 11208–12.

32 Li, J., Fu, J., Cong, Y., Wu, Y., Xue, L.J. and Han, Y.C. (2006) Macroporous fluoropolymeric films templated by silica colloidal assembly: a possible route to super-hydrophobic surfaces. *Applied Surface Science*, **252**, 2229–34.

33 Hikita, M., Tanaka, K., Nakamura, T., Kajiyama, T. and Takahara, A. (2005) Super-liquid-repellent surfaces prepared by colloidal silica nanoparticles covered with fluoroalkyl groups. *Langmuir*, **21**, 7299–302.

34 Onda, T., Shibuichi, S., Satoh, N. and Tsujii, K. (1996) Super-water-repellent fractal surfaces. *Langmuir*, **12**, 2125–7.

35 Guo, Z. and Liu, W. (2007) Biomimic from the superhydrophobic plant leaves in nature: binary structure and unitary structure. *Plant Science*, **172**, 1103–12.

36 Wu, X.D., Zheng, L.J. and Wu, D. (2005) Fabrication of superhydrophobic surfaces from microstructured ZnO-based surfaces via a wet-chemical route. *Langmuir*, **21**, 2665–7.

37 Niu, J.J., Wang, J.N. and Xu, Q.F. (2008) Aligned silicon carbide nanowire crossed nets with high superhydrophobicity. *Langmuir*, **24**, 6918–23.

38 Qu, M., Zhao, G., Wang, Q., Cao, X. and Zhang, J. (2008) Fabrication of superhydrophobic surfaces by a Pt nanowire array on Ti/Si substrates. *Nanotechnology*, **19**, 55707.

39 Ma, M., Hill, R.M., Lowery, J.L., Fridrikh, S.V. and Rutledge, G.C. (2005) Electrospun poly (styrene-block-dimethylsiloxane) block copolymer fibers exhibiting superhydrophobicity. *Langmuir*, **21**, 5549–54.

40 Jin, M., Feng, X., Xi, J., Zhai, J., Cho, K., Feng, L. and Jiang, L. (2005) Superhydrophobic PDMS surface with ultra-low adhesive force. *Macromolecular Rapid Communications*, **26**, 1805–9.

41 Xie, D., Fan, G., Zhao, N., Guo, X., Xu, J., Dong, J., Zhang, L., Zhang, Y. and Han, C. (2004) Facile creation of a bionic super-hydrophobic block copolymer surface. *Advanced Materials*, **16**, 1830–3.

42 Yoshimitsu, Z., Nakajima, A., Watanabe, T. and Hashimoto, K. (2002) Effects of surface structure on the hydrophobicity and sliding behavior of water droplets. *Langmuir*, **18**, 5818–22.

43 Gao, L. and McCarthy, T.J. (2006) The "Lotus Effect" explained: two reasons why two length scales of topography are important. *Langmuir*, **22**, 2966–7.

44 Michael, N. and Bhushan, B. (2007) Hierarchical roughness makes superhydrophobic states stable. *Microelectronic Engineering*, **84**, 382–6.

45 Wang, Z., Ci, L., Ajayan, P.M. and Koratkar, N. (2007) Combined micro/nanoscale surface roughness for enhanced hydrophobic stability in carbon nanotube arrays. *Applied Physics Letters*, **90**, 143117.

46 Jin, M., Feng, X.J., Sun, T., Zhai, J., Li, T. and Jiang, L. (2000) Superhydrophobic aligned polystyrene nanotube films with high adhesive force. *Advanced Materials*, **17**, 1977–81.

47 Cho, W.K. and Choi, I.S. (2008) Fabrication of hairy polymeric films inspired by geckos: wetting and high adhesion properties. *Advanced Functional Materials*, **18**, 1089–96.

48 Lai, Y., Lin, C., Huang, J., Zhuang, H., Sun, L. and Nguyen, T. (2008) Markedly controllable adhesion of superhydrophobic spongelike nanostructure TiO_2 films. *Langmuir*, **24**, 3867–73.

49 Feng, L., Zhang, Y., Xi, J., Zhu, Y., Wang, N., Xia, F. and Jiang, L. (2008) Petal effect: a superhydrophobic state with high adhesive force. *Langmuir*, **24**, 4114–19.

50 Lau, K.K.S., Bico, J., Teo, K.B.K., Chhowalla, M., Amaratunga, G.A.J., Milne, W.I., McKinley, G.H. and Gleason, K.K. (2003) Superhydrophobic carbon nanotube forests. *Nano Letters*, **3**, 1701–5.

51 Xiu, Y., Zhu, L., Hess, D.W. and Wong, C.P. (2007) Hierarchical silicon etched structures for controlled hydrophobicity/superhydrophobicity. *Nano Letters*, **7**, 3388–93.

52 Zheng, Y., Gao, X. and Jiang, L. (2007) Directional adhesion of superhydrophobic butterfly wings. *Soft Matter*, **3**, 178–82.

53 Cao, L.L., Hu, H.H. and Gao, D. (2007) Design and fabrication of micro-textures for inducing a superhydrophobic behavior on hydrophilic materials. *Langmuir*, **23**, 4310–14.

54 Tuteja, A., Choi, W., Ma, M., Mabry, J.M., McKinley, G.H. and Cohen, R.E. (2008) Robust omniphobic surfaces. *Proceedings of the National Academy of Sciences of the United States of America*, **105**, 18200–5.

55 Bormashenko, E., Pogreb, R., Whyman, G., Bormashenko, Y. and Erlich, M. (2007) Vibration-induced Cassie-Wenzel wetting transition on rough surfaces. *Applied Physics Letters*, **90**, 201917.

56 Wang, Z., Carlos, L., Hirsa, A. and Koratkar, N. (2007) Impact dynamics and rebound of water droplets on superhydrophobic carbon nanotube arrays. *Applied Physics Letters*, **91**, 023105.

57 Yarin, A.L. (2006) Drop impact dynamics: splashing, spreading, receding, bouncing. *Annual Review of Fluid Mechanics*, **38**, 159–92.

58 Rioboo, R., Tropea, C. and Marengo, M. (2001) Outcomes from a drop impact on solid surfaces. *Atomization and Sprays*, **11**, 155–65.

59 Jung, Y.C. and Bhushan, B. (2008) Dynamic effects of bouncing water droplets on superhydrophobic surfaces. *Langmuir*, **24**, 6262–9.

60 Rioboo, R., Marengo, M. and Tropea, C. (2002) Time evolution of liquid drop impact onto solid, dry surfaces. *Experimental Fluids*, **33**, 112–24.

61 Reyssat, M., Pepin, A., Marty, F., Chen, Y. and Quéré, D. (2006) Bouncing transitions on microtextured materials. *Europhysics Letters*, **74**, 306–12.

62 Richard, D., Clanet, C. and Quéré, D. (2002) Surface phenomena: contact time of a bouncing drop. *Nature*, **417**, 811.

63 Zheng, Q.-S., Yu, Y. and Zhao, Z.-H. (2005) Effects of hydraulic pressure on the stability and transition of wetting modes of superhydrophobic surfaces. *Langmuir*, **21**, 12207–12.

64 McHale, G., Herbertson, D.L., Elliott, S.J., Shirtcliffe, N.J. and Newton, M.I. (2007) Electrowetting of nonwetting liquids and liquid marbles. *Langmuir*, **23**, 918–24.

65 Sun, C., Zhao, X.W., Han, Y.H. and Gu, Z.Z. (2008) Control of water droplet motion by alteration of roughness gradient on silicon wafer by laser surface treatment. *Thin Solid Films*, **516**, 4059–63.

66 Yu, X., Wang, Z., Jiang, Y. and Zhang, X. (2006) Surface gradient material: from superhydrophobicity to superhydrophilicity. *Langmuir*, **22**, 4483–6.

67 Balaur, E., Macak, J.M., Taveira, L. and Schmuki, P. (2005) Tailoring the wettability of TiO_2 nanotube layers. *Electrochemistry Communications*, **7**, 1066–70.

68 Verplanck, N., Coffinier, Y., Thomy, V. and Boukherroub, R. (2007) Wettability switching techniques on superhydrophobic surfaces. *Nanoscale Research Letters*, **2**, 577–96.

69. Han, J.T., Kim, S. and Karim, A. (2007) UVO-tunable superhydrophobic to superhydrophilic wetting transition on biomimetic nanostructured surfaces. *Langmuir*, **23**, 2608–14.

70. Lim, H.S., Han, J.T., Kwak, D., Jin, M. and Cho, K. (2006) Photoreversibly switchable superhydrophobic surface with erasable and rewritable pattern. *Journal of the American Chemical Society*, **128**, 14458–9.

71. Yu, X., Wang, Z., Jiang, Y. and Zhang, X. (2006) Surface gradient material: from superhydrophobicity to superhydrophilicity. *Langmuir*, **22**, 4483–6.

72. Sun, T., Wang, G., Feng, L., Liu, B., Ma, Y., Jiang, L. and Zhu, D. (2003) Reversible switching between superhydrophilicity and superhydrophobicity. *Angewandte Chemie, International Edition*, **43**, 357–60.

73. Chen, T.-H., Chuang, Y.-J., Chieng, C.-C. and Tseng, F.-G. (2007) A wettability switchable surface by microscale surface morphology change. *Journal of Micromechanics and Microengineering*, **17**, 489–95.

74. Quinn, A., Sedev, R. and Ralston, J. (2003) Influence of the electrical double layer in electrowetting. *Journal of Physical Chemistry B*, **107**, 1163–9.

75. Beni, G. and Hackwood, S. (1981) Electro-wetting displays. *Applied Physics Letters*, **38**, 207.

76. Cho, S.K., Moon, H.J. and Kim, C.J. (2003) Creating, transporting, cutting, and merging liquid droplets by electrowetting-based actuation for digital microfluidic circuits. *Journal of Microelectromechanical Systems*, **12**, 70–80.

77. Teh, S., Lin, R., Hung, L. and Lee, A. (2008) Droplet microfluidics. *Lab on a Chip*, **8**, 198–220.

78. Prins, M.W.J., Welters, W.J.J. and Weekamp, J.W. (2001) Fluid control in multichannel structures by electrocapillary pressure. *Science*, **291**, 277–81.

79. Torkelli, A. (2003) Droplet microfluidics on a planar surface. PhD Thesis, Department of Electrical Engineering, Helsinki University of Technology, Espoo, Finland.

80. McHale, G., Shirtcliffe, N.J. and Newton, M.I. (2004) Contact-angle hysteresis on super-hydrophobic surfaces. *Langmuir*, **20**, 10146–9.

81. Herbertson, D.L., Evans, C.R., Shirtcliffe, N.J., McHale, G. and Newton, M.I. (2006) Electrowetting on superhydrophobic SU-8 patterned surfaces. *Sensors and Actuators A*, **130**, 189–93.

82. Krupenkin, T., Taylor, J.A., Schneider, T.M. and Yang, S. (2004) From rolling ball to complete wetting: the dynamic tuning of liquids on nanostructured surfaces. *Langmuir*, **20**, 3824–7.

83. Verplanck, N., Galopin, E., Camart, J.-C., Thomy, V., Coffinier, Y. and Boukherroub, R. (2007) Reversible electrowetting on superhydrophobic silicon nanowires. *Nano Letters*, **7**, 813–17.

84. Bahadur, V. and Garimella, V. (2007) Electrowetting-based control of static droplet states on rough surfaces. *Langmuir*, **23**, 4918–24.

85. Wang, Z. and Koratkar, N. (2008) Electrically controlled wetting and dewetting on micropillar arrays. *Advanced Science Letters*, **1**, 222–5.

86. Krupenkin, T., Taylor, J.A., Wang, E.N., Kolodner, P., Hodes, M. and Salamon, T.R. (2007) Reversible wetting-dewetting transitions on electrically tunable superhydrophobic nanostructured surfaces. *Langmuir*, **23**, 9128–33.

87. Dhindsa, M.S., Smith, N.R. and Heikenfeld, J. (2006) Reversible electrowetting of vertically aligned superhydrophobic carbon nanofibers. *Langmuir*, **22**, 9030–4.

88. Dzubiella, J. and Hansen, J.P. (2005) Electric-field controlled water and ion permeation of a hydrophobic nanopore. *Journal of Chemical Physics*, **120**, 5001–4.

89. Yum, K. and Yu, M.F. (2005) Surface-mediated liquid transport through molecularly thin liquid films on nanotubes. *Physical Review Letters*, **95**, 18602.

90. Hummer, G., Rasaiah, J.C. and Noworyta, J.P. (2001) Water conduction through the hydrophobic channel of a carbon nanotube. *Nature*, **414**, 188–90.

91. Mann, D.J. and Halls, M.D. (2003) Water alignment and proton conduction inside

carbon nanotubes. *Physical Review Letters*, **90**, 195503.

92 Hinds, B.J., Chopra, N., Rantell, T., Andrews, R., Gavalas, V. and Bachas, L.G. (2003) Aligned multiwalled carbon nanotube membranes. *Science*, **303**, 62–5.

93 Holt, J.K., Park, H.G., Wang, Y., Stadermann, M., Artyukhin, A.B., Grigoropoulos, C.P., Noy, A. and Bakajin, O. (2006) Fast mass transport through sub-2-nanometer carbon nanotubes. *Science*, **312**, 1034–7.

94 Wang, Z., Ci, L., Chen, L., Nayak, S., Ajayan, P.M. and Koratkar, N. (2007) Polarity-dependent electrochemically controlled transport of water through carbon nanotube membranes. *Nano Letters*, **7**, 697–702.

95 Chen, J.Y., Kutana, A., Collier, C.P. and Giapis, K.P. (2005) Electrowetting in carbon nanotube. *Science*, **310**, 1480–3.

96 Ito, T., Sun, L. and Crooks, R.M. (2003) Electrochemical etching of individual multiwall carbon nanotubes. *Electrochemical and Solid State Letters*, **6**, 4–7.

97 Parr, R.G. and Yang, W. (1989) *Density-Functional Theory of Atoms and Molecules*, Oxford University Press.

11
Imaging of Thin Films, and Its Application in the Life Sciences

Silvia Mittler

11.1
Introduction

For many decades, thin films down to monolayer thicknesses have been of scientific and engineering interest, with their great potential and ever-increasing numbers of applications. Both, organic and inorganic materials may be deposited in thin and ultrathin film forms on a huge variety of substrate materials and forms. However, depending on the combination of substrate and thin film material, the deposition process and characterization methods use to monitor the thickness, homogeneity, durability, reproducibility and functionality of such films must be selected with the utmost care.

Among the many technological applications of thin films can be included lubricants [1], electronic components [2, 3], paper protection, food and food packaging [4, 5], solar cells [6], smart windows [7], thermal barriers [8], and very many others. Yet, perhaps the greatest area of application is in biomedicine.

Due to the enormous variety of thin film use, this chapter will concentrate on an area that is important in the life sciences, namely the deposition of organic thin films onto nonorganic substrates such as glasses, glasses covered with a metal, or the metal substrates themselves.

Following a brief description of organic thin film fabrication technology and the methods used for the lateral structuring of thin films, details of the imaging technology for thin films are provided. These will include nonresonant optical methods (e.g., Brewster angle microscopy; BAM [9]), as well as resonant optical evanescent field methods (e.g., surface plasmon resonance; SPR and waveguide resonance microscopy, in its various subtypes [10–13]). Among nonresonant evanescent optical methods are included total internal reflection fluorescence (TIRF) microscopy [14], evanescent field waveguide microscopy/scattering microscopy [15], and evanescent fluorescence microscopy [16]. Confocal and confocal Raman microscopy will also be discussed. Neither electron microscopy [17], nor near-field methods such as atomic force microscopy (AFM) [18], scanning tunneling microscopy (STM) [18], scanning electrochemical microscopy

Nanomaterials for the Life Sciences Vol.5: Nanostructured Thin Films and Surfaces.
Edited by Challa S. S. R. Kumar
Copyright © 2010 WILEY-VCH Verlag GmbH & Co. KGaA, Weinheim
ISBN: 978-3-527-32155-1

(SECM) [19] and secondary ion mass-spectrometry imaging [20] will be discussed at this point.

11.2
Thin Film Preparation Methods

11.2.1
Dip-Coating

The simplest means of depositing a material thinly onto a surface to use a random adsorption process, with the coating material in solution. This approach is very effective for polymers, and generally provides homogeneous, smooth, thin films. It is possible to control the film thickness and internal layer structure by altering the concentration of material in solution, the immersion time, and by using multiple dipping cycles [21, 22].

11.2.2
Spin-Coating

Spin-coating represents a more sophisticated means of coating a material onto a flat surface. Here, a substrate is first mounted on a spinning device; the coating material is dissolved in a solvent and applied to the top of the substrate, so that the entire substrate is covered macroscopically. When the spinner is turned on, most of the material is spun off to the side; however, a thin film of the coating material will be deposited on the substrate, depending on a combination of the spinning speed and the viscosity of the applied solution. Notably, the film thickness will increase with a decreasing spinning speed and an increasing viscosity. Due to the stress applied during the spinning process, polymers can demonstrate an anisotropic orientation in spin-coated samples, with the chains showing a preferential orientation parallel to the substrate, and not necessarily a random coil formation [23–25].

11.2.3
Langmuir–Blodgett (LB) Films

Amphiphilic molecules and some hydrophobic molecules can be spread and oriented at an air–water interface due to hydrophilic/hydrophobic interactions. Typically, a Langmuir–Blodgett (LB) trough is used for this purpose (Figure 11.1a), with the amphiphilic molecules being dissolved in a volatile solvent and applied dropwise to the air–water interface of the trough. As the drops spread and distribute the amphiphilic molecules homogeneously over the available water surface, the molecules become oriented due to the hydrophobic/hydrophilic interaction. The hydrophilic parts of the molecules become located in the water phase, while

(a)
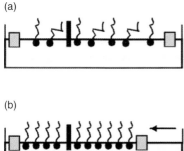

(b)
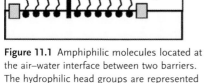

(c)
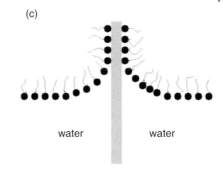

Figure 11.1 Amphiphilic molecules located at the air–water interface between two barriers. The hydrophilic head groups are represented by a circle, the hydrophobic chains by a wiggly line. A Wilhelmy system (black bar) is immersed in the interface to measure the lateral pressure. (a) The system is uncompressed; (b) The system is compressed; (c) Transfer of the monolayer onto a hydrophilic substrate. The hydrophilic head groups align on the substrate surface. Adapted from www.foa.se/surfacebiotech/tt/film.html).

the hydrophobic moieties will point out of the water into the air. Then, when the solvent is evaporated the molecules are pushed together by a barrier system (Figure 11.1b). The lateral pressure (Π) or the surface tension (in mN m^{-1}) can be measured with a so-called "film balance" or "Wilhelmy system." As the film is compressed by the barriers an area–pressure isotherm can be measured, which shows that the molecules are led through a two-dimensional (2-D) phase diagram. Typically, over a large available surface they will first be present in a gas-analogous state, followed by a liquid-analogous state, and finally achieve a solid-analogous state before the film collapses into the third dimension as the available area becomes too small.

When they are in the solid-analogous state, the condensed films can be transferred onto a solid substrate by pushing the substrate (using a film lift) through an air–water interface that carries the monolayer of amphiphilic material (Figure 11.1c). If the substrate is hydrophobic, then the hydrophobic region of the amphiphilic coating material will be in contact with the substrate, while the hydrophilic part of the amphiphilic molecules will form the outer surface of the coated substrate. However, in the case of a hydrophilic substrate the inverse occurs. By using this method it is possible to fabricate a single monolayer thin film, although multilayer films with a controlled internal architecture may also be produced, depending on the sequence of materials that are transferred step-by-step [26–28].

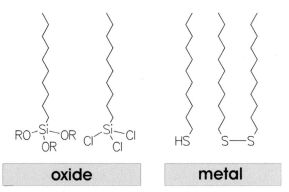

Figure 11.2 Self-assembled monolayer formation via silanes on oxide surfaces, thiols and disulfides on metal surfaces.

11.2.4
Self-Assembled Monolayers

One very clever method of creating monolayers on a substrate is to utilize the self-assembling processes that are driven and stabilized by preferential chemical bonding and van de Waals interactions. This method uses simple "beaker" chemistry, and is applicable to oxide surfaces (silicon wavers and glasses) with silane molecules and on noble metal surfaces (typically gold and silver) using sulfur-containing compounds such as thiols, sulfides, or disulfides (Figure 11.2). A solution of the coating material (often in the form of a thiol) is first prepared in a solvent; the substrate to be coated is then immersed in the solution for a few hours, during which time the thiol moiety binds spontaneously to the metal. With increasing immersion time, the film becomes more dense and the molecules increasingly oriented. As a result, the head groups of the molecules form the new surface, which then carries the physical and chemical characteristics of the head group of the immobilized thiol. A further build-up of layers is possible by using the head group's chemistry. Depending on the length of the self-assembled molecule and the size of the van de Waals interactions, this method can be used to produce dense, almost defect-free, highly oriented and well-controlled ultrathin films [29–31].

11.2.5
Layer-by-Layer Assembly

The layer-by-layer (LBL) technique is a combination of self-assembly and dip-coating, and is based on attractive Coulomb interactions. The charged surface that is required for the LBL method can be prepared using a self-assembled monolayer (SAM) that carries a –COOH or an –NH$_2$ head group. If a negatively charged surface is the starting point, the first layer to be deposited with the LBL method must be positively charged. The first stage is to dissolve the positively charged

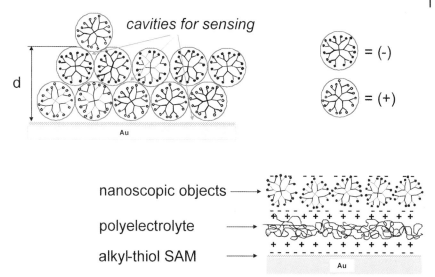

Figure 11.3 Schematic of an example of a Coulomb interaction-driven, layer-by-layer assembly. The upper system is built up by positively and negatively charged dimensionally stable dendrimers, whereas the lower system depicts a structure formed by a flexible polyelectrolyte and a dendrimer with opposite charge.

material to be deposited in a solvent; the negatively charged substrate is then immersed in this solution, which has a fixed pH-value so as to ensure the presence of both charge states needed to attract the positively charged coating molecules to the negative surface. In the next step, the situation is reversed (Figure 11.3), such that the now positively charged substrate is immersed in a solution with a negatively charged molecule. Successive immersions lead to a build-up of the binary material layer structure [27, 32–34].

11.2.6
Polymer Brushes: The "Grafting-From" Approach

The "grafting-from" approach was developed for use with polymers to create the so-called *polymer brush*; this is a thin polymer film where each individual polymer chain is fixed covalently to the substrate at only one end, and nowhere else.

In order to achieve this, an initiator molecule is first immobilized (typically via self-assembly) on a surface (Figure 11.4). The polymerization reaction is then started by adding the monomers to each initiation site. The density of the film depends either on the grafting density, or on the density of the initiation sites at the surface, whereas the film thickness depends on the degree of polymerization (DP) achieved in the polymerization process [35].

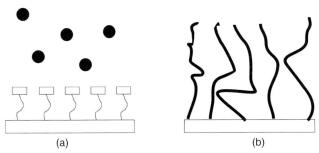

Figure 11.4 Schematic of the synthesis of polymer brushes. (a) A self-assembled monolayer with an initiator (small squares) is immobilized on a substrate, and the monomers (black circles) are added for the polymerization process; (b) The resultant polymer brush.

11.3
Structuring: The Micro- and Nanostructuring of Thin Films

The micro- and nanostructuring of thin films is typically achieved using a lithographic technique, by nanoimprinting, or by a variety of near-field methods. In the case of lithography, three writing media are available: (i) *photons* in photolithography; (ii) *ions* in ion lithography and in focused ion beam (FIB) lithography; and (iii) *electrons* in electron lithography and electron beam lithography.

11.3.1
Photolithography

The foundation of all photolithography is *photochemistry*. Depending on the material system applied, the photon will either destroy a chemical bond and allow (in a further rinsing step) the removal of an illuminated species, or it will form a bond, thereby stabilizing the illuminated material and allowing the nonilluminated material to be rinsed away from the illuminated material (Figure 11.5). Initially, this technology was developed and improved for photoresist lithography in chip production for the electronics industry. The pattern, which should be written into the resist or the thin film, is provided in the form of a "mask" that shields some areas on a sample but allows other areas to be illuminated. A typical device used to perform a patterned illumination using a mask is known as a "mask aligner." Photolithography can also be applied directly to the thin film under investigation. For example, the illumination of a thiol SAM with UV irradiation leads to a chemical reaction at the S–Au bond directly at the substrate surface. As a consequence, weakly bound sulfonates will be formed that can either be washed off or easily replaced in a consecutive SAM-forming step. However, the pattern sizes are diffraction-limited in this process [36–40].

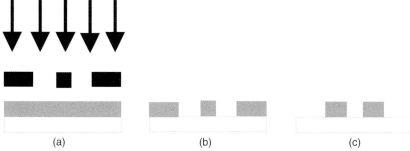

Figure 11.5 Schematic of lithography. (a) The substrate is coated with a thin film (gray, usually referred to as the "resist") which is chemically changed upon irradiation with light or electrons or ions (black arrows). A mask shields part of the thin film; (b) Positive resists, where the chemical reaction creates a more soluble material; (c) A negative resist, where the illumination yield an insoluble thin film.

11.3.2
Ion Lithography and FIB Lithography

In ion lithography, the resist or thin film is destroyed and removed by the impact of the ions on the film-forming molecules. How complete this removal is depends on various parameters, including the ion dose, the film thickness, and the chemistry of the removed material. Two approaches are implemented in ion lithography: (i) a masked approach, where the ions are applied over a larger area (e.g., in an ion accelerator); and (ii) a mask-less FIB approach, where the FIB is steered over the sample so as to manufacture a pattern [41, 42].

11.3.3
Electron Lithography

Electron lithography can also be divided into two approaches: (i) with a broad sample illumination by the electrons and the aid of a mask; and (ii) a mask-less approach (called electron-beam lithography), where an electron beam is steered over the sample. In electron lithography the material that is being laterally structured faces a chemical reaction (typically a bond breakage) upon electron irradiation. This process may be chemically and physically very complicated, and depends on the chemistry of the material to be structured, the electron energy and its dose, as well as many other parameters. Electron-beam lithography with an electron-beam photoresist is a well-established technology [43–45].

11.3.4
Micro-Contact Printing and Nanoimprinting (NIL)

Micro-contact printing (Figure 11.6), and nano- and micro-imprinting are recognized as combination methods. First, a topographic stamp carrying the pattern is

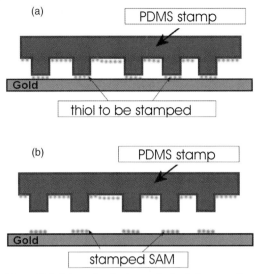

Figure 11.6 Schematic of microcontact printing or soft lithography. (a) A soft stamp brushed with a thiol solution is brought into contact with a gold surface; (b) The resultant patterned gold surface with a stamped SAM and the emptied stamp.

fabricated with the help of a classical lithography method. This stamp is then used in a second step to generate either a pattern or a thin film on a surface. A wide spectrum of methods is available for fabricating the stamp, and for the second pattern-formation step. The stamp materials may be either soft (an example is the silicon rubber polydimethylsiloxane used for soft-stamping thiols onto a metal surface) or hard (as are used for pressure and melt applications). The depth and lateral resolution depend on the stamp and on the second pattern-formation process [46–49].

11.3.5
Near-Field Scanning Methods

Basically, all near-field scanning methods, including AFM, STM, scanning near-field optical microscopy (SNOM) and SCEM, have been implemented in the nanostructuring of surfaces, with both the removal and placement of material having been demonstrated. Frequently, AFM is used in dip-pen nanolithography (DPN), where a nanopatterned SAM on gold is fabricated by supplying (for example) thiols to the substrate via the microscope tip. By using a STM tip under potential, the DPN method can be extended to a writing mechanism that is driven by a potential difference. It is also possible chemically to change, for example, the head groups of a SAM by scanning a metallic AFM tip under a potential over a thin film. Previously, SNOM has been used to implement photochemical reactions in thin films so as to increase the lateral resolution, which is diffraction-limited in classical mask photolithography. These methods typically lead to nanostructures that can be viewed using either the near-field method with which they have been fabricated, or with electron microscopy [50–53].

11.3.6
Other Methods

Other lithographic methods that permit the patterning of surfaces include interference and holographic lithography, flash imprint lithography, and nanosphere lithography [54–58].

11.4
Imaging Technologies

When a structure has a lateral resolution below the optical diffraction limits inherent to optical microscopy, it cannot be imaged by using an optical microscopy technique. Rather, such structures are typically fabricated using electron-beam lithography or with near-field patterning methods, and consequently either a near-field method or electron microscopy will be required to resolve the lateral patterning. Yet, by choosing the correct optical microscopy methodology, ultrathin films (and the very small height contrast within them) can be imaged as long as the lateral structure is above the diffraction limit. Optical microscopy technologies capable of imaging thin films with a height resolution down to ~1 Å, or a height contrast difference of approximately one CH_2 group, are normally based on evanescent optical techniques, or by implementing a special contrast mechanism such as Brewster angle or one- and two-photon confocal microscopy [59, 60].

The imaging technology of near-field methods will not be discussed here. Rather, attention will be focused on special optical microscopy technologies used to image thin films located on an air–water interface, and on glass or metal surfaces.

11.4.1
The Concept of Total Internal Reflection

Before discussing the various microscopy technologies, it might be advantageous to consider the concept of total internal reflection. If a light beam incident on an interface between two transparent dielectric media with refractive indices n_1 (of the medium where the light is propagating in towards the interface) and n_2 (of the medium at which the surface reflection will take place, or into which the beam will be refracted), then Snellius' law of reflection will yield:

$$\frac{\sin\theta_1}{\sin\theta_2} = \frac{n_2}{n_1}$$

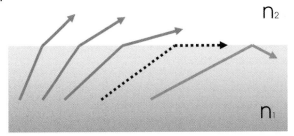

Figure 11.7 Light propagating in the optically thicker medium with refractive index n_1 is reflected and refracted at the interface with the optically thinner medium with a refractive index of n_2. For simplicity, only the diffracted beams are shown. By successively increasing the angle of incidence, the critical angle of total internal reflection is reached. Any further increase in the angle of incidence will result in one reflected beam only.

where θ_1 is the angle of incidence and θ_2 is the angle of refraction. In the case where $n_1 > n_2$, when the light is propagating in an optically thicker medium and is reflected on an optical thinner medium, there exists a critical angle of total internal reflection, $\theta_{critical}$, above which all the light will be reflected without any refraction (Figure 11.7):

$$\theta_{critical} = \arcsin \frac{n_2}{n_1}$$

Upon total internal reflection, an evanescent field is created, which depicts its maximum value at the reflecting interface and decays exponentially into the optically thinner medium. The penetration depth is given by:

$$d = \frac{\lambda}{4\pi\sqrt{n_1^2 \sin^2(\theta_1) - n_2^2}}$$

where λ is the wavelength of the light.

11.4.2
The Concept of Waveguiding

Imagine now a sheet of an optically thick, transparent material enclosed by optically thin, transparent materials (Figure 11.8). A light beam can be trapped in this sheet, when all reflections on the sheet's surface are carried out above the critical angle of total internal reflection; this sheet can be described as a "light guide." In order to form a waveguide, the sheet must be thin, in the order of micrometers. Because destructive interference between different light beams guided in the waveguide under different angles occurs, only a few angles above the critical angle of total internal reflection will lead to a constructive interference, the transport of photons, and the guiding of a wave. These angles are selected by allowing the phase of the light beam being picked up from refection on interface I, along

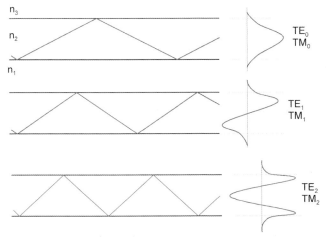

Figure 11.8 Schematic of a three-layer waveguide system with n_2 the refractive index of the guiding layer and n_1 and n_3 the refractive indices of the substrate and the cladding, respectively. In order to assure guiding and total internal reflection, $n_2 > n_{1,3}$. In addition, the internally reflected beams are drawn, and the field distribution occurring along the entire structure for the first three modes. For details of the modes, see the text.

propagation through the waveguide, by the second reflection on interface II and the back-propagation through the waveguide to interface I to be $m\pi$, with m an integer. Therefore, a waveguide allows particular modes to propagate. Moreover, the thicker the waveguide is, the more modes are able to propagate; and conversely, the shorter the wavelength, the more modes are possible. Each mode carries a particular oscillating field distribution with it, and shows the evanescent fields due to the total internal reflection on both outside surfaces (Figure 11.8). Both modes, in s- and p-polarization, are possible. Modes in s-polarization are referred to as "TE" modes and in p-polarization "TM" modes. The complete mode name also carries a number index, for example, TE_1, indicating the number of zero field positions within the mode. Each mode carries a particular wave vector.

This wave vector is larger than the wave vector of freely propagating photons of the same wavelength. Therefore, a coupling medium such as a prism or a grating is necessary to couple freely propagating photons into individual waveguide modes [61, 62].

11.4.3
Brewster Angle Microscopy (BAM)

There is a particular angle, the Brewster angle, under which p-polarized light is not reflected from a surface; however, s-polarized light is reflected. The angular position of the Brewster angle, θ_B, in a reflection spectrum (reflected intensity versus angle of incidence) is defined by the refractive index architecture of the interface:

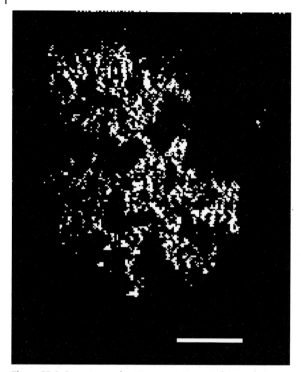

Figure 11.9 Brewster angle microscopy image taken at the Brewster angle of water at room temperature of a dipalmitylphosphatidylethanolamine (DPPE) island on the air–water interface of an LB trough. Scale bar = 2 μm.

$$\theta_B = \arctan\frac{n_2}{n_1} \quad (11.1)$$

where n_1 is the refractive index of the medium in which the Brewster angle is measured, and n_2 is the refractive index of the medium on which the reflection happens. Typically in air, with $n_1 = 1$, Equation 11.1 simplifies to $\theta_B = \arctan n_2$.

Imagine a patterned surface where the pattern is made out of materials with different refractive indices; an example might be a lipid island floating on an air–water interface, as is possible on an LB trough. The lipid island depicts a higher refractive index ($n \approx 1.5$) than the clean water ($n \approx 1.33$) surface. An imaging contrast can be achieved by illuminating this water surface with p-polarized light under the Brewster angle condition of clean water: $\theta_B = \arctan(1.33) \approx 53°$. The clean water surface will not reflect this p-polarized light, and will show dark areas on a screen catching the reflection image, whereas the lipid islands reflect the light and will appear bright on the screen (Figure 11.9). By changing the illumination angle to the Brewster angle of the lipid islands, the contrast will be reversed. This contrast mechanism is implemented in BAM that typically is applied to the characterization of LB films on the air–water interface. Ultraflat solid substrates can also be characterized with this method, although any angular surface irregularities on the substrate will create artifacts in the BAM image. The microscopic set-up uses a laser for illumination and an objective to catch a microscopic image of the interface. Because, both the illumination and the capturing of the image are carried out under an angle to the surface, the images are distorted; in addition, only the central part of the images will be in focus. However, this problem has been overcome by using advanced image-processing software and the recording of multiple images. This microscopy method demonstrates a lateral resolution of 1–2 µm and a height resolution of one monolayer of material, when the correct refractive index contrast is available. Both, static and dynamic experiments can be carried out. Recently, this method was modified for nanosecond pump-probe BAM to image photo-induced phase changes in LB films [9, 63–66].

11.4.4
Resonant Evanescent Methods

11.4.4.1 Surface Plasmon Resonance Microscopy

A surface plasmon (or a plasmon surface polariton) is a longitudinal charge density wave of the free electron gas in a metal, which propagates at a metal–dielectric interface. It exists only in p-polarization, and forms an evanescently decaying H-field in the metal and in the dielectric, with the highest field being located at the interface. The decay length (the length where the field is fallen to 1/e) in the metal is roughly one-tenth of the decay length in the dielectric (for gold–air at $\lambda = 632.8$ nm, 28 nm and 333 nm, respectively). The surface plasmon carries the dispersion:

$$k_x = \frac{\omega}{c}\sqrt{\frac{\varepsilon_1 \varepsilon_2}{\varepsilon_1 + \varepsilon_2}}.$$

The wave vector of the surface plasmon in propagation direction x, k_x, can be described by the dielectric constant of the metal, ε_1, and the dielectric, ε_2, the speed of light, c, and the frequency, ω. A plot of this dispersion (ω versus k_x) shows a dispersion curve that starts at the origin, and increased and bends according to the square-root behavior to a maximum ω-value of ω_p the plasma frequency of the metal

$$\omega_p^2 = \frac{e^2 N}{\varepsilon_0 m_e}$$

where N is the electron density, e is the elementary charge, and m_e is the mass of the electron.

On the other hand, the wave vector of freely propagating photons in the dielectric medium can be written as $k_{\text{photon}} = \frac{\omega}{c}\sqrt{\varepsilon_2}$. This dispersion is represented as the light line in the dispersion plot, which is always positioned at the left (smaller wave vectors) of the surface plasmon dispersion. By comparing both dispersions, it becomes clear that, besides the origin, the wave vector of the surface plasmon is always larger than the wave vector of the photons. In order to excite a surface plasmon with photons by wave vector matching, it is necessary to enhance the wave vector of the photons. This can easily be achieved by introducing a second transparent dielectric, a glass prism, with a higher dielectric constant, ε_3. This leads to a light line with a smaller slope which cuts through the surface plasmon dispersion and allows wave vector matching. Therefore, in the experimental set-up a thin metal film with a thickness of ~50 nm is located at the base of a prism (ε_3). The surface plasmon is excited and detected by turning this prism under illumination (Figure 11.10) and detecting the intensity of the light being reflected from the prism base. This is called an attenuated total reflection (ATR). This reflectivity curve shows, with increasing angle of incidence, θ, on the prism (converted to the angle of incidence within the prism onto its base), first an increase in intensity up to the critical angle of total internal reflection, after which the intensity stays constant, before decreasing; recovery of the reflected intensity then occurs, which is typical SPR behavior. When the light is coupled to the surface plasmon at the correct wave vector due to the prism turning ($k_x = k_{photon} \sin\theta$), the surface plasmon propagates along the metal–air interface, and therefore the photons are missing in the reflected beam. The angle of incidence, θ, where the minimum intensity is found, is called the SPR angle, $\theta_{resonance}$.

If a thin layer is added on top of the 50 nm metal film, the surface plasmon dispersion changes slightly, being moved to higher wave vectors. This dispersion change depends on the optical thickness of the adlayer: $d*n$, where d is the thickness and n is the refractive index. Because of this; the SPR angle will shift to a higher value. This mechanism can be used for generating a contrast in SPR microscopy. If a gold film is laterally patterned with an adlayer, and the prism illuminated at the SPR angle for the clean gold surface over an extended area, then the captured reflection image will show dark areas where the clean gold surfaces are located, and bight areas where the adlayer is located (Figure 11.10a). Moreover, by changing the SPR angle to that of the adlayer, the contrast will be inverted [10].

11.4.4.2 Waveguide Resonance Microscopy

The same experimental set-up as used for ATR is used for waveguide resonance microscopy (Figure 11.10b and c). A prism base with or without the 50 nm of metal carries a waveguide, and on top of this waveguide a laterally patterned film. In this configuration, it is possible to operate with an s-polarized or a p-polarized illumination, leading to TE and TM modes in addition to the SPR when the metal film is present. The waveguide will show its mode structure as very sharp dips in the reflection versus angle of incidence scan. The waveguide modes are also excited by a wave vector match between the incoming photons and the waveguide mode wave vector. The addition of the pattern disturbs the waveguide in such a way that in the areas of the additional material the waveguide wave vector is slightly altered, increased, and will yield a slightly different waveguide mode resonance or slightly

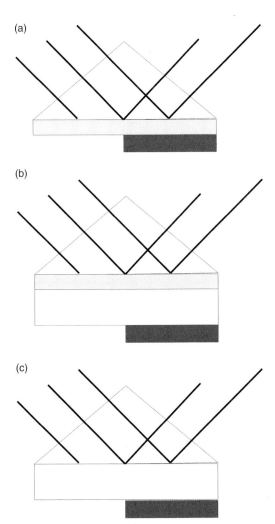

Figure 11.10 The basic principles of (a) resonant surface plasmon and (b,c) resonant waveguide microscopy (gray = metal. white = waveguide; black = adlayer); (b) The TM$_0$ mode is represented by the surface plasmon resonance; (c) Only waveguide modes exist. The angle of incidence is chosen so that the system is in resonance on the left side of the prism base (no adlayer). The surface plasmon or a waveguide mode is excited, and the light is guided away within the left side of the film structure. The light is missing in reflection (missing reflected beam); the microscopy image will therefore be bright for the areas carrying an adlayer (out of the resonance). This is implied by an incoming of three beams, but a reflection of only two beams. For simplicity, the refraction of the beams at the prism interfaces is omitted.

higher coupling angle. With each of the additionally available modes it is then possible to create both an image and an inverse contrast image [11].

11.4.4.3 Surface Plasmon Enhanced Fluorescence Microscopy

As noted above, when a surface plasmon is excited, an evanescent field is present at the interface between the metal layer and the dielectric, typically air or a solvent. This evanescent field can be used for the excitation of fluorescence, and it is possible to combine SPR microscopy with fluorescence microscopy (Figure 11.11a). However, care must be taken here as metals quench fluorescence and hinder the emission of fluorescence photons. Therefore, the fluorescence dye must be placed at a distance from the metal surface, which must be at least the Förster radius. In this case, excitation and emission is possible. For this combined technology the prism must be turned into the SPR angle of the metal with the adlayer

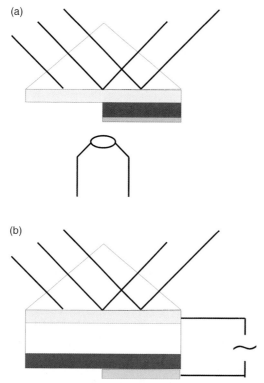

Figure 11.11 (a) The basic principle of surface plasmon-enhanced fluorescence microscopy (gray = metal; black = adlayer; dark gray = chromophores). The fluorescence signal is picked up from the layer system at the prism base. Contrast occurs only when active fluorescence labels are present, even when the entire layer system is in surface plasmon resonance; (b) The electro-optical set-up (white = a voltage applied over a waveguide). Due to the applied potential, in noncentrosymmetric materials the refractive index changes, and therefore so also does the resonance condition for the waveguide mode.

and the fluorescence dye. The evanescent field may then excite the dye, such that it is able to emit. Typically, this microscopy technology is used in sensor applications, where a spotty pattern with different recognition sites is immobilized on the metal film. In the reflection image these dots will be seen as dark dots on a bright background, whereas in the fluorescence microscopy image taken with a microscope objective directly from the metal film, the spots will be seen as bright areas on a dark background. Changing the surface chemistry only very slightly, so that the SPR does not shift but the fluorescence dye is affected (e.g., towards its position in the evanescent field) will yield a change in the fluorescence image, but not in the reflection image. Therefore, this method will lead to an enhanced sensitivity in surface changes [12, 67, 68].

11.4.4.4 Waveguide Resonance Microscopy with Electro-Optical Response

Materials that exhibit a noncentrosymmetric internal structure show an electro-optical or Kerr effect [69]; by applying an electrical potential on such a material the refractive index of the material will be changed. This change can be applied as a contrast mechanism and made visible in waveguide resonance microscopy. The experimental set-up is identical to the prism set-up described above, but in addition a potential is applied, typically with the help of the thin metal layer as one electrode and a second patterned electrode located on top of the layered system (Figure 11.11b) [13, 70, 71].

Other microscope combination methods are available of resonant surface plasmon and resonant waveguide evanescent fields (e.g., SPR Raman microscopy). In this particular method, a grating is implemented to carry out the wave vector match between the photons and the surface plasmon [72–74].

11.4.5
Nonresonant Evanescent Methods

11.4.5.1 Total Internal Reflection Fluorescence (TIRF) Microscopy

As noted above, light under total internal reflection generates an evanescent field in the optically thinner medium. This evanescent field is applied as an illumination and fluorophore excitation source in TIRF microscopy. The first TIRF experiments were carried out implementing a "prism" to establish total internal reflection and the evanescent field. For this, the specimen is placed in a thin cuvette constructed with the prism base and a microscope cover slip. ("Prism" is written here in quotation marks because all types of glass bodies have been used, not necessarily the classical triangular prism form.) The image is taken without image distortion through the cuvette on the opposite side from the "prism," using a classical fluorescence microscopy set-up. This has the disadvantage that the specimen is not accessible for manipulation, and the image is recorded through the entire sample and the cuvette. Huge improvements have been made to this now classical set-up, such that highly specialized microscope objectives are available that help to create the evanescent field at the inner side of a high refractive index cover slip, which forms part of a cuvette and also carries the specimen. The evanescent field-creating objective also records the fluorescence image, with the set-up typically being housed in an inverted fluorescence microscope. With this technology, the specimen is accessible for manipulation from the top and the image can be recorded directly, only through the high refractive index coverslip, from the active area of microscopy [75–77]. Excellent schematics of the prism and the objective-based systems are available on the internet homepages of many microscope manufacturers (e.g., Zeiss or Olympus).

11.4.5.2 Waveguide Scattering Microscopy

When a waveguide mode is propagating through a waveguide, it carries an evanescent field. By using a continuous wave operating laser it is possible to create an evanescent field along the entire waveguide length, and in a width that depends upon the width over which the laser can be coupled homogeneously into the waveguide. A broad illumination can be achieved by combining a spherical lens with a cylinder lens. The evanescent field can be used to illuminate objects located immediately at the surface of the waveguide (Figure 11.12a). Waveguide scattering microscopy can be described as an evanescent field perturbation microscopy. While the waveguide is perfect, and when it shows no surface inhomogeneities, no scattering centers on its surface, no patterned adlayer – and, of course, no scattering centers in its inner layer – the waveguide will operate undisturbed and perfectly. Moreover, the conditions of total internal reflection are maintained, no light is scattered, and the guided light stays within the structure. By viewing such a perfect waveguide from the top or the bottom, it appears dark and the losses of the waveguide are minor. As soon as any additional material is located at the waveguides surface, the waveguide operation is disturbed, the conditions of total internal reflection might be altered locally and, in addition, the photons within the evanescent field can be scattered. Therefore, the waveguide scatters the guided light partly out of its structure. Most of the scattered photons are found in forward propagation direction due to momentum conservation, but under perpendicular, 0° viewing geometry, it is clear that photons escape from the waveguide in this direction and give the waveguide a bright appearance. This mechanism is used to distinguish areas on a waveguide with an adlayer from areas without an adlayer. The contrast between the two different areas can be calculated as:

$$\text{contrast} = \frac{I(\text{bright}) - I(\text{bright})}{I(\text{outside the sample})}$$

where I is the integrated intensity measured over a unit area.

These experiments have been conducted with so called "ion-exchanged waveguides," which show an increase in the penetration depth of the evanescent field and an increase in the integrally guided intensity with increasing mode number. Therefore, the contrast increases with increasing mode number. In addition, it was found that the TE modes showed a higher contrast than TM modes. The scattering intensity is a function of the refractive index difference between the material located on the waveguide which causes the scattering, and the surround-

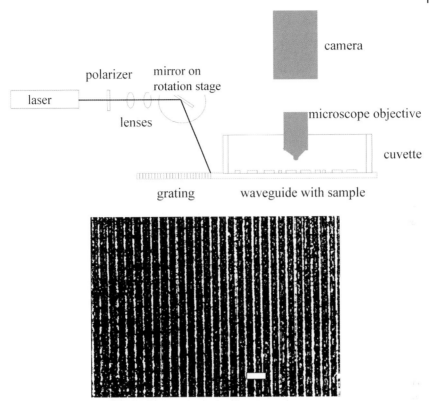

Figure 11.12 Upper: Scheme of waveguide scattering microscopy, a nonresonant method. The scattered photons from the evanescent field are picked up under 0° geometry. Scattering is enhanced in areas on the waveguide where material is located, due to inhomogeneities in the material and to changes in the local total internal reflection conditions. Lower: Waveguide scattering image of a SiO_x grating with a height of 11 nm. Scale bar = 5 μm.

ing material in the cladding. Therefore, the contrast decreases substantially by comparing a sample investigated in air with one immersed in a liquid such as water. The lateral resolution of this method is approximately <1 μm. The smallest height investigated was 11 nm of evaporated SiO_x (Figure 11.12b) [15, 78].

11.4.5.3 Waveguide Evanescent Field Fluorescence Microscopy (WEFFM)

Total internal reflection fluorescence microscopy is demonstrated with the aid of an ion-exchanged waveguide that is operated with a widened continuous wave laser beam to generate an evanescent field over an extended area on the waveguide. The waveguide is mounted in an inverted fluorescence microscope as the lower part of a cuvette. Because the evanescent field is created by the waveguide, this microscopy technology operates without immersion oil and without a high-refractive index coverslip. In addition, the imaging lens can be freely exchanged

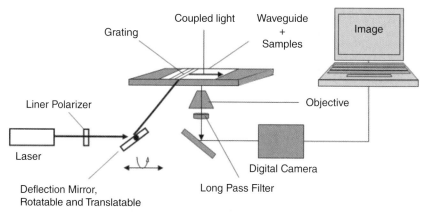

Figure 11.13 Experimental set-up for waveguide evanescent field fluorescence microscopy.

and the magnification altered during an experiment. The specimens are freely accessible from the top so as to allow sample manipulation (Figure 11.13).

In order to learn about the data interpretation of fluorescence intensities emitted from stained specimens, first LB films with systematically changing distances between the waveguide surface and an active fluorescent LB film were investigated. A clear correlation between the intensity at a particular distance from the waveguide surface and the emitted fluorescence intensity was found. The simulated field distribution, and therefore intensity distribution, was seen to correlate with the experimental intensities from the LB films. Osteoclasts were imaged and focal adhesions identified by high-intensity areas in the image. In addition, it is possible by implementing a multimode waveguide and recording images with various modes to measure the distance between the waveguide surface and the dye location (for example, in a cell) if it is not moving. Focal contacts, with separation distances of 10–15 nm, can be distinguished from close contacts, with separation distances of 20–50 nm [16, 79] ; S. Armstrong et al., unpublished results].

11.4.5.4 Confocal Raman Microscopy and One- and Two-Photon Fluorescence Confocal Microscopy

Raman spectra show the energy shift of laser light excitation due to inelastic scattering in samples, which leads to vibrations of chemical bonds within the molecules [80]. Different chemical species consist of different molecular groups, atoms and bonds, so that each molecule can be identified by its fingerprint Raman spectrum. Raman spectroscopy is a nondestructive technique, as only molecular vibrations are excited or annihilated (Stokes- and Anti-Stokes process). In confocal Raman microscopy, the Raman spectra are collected using a high-throughput spectrometer equipped with a CCD camera or a photomultiplier (Figure 11.14). It is possible to generate images by choosing a particular Raman band, a combination of several bands, or by integrating over selected spectral areas. The plane of imaging and depth of imaging within a sample is chosen by the confocal

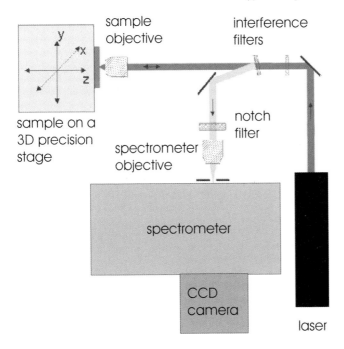

Figure 11.14 Scheme of a Raman confocal microscope.

microscopy technology. Photons are only contributing to the imaging process, which come out of the focal waist (the focal point area) of the illuminating laser. This beam waist can be scanned in all three room directions [81].

In one- or two-photon fluorescence confocal microscopy, the use of a spectrometer is unnecessary, although the filters and/or dichroic mirrors must be able to distinguish between the excitation and emission wavelengths of a particular fluorescent dye. The dye, which is located in the waist of the illuminating laser, is excited either with a single photon in the visible spectrum or with two photons in the near-infrared spectrum, so as to emit one photon in the visible spectrum. These emitted photons contribute to building-up the image. The lateral resolution of this microscopy technology is approximately 200 nm [82]. When images of thin films are taken, it is essential to ensure that the focal waist is scanned along the surface carrying the thin film.

11.5
Application of Thin Films in the Life Sciences

Thin films play a substantial role in all types of biochemical sensors implementing a recognition reaction [83–85], in surface functionalization for biocompatibility, either avoiding or enhancing protein binding and/or cell growth [86–88], in drug delivery [89–92], in bioreactors [92–94], and in cell-surface mimicking [95], among other applications.

11.5.1
Sensors

Sensors based on thin films are typically designed by immobilizing a molecular species on a surface in such a fashion that a molecular moiety, which is able to carry out a specific recognition and binding reaction, is exposed to that very surface. If the molecule is present in an analyte solution, and is recognized and bound, then the appropriate reactions will occur and the thickness and/or refractive index of the thin film will be changed. Such changes may then be recognized and/or quantitated using the above-mentioned resonant technologies [96–98].

One powerful signal enhancement technology is the implementation of chromophores and fluorescence detection in the sensing reaction. To date, a variety of schemes has been established:

- The recognized molecule carries a chromophore, and the fluorescence increase that occurs during binding is then monitored.

- The chromophore is located at the recognizing molecule, and changes its position on binding; this leads to an increase or decrease in fluorescence intensity. An increase can occur if the chromophore moves further away from a metal, as this diminishes any quenching. A decrease can occur if the chromophore changes its position within an evanescent field, to a lower field position.

- Both, the recognizing and the recognized molecules carry a chromophore. The absorption and emission spectra must be chosen in such a way that, upon binding, an energy transfer between the two chromophores can occur, which can be monitored by a fluorescence color change during the binding event [98].

An example of this is a combination of a surface thin film DNA sensor and surface plasmon-enhanced fluorescence microscopy [99]. In a microarray single-stranded catcher, DNA probes were immobilized and organized in individual round surface spots. The single-stranded DNA (ss-DNA) sequences carried 15 thymines as a spacer, to prevent quenching, while quantum dots (QDs) were used as fluorescent probes; these were bound to the ss-DNA target sequences. A simultaneous qualitative analysis of the QD-conjugated analyte DNA strands was obtained as multicolor images.

11.5.2
Surface Functionalization for Biocompatibility

Biocompatibility, fouling and antifouling covers a huge scientific, engineering and technology field, ranging from the acceptance of an implant by a host body to the avoidance of biomass growth on a ship's hull. Here, osteoclasts and osteoblasts – the cells responsible for bone resorption and formation, respectively

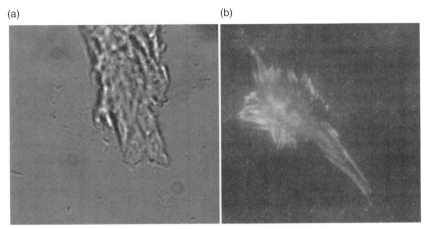

Figure 11.15 (a) Bright-field image and (b) waveguide evanescent field fluorescence microscopy image of an osteoblast (MC3T3-E1) grown on a bare glass waveguide surface. Scale bar = 30 µm (applicable to both images).

[100–103] – provide an excellent example of the link between surface functionalization and biocompatability.

In these studies, 3-mercaptopropanoic acid [MPA; $HS(CH_2)_2COOH$] and octanethiol [OT, $HS(CH_2)_7CH_3$] were applied to gold-coated glass coverslips to produce carboxylic acid-terminated hydrophilic monolayers or methyl-terminated hydrophobic monolayers, respectively. An ultraviolet photo-patterning technique [36] was also implemented to produce surfaces with micrometer-scale regions of well-defined composition. When human osteoblasts were grown on top of the patterned surface, both fluorescence microscopy and confocal fluorescence microscopy showed the cells to be attached almost exclusively (depending on the stripe width) to the hydrophilic areas of the surface, while the hydrophobic regions remained bare [104].

With WEFFM, these types of cell can be imaged in the area of attachment to the substrate. Figure 11.15a shows a bright-field image part of an osteoblast, and Figure 11.15b the corresponding WEFFM image on a clean glass waveguide, without any surface functionalization. Close to the substrate, the cell is clearly located at a slightly "different" position and its focal adhesions or close contacts are visible [16, 79].

11.5.3
Drug Delivery

Historically, drug delivery was discovered almost accidentally in the mid-1960s, when silicone rubber arteriovenous shunts containing a drug were implanted and found to serve as constant-rate drug delivery devices. Since that time, a wide variety of technologies for drug storage and release have been developed. The first macroscopic delivery devices were developed during the 1960s, with microdelivery systems being designed and adopted for clinical purposes during the 1980s. Nanodelivery systems were first considered during the mid-1970s and have subsequently been widely introduced, for example with targeted polymer–drug conjugates and in surface-controlled drug delivery systems [105].

Mammalian cells are distinguished from less-evolved forms of life by the key architectural principle of compartmentalization. Hence, a number of synthetic mimicries of this abundant natural concept have been realized in the form of synthetic particles, and have found application as fully integrated multifunctional vehicles for drug delivery [106]. An example is the thin film fabricated from microcapsules that contain pyrene as a model drug. For this, microcapsules were fabricated using the LBL approach [27, 32–34] on melamine formaldehyde (MF) particles with pyrene and chitosan. Hollow capsules were obtained by dissolving the MF particles in HCl solution; thin films of the microcapsules were then fabricated by drop casting. The capsular thin film was investigated with respect to the amount of pyrene and chitosan layers by using absorption and fluorescence spectroscopy, transmission electron microscopy (TEM), and confocal fluorescence microscopy. It was shown that, as the layer number increased, the fingerprint spectra increased almost linearly, while the drug uptake increased from 0.488 μg per 15 μl MF particles for two layers, to 1.173 μg per 15 μl MF particles for five layers. By using confocal microscopy, the pyrene was shown to be present in the capsular membrane. Release of the entrapped pyrene was demonstrated using a multilayer film of pyrene/chitosan on a quartz substrate that was exposed to NaCl solution at pH 6.5, at room temperature. The amount of pyrene in the saline solution was shown to increase with increasing time. [107]

11.5.4
Bioreactors

Bioreactors have been (and continue to be) implemented for a variety of technological purposes, including tissue engineering [108], biorefineries [109], separation technology [109, 110], and water purification [110, 111].

Typical examples include the membrane bioreactors (MBRs) that are used in waste water treatment [111] and which combine conventional biological wastewater treatment using a suspended biomass with membrane separation. These MBRs represent an attractive alternative to conventional activated sludge treatment using secondary sedimentation, and in fact offer a higher product water quality. Unfortunately, the application of MBRs is constrained by membrane fouling that may be caused by a variety of constituents in the water, including dissolved inorganic or organic compounds, colloids, bacteria, and suspended solids. *Biofouling* is largely attributable to accumulated extracellular materials, rather than to individual bacterial cells. These extracellular materials, which include soluble microbial products and extracellular polymeric substances, consist mainly of polysaccharides, proteins, and other natural organic matter.

Nanofiltration (NF) membranes have the potential for rejecting low-molecular-weight contaminants such as pharmaceuticals, as well as pesticides, that may be hazardous to humans. Commercial NF membranes are typically fabricated from thin-film composites that consist of a charged polyamide polymerized at the interface on a polysulfone ultrafiltration membrane. Whilst the dense nature of the polyamide layer inhibits internal pore fouling, it causes a dramatic reduction in flux. The chemical nature of this interfacial layer renders NF membranes susceptible to fouling by charged species in the feed water. Mayes *et al.* have developed a thin-film composite NF membrane composed of a commercial polyvinylidene fluoride (PVDF) ultrafiltration (UF) membrane coated with an amphiphilic graft copolymer: poly(vinylidene fluoride)-*graft*-poly(oxyethylene) methacrylate (PVDF-*g*-POEM). The microphase separation of the PVDF backbone and short polyethylene oxide (PEO) side chains results in an interconnected network of hydrophilic, charge-neutral, 2 nm-wide "nanochannels" that allow the passage of water and small molecules through the coating. In the study conducted by Mayes and coworkers, the fouling behavior of these new membranes was investigated with bovine serum albumin, sodium alginate, humic acid, and sludge from an aerobic bioreactor with the help of dead-end filtration experiments, AFM measurements, and scanning electron microscopy.

The novel NF membranes with the self-assembling graft copolymer, PVDF-*g*-POEM, exhibited exceptional fouling resistance for a variety of model biofoulant solutions. Typically, the model foulants were filtered out almost completely (99.9%), while the water flux could be completely recovered after membrane rinsing. The antifouling properties of these membranes were attributed to both the nanoscale dimensions of the hydrophilic channels in the self-assembling graft coating, which greatly restrict the size of permeating species, and to the unique properties of PEO, the charge-neutral polymer that hydrogen-bonds with water to create an energetic barrier against the adsorption of biomolecules onto the surface. Their combined antifouling characteristics and high effluent quality suggest that these membranes might represent promising prospective replacements for UF membranes in MBRs, where high concentrations of organic foulants are encountered.

11.5.5
Cell-Surface Mimicking

Lipids, as a major constituent of biological membranes, acting as the outer boundary of cells and organelles. Due to their amphiphilic architecture, lipids assemble *inter alia* into bilayers in water, allowing for a minimization of the contact of hydrophobic chains with the surrounding water. Within biological membranes, lipids do not exhibit any particular order, and are able to diffuse within their membrane plane. Both, peripheral and integral proteins and carbohydrates, may be bound to or incorporated into these approximately 50 Å-thick thin films. These membranes control the inflow and outflow of cells or organelles, much like a "smart" filter. Plasma membranes, for example, are able to block the entry of most toxic substances while simultaneously allowing the selective passage of nutrients, wastes, and metabolites.

Many important biochemical processes occur at a membrane surface by virtue of the presence of membrane proteins. However, a detailed examination of biological membranes reveals them to be complex assemblies of a number of molecular machines that cannot be reassembled piece-by-piece. Moreover, as the intracellular and extracellular networks, such as the cytoskeleton and the extracellular matrix, add to this complexity it becomes difficult to investigate the structures and functions in a direct manner. Consequently, a model system of a biomembrane, with a reduced number of components and parameters to control and with the phospholipid bilayer deposited onto a solid substrate, was designed and used extensively in cell-surface experiments. In fact, over the past 20 years this model has provided much insight into immune reactions and cell adhesion processes.

Supported membranes are prepared by the direct deposition of lipid monolayers or bilayers onto either solid substrates or polymer-coated substrates. A bilayer deposition can be performed using any of four methods:

- The successive transfer of lipid monolayers: for this, a LB trough was used to transfer the lipids monolayer by monolayer from the air–water interface.

- The fusion of lipid vesicles: here, the lipid vesicles were deposited in a suspension onto the substrate. By selecting the correct conditions, the vesicles could rupture and form membrane patches that fused into continuous bilayers.

- Single bilayer spreading: in this approach, a single layer spreading is achieved by depositing a lipid reservoir onto the solid, followed by a second spontaneous spreading.

- Solvent exchange: here, lipid membranes are formed by the exchange of solvents from alcohols.

These supported membranes show the intrinsic "fluid" property required to self-heal local defects, while achieving excellent mechanical stability [95].

At this point, attention will be focused on electrical and optical techniques for investigating cellular signaling reactions in artificial and native membranes immobilized on solid supports [112]. A thiolipid was used to form the first lipid monolayer on a gold electrode; the second monolayer was then formed by incubating the first layer in a solution of the phospholipid which was then diluted stepwise below the critical micellar concentration. *Thiolipids* are phospholipids that carry an SH-group at the end of a spacer at their polar head groups. The hydrated spacer decouples the lipid bilayer from the gold surface, and this results in an aqueous film forming between the electrode and the lipid bilayer, such that the transmembrane proteins can be accommodated. This in turn leads to the creation of GΩ resistance (as investigated using impedance spectroscopy), high-quality membranes with minimal defects. The incorporation of a synthetic channel protein named SLIC (synthetic ligand-gated ion channel) into a preformed tethered lipid layer was reflected by a drop in membrane resistance. A control experiment conducted in another type of artificial biomembrane, the "black lipid membrane," clearly showed single channel formation and rupturing. Selective antibody binding to SLIC in the lipid bilayer increased the membrane resistance as a function of the concentration of the antibody in the aqueous phase.

Large membrane sheets (>100 µm^2) can be detached from the plasma membrane of adherent living cells and transferred onto planar glass surfaces. These supported native cell membranes offer interesting novel possibilities for the investigation of the natural lateral distribution of membrane proteins and lipids, and their interactions during biochemical membrane processes. In order to fabricate these large membrane sheets, human embryonic kidney (HEK) cells were grown, rinsed with buffer, and pressed with a poly-L-lysine-coated coverslip for 3–4 min to create a good contact between the upper membrane of the cells and the glass surface. The glass coverslip was then removed, ripping off the upper cell membranes.

The HEK cells express a variety of representative membrane proteins involved in important signal transduction pathways, including the α1b-adrenergic receptor (α1b-AR) and the neurokinin-1 receptor (NK1R) as prototypical G protein-coupled receptors (GPCRs), G proteins as membrane-anchored proteins, and the serotonin type 3 receptor (5HT3R) as a representative of ligand-gated ion channels. Because the interest here was in ligand-gated ion channels, each of the five subunits of the homopentameric 5HT3R was labeled with a cyan fluorescent protein (CFP), yielding 5HT3R–CFP. The fluorescent analogue of the 5HT3R-specific ligand GR-Cy3 was bound to the membrane. Confocal microscopy indicated that the membrane patches showed a fluorescence of the fusion protein 5HT3R–CFP expressed at the plasma membrane, whereas after incubation with the 5HT3R-specific ligand GR-Cy3, only those receptors located at the periphery of the membrane sheets were labeled.

11.6
Summary

To summarize, optical microscopy technologies in both resonant and nonresonant fashion, offer a wide range for the imaging of thin films and surface features with a high z-resolution and a lateral resolution that is limited by diffraction. In order to extend the lateral resolution, a variety of confocal methods have been developed which show resolutions down to 200 nm. Further resolution may be achieved by using electron microscopy and other near-field methods.

These methods are frequently applied to the study of natural and artificial thin films and interfaces in all fields of science and engineering. Today, the life sciences continue to benefit enormously in terms of both basic and applied research.

References

1 Dahotre, N.B. and Nayak, S. (2005) Nanocoatings for engine application. *Surface and Coating Technology*, 194 (1), 58–67.
2 Dwaber, M., Rabe, K.M. and Scott, J.F. (2005) Physics of thin-film ferroelectric oxides. *Review of Modern Physics*, 77 (4), 1083–130.
3 Lu, W., Xie, P. and Lieber, C.M. (2008) Nanowire transistor performance limits and applications. *IEEE Transactions on Electronic Devices*, 55 (11), 2859–76.
4 Andersson, C. (2008) New ways to enhance the functionality of paperboard by surface treatment – a review. *Packaging Technology and Science*, 21 (6), 339–73.
5 Mendieta-Taboada, O., de Carvalho, R.A. and Sobral, P.J.A. (2008) Dynamic-mechanical analysis: applications in edible film technology. *Quimica Nova*, 31 (2), 384–93.
6 Adachi, M., Jiu, J. and Isoda, S. (2007) Synthesis of morphology-controlled titania nanocrystals and application for dye-sensitized solar cell's. *Current Nanoscience*, 3 (4), 285–95.
7 Niklasson, G.A. and Granqvist, C.G. (2007) Electrochromics for smart windows: thin film of tungsten oxide and nickel oxide, and devices based on these. *Journal of Material Chemistry*, 17 (2), 127–56.
8 Schulz, U., Saruhan, B., Fritscher, K. and Leyens, C. (2004) Review on advanced EB-PVD ceramic topcoats for TBC applications. *International Journal of Applied Ceramic Technology*, 1 (4), 302–15.
9 Baszkin, A. and Norde, W. (2000) *Physical Chemistry of Biological Interfaces*, Marcel Dekker, New York.
10 Hickel, W., Kamp, D. and Knoll, W. (1989) Surface-plasmon microscopy. *Nature*, 339 (6221), 186–
11 Hickel, W. and Knoll, W. (1990) Optical waveguide microscopy. *Applied Physics Letters*, 57 (13), 1286–8.
12 Liebermann, T. and Knoll, W. (2003) Parallel multispot detection of target hybridization to surface-bound probe oligonucleotides of different base mismatch by surface-plasmon field-enhanced fluorescence microscopy. *Langmuir*, 19 (50), 1567–72.
13 Aust, E., Hickel, W., Knobloch, H., Orendi, H. and Knoll, W. (1993) Electrooptical waveguide spectroscopy and waveguide microscopy. *Molecular Crystals and Liquid Crystals*, 227, 49–59.
14 Török, P. and Kao, F.-J. (2007) *Optical Imaging and Microscopy: Techniques and Advanced Systems*, 2nd rev. edn, Springer, Berlin, New York.
15 Thoma, F., Langbein, U. and Mittler-Neher, S. (1997) Waveguide scattering microscopy. *Optics Communications*, 134, 16–20.
16 Hassanzadeh, A., Nitsche, M., Armstrong, S., Dixon, J., Langbein, U. and Mittler, S. (2008) Waveguide evanescent field fluorescence microscopy: thin film fluorescence intensities and its application in cell biology. *Applied Physics Letters*, 92, 233503.
17 Schatten, H. and Pawley, J.B. (eds) (2008) *Biological Low-Voltage Scanning Electron Microscopy*, Springer Science + Business Media, LLC, New York, NY.
18 Magonov, S.N. and Whangbo, M.-H. (1996) *Surface Analysis with STM and AFM: Experimental and Theoretical Aspects of Image Analysis*, VCH Verlag GmbH, Weinheim.
19 Bard, A.J. and Mirkin, M.V. (2001) *Scanning Electrochemical Microscopy*, Marcel Dekker, New York.
20 Leng, Y. (2008) *Materials Characterization: Introduction to Microscopic and Spectroscopic Methods*, John Wiley & Sons, Inc., New York.
21 Scott, E.A., Nichols, M.D., Cordova, L.H., George, B.J., Jun, Y.S. and Elbert, D.L. (2008) Protein adsorption and cell adhesion on nanoscale bioactive coatings formed from poly(ethylene glycol) and albumin microgels. *Biomaterial*, 29 (34), 4481–93.
22 Oliviero, G., Bergese, P., Canavese, G., Chiari, M., Colombi, P., Cretich, M.,

Damin, F., Fiorilli, S., Marasso, S., Ricciardi, C., Rivolo, P. and Depero, L. (2008) A biofunctional polymeric coating for microcantilever molecular recognition. *Analytica Chimica Acta*, 630 (2), 161–7.

23 Mathy, A., Simmrock, H.U. and Bubeck, C. (1991) Optical wave-guiding in thin-films of polyelectrolytes. *Journal of Physics D – Applied Physics*, 24 (6), 1003–8.

24 Keller, S., Blagoi, G., Lillemose, M., Haefliger, D. and Boisen, A. (2008) Processing of thin SU-8 films. *Journal of Micromechanics and Microengineering*, 18 (12), 125020.

25 Mittler-Neher, S., Otomo, A., Stegeman, G.I., Lee, C.Y.-C., Mehta, R., Agrawal, A.K. and Jenekhe, S.A. (1993) Waveguiding in substrate supported and free-standing films of insoluble conjugated polymers. *Applied Physics Letters*, 62, 115–17.

26 Ulman, A. (1991) *An Introduction to Ultrathin Organic Films: From Langmuir-Blodgett to Self-Assembly*, Academic Press, Boston.

27 Ariga, K., Nakanishi, T. and Michinobu, T. (2006) Immobilization of biomaterials to nano-assembled films (self-assembled monolayers, Langmuir-Blodgett films, and layer-by-layer assemblies) and their related functions. *Journal of Nanoscience and Nanotechnology*, 6 (8), 2278–301.

28 Basu, J.K. and Sanyal, M.K. (2002) Ordering and growth of Langmuir-Blodgett films: X-ray scattering studies. *Physics Reports – Review Section of Physics Letters*, 363 (1), 1–84.

29 Ulman, A. (1996) Formation and structure of self-assembled monolayers. *Chemical Reviews*, 96, 1533–54.

30 Schreiber, F. (2000) Structure and growth of self-assembling monolayers. *Progress in Surface Science*, 65, 151–256.

31 Love, J.C., Estroff, L.A., Kriebel, J.K., Nuzzo, R.G. and Whitesides, G.M. (2005) Self-assembled monolayers of thiolates on metals as a form of nanotechnology. *Chemical Reviews*, 105 (4), 1103–69.

32 Ariga, K., Hill, J.P. and Ji, Q.M. (2007) Layer-by-layer assembly as a versatile bottom-up nanofabrication technique for exploratory research and realistic application. *Physical Chemistry Chemical Physics*, 9 (19), 2319–40.

33 Decher, G. and Hong, J.D. (1991) Buildup of ultrathin multilayer films by a self-assembly process. 2. Consecutive adsorption of anionic and cationic bipolar amphiphiles and polyelectrolytes on changed surfaces. *Bericht der Bunsen-Gesellschaft, Physical Chemistry Chemical Physics*, 95 (11), 1430–4.

34 Decher, G. and Hong, J.D. (1991) Macromolecular Chemistry, Buildup of ultrathin multilayer films by a self-assembly process: 1. Consecutive adsorption of anionic and cationic bipolar amphiphiles on charged surfaces. *Macromolecular Symposia*, 46, 321–7.

35 Rühe, J. (1998) Polymers grafted from solid surfaces. *Macromolecular Symposia*, 126, 215–22.

36 Piscevic, D., Tarlov, M. and Knoll, W. (1994) Surface-plasmon microscopy of biomolecular recognition reactions on UV-photopatterned alkanethiol self-assembled monolayers. *Abstracts of Papers of the American Chemical Society*, 208 (Part 1), 237.

37 Li, T.S., Mitsuishi, M. and Miyashita, T. (2004) Ultrathin polymer Langmuir-Blodgett films for microlithography. *Thin Solid Films*, 446 (1), 138–42.

38 Suzuki, K. (2007) *Microlithography: Science and Technology*, 2nd edn, CRC Press, Hoboken.

39 Kunz, C.O., Long, P.C. and Wright, A.N. (1972) Positive and negative photoresist applications of thin films surface-photopolymerized from hexachlorobutadiene. *Polymer Engineering and Science*, 12 (3), 209–12.

40 Benschop, J., Banine, V., Lok, S. and Loopstra, E. (2008) Extreme ultraviolet lithography: status and prospects. *Journal of Vacuum Sciences and Technology*, 26 (6), 2204–7.

41 Vutova, K. and Mladenov, G. (2008) Computer simulation of micro- and nano- structures at electron and ion lithography. *Journal of Optoelectronics and Advances Materials*, 10 (1), 91–7.

42 Gamo, K. and Namba, S. (1991) Recent advances of focused ion-beam technology in maskless deposition and patterning. *Nuclear Instruments and Methods in Physics Research Section B – Beam Interaction with materials and Atoms*, 59 (1), 190–6.

43 Frey, S., Heister, K., Zharnikov, M. and Grunze, M. (2000) Modification of semifluorinated alkanethiolate monolayers by low energy electron irradiation. *Physical Chemistry Chemical Physics*, 2 (9), 1979–87.

44 Zharnikov, M., Frey, S., Heister, K. and Grunze, M. (2000) Modification of alkanethiolate monolayers by low energy electron irradiation: dependence on the substrate material and on the length and isotopic composition of the alkyl chains. *Langmuir*, 16 (6), 2697–705.

45 Marrian, C.R.K. and Tennant, D.M. (2003) Nanofabrication. *Journal of Vacuum Science and Technology A*, 21 (5), S207–15.

46 Schift, H. (2008) Nanoimprint lithography: an old story in modern times? A review. *Journal of Vacuum Science and Technology B*, 26 (2), 458–80.

47 Schift, H. and Kristensen, A. (2007) *Handbook of Nanotechnology*, 2nd edn (ed. B. Bhushan), Springer, Berlin, Chap. A/8, pp. 239–78.

48 Kumar, A. and Whitesides, G.M. (1993) Features of gold having micrometer to centimeter dimensions can be formed through a combination of stamping with an elastomeric stamp and an alkylthiol ink followed by chemical etching. *Applied Physics Letters*, 63 (14), 2002–4.

49 Kumar, A., Biebuyck, H.A. and Whitesides, G.M. (1994) Patterning self-assembled monolayers – application in material science. *Langmuir*, 10 (5), 1498–511.

50 Piner, R.D., Zhu, J., Xu, F., Hong, S.H. and Mirkin, C.A. (1999) "Dip-pen" nanolithography. *Science*, 283 (5402), 661–3.

51 Li, Y., Maynor, B.W. and Liu, J. (2001) Electrochemical AFM "dip-pen" nanolithography. *Journal of the American Chemical Society*, 123 (90), 2105–6.

52 (a) Lercel, M.J., Redinbo, G.F., Graighead, H.G., Sheen, C.W. and Allara, D.L. (1994) Scanning-tunneling microscopy based lithography of octadecanethiol on Au and GaAs. *Applied Physics Letters*, 65 (8), 974–6;
(b) Kim, Y.-T. and Bard, A.J. (1992) Imaging and etching of self-assembled n-octadecanethiol layers on gold with scanning tunneling microscopy. *Langmuir*, 8 (4), 1096–102.

53 Leggett, G.J. (2006) Scanning near-field photolithography – surface photochemistry with nanoscale spatial resolution. *Chemical Society Reviews*, 35 (11), 1150–61.

54 Lasagni, A.F., Acevedo, D.F., Barbero, C.A. and Muecklich, F. (2008) Direct patterning of polystyrene-polymethyl methacrylate copolymer by means of laser interference lithography using UV laser irradiation. *Polymer Engineering and Science*, 48 (12), 2367–72.

55 Zhang, A.P., He, S.L., Kim, K.T., Yoon, Y.K., Burzynski, R., Samoc, M. and Prasad, P.N. (2008) Fabrication of submicron structures in nanoparticle/polymer composite by holographic lithography and reactive ion etching. *Applied Physics Letters*, 93 (20), 203509.

56 Kettle, J., Coppo, P., Lalev, G., Tattershall, C., Dimov, S. and Turner, M.L. (2008) Development and validation of functional imprint material for the step and flash imprint lithography process. *Microelectronic Engineering*, 85 (5–6), 850–2.

57 Deckman, H.W. and Dunsmuir, J.H. (1982) Natural lithography. *Applied Physics Letters*, 41 (4), 377–9.

58 Qian, X., Li, J., Wasserman, D. and Goodhue, W.D. (2008) Uniform InGaAs quantum dot arrays fabricated using nanosphere lithography. *Applied Physics Letters*, 93 (23), 231907.

59 Hibbs, A.R. (2004) *Confocal Microscopy for Biologists*, Kluwer Academic/Plenum Publishers, New York, London.

60 Pawley, J.B. (2006) *Handbook of Biological Confocal Microscopy*, 3rd edn, Springer, New York.

61 Snyder, A.W. and Love, J.D. (1983) *Optical Waveguide Theory*, Chapman & Hall, London, New York.

62 Marcuse, D. (1973) *Integrated Optics*, IEEE Press, New York.

63 Hönig, D. and Möbius, D. (1991) Direct visualization of monolayers at the air-water interface by Brewster angle microscopy. *Journal of Physical Chemistry*, 95 (12), 4590–2.

64 Hönig, D. and Möbius, D. (1992) Brewster-angle microscopy of LB films on solid substrates. *Chemical Physics Letters*, 195 (1), 50–2.

65 Kaerchert, T., Hönig, D. and Möbius, M. (1993) Brewster-angle microscopy – a new method of visualizing the spreading of meibomian lipids. *International Ophthalmology*, 12 (6), 341–8.

66 Hobley, J., Oori, T., Kajimoto, S., Gorelik, S., Hönig, D., Hatanaka, K. and Fukumura, H. (2008) Laser-induced phase change in Langmuir films observed using nanosecond pump-probe Brewster angle microscopy. *Applied Physics A – Material Science and Processing*, 93 (4), 947–54.

67 Feller, B.E., Kellis, J.T. Jr, Cascao-Pereira, L.G., Knoll, W., Robertson, C.R. and Frank, C.W. (2008) Fluorescence quantification for surface plasmon excitation. *Langmuir*, 24 (21), 12303–11.

68 Knoll, W., Kasry, A., Yu, F., Wang, Y., Brunsen, A. and Dostalek, J. (2008) New concepts with surface plasmons and nano-biointerfaces. *Journal of Nonlinear Optical Physics and Materials*, 17 (2), 121–9.

69 Mansuripur, M. (2002) *Classical Optics and Its Applications*, Cambridge University Press, Cambridge, UK, New York.

70 Aust, E.F. and Knoll, W. (1993) Electrooptical waveguide microscopy. *Journal of Applied Physics*, 73 (6), 2705–8.

71 Aust, E.F., Sawodny, M. and Knoll, W. (1994) Surface-plasmon and guided optical wave microscopies. *Scanning*, 16 (6), 353–61.

72 Knoll, W., Philpott, M.R., Swalen, J.D. and Girlando, A. (1982) Surface plasmon enhanced Raman spectra of molecular assemblies. *Journal of Chemical Physics*, 77 (5), 2254–9.

73 Knobloch, H. and Knoll, W. (1991) Raman imaging and Raman spectroscopy with surface plasmon light. *Journal of Chemical Physics*, 94 (2), 835–42.

74 Nemetz, A. and Knoll, W. (1996) Raman spectroscopy and microscopy with plasmon surface polaritons. *Journal of Raman Spectroscopy*, 27 (8), 587–92.

75 Axelrod, D. (1981) Cell-substrate contacts illuminated by total internal reflection fluorescence. *Journal of Cell Biology*, 89 (2), 141–5.

76 Axelrod, D. (1989) Total internal reflection fluorescence microscopy. *Methods in Cell Biology*, 30, 245–70.

77 Axelrod, D. (2008) Total internal reflection fluorescence microscopy, in *Biophysical Tools for Biologists, Vol. 2: In Vivo Techniques*. Academic Press, pp. 169–221.

78 Thoma, F., Armitage, J., Trembley, H., Menges, B., Langbein, U. and Mittler-Neher, S. (1998) Waveguide scattering microscopy in air and water. *SPIE*, 3414, 242–9.

79 Hassanzadeh, A., Armstrong, S., Dixon, S.J. and Mittler, S. (2009) Multimode waveguide evanescent field fluorescence microscopy: measurement of cell-substratum separation distance. *Applied Physics Letters*, 94, 033503.

80 Pelletier, M.J. (1999) *Analytical Applications of Raman Spectroscopy*, Blackwell Science, Oxford, Malden, MA.

81 Hibbs, A.R. (2004) *Confocal Microscopy for Biologists*, Kluwer Academic/Plenum Publishers, New York, London.

82 Govil, A., Pallister, D.M. and Morris, M.D. (1993) 3-dimensional digital confocal Raman microscopy. *Applied Spectroscopy*, 47 (1), 75–9.

83 Göpel, W., Hesse, J. and Zemel, J.N. (1989–1998) *Sensors: A Comprehensive Survey*, Wiley-VCH Verlag GmbH, Weinheim.

84 Davis, F. and Higson, S.P.J. (2005) Structured thin films as functional components within biosensors. *Biosensors and Bioelectronics*, 21 (1), 1–20.

85 Janata, J. and Josowicz, M. (2009) Organic semiconductors in potentiometric gas sensors. *Journal of Solid State Electrochemistry*, 13 (1), 41–9.

86 Han, Y.J., Loo, S.C.J., Lee, J. and Ma, J. (2007) Investigation of the bioactivity and biocompatibility of different glass interfaces with hydroxyapatite, fluorohydroxyapatite and 58S bioactive glass. *Biofactors*, 30 (4), 205–16.

87 Ratner, B.D. (1992) Plasma deposition for biomedical application – a brief review. *Journal of Biomaterials Science – Polymer Edition*, 4 (1), 3–11.

88 Gay, C. (2003) Some fundamentals of adhesion in synthetic adhesives. *Biofouling*, 19 (Suppl.), 53–7.

89 Abbasi, A., Eslamian, M., Heyd, D. and Rousseau, D. (2008) Controlled release of DSBP from genipin-crosslinked gelatin thin films. *Ethical Development and Technology*, 13 (6), 549–57.

90 Zhang, Z.Q., Cao, X.C., Zhao, X.B., Holt, C.M., Lewis, A.L. and Lu, J.R. (2008) Controlled delivery of anti-sense oligodeoxynucleotide from multilayered biocompatible phosphorylcholine polymer films. *Journal of Controlled Release*, 130 (1), 69–76.

91 Jewell, C.A. and Lynn, D.M. (2008) Multilayered polyelectrolyte assemblies as platforms for the delivery of DNA and other nucleic acid-based therapeutics. *Advanced Drug Delivery Reviews*, 60 (9), 979–99.

92 Ariga, A., Nakanishi, T. and Michinobu, T. (2006) Immobilization of biomaterials to nano-assembled films (self-assembled monolayers, Langmuir-Blodgett films, and layer-by-layer assemblies) and their related functions. *Journal of Nanoscience and Nanotechnology*, 6 (8), 2278–301.

93 Adelberg, J.W., Delgado, M.P. and Tomkins, J.P. (2005) Ancymidol and liquid media improve micropropagation of Hemerocallis hybrid cv. "Todd Monroe" on a thin-film "rocker" bioreactor. *Journal of Horticultural Science and Biotechnology*, 80 (6), 774–8.

94 Whang, L.M., Yang, Y.F., Huang, S.J. and Cheng, S.S. (2008) Microbial ecology and performance of nitrifying bacteria in an aerobic membrane bioreactor treating thin-film transistor liquid crystal display wastewater. *Water Science and Technology*, 58 (12), 2365–71.

95 Tanaka, M. (2006) Polymer-supported membranes: physical models of cell surfaces. *MRS Bulletin*, 31 (7), 513–20.

96 Menges, B. and Mittler, S. (2007) Evanescent waves as nanoprobes for surfaces and interfaces: from waveguide technology to sensor application, in *Surface Nanophotonics. Principles and Applications*, Vol. 133 (eds D. Andrews and Z. Gaburro), Springer Series in Optical Science, Springer, Berlin/Heidelberg, pp. 19–47.

97 Knoll, W., Yu, F., Neumann, T., Schiller, S. and Naumann, R. (2003) Supramolecular functional interfacial architectures for biosensor applications. *Physical Chemistry Chemical Physics*, 5 (23), 5169–75.

98 Wiltschi, B., Knoll, W. and Sinner, E.-K. (2006) Binding assays with artificial tethered membranes using surface plasmon resonance. *Methods*, 39 (2), 134–46.

99 Robelek, R., Niu, L.F., Schmid, E.L. and Knoll, W. (2004) Multiplexed hybridization detection of quantum dot-conjugated DAN sequences using surface plasmon enhanced fluorescence microscopy and spectrometry. *Analytical Chemistry*, 76 (20), 6160–5.

100 Jones, F.J. (2001) Teeth and bones: applications of surface science to dental materials and related biomaterials. *Surface Science Reports*, 42 (3–5), 79–205.

101 Rifkin, B.R. and Gay, C.V. (1992) *Biology and Physiology of the Osteoclast*, CRC Press, Boca Raton.

102 Mahapatro, A. and Kulshrestha, A.S. (eds) (2008) *Polymers for Biomedical Applications*, American Chemical Society: Oxford University Press, Washington, DC.

103 Fisher, J.P. (2006) *Tissue Engineering*, Springer, New York, NY.

104 Scotchford, C.A., Cooper, E., Leggett, G.J. and Downes, S. (1998) Growth of human osteoblast-like cells on alkanethiol on gold self-assembled monolayers: the effect of surface

chemistry. *Journal of Biomedical Materials Research*, 41 (3), 431–42.
105 Hoffman, A.S. (2008) The origins and evolution of "controlled" drug delivery systems. *Journal of Controlled Release*, 132 (3), 153–63.
106 Mitragotri, S. and Lahann, J. (2009) Physical approaches to biomaterial design. *Nature Materials*, 8 (1), 15–23.
107 Manna, U. and Patil, S. (2008) Encapsulation of uncharged water-insoluble organic substance in polymeric membrane capsules via layer-by-layer approach. *Journal of Physical Chemistry B*, 112 (42), 13258–62.
108 Depprich, R., Handschel, J., Wiesmann, H.-P., Jäsche-Meyer, J. and Meyer, U. (2008) Use of bioreactors in maxillofacial tissue engineering. *British Journal of Oral and Maxillofacial Surgery*, 46 (5), 349–54.
109 Huang, H.-J., Ramaswamy, S., Tschirner, U.W. and Ramarao, B.V. (2008) A review of separation technologies in current and future biorefineries. *Separation and Purification Technology*, 62 (1), 1–21.
110 Liao, B.-Q., Kraemer, J.T. and Bagley, D.M. (2006) Anaerobic membrane bioreactors: applications and research directions. *Critical Reviews in Environmental Science and Technology*, 36 (6), 489–530.
111 Asatekin, A., Menniti, A., Kang, S.Y., Elimelech, M., Morgenroth, E. and Mayes, A.M. (2006) Antifouling nanofiltration membranes for membrane bioreactors from self-assembling graft copolymers. *Journal of Membrane Science*, 285 (1–2), 81–9.
112 Danelon, C., Terrettaz, S., Guena, O., Koudelka, M. and Vogel, H. (2008) Probing the function of ionotropic and G protein-coupled receptors in surface-confined membranes. *Methods*, 46 (2), 104–15.

12
Structural Characterization Techniques of Molecular Aggregates, Polymer, and Nanoparticle Films

Takeshi Hasegawa

12.1
Introduction

Since surfaces and interfaces provide a field for chemical reactions and light–material interactions, thin films and/or molecular adsorbates on a surface may represent the key to an understanding of the properties of functionalized materials. One key word that is essential when describing functionalized thin materials is the "orientation" of the chemical groups. This refers to the direction of a chemical group, and may often represent the material's characteristics not only in terms of its chemical reactions, but also of its surface and optical properties. In other words, the characteristic properties of a material may be governed by the molecular structure and arrangement within the surface-covering layer. It is crucially important, therefore, to define an appropriate strategy for the molecular design and synthesis of the surface layer. In addition, an understanding of molecular orientation within the surface layer by using appropriate analytical techniques is also important. For the analysis of molecular orientation in thin films, spectroscopy – notably infrared (IR) spectroscopy – are especially useful for acquiring molecular information at the chemical group level, irrespective of the degree of crystallinity of the film. If a thin film has a surface roughness, or if the molecular aggregation is inhomogeneous, then the analysis of surface topography might also important when discussing surface properties, and for this purpose both microscopic and surface chemical techniques will be required. Those techniques that employ vibrational spectroscopy for polymeric thin films, based on these concepts, are described in this chapter.

The concept of molecular orientation, on the other hand, can be extended to light absorption by metal nanoparticles deposited on a solid surface. The collective motion of electron gas density in a metallic material is represented by the plasma frequency, ω_p. In a metal, the electron plasma generates an image whereby oscillation of the electron gas density below ω_p can match the frequency of the incident light to screen (reflect) the light, but not the light frequency above ω_p and the light transmittance of the metal. Consequently, the energy position of ω_p will deter-

Nanomaterials for the Life Sciences Vol.5: Nanostructured Thin Films and Surfaces.
Edited by Challa S. S. R. Kumar
Copyright © 2010 WILEY-VCH Verlag GmbH & Co. KGaA, Weinheim
ISBN: 978-3-527-32155-1

mines the optical properties of the metal, as represented by color. The quantum of plasma is defined as the plasmon; a light-induced plasmon causes "polarity oscillation" coupled with light; thus is referred to as "plasmon polariton," and corresponds to a molecular vibration coupled with light, as observed using IR spectroscopy. Thus, a similar concept of transition moment can be taken into account in order to understand plasmon–polariton absorption in UV-visible absorption spectroscopy. In this chapter, a newly developed method is introduced to analyze the plasmon polariton, namely multiple-angle incidence resolution spectrometry (MAIRS).

12.2
Characterization of Ultrathin Films of Soft Materials

12.2.1
X-Ray Diffraction Analysis

Well-defined atomic arrangements are available in a crystal of metallic and inorganic materials. As these highly ordered arrangements have a repeated structure, analytical techniques based on light diffraction can be employed, including X-ray diffraction (XRD) [1] and low-energy electron diffraction [2]. Both techniques are also useful for the analysis of precise organic molecular arrangements.

The most highly organized assemblies of organic molecules are found in Langmuir–Blodgett (LB) films [3]. An LB film is composed of deposited monolayers that have been transferred from a Langmuir monolayer [3] prepared on an aqueous solution. Typically, the Langmuir monolayer is an assembly of organic molecules, the molecular density of which can be controlled by changing the surface pressure of the monolayer by using a moving barrier placed at the water surface. When a high surface pressure is applied to the Langmuir monolayer, the molecular density is raised, and this induces a highly organized molecular arrangement within the monolayer. Such an organized molecular order is often retained during the LB transfer onto a substrate surface [4, 5], such that an LB film with a high molecular order will be obtained with the wet process. When the LB film is made from highly electron-scattering compounds, such as metallic salts, then even a very thin LB film can be easily analyzed using XRD.

Figure 12.1 shows an XRD pattern of a five-monolayer cadmium stearate LB film deposited on glass [6]. The film has a thickness of approximately 12.5 nm, but clear diffraction peaks are available. The peak locations on the 2θ axis revealed that the bilayer long spacing of the LB film was 5.05 nm. According to Sugi et al., the bilayer spacing, d (nm), of a cadmium salt of an n-fatty acid is related to the number of carbons, n, with a constant, s [7]:

$$d = 0.53 + sn. \tag{12.1}$$

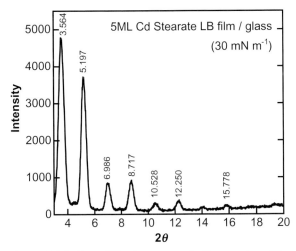

Figure 12.1 An X-ray diffraction pattern of a five-monolayer LB film of cadmium stearate on a glass slide transferred from a Langmuir monolayer at a surface pressure of 30 mN m^{-1}.

The typical value of s for trans-zigzag alkyl chains is known to be 0.254 nm, which corresponds to the distance of the alternate carbon atoms. In the present case, s was calculated to be 0.251 nm from the bilayer spacing (d = 5.05 nm) using n = 18. The tilt angle of the molecule, ϕ, could then be estimated by the two s values:

$$\varphi = \cos^{-1}(0.251/0.254) = 8.6° \tag{12.2}$$

This analytical result strongly suggested that the hydrocarbon chain had a near-perpendicular stance to the substrate. In this manner, the long spacing structure in an LB film along the out-of-plane (OP; surface normal) direction can be analyzed using XRD.

In recent years, XRD analysis of the in-plane (IP; surface parallel) structure by using a grazing-angle incidence geometry has also become available, so that the molecular packing can be discussed in detail. At an early stage of this technique, a very bright X-ray source such as synchrotron orbital radiation was necessary because of the poor signal-to-noise ratio (SNR) [8]. More recently, however, the situation has been greatly improved, such that IP- and OP-XRD measurements are both available using an in-laboratory XRD set-up. Recent progressive studies on LB films using IP-XRD were reported in a series of studies conducted by Nakahara and coworkers [8–10]. Using XRD, Nakahara's group studied the structural changes of a cationic LB film before and after the adsorption of pyranine anions on the film [8–10]. For this, the cationic monolayer was composed of n-octadecylamine (ODA; neutral) [11] and p-phenylenedimethylenebis(octadecyldimethylammonium)dibromide (p-PODB; cationic). Figure 12.2a presents the IP-XRD patterns of the LB films only, and Figure 12.2b the results for LB films associated

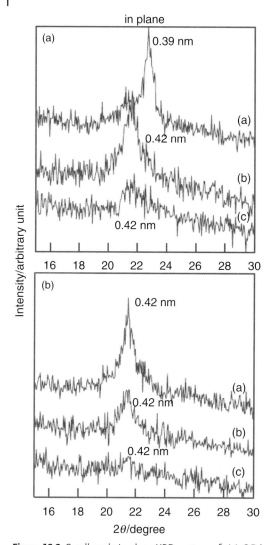

Figure 12.2 Small-angle in-plane XRD pattern of: (a) ODA; (b) ODA: p-PODB = 2:1; and (c) p-PODB LB films. Panels (a) and (b) present results of the LB films adsorbed without and with pyranine, respectively.

with pyranine molecules. Each of the figure panels includes results with different molar ratios of ODA and p-PODB [(a–c) = 1:0, 2:1, and 0:1, respectively]. When no pyranine was adsorbed onto an LB film, the molecular packing structure was seen to depend on the molar ratios. For example, a slightly distorted orthorhombic lattice of an ODA film was changed to an isotropic hexagonal film of a mixture of ODA and p-PODB with a ratio of 2:1. On the other hand, when pyranine was adsorbed onto the LB films via an anion–cation interaction, the LB films exhibited only a single peak at $d_{IP} = 0.42$ nm, which corresponded to an isotropic hexagonal subcell packing. These results clearly suggested that the adsorption of pyranine caused the structure of the LB films to become stabilized. It was also suggested by the authors that the two-dimensional (2-D) spacing of 0.42 nm was necessary in order for the pyranine molecules to penetrate the cationic LB films.

It is clear, therefore, that XRD analysis represents a powerful means of accessing a very fine molecular structure within an ultrathin film of organic compounds. When a polymeric material was studied, however, such an atomic-level structure was often far from the macroscopic character of the material. One of the most essential macroscopic characters of a polymeric material can be captured using a thermodynamic analysis, as represented by differential scanning calorimetry (DSC). In order to bridge the nanoscale and macroscale analyses, combined XRD–DSC measurements may be of great benefit, and these can be carried out simultaneously on an identical sample. Although, the combined XRD–DSC technique would in principle be expected to provide great power, the measurements may yield a number of XRD patterns but with weak DSC peaks for many thin films. Consequently, the best use of these valuable data would be very difficult.

In order to overcome these problems, Hasegawa and coworkers suggested a technique which used an augmented alternative least squares (ALS) regression analysis [12]. This technique is used to retrieve chemically independent spectra with quantity variations of the corresponding compounds, these being decomposed from the collection of spectra based on the classical least squares (CLS) regression technique [13, 14]. Hasegawa and colleagues introduced this technique for the analysis of XRD–DSC data, with CLS being formulated by following relationship:

$$\mathbf{D} = \mathbf{CK} + \mathbf{E}. \tag{12.3}$$

The measured XRD patterns represented by a matrix, **D**, which is composed of a linear combination of pure-component XRD patterns of the independent crystals, **K**, with weighting factors, **C**, except for experimental error, **E**. With the CLS regression model, **C** (or **K**) can be predicted by Equation 12.4 from **D** by using a priori knowledge of **K** (or **C**), no matter what experimental error, **E**, is involved in **D** [12, 15]:

$$\mathbf{C} = \mathbf{D}\mathbf{K}^T(\mathbf{K}\mathbf{K}^T)^{-1} \tag{12.4}$$

where the superscripts T and −1 represent transpose and inverse matrices, respectively. In this manner, **C** or **K** must be known prior to a CLS analysis.

Alternative least squares is an automated decomposition technique based on CLS, which enables the prediction of both **C** and **K** from **D** with no *a priori* knowledge. As **C** and **K** depend on each other quantitatively, a unique solution pair of **C** and **K** is not obtained at any one time. Regardless, the "shapes" of quantity variations in the matrices can be accurately elucidated with ALS, which provides a powerful means of discussing molecular structural changes.

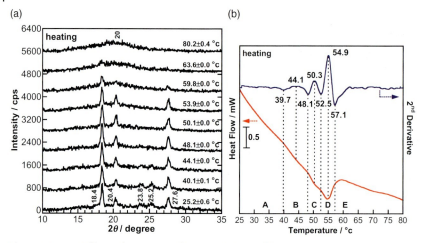

Figure 12.3 (a) Differential scanning calorimetry (DSC) (red) and second-derivative (blue) curves of an LPEI cast film in the first heating process; (b) Selected X-ray diffraction patterns measured during the DSC measurements.

In Figure 12.3 are presented a DSC curve (Figure 12.3a) and selected XRD patterns (Figure 12.3b) of a linear (polyethyleneimine) (LPEI) cast film, following simultaneous measurements on a heating process [12]. As the XRD patterns reflect variations of polymorphism, a few different crystallines will contribute to the measured XRD patterns, which in turn makes the analysis difficult. A total of 21 XRD patterns was then subjected to the augmented ALS analysis (this is an improved ALS as it introduces an additional vector that receives "garbage" information) [12].

The results are presented in Figure 12.4. Three independent XRD patterns are readily yielded extracted from the original 21 patterns, and k_1, k_2, and k_3 correspond perfectly to the XRD patterns of the sesquihydrate crystal, the hemihydrate form, and the amorphous component of LPEI [16, 17]. In addition, quantity variations of the three components are also yielded, as presented in Figure 12.4b. The quantity variations were found to be very consistent with the DSC curve. A second-derivative DSC curve is also plotted for clear recognition of peaks by removing the unnecessary baseline drift (see Figure 12.3), where the peak positions correspond to the zone boundaries in Figure 12.4 (zones A–E), from which the physical meaning of each DSC peak was readily revealed. In this manner, chemometric techniques may sometimes prove to be quite powerful for the quantitative analysis of complicated data.

12.2.2
Infrared Transmission and Reflection Spectroscopy

Infrared spectroscopy is quite useful particularly when the analyte polymer thin film has a poor crystallinity, for which diffraction techniques cannot be used. In

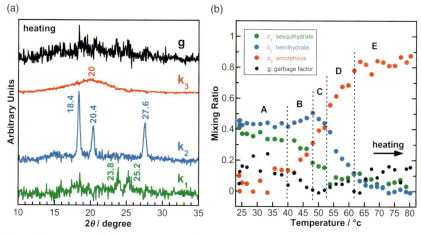

Figure 12.4 (a) X-ray diffraction patterns and (b) quantity variations of the sesquihydrate (green), hemihydrates (blue) and amorphous (red) components deduced by the augmented ALS analysis using a garbage vector (black).

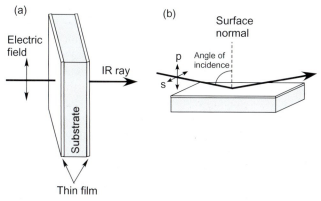

Figure 12.5 Schematics of infrared (a) transmission and (b) reflection–absorption spectrometries.

fact, IR spectroscopy represents a powerful means of elucidating molecular conformation, interactions and orientation, irrespective of the degree of crystallinity.

One IR spectroscopic approach that is used conventionally for molecular orientation analyses in a thin film combines the techniques of IR transmission and reflection–absorption (RA) spectrometries [15, 18] (Figure 12.5). In IR transmission spectrometry, an unpolarized (or sometimes polarized) IR ray is irradiated perpendicularly onto the film surface, and the transmitted ray observed (Figure 12.5a). As the electric field oscillations are perpendicular to the traveling direction of the ray (irrespective of any polarization), the electric fields will always be aligned parallel to the film surface. As a result, molecular vibrations parallel to the film surface (TO modes or Im (ε) [19]) are selectively observed in the IR transmission spectra.

In contrast, the IR RA technique employs a reflection geometry on a metallic substrate covered by an analyte film [15]. Here, an IR ray is irradiated onto the film at a grazing angle (70–85°) from the surface normal, as shown schematically in Figure 12.5b. It has been proven, theoretically, that the surface–normal molecular vibrations (LO modes or Im ($-1/\varepsilon$) [19]) appear selectively in IR RA spectra; this is known as the "surface selection rule" of RA spectrometry, and in this respect the surface selection rules of transmission and RA spectrometries are complementary. Within these rules a molecular orientation analysis can, in theory, be performed using the combination technique [20]. Most importantly, this technique is employed over a wide range of applications.

Kumaran et al. [21] prepared an LB film of a discotic molecular assembly of a phthalocyanine derivative, and subsequently studied the molecular structure and orientation by using the combination technique. Attention was focused on two mutually orthogonal vibrational modes, namely the C–O–C stretching vibration and the C–H bending vibration modes, which are aligned circularly parallel and linearly perpendicular to the discotic column, respectively. Kumaran and colleagues developed a biaxial analytical technique of order parameters for the x, y, and z directions. The tilt angles of a transition moment along the x and y directions were analyzed by two orthogonally polarized IR transmission spectra, whereas the tilt angle from the surface normal was analyzed by measuring an infrared RA spectrum. Whilst the results were clear and acceptable, the technique required an assumption that the microstructure of the thin films was the same on both the gold and silicon substrate (as mentioned above). In fact, this proved to be an analytical limit of the combination technique, although the analytical accuracy was relatively high and the method valuable overall. Of note, however, this limit could be overcome by the use of multiple-angle incidence resolution spectrometry (MAIRS), as described below.

At this point it should also be noted that the RA technique may often be misunderstood, and this has resulted in inappropriate practical applications. In one such example, the IR combination technique was employed in confusing manner to study molecular aggregation structure in a polymer thin film of poly(ethylene oxide) (PEO) [22]. For this, the films were prepared by spreading chloroform solutions of various concentrations on a single-sided "or" double-sided silicon wafer, using the spin-coating technique, the sample being used for both transmission and external-reflection (ER) measurements [22–25]. In the original report, although the ER technique was described as grazing-incidence reflection (GIR), and the authors confusingly used it as the RA technique. Historically, RA has been defined as reflection spectrometry which is limited to a metallic surface, and on which the RA-specific surface selection rule holds [19]. Here, however, a reflection geometry was used for reflection measurements on a semiconductor (nonmetallic) surface, with which it was impossible to retrieve LO modes selectively. In fact, many negative-absorbance bands appeared in the spectra, which were characteristic of

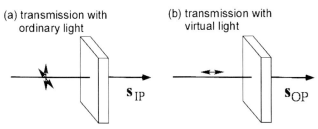

Figure 12.6 Conceptual images of normal-incidence transmission measurements using (a) ordinary light and (b) virtual light.

ER spectrometry, but these were ignored in the author's discussion. Clearly, such spectrometry should be addressed as ER, and not RA. The selection of a correct angle of incidence is very inappropriate; in this case the authors confusingly selected an angle of 75° from the surface normal, as is often the case with RA measurements. Unfortunately, this angle was very close to the Brewster angle for silicon (ca. 73°), and this disturbed the surface selection rule of ER spectrometry, which depends heavily on the Brewster angle. In addition, for the ER measurements, it is important that only a single-sided water is used [25]; otherwise, the three-layer model [26] cannot be considered which, again, disturbs the ER selection rule.

Such confusion of terminology has occasionally caused inevitable problems in practical analyses. Consequently, not only spectroscopists but also IR spectroscopy "users" should pay a careful attention to this issue so as to prevent similar problems occurring in the future.

The IR combination technique also has an intrinsic problem, in that the technique requires two different substrates, namely infrared-transparent and metallic substrates. These totally different two surfaces, in terms of their dielectric properties, would have a major influence on both the structure and electronic properties of an ultrathin film with a monolayer-level thickness. In other words, as identical films cannot be prepared on two different substrates, a direct comparison of the two spectra is not possible.

As these issues proved to be difficult to be overcome by using conventional concepts of optics and spectroscopy, it became necessary to create a totally new analytical concept by introducing the concept of "virtual light" [27]. The term virtual light means that the electric-field oscillations occur in parallel to the traveling direction of the light which is, of course, unavailable in Nature. As shown in Figure 12.6b, however, surface-normal vibrational modes would be selectively observed by using virtual light with a transmission optical geometry. As both optics (Figure 12.6a and b) have a common optical configuration, the measured spectra would, in principle, have a common absorbance scale.

To realize "virtual spectrometry," the intensities of the ordinary and virtual lights are denoted as s_{IP} and s_{OP}, respectively, to build a new measurement theory for the expression of these two lights that interact individually with the IP and OP vibrational modes in a thin film, respectively.

12.2.3
Multiple-Angle Incidence Resolution Spectrometry (MAIRS)

Although the combined technique of transmission and RA spectrometries is useful from a practical aspect, it has one intrinsic problem—namely, that both transparent and metallic surfaces are necessary, and these difference in substrates may influence the molecular structures of monolayer-level thin films that are deposited individually on the two surfaces. In order to overcome this experimental limitation, a virtual light that has an electric field oscillation parallel to the traveling direction of light should be quite useful. In this way, the LO modes could be selectively observed by employing normal-incidence transmission geometry. In other words, the LO-mode observation would be available with the same optical geometry as the transmission measurements.

This suggestion was realized by employing the idea of a "regression equation" [13, 14], a unique concept whereby an observable parameter could be related nondeterministically to another physical parameter. In physics, a physical parameter can be rigorously related to other parameters by the analytical deduction of equations from a set of fundamental data. For example, Beer–Lambert's law [28] can be deduced very precisely from Maxwell's equations [29], where absorbance is rigorously related to the concentration, refractive index, extinction coefficient, and wavelength [25, 29]. In contrast, a regression equation is created by making more observations than the number of undetermined variables in the equation. As a result, the observable parameter cannot strictly be related to the undetermined parameter (this is shown clearly in Equation 12.7). The regression equation, however, creates one parameter which is related to another through a compromise-solution calculation [13, 30] that corresponds to a least-squares solution in multivariate space. The theoretical framework of the regression equation was recently employed to develop a novel spectroscopic technique that would make virtual-light measurements possible.

12.2.3.1 Theoretical Background of MAIRS

In MAIRS, the intensity spectra of transmitted light through the substrate are measured at several angles of incidence, and stored in a matrix, s_{obs} [27, 31, 32]. This matrix is interrelated by using the CLS regression equation to two vectors of IP and OP spectra, which are denoted as s_{IP} and s_{OP}, respectively. It should be noted that a portion of the data involved in s_{obs} is theorized by s_{IP} and s_{OP} (see Figure 12.7), and the remaining part in s_{obs} is left untheorized in $\mathbf{u}$. In this sense, the regression equation is a nondeterministic equation:

$$\mathbf{s}_{obs} = r_{IP}\mathbf{s}_{IP} + r_{OP}\mathbf{s}_{OP} + \mathbf{u} \tag{12.5}$$

The weighting rates of r_{IP} and r_{OP} are functions of the angle of incidence only, which cannot be deduced directly from Maxwell equations [27].

$$r_{IP} = 1 + \cos^2\theta_j + \sin^2\theta_j \tan^2\theta_j, \quad r_{OP} = \tan^2\theta_j \tag{12.6}$$

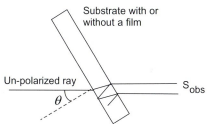

Figure 12.7 Optical scheme of multiple-angle incidence resolution spectrometry (MAIRS) measurements.

With these parameters, a collection of the spectra, **S**, measured at several angles of incidence, θ_j, are related to $\mathbf{s}_{IP}$ and $\mathbf{s}_{OP}$ as follows by the use of the CLS regression framework:

$$\mathbf{S} \equiv \begin{pmatrix} \mathbf{s}_{obs,1} \\ \mathbf{s}_{obs,2} \\ \vdots \end{pmatrix} = \begin{pmatrix} r_{IP,1} & r_{OP,1} \\ r_{IP,2} & r_{OP,2} \\ \vdots & \vdots \end{pmatrix} \begin{pmatrix} \mathbf{s}_{IP} \\ \mathbf{s}_{OP} \end{pmatrix} + \mathbf{U} \equiv \mathbf{R} \begin{pmatrix} \mathbf{s}_{IP} \\ \mathbf{s}_{OP} \end{pmatrix} + \mathbf{U} \tag{12.7}$$

where **S** is a matrix which consists of a collection of "single-beam" spectra (row-wise) as a function of the angle of incidence (column-wise). It should be noted that the single-beam spectra (but not the absorbance spectra) are subjected to the regression analysis.

The experimental requirements for MAIRS are that the substrate must be transparent, and that the sample film must be thin, so that the absorption of the film is sufficiently weak. Otherwise, the normal-incidence transmission model (Figure 12.6) cannot hold. Both, $\mathbf{s}_{IP}$ and $\mathbf{s}_{OP}$ are calculated using Equations 12.7 and 12.8 in the same manner as Equation 12.4.

$$\begin{pmatrix} \mathbf{s}_{IP} \\ \mathbf{s}_{OP} \end{pmatrix} = (\mathbf{R}^T \mathbf{R})^{-1} \mathbf{R}^T \mathbf{S} \tag{12.8}$$

The same analytical procedure is also repeated for the background measurement. As a result, the two sets of $\mathbf{s}_{IP}$ and $\mathbf{s}_{OP}$ are produced, which correspond to single-beam spectra that are virtually measured with the normal and virtual lights. With these single-beam spectra sets, two absorbance spectra ($\mathbf{A}_{IP}$ and $\mathbf{A}_{OP}$) are then calculated, using Equation 12.9:

$$\mathbf{A}_{IP} - \log(\mathbf{s}_{IP}^S / \mathbf{s}_{IP}^B), \quad \mathbf{A}_{OP} - \log(\mathbf{s}_{OP}^S / \mathbf{s}_{OP}^B) \tag{12.9}$$

where the division of the vectors, $(\mathbf{s}_{IP}^S / \mathbf{s}_{IP}^B)$, is calculated as a scalar division at each wavenumber.

The IR-MAIRS technique is "analyst-friendly" on the basis of the simple schematics, that normal-incidence transmission measurements with both ordinary (transverse wave) and virtual (longitudinal wave) lights are performed on a thin film deposited on a transparent substrate (Figure 12.7). Here, the outputs are IP and OP absorbance spectra, which correspond to the conventional transmission and RA spectra, respectively. Today, with the current commercially available MAIRS equipment and software, the results can easily be obtained simply by clicking on a few icons on the computer screen.

12.2.3.2 Molecular Orientation Analysis in Polymer Thin Films by IR-MAIRS

Polymer thin films that involve liquid crystals are aggregates of organic molecules. Most of these films are prepared using spin-coating and dip-coating [33] techniques, followed by thermal annealing; the latter procedure is used to create a greater degree of order in the molecular aggregate [34]. Although, after annealing the molecular orientation may often be improved, the crystallinity may occasionally show no improvement. In this situation, in the case of a thin film with a low crystallinity the technique of XRD cannot be used, although IR-MAIRS may be sufficiently powerful as to reveal the molecular orientation.

A dip-coated film of LPEI on a solid substrate is a good example of the well-ordered film with a low crystallinity. As LPEI is water-soluble, dip-coated films on a solid substrate were prepared from an LPEI aqueous solution. Following the annealing process, the film exhibited no crystallinity, which was confirmed with bright-field XRD using synchrotron radiation. However, the use of IR-MAIRS showed the film to have highly a ordered molecular orientation.

Figure 12.8 presents the IR-MAIRS spectra of an LPEI film that has been dip-coated on a germanium substrate [35]. The OP and IP spectra reflect surface-perpendicular and surface-parallel molecular vibrations, respectively. The band locations and relative band intensities suggest that the LPEI chains are involved in a double-strand helix structure that is found in a thoroughly dried crystal of LPEI [16, 17]. As the dip-coated film exhibited no crystallinity (as noted above), the double-strand helix should be located within a local structure.

In contrast, it has been shown clearly that many bands exhibit MAIRS-dichroism, with some bands being strong in the IP spectrum but weak in the OP

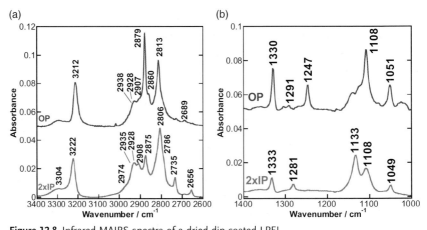

Figure 12.8 Infrared MAIRS spectra of a dried dip-coated LPEI film on a germanium substrate.

spectrum, and *vice versa*. For example, whilst the band at 1247 cm^{-1} appears only in the OP spectrum, the band at 1133 cm^{-1} appears in both spectra, but much more strongly in the IP than in the OP spectrum. This dichroic appearance strongly suggested that the molecules were largely oriented towards the surface; however, the crystallinity was very low, indicating that the LPEI film should be considered as a liquid crystal film.

Following a density functional theory (DFT) calculation, the band at 1247 cm^{-1} was mostly attributed to the CH$_2$ twisting vibration mode, which has a transition moment parallel to the double-strand helix. The appearance of this band only in the OP spectrum strongly suggests, therefore, that the double-strand helix stands perpendicularly on the substrate. The band at 1330 cm^{-1}, which is due to the CH$_2$ wagging vibration mode, is discussed in similar manner, and also supports the perpendicular orientation model of the helix. Although the two bands at 1333 and 1330 cm^{-1} are close to each other in wavenumber, the transition moments are mutually orthogonal [16]; therefore, the separated appearances in the IP and OP spectra are reasonable.

When the band at 1133 cm^{-1} was further studied by Hashida *et al.* [16], using polarized IR spectroscopy, it was shown to have a perpendicular transition moment to the LPEI chain axis. When this chain is involved in a double-strand helix, the mode should be aligned perpendicular to the helix; consequently, when the helix stands perpendicularly on the substrate the mode should appear in the IP spectrum, a point which is in agreement with the MAIRS spectra.

Another key band is the N–H stretching vibration band, which is identified at 3222 and 3212 cm^{-1} in the IP-MAIRS and OP-MAIRS spectra, respectively. As the absorptivity of this mode is weaker, the TO-LO splitting (Berreman's effect) [36, 37] would not be expected, although the large shift between the IP and OP spectra was impressive. If the Berreman effect were to occur, the OP mode should appear at a higher location than the IP mode, a point well proven (on a theoretical basis) by the Lyddane–Sachs–Teller relationship [38]. The possibility of a Berreman effect is, therefore, denied.

If the highly oriented double-strand helix model were true, the factor-group splitting [39] mechanism would represent the most likely possibility of explaining the large shift. This model can be better understood by examining the two schemes shown in Figure 12.9. In the double-strand helix, the atoms are precisely configured, and two hydrogen bonded N–H groups are placed so as to form a cross (as magnified in the figure). In this situation, coupled vibrations yield mutually orthogonal wave functions (normal modes), as well as out-of-phase and in-phase modes. The out-of-phase mode is often located at a higher wavenumber position than the in-phase mode. Thus, the shifted bands in the IR-MAIRS spectra clearly suggest that the double-strand helix stands perpendicular to the surface. Then, if the helix were to incline towards the substrate, both out-of-phase and in-phase modes should appear in the IP and OP spectra. The clear separation of the two modes in the IP and OP spectra additionally concludes that the helix stands perpendicularly to the substrate. In this manner, IR-MAIRS is able clearly to determine molecular orientation, even in a liquid crystal thin film.

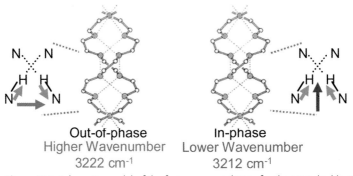

Figure 12.9 Schematic model of the factor group splitting for the LPEI double-strand helix.

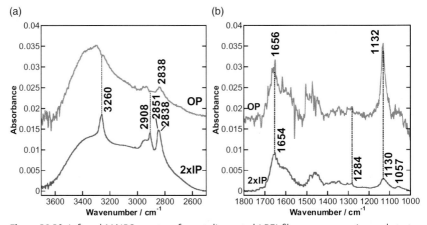

Figure 12.10 Infrared MAIRS spectra of a wet dip-coated LPEI film on a germanium substrate.

The film exhibits another interesting result when it is wetted by the vaporization of water in a beaker. When a water-filled beaker is placed close to the thin film, the film absorbs water rapidly, such that the molecular arrangement changes spontaneously. The IR-MAIRS spectra of the wet LPEI film are shown in Figure 12.10. Due to the hydration, a large broad band of the O–H stretching vibration band centered at $3300\,\text{cm}^{-1}$ dominates the high wavenumber region. The most impressive point, however, is that the N–H stretching vibration band at $3250\,\text{cm}^{-1}$ clearly appears in the IP spectrum only, but is silent in the OP spectrum. This strongly suggests that the N–H group is oriented almost parallel to the film surface.

In a crystal of LPEI, the LPEI chain is known to be stretched out when the anhydrated crystal becomes hydrated [17]. This change is shown schematically in Figure 12.11, if the same change occurs in the dip-coated film without changing the perpendicular molecular stance to the substrate. In the extended molecule model with a perpendicular orientation, each N–H group will direct parallel to the

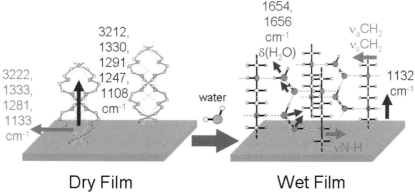

Figure 12.11 Schematics of molecular orientation in the dried and wet films of LPEI.

substrate surface, which is consistent with the IR-MAIRS spectra, and the N–H stretching vibration band will appear only in the IP spectrum. Likewise, both antisymmetric and symmetric CH_2 stretching vibration modes should be aligned parallel to the substrate surface. In fact, the two modes at 2908 and 2851 cm^{-1} appear dominant in the IP spectrum, and in this respect the schematic for both dry and wet dip-coated films should hold true.

The band at 1132 cm^{-1}, which has already been assigned to the C–N stretching vibration mode, appears strongly in the OP spectrum but was diminished in the IP spectrum. As the mode is aligned parallel to the linear molecular chain, the IR-MAIRS spectra may be easily understood when the perpendicular standing mode is taken into account.

Also of interest is the crystalline water in the film. When the molecular arrangement in the "dip-coated film" is compared (analogously) with that in the "crystal," the water molecules should be tilted (as illustrated in Figure 12.11). However, if this model were true, the H_2O scissoring vibration band at approximately 1655 cm^{-1} would be expected to appear in both the IR-MAIRS IP spectrum and the IR-MAIRS OP spectrum. The MAIRS spectra agree with the expectation (see Figure 12.10), which means that the schematic picture would be close to the truth.

In this respect, the LPEI dip-coated film that has poor crystallinity has been revealed clearly to have a perpendicularly standing molecular orientation, which should have been induced by interaction with the substrate surface. The IR-MAIRS technique has also proved to be both powerful and valuable for conducting structural analyses in polymer thin films.

12.2.3.3 Analysis of Metal Thin Films

Unfortunately, the original, above-mentioned MAIRS technique has an experimental limit in that the film-supporting substrate must have a high refractive index (if possible, >3.0). Hence, both silicon and germanium represent suitable substrate materials for MAIRS analysis in the mid-IR region. However, if a low-refractive-index substrate such as calcium fluoride is required, then an alternative technique

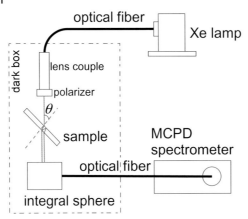

Figure 12.12 Optical set-up of the vis-MAIRS equipment.

using p-polarization, known as p-MAIRS, should be employed. Whilst it is essential that p-MAIRS is optimized experimentally when a substrate with a different refractive index is used, it must be employed to obtain MAIRS spectra in the UV-visible region, as there is no high-refractive-index substrate in the wavelength region. The p-MAIRS technique in the UV-visible region (abbreviated as "vis-MAIRS") represents a powerful new tool for the analysis of surface-plasmon polariton in a metallic thin film or in metal–particle aggregates. For details of the technique, the reader is referred elsewhere [6, 27].

The optical set-up of the visible-MAIRS scheme is shown in Figure 12.12. Here, the incident light is led from a xenon lamp by an optical fiber, and then run through a Gran–Taylor prism to generate p-polarization. The polarized ray is irradiated onto a sample, and the transmitted light led to an integral sphere. The system is arranged so that the multiply reflected light paths in the substrate are all collected by the integral sphere; it is this process of collection that leads to the high level of accuracy of the p-MAIRS analysis. The measured single-beam spectra at the optimal angles of incidence are subjected to a p-MAIRS analysis to yield the final two absorbance spectra.

One sample which has proved to be of great interest for vis-MAIRS is a thin film of metal particles; a typical set-up of a silver thin film evaporated onto a glass substrate is shown in Figure 12.13 [40]. In this situation a rough image of the thickness of the film was less than 10 nm; however, when the size of metal was decreased to nanometer order, the metallic properties (in terms of optical response) underwent a major change which was caused by a large shift in the plasma frequency. When a 5 nm-thick silver-evaporated film was prepared, the film exhibited a purple color that was very different from that of the metal. Such visible absorption is a representative result of the plasmon-polariton, which is a coupled mode of plasmon with the irradiated light. To date, the plasmon–polariton has been analyzed by measuring transmission UV-visible spectra, and the band location has been discussed.

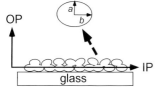

Figure 12.13 A schematic model of a very thin metal-particle film less than 10 nm in thickness.

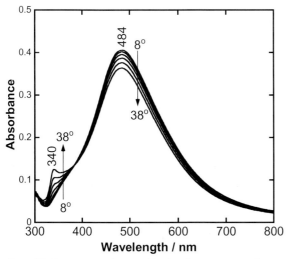

Figure 12.14 Variation of UV-visible absorbance spectra of a 5 nm-thick silver thin film evaporated onto glass.

The conventional UV-visible spectra of a silver film deposited onto glass, and measured at various angles of incidence, are presented in Figure 12.14. Although the film thickness was only about 5 nm, the absorption was very strong such that the absorbance attained was 0.4. When the angle of incidence was small (8°), virtually one band centered at 484 nm dominated the spectrum, but a new band at 340 nm evolved when the angle was increased. These spectra strongly suggested that the plasmon–polariton with a wavelength of 484 nm would lay parallel to the film surface (IP), while the other polariton (which exhibited the 340 nm band) should be perpendicular to the film (OP). Whilst it should be straightforward to obtain a pure IP spectrum by measuring a normal-incidence transmission spectrum, a pure OP spectrum would be difficult to obtain for the metal-particle film; however, vis-MAIRS might represent an effective solution to this problem.

When using the vis-MAIRS technique, two absorption spectra of plasmon–polariton parallel and perpendicular to the film surface are obtained simultaneously from an identical sample. The vis-MAIRS spectra of the same silver film are presented in Figure 12.15 [40]. Here, the IP spectrum showed only one broad band

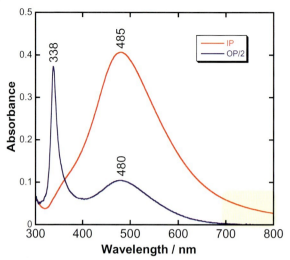

Figure 12.15 Vis-MAIRS spectra of the 5 nm-thick silver thin film evaporated onto glass.

at 485 nm, which was perfectly consistent with the normal-incident transmission spectrum (not shown). On the other hand, the OP spectrum was unique in that almost pure surface-normal polariton absorption was observed in the spectrum. As the absorbance was very large (0.4), some analytical inaccuracy might have been involved (notably at 480 nm), although the characteristic band at 338 nm was clearly extracted.

Based on the results of previous studies which used a simple ellipsoidal metal-particle model [41], the two bands can easily be assigned to plasmon excitation along the long and short axes of the ellipsoidal body. Vis-MAIRS, however, enables further insights to be reached in the metal film. One major benefit of using vis-MAIRS is that the absorption ratio of the IP and OP spectra at a long wavelength can easily be identified. In a long-wavelength region (above 700 nm) the OP spectrum falls to zero absorbance, while the IP spectrum continues to absorb. In the 5 nm-thick silver film, the silver particles were aggregated in the film, whereas few accumulations were apparent in the direction of thickness. It appears, therefore, that the plasmon–polariton wave propagates only in the IP direction, with virtually no propagation expected in the OP direction (see Figure 12.13). Moreover, propagation over a long range causes the energy of the polariton to be lowered, and this should be reflected in a band shift towards a lower wavelength. Therefore, the skirt region over 700 nm is assigned as the long propagating plasmon–polariton along the IP direction due only to the metal-particle aggregates.

Today, following much development, vis-MAIRS is approaching the situation where it is ready to reveal the physical characteristics associated with metallic particles. Most fortuitously, this property is expected to be useful not only for investigations involving physics, but also for applications in both chemistry and medicine, possibly involving the use of quantum dots and metal nanoparticles.

References

1. Fourme, R., Girard, E., Kahn, R., Ascone, I., Mezouar, M., Lin, T. and Johnson, J.E. (2004) State of the art and prospect of macromolecular X-ray crystallography at high hydrostatic pressure. *NATO Science Series II: Mathematics, Physics and Chemistry*, 140, 527–42.
2. Somorjai, G.A. and Rous, P.J. (2008) John Pendry: his contributions to the development of low energy electron diffraction surface crystallography. *Journal of Physics: Condensed Matter*, 20, 304210/1–6.
3. MacRitchie, F. (1990) *Chemistry at Interfaces*, Academic Press, San Diego.
4. Hasegawa, T., Umemura, J. and Takenaka, T. (1992) Fourier transform infrared metal overlayer attenuated total reflection spectra of Langmuir-Blodgett films of 12-hydroxystearic acid and its cadmium salt. *Thin Solid Films*, 210/211, 583–5.
5. Roberts, G.G. (1985) An applied science perspective of Langmuir-Blodgett films. *Advances in Physics*, 34, 475–512.
6. Hasegawa, T., Itoh, Y. and Kasuya, A. (2008) Experimental optimization of p-polarized MAIR spectrometry performed on a Fourier transform infrared spectrometer. *Analytical Sciences*, 24, 105–9.
7. Sugi, M., Fukui, T., Iijima, S. and Iriyama, K. (1979) Electrical properties of Langmuir multilayer films with and without dye-sensitizers. *Bulletin of the Electrotechnical Laboratory*, 43, 625–45.
8. Misawa, Y., Kubo, Y., Tokita, S., Ohkuma, H. and Nakahara, H. (2004) An isothiouronium-derived organized monolayer at the air-water interface: design of film-based anion sensor systems for $H_2PO_4^-$. *Chemistry Letters*, 33, 1118–19.
9. Tsukamoto, O., Villeneuve, M., Sakamoto, A. and Nakahara, H. (2007) Change in the orientation and packing upon adsorption of pyranine molecules onto cationic Langmuir monolayers and Langmuir-Blodgett films. *Bulletin of the Chemical Society of Japan*, 80, 1723–30.
10. Murata, M., Villeneuve, M. and Nakahara, H. (2005) Control of two-dimensional aggregates of a long-chain merocyanine by change of molecular environments in monolayer assemblies. *Chemical Physics Letters*, 405, 416–21.
11. Bruylants, G., Wouters, J. and Michaux, C. (2005) Differential scanning calorimetry in life science: thermodynamics, stability, molecular recognition and application in drug design. *Current Medicinal Chemistry*, 12, 2011–20.
12. Tauler, R., Smilde, A.K. and Kowalski, B.J. (1995) Selectivity, local rank, three-way data analysis and ambiguity in multivariate curve resolution. *Journal of Chemometrics*, 9, 31–58.
13. Hasegawa, T. (2002) Principal component regression and partial least squares modelings, in *Handbook of Vibrational Spectroscopy*, Vol. 3 (eds P.R. Griffiths and J. Chalmers), John Wiley & Sons, Inc., New York, pp. 2293–312.
14. Wehlburg, C.M., Haaland, D.M., Melgaard, D.K. and Martin, L.E. (2002) New hybrid algorithm for maintaining multivariate quantitative calibrations of a near-infrared spectrometer. *Applied Spectroscopy*, 56, 605–14.
15. Umemura, J. (2002) Reflection–absorption spectroscopy of thin films on metallic substrates, in *Handbook of Vibrational Spectroscopy*, Vol. 2 (eds P.R. Griffiths and J. Chalmers), John Wiley & Sons, Inc., New York, pp. 982–98.
16. Hashida, T., Tashiro, K., Aoshima, S. and Inaki, Y. (2002) Structural investigation on water-induced phase transitions of poly(ethylene imine). 1. Time-resolved infrared spectral measurements in the hydration process. *Macromolecules*, 35, 4330–6.
17. Chatani, Y., Tadokoro, H., Saegusa, T. and Ikeda, H. (1981) Structural studies of poly(ethylenimine). 1. Structures of two hydrates of poly(ethylenimine): sesquihydrate and dihydrate. *Macromolecules*, 14, 315–21.
18. Mirabella, F.M. (1998) *Modern Techniques in Applied Molecular Spectroscopy*, John

Wiley & Sons, Inc., New York, pp. 83–102.
19. Ibach, H. and Lüth, H. (1996) *Solid-State Physics: An Introduction to Theory & Experiment*, Springer, Berlin.
20. Umemura, J., Kamata, T., Kawai, T. and Takenaka, T. (1990) Quantitative evaluation of molecular orientation in thin Langmuir-Blodgett films by FT-IR transmission and reflection-absorption spectroscopy. *Journal of Physical Chemistry*, 94, 62–7.
21. Kumaran, N., Donley, C.L., Mendes, S.B. and Armstrong, N.R. (2008) Characterization of the angular orientation distribution of discotic molecules in thin-film assemblies: combinations of polarized transmission and reflection-absorption infrared spectroscopies. *Journal of Physical Chemistry C*, 112, 4971–7.
22. Schönherr, H. and Frank, C.W. (2003) Ultrathin films of poly(ethylene oxides) on oxidized silicon. 1. Spectroscopic characterization of film structure and crystallization kinetics. *Macromolecules*, 36, 1188–98.
23. Kattner, J. and Hoffmann, H. (2002) External-reflection spectroscopy of thin films on dielectric substrates, in *Handbook of Vibrational Spectroscopy*, Vol. 2 (eds P.R. Griffiths and J. Chalmers), John Wiley & Sons, Inc., New York, pp. 1009–27.
24. Hasegawa, T., Umemura, J. and Takenaka, T. (1993) Infrared external reflection study of molecular orientation in thin Langmuir–Blodgett films. *Journal of Physical Chemistry*, 97, 9009–12.
25. Hasegawa, T., Takeda, S., Kawaguchi, A. and Umemura, J. (1995) Quantitative analysis of uniaxial molecular orientation in Langmuir-Blodgett films by infrared reflection spectroscopy. *Langmuir*, 11, 1236–43.
26. Yeh, P. (1998) *Optical Waves in Layered Media*, John Wiley & Sons, Inc., New Jersey, pp. 102–17.
27. Hasegawa, T. (2002) A novel measurement technique of pure out-of-plane vibrational modes in thin films on a nonmetallic material with no polarizer. *Journal of Physical Chemistry B*, 106, 4112–15.
28. Griffiths, P.R. (2002) Beer's law, in *Handbook of Vibrational Spectroscopy*, Vol. 1 (eds P.R. Griffiths and J. Chalmers), John Wiley & Sons, Inc., New York, pp. 2225–34.
29. Milosevic, M. and Berets, S.L. (1993) Applications of the theory of optical spectroscopy to numerical simulations. *Applied Spectroscopy*, 47, 566–74.
30. Kramer, R. (1998) *Chemometric Techniques for Quantitative Analysis*, Marcel Dekker, New York.
31. Hasegawa, T. (2007) A new spectroscopic tool for surface layer analysis: multiple-angle incidence resolution spectrometry. *Analytical and Bioanalytical Chemistry*, 388, 7–15.
32. Hasegawa, T. (2008) A new approach to analysis of molecular structure in thin films: infrared multiple-angle incidence resolution spectrometry. *Applied Spectroscopy Reviews*, 43, 181–201.
33. Norrman, K., Ghanbari-Siahkali, A. and Larsen, N.B. (2005) Studies of spin-coated polymer films. *Annual Reports on the Progress of Chemistry, Section C*, 101, 174–201.
34. Matsunaga, M., Suzuki, T., Yamamoto, K. and Hasegawa, T. (2008) Molecular structure analysis in a dip-coated thin film of poly (2-perfluoro octylethyl acrylate) by infrared multiple-angle incidence resolution spectrometry. *Macromolecules*, 41, 5780–4.
35. Kakuda, H., Okada, T. and Hasegawa, T. (2008) anisotropic molecular structure in dip-coated films of linear poly(ethylene imine) studied by infrared multiple-angle incidence resolution spectrometry. *Journal of Physical Chemistry B*, 112, 12940–5.
36. Berreman, D.W. (1963) Infrared absorption at longitudinal optic frequency in cubic crystal films. *Physical Review*, 130, 2193–8.
37. Hasegawa, T., Nishijo, J., Umemura, J. and Theiss, W. (2001) Simultaneous evaluation of molecular-orientation and optical parameters in ultrathin films by oscillators-model simulation and infrared external reflection spectroscopy. *Journal of Physical Chemistry B*, 105, 11178–85.
38. Lyddane, R.H., Sachs, R.G. and Teller, E. (1941) The polar vibrations of alkali halides. *Physical Review*, 59, 673–6.

39 Tasumi, M. and Shimanouchi, T. (1965) Crystal vibrations and intermolecular forces of polymethylene crystals. *Journal of Physical Chemistry*, 43, 1245–58.
40 Hasegawa, T., Itoh, Y. and Kasuya, A. (2008) Development of UV-visible multiple-angle incidence resolution spectrometry and application study of Anisotropic surface-plasmon excitation in a silver thin film on a glass substrate. *Analytical Chemistry*, 80, 5630–4.
41 Zhao, L., Kelly, K.L. and Schatz, G.C. (2003) The extinction spectra of silver nanoparticle arrays: influence of array structure on plasmon resonance wavelength and width. *Journal of Physical Chemistry B*, 107, 7343–50.

Index

a
adhesion
– adaptive 122
– bacterial 3, 10, 13f.
– cell 3, 9, 15ff.
– cell membrane 123
– focal 22, 116, 160
– microbial 3
– plasma albumin 7
– platelet 7ff.
– platelet–platelet 8
– -promoter 12
ADSA (axisymmetric drop shape analysis), see contact angle technique
adsorption
– activation energy barrier 60
– BSA 3f.
– dissociative 266
– enthalpy 266
– equilibrium 212
– fibrinogen 3f., 7, 9, 21
– frequency 213
– isotherm 236
– lysozyme 3f., 21
– molecular 266
– nonspecific 7f., 15, 56f., 60, 66, 75, 180f.
– physical 62
– protein 2ff.
– selective 67, 243
AFM (atomic force microscopy)
– cell–polymer interaction 86f.
– coated tip 313f.
– dip-pen nanolithography (DPN) 304ff.
– ELP dot array 130
– nanografting 316
– nanoindentation 22, 257, 286
– nanopatterned 90
– polymer brush gradient 72
– scratching 316
– silicon nitride tip 313
– single-molecule conductivity 190
– supported lipid bilayers 210, 282
– stimuli-responsive polymer film 112
– /STM 190
– tapping mode 130
– zirconia nanofilms 252f., 258
alginate (ALG) 22, 108f.
anchimeric effects 79
antibacterial activity 10ff.
antifouling 4, 74
– bacterial 122
– glycocalyx 4
– phosphorylcholine 7
antithrombogenic 6f., 18
apatite 263f., 266
ATRP (atom transfer radical polymerization), see polymerization
AuNPs
– aggregation 176, 178ff.
– antibody-modified 181f.
– application, see microarray
– core–shell 177, 182
– DNA-modified probes 179f.
– extinction coefficient 175
– functionalization 178ff.
– grain sizes 178f.
– shape 177, 182
– synthesis 177ff.

b
bacteria
– colonization 13f.
– Gram-negative 10ff.
– Gram-postive 10ff.
– viable 13f.

binding
- constant 83
- nonspecific 314f.
bioactivity 109f.
- enzyme 110
- ionic compounds 127
- molecules 125
- zirconia nanofilms 263ff.
bioconjugates 57, 60, 63
- intelligent 109f.
- polymer surfaces 79ff.
biodegradation 12, 18, 111
biofilm
- control 122ff.
- formation 115
biofouling 29, 123 384, 386
biofunctionalization 55ff.
biointerfaces 55ff.
- applications 59
- fabrication 74f.
- functional 56f., 64, 270
- nanostructured 269f.
- polymer-based 58ff.
- zirconia nanofilms 269f.
biomacromolecules 109
- diffusion rate 279
biomarkers 179
bioreactivity 74
bioreactors, *see* membrane
biorecognition 310
biosensors 16, 55f., 58, 384
- AuNP array 188ff.
- DNA 76, 188ff.
- hydrogen peroxide 270
- patterned polymeric thin films 89f.
- thermal stability 271
- zirconia nanofilms 269ff.
blood
- clotting 7f.
- coagulation 8, 216
- compatibility 6, 8, 76
- testing algorithms, *see* diagnostics
- vessels 9
BMA (*n*-butyl methacrylate) 119f.
Boltzmann constant 187
bonding
- covalent 313
- nonspecific 319
bone
- -bonding rate 149
- cell metabolism 164
- graft 150
- growth 149

- marrow mesenchymal stem cells (BMMSCs) 267ff.
- mineralization 229
Brewster angle microscopy (BAM) 363, 373ff.
BSA (bovine serum albumin) 3f., 322
byproducts 77, 80

c

calcium phosphate (CaP) 149, 154f., 164
capillarity 332
captive bubble method 113
carbon nanotubes (CNTs) 8
- buckypaper 293
- controlled wettability 344, 350f.
- multiwalled (MWCNTs) 336f., 340f., 350f.
- single-walled (SWCNTs) 24, 292ff., 351
- water droplet impact 340f.
Cassie equation 334f., 337f.
Cassie–Baxter (CB) equation 334
catalyst 68
- AuNPs 182f.
- bio- 270
- copper 82, 85
- decomposition 9
cell
- activity 116
- apoptosis 23, 56f.
- attachment 115f.
- –cell junctions 25
- culture 14, 18, 104, 115, 121
- dedifferentiation 55f.
- detachment 25, 117, 119
- differentiation 55f., 165, 268
- encapsulation 30
- growth 16f., 21, 23, 57
- heterotypic cocultures 26
- human osteoblast-like (HOB) 160, 165
- junction 115f.
- lift-off 116
- membrane 115, 160
- migration 56
- proliferation 16, 22, 56, 115, 120f., 150, 268
- recovery efficiency 116
cell sheet engineering 25f., 28, 111, 115ff.
- 3-D cell sheet tissue engineering 121
- transplantation 121
cell
- sheet harvesting 25f.
- sorting 123
- spreading 15, 21, 57, 166
- stem 24

Index

- – surface coverage 68
- - surface mimiking 387f.
- - viability 20, 23, 35
- ceramic
- – bioactive 149
- – bioinert 263
- – glass- 149
- – nanocoatings 149ff.
- – titania 203ff.
- – zirconia 251ff.
- chemical vapor deposition (CVD), *see* deposition
- chemiluminescence 320
- chemistry
- – analytical 178
- – click 58, 82, 84
- – coupling 60
- – EDC/NHS 86
- – interfacial 112f., 204
- – NHS 65f., 74, 81f.
- CHI (chitosan) 3, 14, 18, 22, 108
- chromatograhic
- – applications 128
- – separation 124ff.
- chromatorahpy 124ff.
- – HPLC (high-performance liquid chromatograhy) 124
- – thermoresponsive chromatograhy columns 125
- CLS (classical least square) regression technique 401, 406
- CLSM (confocal laser scanning microscopy) 89
- – stimuli-responsive polymer film 112
- CMA (choline methacrylate) 16
- coatings, *see* thin films
- collagen
- – calcified titania surface 229ff.
- – fibrillar 226
- – hydroxylated titania surface 228f.
- – oxidized titania surface 228
- – titania surface 227ff.
- composite
- – AuNPs/polyelectrolyte multilayer 290ff.
- – doped zirconia–alumina 263
- – nanofibrous films 295ff.
- – nanostrand 297
- – organoceramic 259
- – polymer/CNT 289
- – silver/polymer 12
- conjugation
- – amphiphilic 9
- – nanoparticle–nanorod 182
- – PEG–DOPA–lysine 15

- – site-specific 109f.
- – SPEO–diisocyanate 9
- contact angle technique 22, 112f., 333ff.
- – apparent 335, 346
- – critical equilibrium 335, 338
- – droplet's dynamic 340f.
- – fibronectin in aqueous solution 236
- – hysteresis 345
- – spreading 345
- – static 336, 340f.
- coupling
- – covalent 3, 80ff.
- – moities 62f.
- – noncovalent 83
- – polymer surfaces 79ff.
- crosslinking
- – AuNPs 181
- – chemical 18
- – covalently 33
- – density 22, 57
- – interlayer 24
- – lysine 110f.
- CVD, *see* deposition
- cyclic voltammetric scanning 270, 272
- cytocompatible 9
- cytophilic 21, 23
- cytophobic 21
- cytotoxicity 16, 20, 111
- – polyelectrolyte 25
- – zirconia 267

d

- Debye length 61, 213
- dendritic
- – cell immobilization 17
- – polyglycerol monolayer 5
- deposition
- – aerosol-gas 183
- – anodic oxidation 251, 260f.
- – biomimetic 150
- – cathodic arc plasma 251f., 260f.
- – chemical vapor deposition (CVD) 10, 36, 64, 75ff.
- – dip-coating 257, 364, 408, 410f.
- – electrochemical 251, 257ff.
- – electron beam 286
- – electrophoretic 150, 230
- – initiated chemical vapor deposition (iCVD) 13, 29, 36
- – layer-by-layer (LbL) 9, 11, 22, 34, 64
- – liquid-phase 12
- – low-pressure glow discharge 75f., 107
- – micro-arc oxidation (MAO) 260ff.

– plasma 4, 12, 16, 75ff.
– plasma-enhanced CVD (PE-CVD) 89
– plasma spraying 254f., 263, 268f.
– pulsed laser 150
– sol–gel 150, 251, 255ff.
– spin-coating 229, 364
– sputtering, see magnetron sputtering
Derjaguin–Landau–Verwey–Overbeek (DLVO) theory 213
detection
– antigens 321
– cancer 321f.
– colorimetric 180f.
– cytokine test 321
– enzyme-linked immunosorbent assay (ELISA) 320f.
– fluorescence-type 180
– immunoassay techniques 181f.
– label-free 60, 188, 194
– limit of 303
– radioallergosorbent test (CAP/RAST) 320f.
– single nucleotide polymorphism (SNP) 193f.
– tests 319ff.
– viruses 318ff.
DFT (density functional theory) 409
diagnostics
– allergic disease 320f.
– blood testing algorithms 316f.
– medical 317ff.
– virus array generation 318ff.
differential interference contrast microscopy (DIC) 319f.
differential scanning calorimetry (DSC) 401f.
diffusion
– interlayer 33, 259
– Knudsen diffusion model 279
– molecular 279
– oxygen ion 260
– polymer 65
– rate 124
– stimuli-modulated 123
– water molecules 120
dissociation 76
– mechanical 115f.
DMSO (dimethylsulfoxide) 3
DNA, see microarrays
DOPA (dihydroxyphenylalanine) 3, 15
droplet, see water droplet
drug delivery
– applications 25, 33f., 385f.
– systems 33, 104

drug development
– drug metabolizing enzyme–toxicology assay chip (MetaChip) 323f.
– dip-pen nanolithography 316ff.
– micro/nano bioarrays 317ff.
drug
– in-situ reloading 115
– loading 115
– release 114f., 124
– screening 86, 111, 318, 323
– testing 318
DTT (dithiothreitol) 6
dynamic light scattering 182

e
ECM (extracellular matrix) 4, 17, 25f., 57
– disruption 115
– proteins on titania surface 226ff.
EDC ((ethyl-3-(3-dimethyl aminopropyl) carbidiimide)) 86
EDL (electrical double layer) 213, 216, 345
EDX (energy-dispersive X-ray spectroscopy)
– zirconia nanofilms 264
elastic modulus, see Young's modulus
electroactive linkage groups 27f.
electrode
– DNA-modiefied 272
– glassy carbon (GC) 271
– pyrolytic graphite (PG) 271
electrokinetic measurements
– stimuli-responsive polymer film 112
electron beam irradiation 107, 121
electrospinning 292
electrowetting-on-dielectric (EWOD) phenomenon, see wettability switching
ellipsometry
– in situ 119
– stimuli-responsive polymer film 112
ELP, see polypeptide
encapsulating 35f.
energy management 331
etching 75, 309, 346
ETFE (ethylene tetrafluoroethylene) 71
evanescent field waveguide microscopy/scattering, see waveguide
evaporation 256, 285

f
FBS (fetal bovine serum) 22
fluorescence-activated cell sorting (FACS) 123
fluorescence microscopy 319f.
– evanescent 363

– one-photon confocal   371, 382f.
– two-photon confocal   371, 382f.
fluorescence spectroscopy   176
free-standing thin films   277ff.
– black lipid membranes (BLM)   282f.
– CNTs   279, 289, 292f.
– common black film   284
– dried foam film (DFF)   284ff.
– equilibrium   284
– foam   283ff.
– ionic liquids   286ff.
– mechanical stability   286, 289f.
– mesoporous   289
– nanofibrous   292ff.
– Newton black film   283f., 286f.
– partitions   277f.
– SLPs, see supported lipid bilayers
– solid surfaces   289f.
– thermal stability   286
– ultrathin   288f.
friction coefficient   15, 115
functional groups   63f., 72f., 120f.
– reactive   109

g

gamma-irradiation   107
gene delivery   36ff.
Gibbs–Marangoni effect   284f.
glass transition temperature   58
grafting
– covalently   26, 64
– density   11, 65f., 112, 117f., 124, 126
– electro-   58
– electron beam   117f.
– ELPs   111
– -from approach   68f., 367f.
– PEG   59, 67
– photo-   107
– PNIPAAm   107
– polymer   25, 59, 63ff.
– /polymerization   14
– -to approach   65f., 125
growth factor
– insulin (INS)   120
– nerve (NGF)   29
GSH (glutathione)   9
GSNO (S-nitrosoglutathione)   9

h

Hagen–Poiseuille equation   279
HCECs (human corneal epithelial cells)   16
high throughput   317, 323
hybridization reaction   180f., 186, 192, 194, 270ff.

– bridged   194
– sandwich   180, 189
hydration   117, 119
– de-   118
– spontaneous   25
– surface   122
– temperature dependent   126
– thickness   13
hydrogel   3f., 15, 29, 55, 84f., 117f.
– collapsed   118, 122
– ionic   107
– PNIPAAm-based   106, 112, 114
– swelling/deswelling   107, 112, 114
hydrogen
– abstraction   4
– bonding   336
hydrolysis   24
– NHS ester   82
hydroxyapatite (HA)   149ff.
– acellular testing   154ff.
– carbonate-containing   266
– grain growth   154f.
– nanocrystals   230f.
– osteointegration   206
– silicon incorporation   149
– silicon-substituted, see SiHA

i

imaging
– applications   383ff.
– technologies   371ff.
– thin films   363ff.
immobilization
– analyte   62
– antibiotics   12
– antibody   62
– bioactive molecules   15, 60, 111, 270, 273
– covalent   14, 63, 78, 310
– DNA   60, 64, 81f., 271f.
– dendritic cells   17
– electrode   176
– electrostatic   310
– ELP   129
– enzyme   270
– functionalited PEG   65
– hydrogel   106f., 120
– hydrophobic   310
– macromolecules   120
– nonspecific   310
– peptides   58
– proteins   58
– PS   69
– selective   15

immunoassay techniques, *see* detection method
immunofluorescent staining   87
immunoreaction efficiency   89f.
impedance spectroscopy   388
– AuNPs   190f.
inclination angle   332, 337
initiater, *see* surfaces
interaction
– anion–cation   400
– antigen–antibody   89
– attractive   60, 213, 366
– AuNP–analyte–AuNP   176
– AuNP–DNA   186
– biotin–streptavidin   83f.
– cell–biomaterial   31
– cell–cell   131
– cell–cell junction   116
– cell–ECM   104
– cell–implant   268
– cell–polymer   86ff.
– cell–protein   131
– cell–surface   16, 55f., 89, 117
– double layer   213, 216
– electrostatic   6, 73f., 86, 126f., 185, 213, 215ff.
– energy   213ff.
– fibronectin–titania   236ff.
– heterotypic cell   121
– hydration   213, 215
– hydrophobic   11, 126f.
– ionic   8
– nanoparticle–substrate   184
– nonspecific   59, 273, 322
– physicochemical   116
– polymer–surface   65
– protein–drug   318
– protein–protein   318, 322
– protein–surface   6, 131
– repulsive   213, 216, 284
– selective   315
– solid–liquid–gas   214
– solid–water–liquid   214
– steric   61, 70
– van der Waals   213ff.
interface
– adsorption–desorption   270f.
– air–water   285, 289, 291f., 364f.
– architectures   62
– bio-   55
– bone/implant   154
– liquid–air   333
– liquid–gas–solid   337
– liquid–vapor   333, 339
– metal–air   376
– natural   55
– oil/water   292
– protein molecules–inorganic materials   272
– recognition   270
– soild–air   333
– solid–fluid   113
– solid–liquid   113, 122, 333, 336
– tissue–implant   2
interfacial
– energy   346
– tension   266
– water   205
ion exchange groups   127
ionic
– strength   107, 110, 208f.
– substitution   149
IRS (infrared spectroscopy)   397f., 402ff.
– collagen/HA nanocrystals   231f.
– fibronectin   238, 240
– hydroxyapatide (HA)   152ff.
– -MAIRS   408ff.
– reflection–absorption (RA)   402ff.
– transmission   402ff.
IRRAS (infrared reflection adsorption spectroscopy)
– PC headgroup structure   217ff.
– polarization modulation- (PM-IRRA)   221
isoelectric point   266
ITO (indium tin oxide)   183, 293

l

lab-on-chip (LOC) devices   109, 128ff.
– microarrays   323ff.
label-free, *see* detection
Langmuir–Blodgett (LB) method   207, 220, 364f., 374f., 382, 398ff.
Langmuir–Schaefer (LS) method   207
layer-by-layer (LbL) assembly   9, 11, 22, 366f.
– free-standing thin film   289f.
– hydrogen-bonding   34
– PEM thin films   18ff.
– zirconia nanofilms   272
ligand exchange technique   185f.
lithography   303f., 368ff.
– applications   306f., 316ff.
– dip-pen (DPN)   303ff.
– electrochemical   307ff.
– electron beam   303f., 368f.
– focused ion beam (FIB)   368f.
– holographic   371
– interference   371

– ion lithography 368f.
– medical diagnostics 316ff.
– micro/nano-contact printing 56, 304f., 369f.
– nanoimprinting 303, 368f.
– nanolithography 303ff.
– parameters 305f.
– photolithography 368f.
– thermal DPN 308f.
Lotus leaf effect 331, 335, 337
lower critical solution temperature (LCST) 25f., 105ff.
– reversible 109
– salting-in 106
– salting-out 106

m

magnetron sputtering 150ff.
– RF 262
– SiHA coatings, see hydroxyapatide
– system 151
– zirconia 251, 262f.
MAIRS (multiple-angle incidence resolution spectroscopy) 398, 404, 406ff.
– metal thin films 411ff.
membrane
– bacterial cell 11
– bio- 281f., 388f.
– bioreactors (MBR) 386f.
– black lipid (BLM) 282
– chip 280
– CNT 350ff.
– -disrupting activity 10
– filtration 292, 295
– flux 279
– grafted 120
– nanofiltration (NF) 387
– nanoseparation 278f., 298f.
– nuclear 281
– permeation 278f., 281
– polymeric composite 124
– porous 120, 124
– proteins 25, 389
– resistance 389
– reverse osmosis 278
– semipermeable 278
– supported 388
– ultrafiltration 279f., 387
– ultrathin porous nanocrystalline silicon 280
microarray
– aligned MWCNTs 350
– aspect ratio 346ff.
– AuNP 176, 179f., 182ff.

– -based algorithms 316f.
– bioarray 313f., 316f.
– biochips 85, 89, 306
– DNA 59f., 82, 180, 269
– fabrication 318
– gas sensing 186
– high-density 86, 313f.
– microchannel network 311f.
– oligonucleotide 89f.
– patterning 317ff.
– protein 59, 319, 321
– protein/DNA 321
– sandwich 321
– silicon electrode 23
microfluidics devices 77, 128f., 331
Mie theory 179
molecular
– aggregates 397ff.
– loading 84
– mobility 114
molecules
– amphiphilic 365
– branched macro- 84
– denatured 58
– multi-armed 3
– phospholipids 7
multifunctional copolymers 63f.
multilayer
– hydrogen-bonded 11, 34f.
– nanofilms 22
– polyelectrolyte-free 24
myotubes 22

n

nanobioarray 303, 316ff.
– patterning 317ff.
– protein 319, 321
– sandwich 321
nanocoatings, see thin fims
nanocomposites, see composite
nanoembossing technique 31, 121
nanofiber
– metal hydroxide nanostrands 293ff.
nanofilms, see thin films
nanofountain probe method 311f.
nanogap technique 190f.
– break junction method 190
nanolithography, see lithography
nanoparticles 36
– core–shell 177, 182, 298
– dip-pen patterning 307
– free-standing thin film 290
– gold, see AuNPs
– magnetic 307

– platinum (Pt)   178
– polymer   124
– self-assembled   291
– semiconductor   175
– silica   12
– silver   11f.
– spherical metal   177
– zirconia   252, 269
nanopatterning, *see* patterning
nanostrands, *see* nanofiber
nanotube
– aspect ratio   260
– double-walled (DWNT)   279
– zirconia   260, 269
nanowires
– electrowetting   346
– zirconia   269
NEXAFS (near edge X-ray absorption fine structure)   11
– fibronectin   240
NHS (*N*-hydroxysuccinimide)   3
– chemistry   65f., 74, 81f.
– –PEG-aminosilane   3
nuclear pore complex   281

o

optical microscopy   31
– wide-field   68
optically transparent   31
optical waveguide lightmode spectroscopy (OWLS)   74

p

PAA (poly(acrylic acid))   19f., 34, 89
PAH (poly(allylamine hydrochloride))   6, 19ff.
passivation   314
– gas   346
– layer   60, 346
patterning
– allergy   320
– antibodies   321
– bio-   303
– 2-D   89, 309
– 3-D   309
– direct   310ff.
– in situ photo-   130
– indirect   310f., 314ff.
– micro-   57, 60, 84, 317ff.
– nano-   57, 60, 84, 317ff.
– rate   311
– surfaces   26, 31ff.
– virus   319f.
PBS (phosphate-buffered saline)   4, 16, 130

PC (phosphatidylcholine)   16, 206ff.
– structure on titania surface   217ff.
PCP (paracyclophane)   10
PCR (polymerase chain reaction), *see* reaction
PDDA (poly(diallyldimethylammonium))   259
PDMS (poly(dimethyl siloxane))   8, 12, 14, 77, 81
PEG (poly(ethylene glycol))
– grafts   6
– hydrophilicity   60, 86
– packing density   13, 17, 72
– radii of gyration   72
– side chains   30
PEIs (poly(ethylene imines))   11
PEM (polyelectrolyte multilayer)   5f., 9, 18ff.
– PAH/PAA   19ff.
– PLL/PGA   20, 22
– polysaccharide–polypeptide   22
– PSS/PAH   19ff.
– single-walled CNT   24
PEO (poly(ethylene oxide))   89
permeation
– controllable   124
– stimuli-modulated   123
PET (poly(ethylene terephthalate))   14, 29
PGA (poly(L-glutamic acid))   20, 22
pH   73, 80
– -responsive polymers   86, 107f.
– supported vesicles layer   211ff.
phase transition   110, 115, 122, 126, 129
photochemically   55, 63
– reactive species   65f.
photolithography, *see* lithography
physical entrapment   310
physiosorption   8, 70
– collagen on titania surfaces   243
– multifunctional copolymers   63f., 72f.
– nonspecific   58
PLA (poly(lactic acid))   31ff.
plasma, *see* deposition
PLCL (poly(L-lactide-*co*-ε-caprolactone))   86f.
PLGA (poly(DL-lactide-*co*-glycolide))   29
PLL (poly(L-lysine))   20, 22, 30, 66
PLL-g-PEG ((poly(L-lysine)-g-poly(ethylene glycol))   73ff.
PMMA (poly(methyl methacrylate))   8, 81
PNIPAAm ((poly(*N*-isopropylacrylamide))   105f.
polyacids, *see* polymer

polyelectrolyte 6, 107
– catechol-functionalized 12
– multilayer, see PEM
poly(ether urethane) 9
polymer
– aggregation 124
– antirhombogenitic 9
– biocompatibility 111
– –biomolecule systems 109
– block co- 10, 14, 34, 63f., 66f.
polymer brushes 3f., 8, 11f., 66f., 367f.
– biohybrid hydrophillic 86
– biomacromolecular 71
– dual-function 11, 15
– gradient 71f.
– homo- 15
– surface-grafted PEGylated 15
– surface-initiated polymerization (SIP) 66ff.
– synthesis techniques 71
– zwitterionic 13, 16
polymer
– cationic 10f.
– chain mobility 117
– crosslinked 3f., 56, 58f., 107
– dehydration 110
– dual-responsive 11, 15, 108f.
– end-functionalized 64f., 67, 69
– flexibility 5
– hydrophilicity 60, 86, 105f., 109f.
– hydrophobicity 86, 105f., 109f.
– hyperbranched 4f., 16f.
– linear 56, 58f.
– multiresponsive 11, 15, 108f.
– pendent proups 107
– polyacids 107f.
– polydispersity 68f.
– star 59, 64, 90
– stimuli-responsive 103ff.
– surface modification 79ff.
– swelling ratio 107, 112, 114
– swelling/shrinking behavior 12, 77, 107f., 128
– thermoresponsive 25f., 105ff.
– triblock co- 15
– zwitterionic 4f., 16
polymeric surface modifiers (PSMs) 11
polymerization 26, 56
– ATRP (atom transfer radical polymerization) 12, 15f., 69f., 112, 126
– co- 106, 118
– condensation 16
– controlled radical 68f.
– CVD 77f., 82

– electron beam 117
– free-radical co- 27, 118, 122
– in situ 129
– living cationic 69
– nitroxid-mediated (NMP) 69
– photo- 29f., 86
– plasma 75, 113
– responsive bio- 110f.
– reversible addition-fragmentation chain-transfer (RAFT) 69, 112, 126
– ring-opening (ROP) 69
– ring-opening metathesis (ROMP) 69
– surface-initiated (SIP) 58, 63, 66ff.
– surface-initiated ATRP (SI-ATRP) 69f., 118
– UV-initiated 14, 71
polypeptide
– -based biopolymers 110
– elastin-like (ELP) 110f., 121, 129f.
porosity
– CaP structure 164
– SiHA coatings 155
– temperature dependent 128
PP (poly(propylene)) 123
precipitation 107
– apatite 266f.
– AuNPs 180
– CaP 164
– SiHA 155
printing
– inkjet 303f.
– micro/nano-contact, see lithography
protein
– adsorption 2ff.
– bioactive 12
– diameter 61
– His-tagged 316
– membrane-spanning 56
– plasma 2
– purification 111
– repellency 4, 18
– resistance 60f.
– transmembrane 56
PSS (poly(styrene)) 14, 65
PSS (poly(styrene sulfonate)) 5f., 19ff.
PTFE (poly(tetrafluoroethylene)) 12
PTMG (poly(tetramethylene glycol)) 17
purification 111, 311, 315
pyrolysis 78
PVA (poly(vinyl alcohol)) 86

q

QCMB (quarz crystal microbalance) 176, 186, 207, 209
– collagen on titania surface 229

– frequency   207
quantum dots (QD)   318, 321f.

r
radical   27, 71
– formation   76
Raman microscopy
– confocal   363, 382f.
Raman spectroscopy (RS)
– fibronectin   238f.
– stimuli-responsive polymer film   112
reaction
– activation–deactivation   69
– addition reactions   24, 80, 82f., 85
– conjugation   65f., 109f., 182
– 1,3-dipolar click   24
– hybridization   180f., 186, 192, 194, 270ff.
– PCR (polymerase chain reaction)   325
– substitution   80f.
refractive index   213, 374f.
release matrices   114f.
replication process
– two-step   88
resonant optical evanescent field methods
– surface plasmon resonance, see SPR
– waveguide resonance microscopy   363
retention time   125, 127
Reynolds number   341
RGD (Arg-Gly-Asp)   13, 88, 120
RGDS (Arg-Gly-Asp-Ser)   27f., 120

s
SAM (self-assembled monolayers)   2, 28, 366
– dip-pen nanolithography   309
– initiator-functionalized   68f.
– PEGylated   2, 4f., 16f.
– silane   126
– zirconia   261
SCEM (scanning electrochemical microscopy)   364, 370
secondary ion-mass spectrometry imaging   364
SEM (scanning electron microscopy)   32
– AuNPs   183
– dried foam film   285
– metal hydroxide nanostrands   294
– nanocomposite fibers   295, 297f.
– zirconia nanofilms   252f., 261
sensor
– chemiresistor   187f.
– electrical   186, 191ff.

– gas   186
– sensitivity   193f.
SERS (surface-enhanced Raman scattering)   176
SFG (sum frequency generation) vibrational spectroscopy   112f.
signal-to-noise ratio (SNR)   192, 399
– AuNP array   192
signal transduction   56
SiHA (silicon-substituted hydroxyapatide)   150ff.
– acellular testing   154ff.
– crystallization   160
– mineralization   167
– nucleation   160
simulated body fluid (SBF)   154f., 263ff.
single nucleotide polymorphism (SNP), see detection methods
SLIC (synthetic ligand-gated ion channel)   389
sliding angle   337
SNOM (scanning near-field optical microscopy)   370
spectrophotometric analysis   212f.
SPEO (staeryl poly(ethylene oxide))   9
SP(surface plasmon)-enhanced fluorescence microscopy   378f.
SPR (surface plasmon resonance)   21, 60, 363, 375ff.
– angle   376
– attenuated total reflection (ATR)   376
– AuNPs   179, 182f.
– DNA binding   77
– in situ   76
– stimuli-responsive polymer film   112
STM (scanning tunneling microscopy)   190, 363, 370
Stone–Wales effect   354
streptavidin (SA)   30f.
substrate
– biocompatibility   15
– biofunctionality   15
– conductive   183
– glass   118, 176, 183, 308
– gold   314
– hydrophobic   4
– indium tin oxide (ITO)   183
– mica   309, 314
– multilayer   315
– PDMS   12
– porous   288
– quartz   183
– silicon   31, 118, 251, 280, 307ff.
– silicon oxide   313

– stainless-steel   12, 18, 259
– tissue engineering   14ff.
– titanium   155, 158, 160ff.
superhydrophobicity, see surface
supported lipid bilayers (SLBs)
– composition   208ff.
– specific peptides   210
– free-standing thin films   282ff.
supported vesicles layer (SVL), see supported lipid bilayers
surface
– amine-functionalized   3
– area   266
– area to volume ratio   331
– beads-on-strings   339
– biofunctional   56ff.
– bioinert   70
– BSA-passivated   7
– cell-repellant   16
– charge   22f., 60
– charge density   127
– dendrinized   16
– energy   333, 336, 342
– fluorinated   28f.
– hydrophilic   2f., 104
– immobilization   12, 15, 27f., 63f., 78f.
– initiator   68f.
– low-friction   15
– micro/nanoscale roughness   337ff.
– multistep modification   64
– nonfouling   65
– non-PEGylated   3f.
– PAA/PEO   90
– passivation   64
– patterned   26, 31ff.
– PEGylated   2ff.
– PET   14
– polystyrene   7
– protein-absorbing   314
– PTFE   12
– reentrant structure   338f.
– roughness   31, 55, 229, 332, 334f., 342ff.
– self-cleaning   331f., 335, 338
– single step modification   64
– stimuli-responsive   103ff.
– superhydrophobicity   332f., 335ff.
– tension   236, 283, 333, 345
– titanium   12ff.
– -to-volume ratio   269
– total free energy   266
– two-scale roughness   336ff.
– unitary roughness   336
surfactant   74, 283
– antimicrobial   123

– catanionic   292
– layer   284f.
– nonionic   286, 288
– polymer   288f.

t

Tabor–Winterton (TW) approximation   213f.
TCPS (tissue culture grade polystyrene)   3, 26f.
TEM (transmission electron microscopy)   194
– collagen on titania surface   230
– dried foam film   284
– high resolution (HRTEM)   264, 298
– nanocomposite fibers   295f., 298
– zirconia nanofilms   261, 264
templates   71, 184
thermodynamic
– analysis   266
– control   60
thin films
– amorphous   151f.
– amphiphilic block copylymer   3
– antibacterial   10ff.
– antibiotic-conjugated polymer   12f.
– antimicrobial   10, 13
– antithrombogenic   6f.
– applications   383ff.
– Au nanofilms   175ff.
– bactericidal   11
– bioactive   18, 28f.
– biocatalytic   8
– biocompatible   2, 13, 29, 120, 383ff.
– biodegradable   12, 18, 24
– biomimetic antibacterial   13
– building additive   9
– clot-lyzing   8
– conductivity   29, 185ff.
– crystallinity   151f., 155
– 2-D (two-dimensional)   183, 185, 187, 309
– 3-D (three-dimensional)   66, 88f., 183, 188, 309
– dip-coated   9, 72
– DNA–polycation multilayer   36ff.
– dual-function antibacterial   11, 15
– electrical insulating   29
– electroactive   27f.
– foam   277, 285ff.
– heparin-mimetic   8
– hydrogel   3f., 15, 29, 55
– hydrophilic   2f.
– hydrophobic   22f.
– hyperbranched polymers   4f.

– imaging 363ff.
– imaging technologies 371ff.
– lysine-based anticlotting 8
– mechanical stability 14, 18, 31, 57, 277, 282, 286, 289f.
– membrane-mimetic 7
– metal 411ff.
– multilayer 5f., 11
– multipotent covalent coatings 74f.
– nanocomposite 8
– nanocomposite polymer 11f.
– nonbiocidal 13
– patterned 31ff.
– PEGylated 2f., 15, 60f.
– polycrystalline 151f.
– polyelectrolyte multilayer, see PEM
– polysaccharide-based 18
– polyurethane 9f., 16f.
– polyurethane–hyaluronic 9
– porosity 128, 155, 164, 256, 259, 262
– preparation methods 364ff.
– protein-repellant 2f., 58
– protein-resistant polymer 4f., 6
– (PSS/PAH)$_n$ 5f.
– resistivity 184, 192f.
– RGD-functionalized 13
– self-standing, see free-standing nanostructured
– stability 16, 65
– stiffness 22
– stimuli-responsive polymer 103ff.
– structuring 368ff.
– tantalum 18
– temperature-responsive polymer 25ff.
– thickness 11, 22, 61, 65, 117f., 257, 259
– TiO$_2$, see titania
– ultrathin 58, 256, 281f., 284, 288, 398ff.
– ZrO$_2$, see zirconia nanofilms
– zwitterionic 4, 6f., 15f., 22
thiol
– binder 183f.
– cationic 185
– goups 178, 183, 186
TIRF (total internal reflection fluorescence microscopy) 363, 379f.
tissue engineering
– applications 14, 22
– scaffolds 17, 24
tissue reconstruction 25
titania 203ff.
– 3-D (three-dimensional) lattice 206
– nanolayer 204
– lipid bilayers 206ff.
– surface 205ff.

titanium implant 203ff.
ToF-SIMS (time-of-flight-secondary ion mass spectroscopy)
– cell lift-off 116
– stimuli-responsive polymer film 112
total hip replacements (THRs) 251
TPA (tissue-plasminogen activator) 8
transition temperature 109ff.
tunneling 186f.

u

UV
– extreme (EUV) 71
– -induced grafting/polymerization 14, 71
– irradiation 29, 368
– -visible spectra 179, 181, 185, 298, 413

v

vesicles, see supported vesicles layer

w

waveguide
– evanescent field fluorescence microscopy (WEFFM) 381f.
– resonant microscopy 376f., 379
– scattering microscopy 380f.
water droplet
– Cassie state 334ff.
– decelerating 342
– expulsion 349
– Gibbs free energy 335
– impact dynamics 340ff.
– shape 335, 337, 351ff.
– spreading 341f.
– water 336f.
– Wenzel state 334ff.
water
– -harvesting system 332
– -repellency 331f.
– striders 332
Wenzel equation 334f., 337
wettability
– heterogeneous 334
– homogeneous 334
– specific 333
– surface 25, 107, 166, 331ff.
– switching 333, 344ff.
wettability
– temperature-dependent 117, 125f.
– thermoresponsive films 113f.
– water
wetting
– complete 333
– complete drying 333

– dynamic   337
– electrochemically controlled   350ff.
– electrowetting   345ff.
– polarity-dependent   350ff.
– surface wetting induced by morphology (SWIM) changes   344
Wilhelmy system   365

x

XPS (X-ray photoelectron spectroscopy)   11, 65, 116
– collagen/HA nanocrystals   231
– collagen on titania surface   229
– fibronectin titania surface   234ff.
XRD (X-ray diffraction)
– -DSC   401f.
– grazing-angle incidence   399
– IP- (in-plane)   399f.
– OP- (out-of-plane)   399
– SiHA coatings   152, 154f., 158
– small-angle   400
– zirconia nanofilms   252f.
– ultrathin films   398ff.

y

Young–Laplace equation   283
Young's equation   333
Young's modulus   22, 57, 257, 286, 290f.

z

zirconia
– doped   257ff.
– gel   266, 270
– monoclinic phase   252, 266f.
– polycrystalline   263, 265
– tetragonal phase   252, 257, 266f.
zirconia nanofilms   251ff.
– applications   269ff.
– bioactivity   263ff.
– biocompatibility   265
– cell behavior   267fff.
– crystallization   255
– nanostructured   254f., 262
– porous   256, 259, 262f.
– preparation, *see* deposition